Praise for
The Age of Surveillance Capitalism

"An intensively researched, engagingly written chronicle of surveillance capitalism's origins and its deleterious prospects for our society. . . . This is the rare book that we should trust to lead us down the long hard road of understanding."
—Jacob Silverman, *New York Times Book Review*

"An original and often brilliant work, and it arrives at a crucial moment, when the public and its elected representatives are at last grappling with the extraordinary power of digital media and the companies that control it. Like another recent masterwork of economic analysis, Thomas Piketty's 2013 *Capital in the Twenty-First Century*, the book challenges assumptions, raises uncomfortable questions about the present and future, and stakes out ground for a necessary and overdue debate. Shoshana Zuboff has aimed an unsparing light onto the shadowy new landscape of our lives. The picture is not pretty."
—Nicholas Carr, *Los Angeles Review of Books*

"Zuboff's book is a brilliant, arresting analysis of the digital economy and a plea for a social awakening about the enormity of the changes that technology is imposing on political and social life." —Paul Starr, *Foreign Affairs*

"A book that no tech industry official will want the American public to read. . . . Insanely brilliant, deeply unsettling." —*Pittsburgh Post-Gazette*

"Groundbreaking, magisterial . . . unmissable. As we grope around in the darkness trying to grasp the contours of our digital era, *The Age of Surveillance Capitalism* shines a searing light on how this latest revolution is transforming our economy, politics, society—and lives." —*Financial Times*

"The most ambitious attempt yet to paint the bigger picture and to explain how the effects of digitization that we are now experiencing as individuals and citizens have come about. . . . A continuation of a tradition that includes Adam Smith, Max Weber, Karl Polanyi and—dare I say it—Karl Marx. . . . If we fail to tame the new capitalist mutant rampaging through our societies then we will only have ourselves to blame, for we can no longer plead ignorance. A striking and illuminating book." —John Naughton, *Observer*

"A bold, important book. . . . Combining in-depth technical understanding and a broad, humanistic scope, Zuboff has written what may prove to be the first definitive account of the economic—and thus social and political—condition of our age." —James Bridle, *Guardian*

"Comprehensive and impassioned . . . an important book."
—Bryan Appleyard, *Sunday Times* (London)

"It's quite possible that the single most important book about politics, economics, culture and society in this century is Shoshana Zuboff's *The Age of Surveillance Capitalism: The Fight for a Human Future at the New Frontier of Power*. She explains with far more power than anyone has done before the emergence of a whole new form of capitalism based on the expropriation of the personal data we freely give to vast corporations. It's the *Das Kapital* for our times."

—Fintan O'Toole, *Irish Times*

"Groundbreaking. . . . Aiming to apply Marx's account of surplus value in a time when capital is accumulated through knowledge-based technology, she has given us an illuminating critical perspective on the regime of surveillance under which we all now live." —John Gray, *New Statesman*

"An exceptional and necessary book about the information civilization we have become." —David Patrikarakos, *Literary Review*

"This book's major contribution is to give a name to what's happening, to put it in cultural and historical perspective, and to ask us to pause long enough to think about the future and how it might be different from today."

—Frank Rose, *Wall Street Journal*

"Zuboff's blow-by-blow accounts of the key players ignoring, mocking and finally riding roughshod over even governmental efforts to stop them are consistently shocking." —Katrina Gulliver, *Times Literary Supplement*

"*The Age of Surveillance Capitalism* has been compared to seminal works from Adam Smith's *The Wealth of Nations* to Thomas Piketty's *Capital in the Twenty-First Century*, and with good reason." —*Changeboard Magazine*

"If a book's importance is gauged by how effectively it describes the world we're in, and how much potential it has to change said world, then in my view it's easily the most important book to be published this century." —Zadie Smith

"Shoshana Zuboff's *The Age of Surveillance Capitalism*—the most important book I have ever read on the intersection of technology, politics, and society— offers a radical reinterpretation of the changes the tech industry has wrought."

—Micah Sifry, *The American Prospect*

"A step-by-step account of the building of the digital iron cage."

—Alex Ross, NewYorker.com

"A warning bell, sounded clearly for both the people in danger and of those with the power to do something to keep them safe . . . a truly sobering shock to the system, a call for ordinary people to re-assert control before it's too late."

—*The National* (UAE)

"Thorough and scholarly evisceration of technology companies' data-gathering efforts." —*Washington Post*

"I feel like I am just a little bit safer from the predators of our time just for having it in my grasp." —Anand Giridharadas, author of *Winner Take All*

"Shoshana Zuboff's *The Age of Surveillance Capitalism* is already drawing comparisons to seminal socioeconomic investigations like Rachel Carson's *Silent Spring* and Karl Marx's *Das Kapital*. Zuboff's book deserves these comparisons and more: Like the former, it's an alarming exposé about how business interests have poisoned our world, and like the latter, it provides a framework to understand and combat that poison. But *The Age of Surveillance Capitalism*, named for the now-popular term Zuboff herself coined five years ago, is also a masterwork of horror. It's hard to recall a book that left me as haunted as Zuboff's, with its descriptions of the gothic algorithmic daemons that follow us at nearly every instant of every hour of every day to suck us dry of metadata. Even those who've made an effort to track the technology that tracks us over the last decade or so will be chilled to their core by Zuboff, unable to look at their surroundings the same way." —Sam Biddle, *The Intercept*

"A wake-up call about how tech companies monetize every moment of our lives—and threaten our free will in the process." —*Wired UK*

"One of the most important criticisms of the power of Big Tech."
—Rana Foroohar, *Financial Times*

"Chilling and essential." —*Globe and Mail*

"The ominous relationship between modern capitalism and digital technology is put under timely scrutiny by Zuboff, whose *In the Age of the Smart Machine* marked her out as a chief voice concerning the information age and its impending risks." —*Irish Independent*

"Zuboff thoughtfully examines the economic and philosophical implications of surveillance capitalism; warns that our children, in their ceaseless quest for connectivity, are harbingers of what lies ahead; and urges public outrage over the theft of our humanity. . . . A big, sprawling, and alarming case for 'the darkening of the digital dream.'" —*Kirkus Reviews*

"Zuboff powerfully argues that the digital revolution has put both our privacy and freedom in danger. . . . Essential reading." —*Washington Book Review*

"*The Age of Surveillance Capitalism* is one of the most important books in recent years on the malaise afflicting the modern techno-savvy age that has brought in its wake not just the blitzkrieg of information with no measurable increase in knowledge or wisdom, but given rise to systems that aim at behavior control, threatening human nature with serious consequences." —*The Hindu*

"Shoshana Zuboff is now finally getting universal recognition as the profound, eloquent, and erudite prophet of our time—a reputation she has long deserved."
—Giovanni Buttarelli, European Data Protection Supervisor

THE AGE OF SURVEILLANCE CAPITALISM

*The Fight for a Human Future at
the New Frontier of Power*

SHOSHANA ZUBOFF

PUBLICAFFAIRS
NEW YORK

Cover design by Pete Garceau
Cover copyright © 2019 Hachette Book Group, Inc.

PublicAffairs
Hachette Book Group
1290 Avenue of the Americas, New York, NY 10104
www.publicaffairsbooks.com
@Public_Affairs

Printed in the United States of America

First Edition: January 2019

Published by PublicAffairs, an imprint of Perseus Books, LLC, a subsidiary of Hachette Book Group, Inc. The PublicAffairs name and logo is a trademark of the Hachette Book Group.

The Hachette Speakers Bureau provides a wide range of authors for speaking events. To find out more, go to www.hachettespeakersbureau.com or call (866) 376-6591.

Sonnets from China, copyright © 1945 by W. H. Auden, renewed 1973 by The Estate of W. H. Auden; and "We Too Had Known Golden Hours," copyright © 1951 by W. H. Auden and renewed 1979 by The Estate of W.H. Auden; from *W. H. Auden Collected Poems* by W. H. Auden. Used by permission of Random House, an imprint and division of Penguin Random House LLC. All rights reserved.

Excerpt(s) from *Social Physics: How Good Ideas Spread—The Lessons from a New Science* by Alex Pentland, copyright © 2014 by Alex Pentland. Used by permission of Penguin Press, an imprint of Penguin Publishing Group, a division of Penguin Random House LLC. All rights reserved.

The publisher is not responsible for websites (or their content) that are not owned by the publisher.

Print book interior design by Six Red Marbles Inc.

Library of Congress Cataloging-in-Publication Data
Names: Zuboff, Shoshana, 1951– author.
Title: The age of surveillance capitalism : the fight for a human future at the new frontier of power / Shoshana Zuboff.
Description: First edition. | New York : PublicAffairs, 2018. | Includes bibliographical references and index.
Identifiers: LCCN 2018003901 (print) | LCCN 2018039998 (ebook) | ISBN 9781610395700 (ebook) | ISBN 9781610395694 (hardcover)
Subjects: LCSH: Consumer behavior—Data processing. | Consumer profiling—Data processing. | Information technology—Social aspects.
Classification: LCC HF5415.32 (ebook) | LCC HF5415.32 .Z83 2018 (print) | DDC 306.3–dc23
LC record available at https://lccn.loc.gov/2018003901

ISBNs: 978-1-61039-569-4 (hardcover), 978-1-61039-570-0 (ebook)

LSC-C

Printing 7, 2020

This book is dedicated to the past and the future:
In memory of my Beloved, Jim Maxmin.
In memory of my courageous friend, Frank Schirrmacher.
In honor of my children,
Chloe Sophia Maxmin and Jacob Raphael Maxmin—
I write to fortify your futures and
the moral cause of your generation.

Chilled by the Present, its gloom and its noise,
On waking we sigh for an ancient South,
A warm nude age of instinctive poise,
A taste of joy in an innocent mouth.

At night in our huts we dream of a part
In the balls of the Future: each ritual maze
Has a musical plan, and a musical heart
Can faultlessly follow its faultless ways.

We envy streams and houses that are sure,
But, doubtful, articled to error, we
Were never nude and calm as a great door,

And never will be faultless like our fountains:
We live in freedom by necessity,
A mountain people dwelling among mountains.

—W. H. AUDEN

SONNETS FROM CHINA, XVIII

THE DEFINITION

Sur-veil-lance Cap-i-tal-ism, n.

1. A new economic order that claims human experience as free raw material for hidden commercial practices of extraction, prediction, and sales; **2.** A parasitic economic logic in which the production of goods and services is subordinated to a new global architecture of behavioral modification; **3.** A rogue mutation of capitalism marked by concentrations of wealth, knowledge, and power unprecedented in human history; **4.** The foundational framework of a surveillance economy; **5.** As significant a threat to human nature in the twenty-first century as industrial capitalism was to the natural world in the nineteenth and twentieth; **6.** The origin of a new instrumentarian power that asserts dominance over society and presents startling challenges to market democracy; **7.** A movement that aims to impose a new collective order based on total certainty; **8.** An expropriation of critical human rights that is best understood as a coup from above: an overthrow of the people's sovereignty.

CONTENTS

PART III

INSTRUMENTARIAN POWER FOR A THIRD MODERNITY

CONCLUSION

INTRODUCTION

CHAPTER ONE

HOME OR EXILE IN THE DIGITAL FUTURE

I saw him crying, shedding floods of tears upon
Calypso's island, in her chambers.
She traps him there; he cannot go back home.

—HOMER, *THE ODYSSEY*

I. The Oldest Questions

"Are we all going to be working for a smart machine, or will we have smart people around the machine?" The question was posed to me in 1981 by a young paper mill manager sometime between the fried catfish and the pecan pie on my first night in the small southern town that was home to his mammoth plant and would become my home periodically for the next six years. On that rainy night his words flooded my brain, drowning out the quickening *tap tap tap* of raindrops on the awning above our table. I recognized the oldest political questions: Home or exile? Lord or subject? Master or slave? These are eternal themes of knowledge, authority, and power that can never be settled for all time. There is no end of history; each generation must assert its will and imagination as new threats require us to retry the case in every age.

Perhaps because there was no one else to ask, the plant manager's voice was weighted with urgency and frustration: "What's it gonna be? Which way are we supposed to go? I must know now. There is no time to spare." I wanted the answers, too, and so I began the project that thirty years ago became my first book, *In the Age of the Smart Machine: The Future of Work and Power.* That work turned out to be the opening chapter in what became a lifelong quest to answer the question "Can the digital future be our home?"

It has been many years since that warm southern evening, but the oldest questions have come roaring back with a vengeance. The digital realm is overtaking and redefining everything familiar even before we have had a chance to ponder and decide. We celebrate the networked world for the many ways in which it enriches our capabilities and prospects, but it has birthed whole new territories of anxiety, danger, and violence as the sense of a predictable future slips away.

When we ask the oldest questions now, billions of people from every social strata, generation, and society must answer. Information and communications technologies are more widespread than electricity, reaching three billion of the world's seven billion people.[1] The entangled dilemmas of knowledge, authority, and power are no longer confined to workplaces as they were in the 1980s. Now their roots run deep through the necessities of daily life, mediating nearly every form of social participation.[2]

Just a moment ago, it still seemed reasonable to focus our concerns on the challenges of an information workplace or an information society. Now the oldest questions must be addressed to the widest possible frame, which is best defined as "civilization" or, more specifically, *information civilization.* Will this emerging civilization be a place that we can call home?

All creatures orient to home. It is the point of origin from which every species sets its bearings. Without our bearings, there is no way to navigate unknown territory; without our bearings, we are lost. I am reminded of this each spring when the same pair of loons returns from their distant travels to the cove below our window. Their haunting cries of homecoming, renewal, connection, and safeguard lull us to sleep at night, knowing that we too are in our place. Green turtles hatch and go down to the sea, where they travel many thousands of miles, sometimes for ten years or twenty. When ready to lay their eggs, they retrace their journey back to the very patch of beach where they were born. Some birds annually fly for thousands of miles, losing as much as half their body weight, in order to mate in their birthplace. Birds, bees, butterflies...nests, holes, trees, lakes, hives, hills, shores, and hollows... nearly every creature shares some version of this deep attachment to a place in which life has been known to flourish, the kind of place we call *home.*

It is in the nature of human attachment that every journey and expulsion sets into motion the search for home. That *nostos,* finding home, is among our most profound needs is evident by the price we are willing to pay for it. There is a universally shared ache to return to the place we left behind or to found a new home in which our hopes for the future can nest and grow. We still recount the travails of Odysseus and recall what human beings will endure for the sake of reaching our own shores and entering our own gates.

Because our brains are larger than those of birds and sea turtles, we know that it is not always possible, or even desirable, to return to the same patch of earth. Home need not always correspond to a single dwelling or place. We can choose its form and location but not its meaning. Home is where we know and where we are known, where we love and are beloved. Home is mastery, voice, relationship, and sanctuary: part freedom, part flourishing... part refuge, part prospect.

The sense of home slipping away provokes an unbearable yearning. The Portuguese have a name for this feeling: *saudade,* a word said to capture the homesickness and longing of separation from the homeland among emigrants across the centuries. Now the disruptions of the twenty-first century have turned these exquisite anxieties and longings of dislocation into a universal story that engulfs each one of us.[3]

II. Requiem for a Home

In 2000 a group of computer scientists and engineers at Georgia Tech collaborated on a project called the "Aware Home."[4] It was meant to be a "living laboratory" for the study of "ubiquitous computing." They imagined a "human-home symbiosis" in which many animate and inanimate processes would be captured by an elaborate network of "context aware sensors" embedded in the house and by wearable computers worn by the home's occupants. The design called for an "automated wireless collaboration" between the platform that hosted personal information from the occupants' wearables and a second one that hosted the environmental information from the sensors.

There were three working assumptions: first, the scientists and engineers understood that the new data systems would produce an entirely new knowledge domain. Second, it was assumed that the rights to that new knowledge and the power to use it to improve one's life would belong exclusively to the people who live in the house. Third, the team assumed that for all of its digital wizardry, the Aware Home would take its place as a modern incarnation of the ancient conventions that understand "home" as the private sanctuary of those who dwell within its walls.

All of this was expressed in the engineering plan. It emphasized trust, simplicity, the sovereignty of the individual, and the inviolability of the home as a private domain. The Aware Home information system was imagined as a simple "closed loop" with only two nodes and controlled entirely by the home's occupants. Because the house would be "constantly monitoring the occupants' whereabouts and activities...even tracing its inhabitants' medical conditions," the team concluded, "there is a clear need to give the occupants knowledge and control of the distribution of this information." All the information was to be stored on the occupants' wearable computers "to insure the privacy of an individual's information."

By 2018, the global "smart-home" market was valued at $36 billion and expected to reach $151 billion by 2023.[5] The numbers betray an earthquake beneath their surface. Consider just one smart-home device: the Nest thermostat, which was made by a company that was owned by Alphabet, the Google holding company, and then merged with Google in 2018.[6] The Nest thermostat does many things imagined in the Aware Home. It collects data about its uses and environment. It uses motion sensors and computation to "learn" the behaviors of a home's inhabitants. Nest's apps can gather data from other connected products such as cars, ovens, fitness trackers, and beds.[7] Such systems can, for example, trigger lights if an anomalous motion is detected, signal video and audio recording, and even send notifications to homeowners or others. As a result of the merger with Google, the thermostat, like other Nest products, will be built with Google's artificial intelligence capabilities, including its personal digital "assistant."[8] Like the Aware Home, the thermostat and its brethren devices create immense new stores of knowledge and therefore new power—but for whom?

Wi-Fi–enabled and networked, the thermostat's intricate, personalized data stores are uploaded to Google's servers. Each thermostat comes with a "privacy policy," a "terms-of-service agreement," and an "end-user licensing agreement." These reveal oppressive privacy and security consequences in which sensitive household and personal information are shared with other smart devices, unnamed personnel, and third parties for the purposes of predictive analyses and sales to other unspecified parties. Nest takes little responsibility for the security of the information it collects and none for how the other companies in its ecosystem will put those data to use.[9] A detailed analysis of Nest's policies by two University of London scholars concluded that were one to enter into the Nest ecosystem of connected devices and apps, each with their own equally burdensome and audacious terms, the purchase of a single home thermostat would entail the need to review nearly a thousand so-called contracts.[10]

Should the customer refuse to agree to Nest's stipulations, the terms of service indicate that the functionality and security of the thermostat will be deeply compromised, no longer supported by the necessary updates meant to ensure its reliability and safety. The consequences can range from frozen pipes to failed smoke alarms to an easily hacked internal home system.[11]

By 2018, the assumptions of the Aware Home were gone with the wind. Where did they go? What was that wind? The Aware Home, like many other visionary projects, imagined a digital future that empowers individuals to lead more-effective lives. What is most critical is that in the year 2000 this vision naturally assumed an unwavering commitment to the privacy of individual experience. Should an individual choose to render her experience digitally, then she would exercise exclusive rights to the knowledge garnered from such data, as well as exclusive rights to decide how such knowledge might be put to use. Today these rights to privacy, knowledge, and application have been usurped by a bold market venture powered by unilateral claims to others' experience and the knowledge that flows from it. What does this sea change mean for us, for our children, for our democracies, and for the very possibility of a human future in a digital world? This book aims to answer these questions. It is about the darkening of the digital dream and its rapid mutation into a voracious and utterly novel commercial project that I call *surveillance capitalism*.

III. *What Is Surveillance Capitalism?*

Surveillance capitalism unilaterally claims human experience as free raw material for translation into behavioral data. Although some of these data are applied to product or service improvement, the rest are declared as a proprietary *behavioral surplus,* fed into advanced manufacturing processes known as "machine intelligence," and fabricated into *prediction products* that anticipate what you will do now, soon, and later. Finally, these prediction products are traded in a new kind of marketplace for behavioral predictions that I call *behavioral futures markets.* Surveillance capitalists have grown immensely wealthy from these trading operations, for many companies are eager to lay bets on our future behavior.

As we shall see in the coming chapters, the competitive dynamics of these new markets drive surveillance capitalists to acquire ever-more-predictive sources of behavioral surplus: our voices, personalities, and emotions. Eventually, surveillance capitalists discovered that the most-predictive behavioral data come from intervening in the state of play in order to nudge, coax, tune, and herd behavior toward profitable outcomes. Competitive pressures produced this shift, in which automated machine processes not only *know* our behavior but also *shape* our behavior at scale. With this reorientation from knowledge to power, it is no longer enough to automate information flows *about us;* the goal now is to *automate us.* In this phase of surveillance capitalism's evolution, the means of production are subordinated to an increasingly complex and comprehensive "means of behavioral modification." In this way, surveillance capitalism births a new species of power that I call *instrumentarianism.* Instrumentarian power knows and shapes human behavior toward others' ends. Instead of armaments and armies, it works its will through the automated medium of an increasingly ubiquitous computational architecture of "smart" networked devices, things, and spaces.

In the coming chapters we will follow the growth and dissemination of these operations and the instrumentarian power that sustains them. Indeed, it has become difficult to escape this bold market project, whose tentacles reach from the gentle herding of innocent Pokémon Go players to eat, drink, and purchase in the restaurants, bars, fast-food joints, and shops that pay to play in its behavioral futures markets to the ruthless expropriation of surplus from Facebook profiles for the purposes of shaping individual behavior, whether it's buying pimple cream at 5:45 P.M. on Friday, clicking "yes" on an offer of new running shoes as the endorphins race through your brain after your long Sunday morning run, or voting next week. Just as industrial capitalism was driven to the continuous intensification of the means of production, so surveillance capitalists and their market players are now locked into the continuous intensification of the means of behavioral modification and the gathering might of instrumentarian power.

Surveillance capitalism runs contrary to the early digital dream, consigning the Aware Home to ancient history. Instead, it strips away the illusion that the networked form has some kind of indigenous moral content, that being "connected" is somehow intrinsically pro-social, innately inclusive, or naturally tending toward the democratization of knowledge. Digital connection is now a means to others' commercial ends. At its core, surveillance capitalism is parasitic and self-referential. It revives Karl Marx's old image of capitalism as a vampire that feeds on labor, but with an unexpected turn. Instead of labor, surveillance capitalism feeds on every aspect of every human's experience.

Google invented and perfected surveillance capitalism in much the same way that a century ago General Motors invented and perfected managerial capitalism. Google was the pioneer of surveillance capitalism in thought and practice, the deep pocket for research and development, and the trailblazer in experimentation and implementation, but it is no longer the only actor on this path. Surveillance capitalism quickly spread to Facebook and later to Microsoft. Evidence suggests that Amazon has veered in this direction, and it is a constant challenge to Apple, both as an external threat and as a source of internal debate and conflict.

As the pioneer of surveillance capitalism, Google launched an unprecedented market operation into the unmapped spaces of the internet, where it faced few impediments from law or competitors, like an invasive species in a landscape free of natural predators. Its leaders drove the systemic coherence of their businesses at a breakneck pace that neither public institutions nor individuals could follow. Google also benefited from historical events when a national security apparatus galvanized by the attacks of 9/11 was inclined to nurture, mimic, shelter, and appropriate surveillance capitalism's emergent capabilities for the sake of total knowledge and its promise of certainty.

Surveillance capitalists quickly realized that they could do anything they wanted, and they did. They dressed in the fashions of advocacy and emancipation, appealing to and exploiting contemporary anxieties, while the real action was hidden offstage. Theirs was an invisibility cloak woven in equal measure to the rhetoric of the empowering web, the ability to move swiftly, the confidence of vast revenue streams, and the wild, undefended nature of the territory they would conquer and claim. They were protected by the inherent illegibility of the automated processes that they rule, the ignorance that these processes breed, and the sense of inevitability that they foster.

Surveillance capitalism is no longer confined to the competitive dramas of the large internet companies, where behavioral futures markets were first aimed at online advertising. Its mechanisms and economic imperatives have become the default model for most internet-based businesses. Eventually, competitive pressure drove expansion into the offline world, where the same foundational mechanisms that expropriate your online browsing, likes, and clicks are trained on your run in the park, breakfast conversation, or hunt for a parking space. Today's prediction products are traded in behavioral futures markets that extend beyond targeted online ads to many other sectors, including insurance, retail, finance, and an ever-widening range of goods and services companies determined to participate in these new and profitable markets. Whether it's a "smart" home device, what the insurance companies call "behavioral underwriting," or any one of thousands of other transactions, we now pay for our own domination.

Surveillance capitalism's products and services are not the objects of a value exchange. They do not establish constructive producer-consumer reciprocities. Instead, they are the "hooks" that lure users into their extractive operations in which our personal experiences are scraped and packaged as the means to others' ends. We are not surveillance capitalism's "customers." Although the saying tells us "If it's free, then you are the product," that is also incorrect. We are the sources of surveillance capitalism's crucial surplus: the objects of a technologically advanced and increasingly inescapable raw-material-extraction operation. Surveillance capitalism's actual customers are the enterprises that trade in its markets for future behavior.

This logic turns ordinary life into the daily renewal of a twenty-first-century Faustian compact. "Faustian" because it is nearly impossible to tear ourselves away, despite the fact that what we must give in return will destroy life as we have known it. Consider that the internet has become essential for social participation, that the internet is now saturated with commerce, and that commerce is now subordinated to surveillance capitalism. Our dependency is at the heart of the commercial surveillance project, in which our felt needs for effective life vie against the inclination to resist its bold incursions. This conflict produces a psychic numbing that inures us to the realities of being tracked, parsed, mined, and modified. It disposes us to rationalize the situation in resigned cynicism, create excuses that operate like defense mechanisms ("I have nothing to hide"), or find other ways to stick our heads in the sand, choosing ignorance out of frustration and helplessness.[12] In this way, surveillance capitalism imposes a fundamentally illegitimate choice that twenty-first-century individuals should not have to make, and its normalization leaves us singing in our chains.[13]

Surveillance capitalism operates through unprecedented asymmetries in knowledge and the power that accrues to knowledge. Surveillance capitalists know everything *about us,* whereas their operations are designed to be unknowable *to us.* They accumulate vast domains of new knowledge *from us,* but not *for us.* They predict our futures for the sake of others' gain, not ours. As long as surveillance capitalism and its behavioral futures markets are allowed to thrive, ownership of the new means of behavioral modification eclipses ownership of the means of production as the fountainhead of capitalist wealth and power in the twenty-first century.

These facts and their consequences for our individual lives, our societies, our democracies, and our emerging information civilization are examined in detail in the coming chapters. The evidence and reasoning employed here suggest that surveillance capitalism is a rogue force driven by novel economic imperatives that disregard social norms and nullify the elemental rights associated with individual autonomy that are essential to the very possibility of a democratic society.

Just as industrial civilization flourished at the expense of nature and now threatens to cost us the Earth, an information civilization shaped by surveillance capitalism and its new instrumentarian power will thrive at the expense of human nature and will threaten to cost us our humanity. The industrial legacy of climate chaos fills us with dismay, remorse, and fear. As surveillance capitalism becomes the dominant form of information capitalism in our time, what fresh legacy of damage and regret will be mourned by future generations? By the time you read these words, the reach of this new form will have grown as more sectors, firms, startups, app developers, and investors mobilize around this one plausible version of information capitalism. This mobilization and the resistance it engenders will define a key battleground upon which the possibility of a human future at the new frontier of power will be contested.

IV. The Unprecedented

One explanation for surveillance capitalism's many triumphs floats above them all: it is *unprecedented*. The unprecedented is necessarily unrecognizable. When we encounter something unprecedented, we automatically interpret it through the lenses of familiar categories, thereby rendering invisible precisely that which is unprecedented. A classic example is the notion of the "horseless carriage" to which people reverted when confronted with the unprecedented facts of the automobile. A tragic illustration is the encounter between indigenous people and the first Spanish conquerors. When the Taínos of the pre-Columbian Caribbean islands first laid eyes on the sweating, bearded Spanish soldiers trudging across the sand in their brocade and armor, how could they possibly have recognized the meaning and portent of that moment? Unable to imagine their own destruction, they reckoned that those strange creatures were gods and welcomed them with intricate rituals of hospitality. This is how the unprecedented reliably confounds understanding; existing lenses illuminate the familiar, thus obscuring the original by turning the unprecedented into an extension of the past. This contributes to the normalization of the abnormal, which makes fighting the unprecedented even more of an uphill climb.

On a stormy night some years ago, our home was struck by lightning, and I learned a powerful lesson in the comprehension-defying power of the unprecedented. Within moments of the strike, thick black smoke drifted up the staircase from the lower level of the house and toward the living room. As we mobilized and called the fire department, I believed that I had just a minute or two to do something useful before rushing out to join my family. First, I ran upstairs and closed all the bedroom doors to protect them from smoke damage. Next, I tore back downstairs to the living room, where I gathered up as many of our family photo albums as I could carry and set them outside on a covered porch for safety. The smoke was just about to reach me when the fire marshal arrived to grab me by the shoulder and yank me out the door. We stood in the driving rain, where, to our astonishment, we watched the house explode in flames.

I learned many things from the fire, but among the most important was the unrecognizability of the unprecedented. In that early phase of crisis, I could imagine our home scarred by smoke damage, but I could not imagine its disappearance. I grasped what was happening through the lens of past experience, envisioning a distressing but ultimately manageable detour that would lead back to the status quo. Unable to distinguish the unprecedented, all I could do was to close doors to rooms that would no longer exist and seek safety on a porch that was fated to vanish. I was blind to conditions that were unprecedented in my experience.

I began to study the emergence of what I would eventually call surveillance capitalism in 2006, interviewing entrepreneurs and staff in a range of tech companies in the US and the UK. For several years I thought that the unexpected and disturbing practices that I documented were detours from the main road: management oversights or failures of judgment and contextual understanding.

My field data were destroyed in the fire that night, and by the time I picked up the thread again early in 2011, it was clear to me that my old horseless-carriage lenses could not explain or excuse what was taking shape. I had lost many details hidden in the brush, but the profiles of the trees stood out more clearly than before: information capitalism had taken a decisive turn toward a new logic of accumulation, with its own original operational mechanisms, economic imperatives, and markets. I could see that this new form had broken away from the norms and practices that define the history of capitalism and in that process something startling and unprecedented had emerged.

Of course, the emergence of the unprecedented in economic history cannot be compared to a house fire. The portents of a catastrophic fire were unprecedented in my experience, but they were not original. In contrast, surveillance capitalism is a new actor in history, both original and sui generis. It is of its own kind and unlike anything else: a distinct new planet with its own physics of time and space, its sixty-seven-hour days, emerald sky, inverted mountain ranges, and dry water.

Nonetheless, the danger of closing doors to rooms that will no longer exist is very real. The unprecedented nature of surveillance capitalism has enabled it to elude systematic contest because it cannot be adequately grasped with our existing concepts. We rely on categories such as "monopoly" or "privacy" to contest surveillance capitalist practices. And although these issues are vital, and even when surveillance capitalist operations are also monopolistic and a threat to privacy, the existing categories nevertheless fall short in identifying and contesting the most crucial and unprecedented facts of this new regime.

Will surveillance capitalism continue on its current trajectory to become the dominant logic of accumulation of our age, or, in the fullness of time, will we judge it to have been a toothed bird: A fearsome but ultimately doomed dead end in capitalism's longer journey? If it is to be doomed, then what will make it so? What will an effective vaccine entail?

Every vaccine begins in careful knowledge of the enemy disease. This book is a journey to encounter what is strange, original, and even unimaginable in surveillance capitalism. It is animated by the conviction that fresh observation, analysis, and new naming are required if we are to grasp the unprecedented as a necessary prelude to effective contest. The chapters that follow will examine the specific conditions that allowed surveillance capitalism to root and flourish as well as the "laws of motion" that drive the action and expansion of this market form: its foundational mechanisms, economic imperatives, economies of supply, construction of power, and principles of social ordering. Let's close doors, but let's make sure that they are the right ones.

V. The Puppet Master, Not the Puppet

Our effort to confront the unprecedented begins with the recognition that *we hunt the puppet master, not the puppet*. A first challenge to comprehension is the confusion between surveillance capitalism and the technologies it employs. Surveillance capitalism is not technology; it is a logic that imbues technology and commands it into action. Surveillance capitalism is a market form that is unimaginable outside the digital milieu, but it is not the same as the "digital." As we saw in the story of the Aware Home, and as we shall see again in Chapter 2, the digital can take many forms depending upon the social and economic logics that bring it to life. It is capitalism that assigns the price tag of subjugation and helplessness, not the technology.

That surveillance capitalism is a logic in action and not a technology is a vital point because surveillance capitalists want us to think that their practices are inevitable expressions of the technologies they employ. For example, in 2009 the public first became aware that Google maintains our search histories indefinitely: data that are available as raw-material supplies are also available to intelligence and law-enforcement agencies. When questioned about these practices, the corporation's former CEO Eric Schmidt mused, "The reality is that search engines including Google do retain this information for some time."[14]

In truth, search engines do not retain, but surveillance capitalism does. Schmidt's statement is a classic of misdirection that bewilders the public by conflating commercial imperatives and technological necessity. It camouflages the concrete practices of surveillance capitalism and the specific choices that impel Google's brand of search into action. Most significantly, it makes surveillance capitalism's practices appear to be inevitable when they are actually meticulously calculated and lavishly funded means to self-dealing commercial ends. We will examine this notion of "inevitabilism" in depth in Chapter 7. For now, suffice to say that despite all the futuristic sophistication of digital innovation, the message of the surveillance capitalist companies barely differs from the themes once glorified in the motto of the 1933 Chicago World's Fair: "Science Finds—Industry Applies—Man Conforms."

In order to challenge such claims of technological inevitability, we must establish our bearings. We cannot evaluate the current trajectory of information civilization without a clear appreciation that technology is not and never can be a thing in itself, isolated from economics and society. This means that technological inevitability does not exist. Technologies are always economic means, not ends in themselves: in modern times, technology's DNA comes already patterned by what the sociologist Max Weber called the "economic orientation."

Economic ends, Weber observed, are always intrinsic to technology's development and deployment. "Economic action" determines objectives, whereas technology provides "appropriate *means*." In Weber's framing, "The fact that what is called the technological development of modern times has been so largely oriented economically to profit-making is one of the fundamental facts of the history of technology."[15] In a modern capitalist society, technology was, is, and always will be an expression of the economic objectives that direct it into action. A worthwhile exercise would be to delete the word "technology" from our vocabularies in order to see how quickly capitalism's objectives are exposed.

Surveillance capitalism employs many technologies, but it cannot be equated with any technology. Its operations may employ platforms, but these operations are not the same as platforms. It employs machine intelligence, but it cannot be reduced to those machines. It produces and relies on algorithms, but it is not the same as algorithms. Surveillance capitalism's unique economic imperatives are the puppet masters that hide behind the curtain orienting the machines and summoning them to action. These imperatives, to indulge another metaphor, are like the body's soft tissues that cannot be seen in an X-ray but do the real work of binding muscle and bone. We are not alone in falling prey to the technology illusion. It is an enduring theme of social thought, as old as the Trojan horse. Despite this, each generation stumbles into the quicksand of forgetting that technology is an expression of other interests. In modern times this means the interests of capital, and in our time it is surveillance capital that commands the digital milieu and directs our trajectory toward the future. Our aim in this book is to discern the laws of surveillance capitalism that animate today's exotic Trojan horses, returning us to age-old questions as they bear down on our lives, our societies, and our civilization.

We have stood at this kind of precipice before. "We've stumbled along for a while, trying to run a new civilization in old ways, but we've got to start to make this world over." It was 1912 when Thomas Edison laid out his vision for a new industrial civilization in a letter to Henry Ford. Edison worried that industrialism's potential to serve the progress of humanity would be thwarted by the stubborn power of the robber barons and the monopolist economics that ruled their kingdoms. He decried the "wastefulness" and "cruelty" of US capitalism: "Our production, our factory laws, our charities, our relations between capital and labor, our distribution—all wrong, out of gear." Both Edison and Ford understood that the modern industrial civilization for which they harbored such hope was careening toward a darkness marked by misery for the many and prosperity for the few.

Most important for our conversation, Edison and Ford understood that the moral life of industrial civilization would be shaped by the practices of capitalism that rose to dominance in their time. They believed that America, and eventually the world, would have to fashion a new, more rational capitalism in order to avert a future of misery and conflict. Everything, as Edison suggested, would have to be reinvented: new technologies, yes, but these would have to reflect new ways of understanding and fulfilling people's needs; a new economic model that could turn those new practices into profit; and a new social contract that could sustain it all. A new century had dawned, but the evolution of capitalism, like the churning of civilizations, did not obey the calendar or the clock. It was 1912, and still the nineteenth century refused to relinquish its claim on the twentieth.

The same can be said of our time. As I write these words, we are nearing the end of the second decade of the twenty-first century, but the economic and social contests of the twentieth continue to tear us apart. These contests are the stage upon which surveillance capitalism made its debut and rose to stardom as the author of a new chapter in the long saga of capitalism's evolution. This is the dramatic context to which we will turn in the opening pages of Part I: the place upon which we must stand in order to evaluate our subject in its rightful context. Surveillance capitalism is not an accident of overzealous technologists, but rather a rogue capitalism that learned to cunningly exploit its historical conditions to ensure and defend its success.

VI. The Outline, Themes, and Sources of this Book

This book is intended as an initial mapping of a terra incognita, a first foray that I hope will pave the way for more explorers. The effort to understand surveillance capitalism and its consequences has dictated a path of exploration that crosses many disciplines and historical periods. My aim has been to develop the concepts and frameworks that enable us to see the pattern in what have appeared to be disparate concepts, phenomena, and fragments of rhetoric and practice, as each new point on the map contributes to materializing the puppet master in flesh and bone.

Many of the points on this map are necessarily drawn from fast-moving currents in turbulent times. In making sense of contemporary developments, my method has been to isolate the deeper pattern in the welter of technological detail and corporate rhetoric. The test of my efficacy will be in how well this map and its concepts illuminate the unprecedented and empower us with a more cogent and comprehensive understanding of the rapid flow of events that boil around us as surveillance capitalism pursues its long game of economic and social domination.

The Age of Surveillance Capitalism has four parts. Each presents four to five chapters as well as a final chapter intended as a coda that reflects on and conceptualizes the meaning of what has gone before. Part I addresses the foundations of surveillance capitalism: its origins and early elaboration. We begin in Chapter 2 by setting the stage upon which surveillance capitalism made its debut and achieved success. This stage setting is important because I fear that we have contented ourselves for too long with superficial explanations of the rapid rise and general acceptance of the practices associated with surveillance capitalism. For example, we have credited notions such as "convenience" or the fact that many of its services are "free." Instead, Chapter 2 explores the social conditions that summoned the digital into our everyday lives and enabled surveillance capitalism to root and flourish. I describe the "collision" between the centuries-old historical processes of individualization that shape our experience as self-determining individuals and the harsh social habitat produced by a decades-old regime of neoliberal market economics in which our sense of self-worth and needs for self-determination are routinely thwarted. The pain and frustration of this contradiction are the condition that sent us careening toward the internet for sustenance and ultimately bent us to surveillance capitalism's draconian quid pro quo.

Part I moves on to a close examination of surveillance capitalism's invention and early elaboration at Google, beginning with the discovery and early development of what would become its foundational mechanisms, economic imperatives, and "laws of motion." For all of Google's technological prowess and computational talent, the real credit for its success goes to the radical social relations that the company declared as facts, beginning with its disregard for the boundaries of private human experience and the moral integrity of the autonomous individual. Instead, surveillance capitalists asserted their right to invade at will, usurping individual decision rights in favor of unilateral surveillance and the self-authorized extraction of human experience for others' profit. These invasive claims were nurtured by the absence of law to impede their progress, the mutuality of interests between the fledgling surveillance capitalists and state intelligence agencies, and the tenacity with which the corporation defended its new territories. Eventually, Google codified a tactical playbook on the strength of which its surveillance capitalist operations were successfully institutionalized as the dominant form of information capitalism, drawing new competitors eager to participate in the race for surveillance revenues. On the strength of these achievements, Google and its expanding universe of competitors enjoy extraordinary new asymmetries of knowledge and power, unprecedented in the human story. I argue that the significance of these developments is best understood as the privatization of the *division of learning in society,* the critical axis of social order in the twenty-first century.

Part II traces the migration of surveillance capitalism from the online environment to the real world, a consequence of the competition for prediction products that approximate certainty. Here we explore this new *reality business,* as all aspects of human experience are claimed as raw-material supplies and targeted for rendering into behavioral data. Much of this new work is accomplished under the banner of "personalization," a camouflage for aggressive extraction operations that mine the intimate depths of everyday life. As competition intensifies, surveillance capitalists learn that extracting human experience is not enough. The most-predictive raw-material supplies come from intervening in our experience to shape our behavior in ways that favor surveillance capitalists' commercial outcomes. New automated protocols are designed to influence and modify human behavior at scale as the means of production is subordinated to a new and more complex *means of behavior modification.* We see these new protocols at work in Facebook's contagion experiments and the Google-incubated augmented reality "game" Pokémon Go. The evidence of our psychic numbing is that only a few decades ago US society denounced mass behavior-modification techniques as unacceptable threats to individual autonomy and the democratic order. Today the same practices meet little resistance or even discussion as they are routinely and pervasively deployed in the march toward surveillance revenues. Finally, I consider surveillance capitalism's operations as a challenge to the elemental *right to the future tense,* which accounts for the individual's ability to imagine, intend, promise, and construct a future. It is an essential condition of free will and, more poignantly, of the inner resources from which we draw *the will to will.* I ask and answer the question *How did they get away with it?* Part II ends with a meditation on our once and future history. *If industrial capitalism dangerously disrupted nature, what havoc might surveillance capitalism wreak on human nature?*

Part III examines the rise of instrumentarian power; its expression in a ubiquitous sensate, networked, computational infrastructure that I call *Big Other;* and the novel and deeply antidemocratic vision of society and social relations that these produce. I argue that instrumentarianism is an unprecedented species of power that has defied comprehension in part because it has been subjected to the "horseless-carriage" syndrome. Instrumentarian power has been viewed through the old lenses of totalitarianism, obscuring what is different and dangerous. Totalitarianism was a transformation of the state into a project of total possession. Instrumentarianism and its materialization in Big Other signal the transformation of the market into a project of total certainty, an undertaking that is unimaginable outside the digital milieu and the logic of surveillance capitalism. In naming and analyzing instrumentarian power, I explore its intellectual origins in early theoretical physics and its later expression in the work of the radical behaviorist B. F. Skinner.

Part III follows surveillance capitalism into a second phase change. The first was the migration from the virtual to the real world. The second is a shift of focus from the real world to the social world, as society itself becomes the new object of extraction and control. Just as industrial society was imagined as a well-functioning machine, instrumentarian society is imagined as a human simulation of machine learning systems: a confluent hive mind in which each element learns and operates in concert with every other element. In the model of machine confluence, the "freedom" of each individual machine is subordinated to the knowledge of the system as a whole. Instrumentarian power aims to organize, herd, and tune society to achieve a similar *social confluence,* in which group pressure and computational certainty replace politics and democracy, extinguishing the felt reality and social function of an individualized existence. The youngest members of our societies already experience many of these destructive dynamics in their attachment to social media, the first global experiment in the human hive. I consider the implications of these developments for a second elemental right: *the right to sanctuary.* The human need for a space of inviolable refuge has persisted in civilized societies from ancient times but is now under attack as surveillance capital creates a world of "no exit" with profound implications for the human future at this new frontier of power.

In the final chapter I conclude that surveillance capitalism departs from the history of market capitalism in surprising ways, demanding both unimpeded freedom *and* total knowledge, abandoning capitalism's reciprocities with people and society, and imposing a totalizing collectivist vision of life in the hive, with surveillance capitalists and their data priesthood in charge of oversight and control. Surveillance capitalism and its rapidly accumulating instrumentarian power exceed the historical norms of capitalist ambitions, claiming dominion over human, societal, and political territories that range far beyond the conventional institutional terrain of the private firm or the market. As a result, surveillance capitalism is best described as a *coup from above,* not an overthrow of the state but rather an overthrow of the people's sovereignty and a prominent force in the perilous drift toward democratic deconsolidation that now threatens Western liberal democracies. Only "we the people" can reverse this course, first by naming the unprecedented, then by mobilizing new forms of collaborative action: the crucial friction that reasserts the primacy of a flourishing human future as the foundation of our information civilization. *If the digital future is to be our home, then it is we who must make it so.*

My methods combine those of a social scientist inclined toward theory, history, philosophy, and qualitative research with those of an essayist: an unusual but intentional approach. As an essayist, I occasionally draw upon my own experiences. I do this because the tendency toward psychic numbing is increased when we regard the critical issues examined here as just so many abstractions attached to technological and economic forces beyond our reach. We cannot fully reckon with the gravity of surveillance capitalism and its consequences unless we can trace the scars they carve into the flesh of our daily lives.

As a social scientist, I have been drawn to earlier theorists who encountered the unprecedented in their time. Reading from this perspective, I developed a fresh appreciation for the intellectual courage and pioneering insights of classic texts, in which authors such as Durkheim, Marx, and Weber boldly theorized industrial capitalism and industrial society as it rapidly constructed itself in their midst during the nineteenth and early twentieth centuries. My work here has also been inspired by mid-twentieth-century thinkers such as Hannah Arendt, Theodor Adorno, Karl Polanyi, Jean-Paul Sartre, and Stanley Milgram, who struggled to name the unprecedented in their time as they faced the comprehension-defying phenomena of totalitarianism and labored to grasp their trail of consequence for the prospects of humanity. My work has also been deeply informed by the many insights of visionary scholars, technology critics, and committed investigative journalists who have done so much to illuminate key points on the map that emerges here.

During the last seven years I have focused closely on the top surveillance capitalist firms and their growing ecosystems of customers, consultants, and competitors, all of it informed by the larger context of technology and data science that defines the Silicon Valley zeitgeist. This raises another important distinction. Just as surveillance capitalism is not the same as technology, this new logic of accumulation cannot be reduced to any single company or group of companies. The top five internet companies—Apple, Google, Amazon, Microsoft, and Facebook—are often regarded as a single entity with similar strategies and interests, but when it comes to surveillance capitalism, this is not the case.

First, it is necessary to distinguish between capitalism and surveillance capitalism. As I discuss in more detail in Chapter 3, that line is defined in part by the purposes and methods of data collection. When a firm collects behavioral data with permission and solely as a means to product or service improvement, it is committing capitalism but not surveillance capitalism. Each of the top five tech companies practices capitalism, but they are not all pure surveillance capitalists, at least not now.

For example, Apple has so far drawn a line, pledging to abstain from many of the practices that I locate in the surveillance capitalist regime. Its behavior in this regard is not perfect, the line is sometimes blurred, and Apple might well change or contradict its orientation. Amazon once prided itself on its customer alignment and the virtuous circle between data collection and service improvement. Both firms derive revenues from physical and digital products and therefore experience less financial pressure to chase surveillance revenues than the pure data companies. As we see in Chapter 9, however, Amazon appears to be migrating toward surveillance capitalism, with its new emphasis on "personalized" services and third-party revenues.

Whether or not a corporation has fully migrated to surveillance capitalism says nothing about other vital issues raised by its operations, from monopolistic and anticompetitive practices in the case of Amazon to pricing, tax strategies, and employment policies at Apple. Nor are there any guarantees for the future. Time will tell if Apple succumbs to surveillance capitalism, holds the line, or perhaps even expands its ambitions to anchor an effective alternative trajectory to a human future aligned with the ideals of individual autonomy and the deepest values of a democratic society.

One important implication of these distinctions is that even when our societies address capitalist harms produced by the tech companies, such as those related to monopoly or privacy, those actions do not ipso facto interrupt a firm's commitment to and continued elaboration of surveillance capitalism. For example, calls to break up Google or Facebook on monopoly grounds could easily result in establishing multiple surveillance capitalist firms, though at a diminished scale, and thus clear the way for more surveillance capitalist competitors. Similarly, reducing Google and Facebook's duopoly in online advertising does not reduce the reach of surveillance capitalism if online advertising market share is simply spread over five surveillance capitalist firms or fifty, instead of two. Throughout this book I focus on the unprecedented aspects of surveillance capitalist operations that must be contested and interrupted if this market form is to be contained and vanquished.

My focus in these pages tends toward Google, Facebook, and Microsoft. The aim here is not a comprehensive critique of these companies as such. Instead, I view them as the petri dishes in which the DNA of surveillance capitalism is best examined. As I suggested earlier, my goal is to map a new logic and its operations, not a company or its technologies. I move across the boundaries of these and other companies in order to compile the insights that can flesh out the map, just as earlier observers moved across many examples to grasp the once-new logics of managerial capitalism and mass production. It is also the case that surveillance capitalism was invented in the United States: in Silicon Valley and at Google. This makes it an American invention, which, like mass production, became a global reality. For this reason, much of this text focuses on developments in the US, although the consequences of these developments belong to the world.

In studying the surveillance capitalist practices of Google, Facebook, Microsoft, and other corporations, I have paid close attention to interviews, patents, earnings calls, speeches, conferences, videos, and company programs and policies. In addition, between 2012 and 2015 I interviewed 52 data scientists from 19 different companies with a combined 586 years of experience in high-technology corporations and startups, primarily in Silicon Valley. These interviews were conducted as I developed my "ground truth" understanding of surveillance capitalism and its material infrastructure. Early on I approached a small number of highly respected data scientists, senior software developers, and specialists in the "internet of things." My interview sample grew as scientists introduced me to their colleagues. The interviews, sometimes over many hours, were conducted with the promise of confidentiality and anonymity, but my gratitude toward them is personal, and I publicly declare it here.

Finally, throughout this book you will read excerpts from W. H. Auden's *Sonnets from China*, along with the entirety of Sonnet XVIII. This cycle of Auden's poems is dear to me, a poignant exploration of humanity's mythic history, the perennial struggle against violence and domination, and the transcendent power of the human spirit and its relentless claim on the future.

PART I
THE FOUNDATIONS OF SURVEILLANCE CAPITALISM

CHAPTER TWO

AUGUST 9, 2011: SETTING THE STAGE FOR SURVEILLANCE CAPITALISM

The dangers and the punishments grew greater,
And the way back by angels was defended
Against the poet and the legislator.

—W. H. AUDEN
SONNETS FROM CHINA, II

On August 9, 2011, three events separated by thousands of miles captured the bountiful prospects and gathering dangers of our emerging information civilization. First, Silicon Valley pioneer Apple promised a digital dream of new solutions to old economic and social problems, and finally surpassed Exxon Mobil as the world's most highly capitalized corporation. Second, a fatal police shooting in London sparked extensive rioting across the city, engulfing the country in a wave of violent protests. A decade of explosive digital growth had failed to mitigate the punishing austerity of neoliberal economics and the extreme inequality that it produced. Too many people had come to feel excluded from the future, embracing rage and violence as their only remedies. Third, Spanish citizens asserted their rights to a human future when they challenged Google by demanding "the right to be forgotten." This milestone alerted the world to how quickly the cherished dreams of a more just and democratic digital future were shading into nightmare, and it foreshadowed a global political contest over the fusion of digital capabilities and capitalist ambitions. We relive that August day every day as in some ancient fable, doomed to retrace this looping path until the soul of our information civilization is finally shaped by democratic action, private power, ignorance, or drift.

I. The Apple Hack

Apple thundered onto the music scene in the midst of a pitched battle between demand and supply. On one side were young people whose enthusiasm for Napster and other forms of music file sharing expressed a new quality of demand: consumption my way, what I want, when I want it, where I want it. On the other side were music-industry executives who chose to instill fear and to crush that demand by hunting down and prosecuting some of Napster's most-ardent users. Apple bridged the divide with a commercially and legally viable solution that aligned the company with the changing needs of individuals while working with industry incumbents. Napster hacked the music industry, but Apple appeared to have hacked capitalism.

It is easy to forget just how dramatic Apple's hack really was. The company's profits soared largely on the strength of its iPod/iTunes/iPhone sales. *Bloomberg Businessweek* described Wall Street analysts as "befuddled" by this mysterious Apple "miracle." As one gushed, "We can't even model out some of the possibilities.... It's like a religion."[1] Even today the figures are staggering: three days after the launch of the Windows-compatible iTunes platform in October 2003, listeners downloaded a million copies of the free iTunes software and paid for a million songs, prompting Steve Jobs to announce, "In less than one week we've broken every record and become the largest online music company in the world."[2] Within a month there were five million downloads, then ten million three months later, then twenty-five million three months after that. Four and a half years later, in January 2007, that number rose to two billion, and six years later, in 2013, it was 25 billion. In 2008 Apple surpassed Walmart as the world's largest music retailer. iPod sales were similarly spectacular, exploding from 1 million units per month after the music store's launch to 100 million less than four years later, when Apple subsumed the iPod's functions in its revolutionary iPhone, which drove another step-function of growth. A 2017 study of stock market returns concluded that Apple had generated more profit for investors than any other US company in the previous century.[3]

One hundred years before the iPod, mass production provided the gateway to a new era when it revealed a parallel universe of economic value hidden in new and still poorly understood mass consumers who wanted goods, but at a price they could afford. Henry Ford reduced the price of an automobile by 60 percent with a revolutionary industrial logic that combined high volume and low unit cost. He called it "mass production," summarized in his famous maxim "You can have any color car you want so long as it's black."

Later, GM's Alfred Sloan expounded on that principle: "By the time we have a product to show them [consumers], we are necessarily committed to selling that product because of the tremendous investment involved in bringing it to market."[4] The music industry's business model was built on telling its consumers what they would buy, just like Ford and Sloan. Executives invested in the production and distribution of CDs, and it was the CD that customers would have to purchase.

Henry Ford was among the first to strike gold by tapping into the new mass consumption with the Model T. As in the case of the iPod, Ford's Model T factory was pressed to meet the immediate explosion of demand. Mass production could be applied to anything, and it was. It changed the framework of production as it diffused throughout the economy and around the world, and it established the dominance of a new mass-production capitalism as the basis for wealth creation in the twentieth century.

The iPod/iTunes innovations flipped this century-old industrial logic, leveraging the new capabilities of digital technologies to *invert* the consumption experience. Apple rewrote the relationship between listeners and their music with a distinct commercial logic that, while familiar to us now, was also experienced as revolutionary when first introduced.

The Apple inversion depended on a few key elements. Digitalization made it possible to *rescue* valued assets—in this case, songs—from the institutional spaces in which they were trapped. The costly institutional procedures that Sloan had described were eliminated in favor of a direct route to listeners. In the case of the CD, for example, Apple *bypassed* the physical production of the product along with its packaging, inventory, storage, marketing, transportation, distribution, and physical retailing. The combination of the iTunes platform and the iPod device made it possible for listeners to continuously *reconfigure* their songs at will. No two iPods were the same, and an iPod one week was different from the same iPod another week, as listeners decided and re-decided the dynamic pattern. It was an excruciating development for the music industry and its satellites—retailers, marketers, etc.—but it was exactly what the new listeners wanted.

How should we understand this success? Apple's "miracle" is typically credited to its design and marketing genius. Consumers' eagerness to have "what I want, when, where, and how I want it" is taken as evidence of the demand for "convenience" and sometimes even written off as narcissism or petulance. In my view, these explanations pale against the unprecedented magnitude of Apple's accomplishments. We have contented ourselves for too long with superficial explanations of Apple's unprecedented fusion of capitalism and the digital rather than digging deeper into the historical forces that summoned this new form to life.

Just as Ford tapped into a new mass consumption, Apple was among the first to experience explosive commercial success by tapping into a new society of individuals and their demand for individualized consumption. The inversion implied a larger story of a commercial reformation in which the digital era finally offered the tools to shift the focus of consumption from the mass to the individual, liberating and reconfiguring capitalism's operations and assets. It promised something utterly new, urgently necessary, and operationally impossible outside the networked spaces of the digital. Its implicit promise of an advocacy-oriented alignment with our new needs and values was a confirmation of our inner sense of dignity and worth, ratifying the feeling that we matter. In offering consumers respite from an institutional world that was indifferent to their individual needs, it opened the door to the possibility of a new rational capitalism able to reunite supply and demand by connecting us to what we really want in exactly the ways that we choose.

As I shall argue in the coming chapters, the same historical conditions that sent the iPod on its wild ride summoned the emancipatory promise of the internet into our everyday lives as we sought remedies for inequality and exclusion. Of most significance for our story, these same conditions would provide important shelter for surveillance capitalism's ability to root and flourish. More precisely, the Apple miracle and surveillance capitalism each owes its success to the destructive collision of two opposing historical forces. One vector belongs to the longer history of modernization and the centuries-long societal shift from the mass to the individual. The opposing vector belongs to the decades-long elaboration and implementation of the neoliberal economic paradigm: its political economics, its transformation of society, and especially its aim to reverse, subdue, impede, and even destroy the individual urge toward psychological self-determination and moral agency. The next sections briefly sketch the basic contours of this collision, establishing terms of reference that we will return to throughout the coming chapters as we explore surveillance capitalism's rapid rise to dominance.

II. The Two Modernities

Capitalism evolves in response to the needs of people in a time and place. Henry Ford was clear on this point: "Mass production begins in the perception of a public need."[5] At a time when the Detroit automobile manufacturers were preoccupied with luxury vehicles, Ford stood alone in his recognition of a nation of newly modernizing individuals—farmers, wage earners, and shopkeepers—who had little and wanted much, but at a price they could afford. Their "demand" issued from the same conditions of existence that summoned Ford and his men as they discovered the transformational power of a new logic of standardized, high-volume, low-unit-cost production. Ford's famous "five-dollar day" was emblematic of a systemic logic of reciprocity. In paying assembly-line workers higher wages than anyone had yet imagined, he recognized that the whole enterprise of mass production rested upon a thriving population of mass consumers.

Although the market form and its bosses had many failings and produced many violent facts, its populations of newly modernizing individuals were valued as the necessary sources of customers and employees. It depended upon its communities in ways that would eventually lead to a range of institutionalized reciprocities. On the outside the drama of access to affordable goods and services was bound by democratic measures and methods of oversight that asserted and protected the rights and safety of workers and consumers. On the inside were durable employment systems, career ladders, and steady increases in wages and benefits.[6] Indeed, considered from the vantage point of the last forty years, during which this market form was systematically deconstructed, its reciprocity with the social order, however vexed and imperfect, appears to have been one of its most-salient features.

The implication is that new market forms are most productive when they are shaped by an allegiance to the actual demands and mentalities of people. The great sociologist Emile Durkheim made this point at the dawn of the twentieth century, and his insight will be a touchstone for us throughout this book. Observing the dramatic upheavals of industrialization in his time—factories, specialization, the complex division of labor—Durkheim understood that although economists could describe these developments, they could not grasp their *cause*. He argued that these sweeping changes were "caused" by the changing needs of people and that economists were (and remain) systematically blind to these social facts:

> The division of labor appears to us otherwise than it does to economists. For them, it essentially consists in greater production. For us, this greater productivity is only a necessary consequence, a repercussion of the phenomenon. If we specialize, it is not to produce more, but it is to enable us to live in the *new conditions of existence* that have been made for us.[7]

The sociologist identified the perennial human quest to live effectively in our "conditions of existence" as the invisible causal power that summons the division of labor, technologies, work organization, capitalism, and ultimately civilization itself. Each is forged in the same crucible of human need that is produced by what Durkheim called the always intensifying "violence of the struggle" for effective life: "If work becomes more divided," it is because the "struggle for existence is more acute."[8] The rationality of capitalism reflects this alignment, however imperfect, with the needs that people experience as they try to live their lives effectively, struggling with the conditions of existence that they encounter in their time and place.

When we look through this lens, we can see that those eager customers for Ford's incredible Model T and the new consumers of iPods and iPhones are expressions of the conditions of existence that characterized their era. In fact, each is the fruit of distinct phases of a centuries-long process known as "individualization" that is the human signature of the modern era. Ford's mass consumers were members of what has been called the "first modernity,"[9] but the new conditions of the "second modernity" produced a new kind of individual for whom the Apple inversion, and the many digital innovations that followed, would become essential. This second modernity summoned the likes of Google and Facebook into our lives, and, in an unexpected twist, helped to enable the surveillance capitalism that would follow.

What are these modernities and how do they matter to our story? The advent of the individual as the locus of moral agency and choice initially occurred in the West, where the conditions for this emergence first took hold. First let's establish that the concept of "individualization" should not be confused with the neoliberal ideology of "individualism" that shifts all responsibility for success or failure to a mythical, atomized, isolated individual, doomed to a life of perpetual competition and disconnected from relationships, community, and society. Neither does it refer to the psychological process of "individuation" that is associated with the lifelong exploration of self-development. Instead, individualization is a consequence of long-term processes of modernization.[10]

Until the last few minutes of human history, each life was foretold in blood and geography, sex and kin, rank and religion. I am my mother's daughter. I am my father's son. The sense of the human being as an *individual* emerged gradually over centuries, clawed from this ancient vise. Around two hundred years ago, we embarked upon the first modern road where life was no longer handed down one generation to the next according to the traditions of village and clan. This "first modernity" marks the time when life became "individualized" for great numbers of people as they separated from traditional norms, meanings, and rules.[11] That meant each life became an open-ended reality to be discovered rather than a certainty to be enacted. Even where the traditional world remains intact for many people today, it can no longer be experienced as the only possible story.

I often think about the courage of my great-grandparents. What mixture of sadness, terror, and exhilaration did they feel when in 1908, determined to escape the torments of the Cossacks in their tiny village outside of Kiev, they packed their five children, including my four-year-old grandfather Max, and all their belongings into a wagon and pointed the horses toward a steamer bound for America? Like millions of other pioneers of this first modernity, they escaped a still-feudal world and found themselves improvising a profoundly new kind of life. Max would later marry Sophie and build a family far from the rhythms of the villages that birthed them. The Spanish poet Antonio Machado captured the exhilaration and daring of these first-modernity individuals in his famous song: "Traveler, there is no road; the road is made as you go." This is what "search" has meant: a journey of exploration and self-creation, not an instant swipe to already composed answers.

Still, the new industrial society retained many of the hierarchical motifs of the older feudal world in its patterns of affiliation based on class, race, occupation, religion, ethnicity, sex, and the leviathans of mass society: its corporations, workplaces, unions, churches, political parties, civic groups, and school systems. This new world order of the mass and its bureaucratic logic of concentration, centralization, standardization, and administration still provided solid anchors, guidelines, and goals for each life.

Compared to their parents and all the generations before, Sophie and Max had to make things up on their own, but not everything. Sophie knew she would raise the family. Max knew he would earn their living. You adapted to what the world had on offer, and you followed the rules. Nor did anyone ask your opinion or listen if you spoke. You were expected to do what you were *supposed* to do, and little by little you made your way. You raised a nice family, and eventually you'd have a house, car, washing machine, and refrigerator. Mass production pioneers like Henry Ford and Alfred Sloan had found a way to get you these things at a price you could afford.

If there was anxiety, it reflected the necessity of living up to the requirements of one's roles. One was expected to suppress any sense of self that spilled over the edges of the given social role, even at considerable psychic cost. Socialization and adaptation were the materials of a psychology and sociology that regarded the nuclear family as the "factory" for the "production of personalities" ready-made for conformity to the social norms of mass society.[12] Those "factories" also produced a great deal of pain: the feminine mystique, closeted homosexuals, church-going atheists, and back-alley abortions. Eventually, though, they even produced people like you and me.

When I set out on the open road, there were few answers, nothing to emulate, no compass to follow except for the values and dreams that I carried inside me. I was not alone; the road was filled with so many others on the same kind of journey. The first modernity birthed us, but we brought a new mentality to life: a "second modernity."[13] What began as a modern migration from traditional lifeways bloomed into a new society of people born to a sense of psychological individuality, with its double-edged birthright of liberation and necessity. We experience both the right and the requirement to choose our own lives. No longer content to be anonymous members of the mass, we feel our entitlement to self-determination, an obvious truth to us that would have been an impossible act of hubris for Sophie and Max. This mentality is an extraordinary achievement of the human spirit, even as it can be a life sentence to uncertainty, anxiety, and stress.

Since the second half of the twentieth century, the individualization story has taken this new turn toward a "second modernity." Industrialization modernity and the practices of mass production capitalism at its core produced more wealth than had ever been imagined possible. Where democratic politics, distributional policies, access to education and health care, and strong civil society institutions complemented that wealth, a new "society of individuals" first began to emerge. Hundreds of millions of people gained access to experiences that had once been the preserve of a tiny elite: university education, travel, improved life expectancy, disposable income, rising standards of living, broad access to consumer goods, varied communication and information flows, and specialized, intellectually demanding work.

The hierarchical social compact and mass society of the first modernity promised predictable rewards, but their very success was the knife that cut us loose and sent us tumbling onto the shores of the second modernity, propelling us toward more-intricate and richly patterned lives. Education and knowledge work increased mastery of language and thought, the tools with which we create personal meaning and form our own opinions. Communication, information, consumption, and travel stimulated individual self-consciousness and imaginative capabilities, informing perspectives, values, and attitudes in ways that could no longer be contained by predefined roles or group identity. Improved health and longer life spans provided the time for a self-life to deepen and mature, fortifying the legitimacy of personal identity over and against a priori social norms.

Even when we revert to traditional roles, these are choices now rather than absolute truths imposed at birth. As the great clinician of identity, Erik Erikson, once described it, "The patient of today suffers most under the problem of what he should believe and who he should—or...might—be or become; while the patient of early psychoanalysis suffered most under inhibitions which prevented him from being what and who he thought he knew he was."[14] This new mentality has been most pronounced in wealthier countries, but research shows significant pluralities of second-modernity individuals in nearly every region of the world.[15]

The first modernity suppressed the growth and expression of self in favor of collective solutions, but by the second modernity, the self is all we have. The new sense of psychological sovereignty broke upon the world long before the internet appeared to amplify its claims. We learn through trial and error how to stitch together our lives. Nothing is given. Everything must be reviewed, renegotiated, and reconstructed on the terms that make sense to us: family, religion, sex, gender, morality, marriage, community, love, nature, social connections, political participation, career, food...

Indeed, it was this new mentality and its demands that summoned the internet and the burgeoning information apparatus into our everyday lives. The burdens of life without a fixed destiny turned us toward the empowering information-rich resources of the new digital milieu as it offered new ways to amplify our voices and forge our own chosen patterns of connection. So profound is this phenomenon that one can say without exaggeration that the individual as the author of his or her own life is the protagonist of our age, whether we experience this fact as emancipation or affliction.[16]

Western modernity had formed around a canon of principles and laws that confer inviolable individual rights and acknowledge the sanctity of each individual life.[17] However, it was not until the second modernity that felt experience began to catch up with formal law. This felt truth has been expressed in new demands to make *actual* in everyday life what is already established in law.[18]

In spite of its liberating potential, the second modernity was slated to become a hard place to live, and our conditions of existence today reflect this trouble. Some of the challenges of the second modernity arise from the inevitable costs associated with the creation and sustenance of one's own life, but second-modernity instability is also the result of institutionalized shifts in economic and social policies and practices associated with the neoliberal paradigm and its rise to dominance. This far-reaching paradigm has been aimed at containing, rechanneling, and reversing the secular wave of second-modernity claims to self-determination and the habitats in which those claims can thrive. We live in this *collision* between a centuries-old story of modernization and a decades-old story of economic violence that thwarts our pursuit of effective life.

There is a rich and compelling literature that documents this turning point in economic history, and my aim here is simply to call attention to some of the themes in this larger narrative that are vital to our understanding of the collision: the condition of existence that summoned both the Apple "miracle" and surveillance capitalism's subsequent gestation and growth.[19]

III. The Neoliberal Habitat

The mid-1970s saw the postwar economic order under siege from stagnation, inflation, and sharply reduced growth, most markedly in the US and the UK. There were also new pressures on the political order as second-modernity individuals—especially students, young workers, African Americans, women, Latinos, and other marginalized groups—mobilized around demands for equal rights, voice, and participation. In the US the Vietnam War was a focal point of social unrest, and the corruption exposed by the Watergate scandal triggered public insistence on political reform. In the UK inflation had strained industrial relations beyond the breaking point. In both countries the specter of apparently intractable economic decay combined with vocal new demands on the democratic social compact produced confusion, anxiety, and desperation among elected officials ill-equipped to judge why once-reliable Keynesian policies had failed to reverse the course.

Neoliberal economists had been waiting in the wings for this opportunity, and their ideas flowed into the "policy vacuum" that now bedeviled both governments.[20] Led by the Austrian economist Friedrich Hayek, fresh from his 1974 Nobel Prize, and his American counterpart Milton Friedman, who received the Nobel two years later, they had honed their radical free-market economic theory, political ideology, and pragmatic agenda throughout the postwar period at the fringe of their profession, under the shadow of Keynesian domination, and now their time had come.[21]

The free-market creed originated in Europe as a sweeping defense against the threat of totalitarian and communist collectivist ideologies. It aimed to revive acceptance of a self-regulating market as a natural force of such complexity and perfection that it demanded radical freedom from all forms of state oversight. Hayek explained the necessity of absolute individual and collective submission to the exacting disciplines of the market as an unknowable "extended order" that supersedes the legitimate political authority vested in the state: "Modern economics explains how such an extended order... constitutes an information-gathering process... that no central planning agency, let alone any individual, could know as a whole, possess, or control...."[22] Hayek and his ideological brethren insisted on a capitalism stripped down to its raw core, unimpeded by any other force and impervious to any external authority. Inequality of wealth and rights was accepted and even celebrated as a necessary feature of a successful market system and as a force for progress.[23] Hayek's ideology provided the intellectual superstructure and legitimation for a new theory of the firm that became another crucial antecedent to the surveillance capitalist corporation: its structure, moral content, and relationship to society.

The new conception was operationalized by economists Michael Jensen and William Meckling. Leaning heavily on Hayek's work, the two scholars took an ax to the pro-social principles of the twentieth-century corporation, an ax that became known as the "shareholder value movement." In 1976 Jensen and Meckling published a landmark article in which they reinterpreted the manager as a sort of parasite feeding off the host of ownership: unavoidable, perhaps, but nonetheless an obstacle to shareholder wealth. They boldly argued that the structural disconnect between owners and managers "can result in the value of the firm being substantially lower than it otherwise could be."[24] If managers suboptimized the value of the firm to its owners in favor of their own preferences and comfort, it was only rational for them to do so. The solution, these economists argued, was to assert the market's signal of value, the share price, as the basis for a new incentive structure intended to finally and decisively align managerial behavior with owners' interests. Managers who failed to bend to the ineffable signals of Hayek's "extended order" would quickly become prey to the "barbarians at the gate" in a new and vicious hunt for unrealized market value.

In the "crisis of democracy" zeitgeist, the neoliberal vision and its reversion to market metrics was deeply attractive to politicians and policy makers, both as the means to evade political ownership of tough economic choices and because it promised to impose a new kind of *order* where disorder was feared.[25] The absolute authority of market forces would be enshrined as the ultimate source of imperative control, displacing democratic contest and deliberation with an ideology of atomized individuals sentenced to perpetual competition for scarce resources. The disciplines of competitive markets promised to quiet unruly individuals and even transform them back into subjects too preoccupied with survival to complain.

As the old collectivist enemies had receded, new ones took their place: state regulation and oversight, social legislation and welfare policies, labor unions and the institutions of collective bargaining, and the principles of democratic politics. Indeed, all these were to be replaced by the market's version of truth, and competition would be the solution to growth. The new aims would be achieved through supply-side reforms, including deregulation, privatization, and lower taxes.

Thirty-five years before Hayek and Friedman's ascendance, the great historian Karl Polanyi wrote eloquently on the rise of the market economy. Polanyi's studies led him to conclude that the operations of a self-regulating market are profoundly destructive when allowed to run free of countervailing laws and policies. He described the *double movement*: "a network of measures and policies...integrated into powerful institutions designed to check the action of the market relative to labor, land, and money."[26]

The double movement, Polanyi argued, supports the market form while tethering it to society: balancing, moderating, and mitigating its destructive excesses. Polanyi observed that such countermeasures emerged spontaneously in every European society during the second half of the nineteenth century. Each constructed legislative, regulatory, and institutional solutions to oversee contested new arenas such as workers' compensation, factory inspection, municipal trading, public utilities, food safety, child labor, and public safety.

In the US the double movement was achieved through decades of social contest that harnessed industrial production, however imperfectly, to society's needs. It appeared in the trust busting, civil society, and legislative reforms of the Progressive Era. Later it was elaborated in the legislative, juridical, social, and tax initiatives of the New Deal and the institutionalization of Keynesian economics during the post–World War II era: labor market, tax, and social welfare policies that ultimately increased economic and social equality.[27] The double movement was further developed in the legislative initiatives of the Great Society, especially civil rights law and landmark environmental legislation. Many scholars credit such countermeasures with the success of market democracy in the US and Europe, a political economics that proved far more adaptive in its ability to produce reciprocities of demand and supply than either leftist theorists or even Polanyi had imagined, and by mid-century the large corporation appeared to be a deeply rooted and durable modern social institution.[28]

The double movement was scheduled for demolition under the neoliberal flag, and implementation began immediately. In 1976, the same year that Jensen and Meckling published their pathbreaking analysis, President Jimmy Carter initiated the first significant efforts to radically align the corporation with Wall Street's market metrics, targeting the airline, transportation, and financial sectors with a bold program of deregulation. What began as a "ripple" turned into "a tidal wave that washed away controls from large segments of the economy in the last two decades of the twentieth century."[29] The implementation that began with Carter would define the Reagan and Thatcher eras, virtually every subsequent US presidency, and much of the rest of the world, as the new fiscal and social policies spread to Europe and other regions in varying degrees.[30]

Thus began the disaggregation and diminishment of the US public firm.[31] The public corporation as a social institution was reinterpreted as a costly error, and its long-standing reciprocities with customers and employees were recast as destructive violations of market efficiency. Financial carrots and sticks persuaded executives to dismember and shrink their companies, and the logic of capitalism shifted from the profitable production of goods and services to increasingly exotic forms of financial speculation. The disciplines imposed by the new market operations stripped capitalism down to its raw core, and by 1989 Jensen confidently proclaimed the "eclipse of the public corporation."[32]

By the turn of the century, as the foundational mechanisms of surveillance capitalism were just beginning to take shape, "shareholder value maximization" was widely accepted as the "objective function" of the firm.[33] These principles, culled from a once-extremist philosophy, were canonized as standard practice across commercial, financial, and legal domains.[34] By 2000, US public corporations employed fewer than half as many Americans as they did in 1970.[35] In 2009 there were only half as many public firms as in 1997. The public corporation had become "unnecessary for production, unsuited for stable employment and the provision of social welfare services, and incapable of proving a reliable long-term return on investment."[36] In this process the cult of the "entrepreneur" would rise to near-mythic prominence as the perfect union of ownership and management, replacing the rich existential possibilities of the second modernity with a single glorified template of audacity, competitive cunning, dominance, and wealth.

IV. The Instability of the Second Modernity

On August 9, 2011, around the same time that cheers erupted in Apple's conference room, 16,000 police officers flooded the streets of London, determined to quell "the most widespread and prolonged breakdown of order in London's history since the Gordon riot of 1780."[37] The rioting had begun four nights earlier when a peaceful vigil triggered by the police shooting of a young man suddenly turned violent. In the days that followed, the number of rioters swelled as looting and arson spread to twenty-two of London's thirty-two boroughs and other major cities across Britain.[38] Over four days of street action, thousands of people caused property damage of over $50 million, and 3,000 people were arrested.

Even as Apple's ascension appeared to ratify the claims of second-modernity individuals, the streets of London told the grim legacy of a three-decade experiment in economic growth through exclusion. One week after the rioting, an article by sociologist Saskia Sassen in the *Daily Beast* observed that "if there's one underlying condition, it has to do with the unemployment and bitter poverty among people who desire to be part of the middle class and who are keenly aware of the sharp inequality between themselves and their country's wealthy elite. These are in many ways social revolutions with a small 'r,' protests against social conditions that have become unbearable."[39]

What were the social conditions that had become so *unbearable?* Many analysts agreed that the tragedy of Britain's riots was set into motion by neoliberalism's successful transformation of society: a program that was most comprehensively executed in the UK and the US. Indeed, research from the London School of Economics based on interviews with 270 people who had participated in the rioting reported on the predominant theme of inequality: "no job, no money."[40] The terms of reference in nearly every study sound the same drumbeat: lack of opportunity, lack of access to education, marginalization, deprivation, grievance, hopelessness.[41] And although the London riots differed substantially from other protests that preceded and followed, most notably the Indignados movement that began with a large-scale public mobilization in Madrid in May 2011 and the Occupy movement that would emerge on September 17 in Wall Street's Zuccotti Park, they shared a point of origin in the themes of economic inequality and exclusion.[42]

The US, the UK, and most of Europe entered the second decade of the twenty-first century facing economic and social inequalities more extreme than anything since the Gilded Age and comparable to some of the world's poorest countries.[43] Despite a decade of explosive digital growth that included the Apple miracle and the penetration of the internet into everyday life, dangerous social divisions suggested an even more stratified and antidemocratic future. "In the age of new consensus financial policy stabilization," one US economist wrote, "the economy has witnessed the largest transfer of income to the top in history."[44] A sobering 2016 report from the International Monetary Fund warned of instability, concluding that the global trends toward neoliberalism "have not delivered as expected." Instead, inequality had significantly diminished "the level and the durability of growth" while increasing volatility and creating permanent vulnerability to economic crisis.[45]

The quest for effective life had been driven to the breaking point under the aegis of market freedom. Two years after the North London riots, research in the UK showed that by 2013, poverty fueled by lack of education and unemployment already excluded nearly a third of the population from routine social participation.[46] Another UK report concluded, "Workers on low and middle incomes are experiencing the biggest decline in their living standards since reliable records began in the mid-19th Century."[47] By 2015, austerity measures had eliminated 19 percent, or 18 billion pounds, from the budgets of local authorities, had forced an 8 percent cut in child protection spending, and had caused 150,000 pensioners to no longer enjoy access to vital services.[48] By 2014 nearly half of the US population lived in functional poverty, with the highest wage in the bottom half of earners at about $34,000.[49] A 2012 US Department of Agriculture survey showed that close to 49 million people lived in "food-insecure" households.[50]

In *Capital in the Twenty-First Century,* the French economist Thomas Piketty integrated years of income data to derive a general law of accumulation: the rate of return on capital tends to exceed the rate of economic growth. This tendency, summarized as $r > g$, is a dynamic that produces ever-more-extreme income divergence and with it a range of antidemocratic social consequences long predicted as harbingers of an eventual crisis of capitalism. In this context, Piketty cites the ways in which financial elites use their outsized earnings to fund a cycle of political capture that protects their interests from political challenge.[51] Indeed, a 2015 *New York Times* report concluded that 158 US families and their corporations provided almost half ($176 million) of all the money that was raised by both political parties in support of presidential candidates in 2016, primarily in support of "Republican candidates who have pledged to pare regulations, cut taxes...and shrink entitlements."[52] Historians, investigative journalists, economists, and political scientists have analyzed the intricate facts of a turn toward oligarchy, shining a light on the systematic campaigns of public influence and political capture that helped drive and preserve an extreme free-market agenda at the expense of democracy.[53]

A précis of Piketty's extensive research may be stated simply: *capitalism should not be eaten raw.* Capitalism, like sausage, is meant to be cooked by a democratic society and its institutions because raw capitalism is antisocial. As Piketty warns, "A market economy...if left to itself...contains powerful

forces of divergence, which are potentially threatening to democratic societies and to the values of social justice on which they are based."[54] Many scholars have taken to describing these new conditions as *neofeudalism,* marked
by the consolidation of elite wealth and power far beyond the control of ordinary people and the mechanisms of democratic consent.[55] Piketty calls it
a return to "patrimonial capitalism," a reversion to a premodern society in
which one's life chances depend upon inherited wealth rather than merito-
cratic achievement.[56]

We now have the tools to grasp the collision in all of its destructive complexity: *what is unbearable is that economic and social inequalities have reverted
to the preindustrial "feudal" pattern but that we, the people, have not.* We are
not illiterate peasants, serfs, or slaves. Whether "middle class" or "marginalized," we share the collective historical condition of individualized persons with
complex social experiences and opinions. We are hundreds of millions or even
billions of second-modernity people whom history has freed both from the
once-immutable facts of a destiny told at birth and from the conditions of
mass society. We know ourselves to be worthy of dignity and the opportunity
to live an effective life. This is existential toothpaste that, once liberated, cannot be squeezed back into the tube. Like a detonation's rippling sound waves
of destruction, the reverberations of pain and anger that have come to define
our era arise from this poisonous collision between inequality's facts and inequality's feelings.[57]

Back in 2011, those 270 interviews of London participants in the riots also
reflected the scars of this collision. "They expressed it in different ways," the
report concludes, "but at heart what the rioters talked about was a pervasive
sense of injustice. For some, this was economic—the lack of a job, money, or
opportunity. For others it was more broadly social, not just the absence of material things, but how they felt they were treated compared with others...." The
"sense of being invisible" was "widespread." As one woman explained, "The
young these days *need to be heard.* It's got to be justice for them." And a young
man reflected, "When no one cares about you you're gonna eventually make
them care, you're gonna cause a disturbance."[58] Other analyses cite "the denial
of dignity" expressed in the wordless anger of the North London rampage.[59]

When the Occupy movement erupted on another continent far from
London's beleaguered neighborhoods, it appeared to have little in common

with the violent eruptions that August. The 99 percent that Occupy intended to represent is not marginalized; on the contrary, the very legitimacy of Occupy was its claim to supermajority status. Nevertheless, Occupy revealed a similar conflict between inequality's facts and inequality's feelings, expressed in a creatively individualized political culture that insisted on "direct democracy" and "horizontal leadership."[60] Some analysts concluded that it was this conflict that ultimately crippled the movement, with its "inner core" of leaders unwilling to compromise their highly individualized approach in favor of the strategies and tactics required for a durable mass movement.[61] However, one thing is certain: there were no serfs in Zuccotti Park. On the contrary, as one close observer of the movement ruminated, "What is different is that from the start very large sections of we, the people, proved to be wiser than our rulers. We saw further and proved to have better judgment, thus reversing the traditional legitimacy of our elite governance that those in charge know better than the unwashed."[62]

This is the existential contradiction of the second modernity that defines our conditions of existence: we want to exercise control over our own lives, but everywhere that control is thwarted. Individualization has sent each one of us on the prowl for the resources we need to ensure effective life, but at each turn we are forced to do battle with an economics and politics from whose vantage point we are but ciphers. We live in the knowledge that our lives have unique value, but we are treated as invisible. As the rewards of late-stage financial capitalism slip beyond our grasp, we are left to contemplate the future in a bewilderment that erupts into violence with increasing frequency. Our expectations of psychological self-determination are the grounds upon which our dreams unfold, so the losses we experience in the slow burn of rising inequality, exclusion, pervasive competition, and degrading stratification are not only economic. They slice us to the quick in dismay and bitterness because we know ourselves to be worthy of individual dignity and the right to a life on our own terms.

The deepest contradiction of our time, the social philosopher Zygmunt Bauman wrote, is "the yawning gap between the right of self-assertion and the capacity to control the social settings which render such self-assertion feasible. It is from that abysmal gap that the most poisonous effluvia contaminating the lives of contemporary individuals emanate." Any new chapter in the centuries-old story of human emancipation, he insisted, must begin here. Can the instability of the second modernity give way to a new synthesis: a *third modernity* that transcends the collision, offering a genuine path to a flourishing and effective life for the many, not just the few? What role will information capitalism play?

V. A Third Modernity

Apple once launched itself into that "abysmal gap," and for a time it seemed that the company's fusion of capitalism and the digital might set a new course toward a third modernity. The promise of an advocacy-oriented digital capitalism during the first decade of our century galvanized second-modernity populations around the world. New companies such as Google and Facebook appeared to bring the promise of the inversion to life in new domains of critical importance, rescuing information and people from the old institutional confines, enabling us to find what and whom we wanted, when and how we wanted to search or connect.

The Apple inversion implied trustworthy relationships of advocacy and reciprocity embedded in an alignment of commercial operations with consumers' genuine interests. It held out the promise of a new digital market form that might transcend the collision: an early intimation of a third-modernity capitalism summoned by the self-determining aspirations of individuals and indigenous to the digital milieu. The opportunity for "my life, my way, at a price I can afford" was the human promise that quickly lodged at the very heart of the commercial digital project, from iPhones to one-click ordering to massive open online courses to on-demand services to hundreds of thousands of web-based enterprises, apps, and devices.

There were missteps, shortfalls, and vulnerabilities, to be sure. The potential significance of Apple's tacit new logic was never fully grasped, even by the company itself. Instead, the corporation produced a steady stream of contradictions that signaled business as usual. Apple was criticized for extractive pricing policies, offshoring jobs, exploiting its retail staff, abrogating responsibility for factory conditions, colluding to depress wages via illicit noncompete agreements in employee recruitment, institutionalized tax evasion, and a lack of environmental stewardship—just to name a few of the violations that seemed to negate the implicit social contract of its own unique logic.

When it comes to genuine economic mutation, there is always a tension between the new features of the form and its mother ship. A combination of old and new is reconfigured in an unprecedented pattern. Occasionally, the elements of a mutation find the right environment in which to be "selected" for propagation. This is when the new form stands a chance of becoming fully institutionalized and establishes its unique migratory path toward the future. But it's even more likely that potential mutations meet their fate in "transition failure," drawn back by the gravitational pull of established practices.[63]

Was the Apple inversion a powerful new economic mutation running the gauntlet of trial and error on its way to fulfilling the needs of a new age, or was it a case of transition failure? In our enthusiasm and growing dependency on technology, we tended to forget that the same forces of capital from which we had fled in the "real" world were rapidly claiming ownership of the wider digital sphere. This left us vulnerable and caught unawares when the early promise of information capitalism took a darker turn. We celebrated the promise of "help is on the way" while troubling questions broke through the haze with increasing regularity, each one followed by a predictable eruption of dismay and anger.

Why did Google's Gmail, launched in 2004, scan private correspondence to generate advertising? As soon as the first Gmail user saw the first ad targeted to the content of her private correspondence, public reaction was swift. Many were repelled and outraged; others were confused. As Google chronicler Steven Levy put it, "By serving ads related to content, Google seemed almost to be reveling in the fact that users' privacy was at the mercy of the policies and trustworthiness of the company that owned the servers. And since those ads made profits, Google was making it clear that it would exploit the situation."[64]

In 2007 Facebook launched Beacon, touting it as "a new way to socially distribute information." Beacon enabled Facebook advertisers to track users across the internet, disclosing users' purchases to their personal networks without permission. Most people were outraged by the company's audacity, both in tracking them online and in usurping their ability to control the disclosure of their own facts. Facebook founder Mark Zuckerberg shut the program down under duress, but by 2010 he declared that privacy was no longer a social norm and then congratulated himself for relaxing the company's "privacy policies" to reflect this self-interested assertion of a new social condition.[65] Zuckerberg had apparently never read user Jonathan Trenn's rendering of his Beacon experience:

> I purchased a diamond engagement ring set from overstock in preparation for a New Year's surprise for my girlfriend.... Within hours, I received a shocking call from one of my best friends of surprise and "congratulations" for getting engaged.(!!!) Imagine my horror when I learned that overstock had published the details of my purchase (including a link to the item and its price) on my public Facebook newsfeed, as well as notifications to all of my friends. ALL OF MY FRIENDS, including my girlfriend, and all of her friends, etc.... ALL OF THIS WAS WITHOUT MY CONSENT OR KNOWLEDGE. I am totally distressed that my surprise was ruined, and what was meant to be something special and a lifetime memory for my girlfriend and I was destroyed by a totally underhanded and infuriating privacy invasion. I want to wring the neck of the folks at overstock and facebook who thought that this was a good idea. It sets a terrible precedent on the net, and I feel that it ruined a part of my life.[66]

Among the many violations of advocacy expectations, ubiquitous "terms-of-service agreements" were among the most pernicious.[67] Legal experts call these "contracts of adhesion" because they impose take-it-or-leave-it conditions on users that stick to them whether they like it or not. Online "contracts" such as terms-of-service or terms-of-use agreements are also referred to as "click-wrap" because, as a great deal of research shows, most people get wrapped in these oppressive contract terms by simply clicking on the box that says "I agree" without ever reading the agreement.[68] In many cases, simply browsing a website obligates you to its terms-of-service agreement even if you don't know it. Scholars point out that these digital documents are excessively long and complex in part to discourage users from actually reading the terms, safe in the knowledge that most courts have upheld the legitimacy of click-wrap agreements despite the obvious lack of meaningful consent.[69] US Supreme Court Chief Justice John Roberts admitted that he "doesn't read the computer fine print."[70] Adding insult to injury, terms of service can be altered unilaterally by the firm at any time, without specific user knowledge or consent, and the terms typically implicate other companies (partners, suppliers, marketers, advertising intermediaries, etc.) without stating or accepting responsibility for *their* terms of service. These "contracts" impose an unwinnable infinite regress upon the user that law professor Nancy Kim describes as "sadistic."

Legal scholar Margaret Radin observes the Alice-in-Wonderland quality of such "contracts." Indeed, the sacred notions of "agreement" and "promise" so critical to the evolution of the institution of contract since Roman times have devolved to a "talismanic" signal "merely indicating that the firm deploying the boilerplate wants the recipient to be bound."[71] Radin calls this "private eminent domain," a unilateral seizure of rights without consent. She regards such "contracts" as a moral and democratic "degradation" of the rule of law and the institution of contract, a perversion that restructures the rights of users granted through democratic processes, "substituting for them the system that the firm wishes to impose.... Recipients must enter a legal universe of the firm's devising in order to engage in transactions with the firm."[72]

The digital milieu has been essential to these degradations. Kim points out that paper documents once imposed natural restraints on contracting behavior simply by virtue of their cost to produce, distribute, and archive. Paper contracts require a physical signature, limiting the burden a firm is likely to impose on a customer by requiring her to read multiple pages of fine print. Digital terms, in contrast, are "weightless." They can be expanded, reproduced, distributed, and archived at no additional cost. Once firms understood that the courts were disposed to validate their click-wrap and browse-wrap agreements, there was nothing to stop them from expanding the reach of these degraded contracts "to extract from consumers additional benefits unrelated to the transaction."[73] This coincided with the discovery of behavioral surplus that we examine in Chapter 3, as terms-of-service agreements were extended to include baroque and perverse "privacy policies," establishing another infinite regress of these terms of expropriation. Even the former Federal Trade Commission Chairperson Jon Leibowitz publicly stated, "We all agree that consumers don't read privacy policies."[74] In 2008 two Carnegie Mellon professors calculated that a reasonable reading of all the privacy policies that one encounters in a year would require 76 full workdays at a national opportunity cost of $781 billion.[75] The numbers are much higher today. Still, most users remain unaware of these "rapacious" terms that, as Kim puts it, allow firms "to acquire rights without bargaining and to stealthily establish and embed practices before users, and regulators, realize what has happened."[76]

At first, it had seemed that the new internet companies had simply failed to grasp the moral, social, and institutional requirements of their own economic logic. But with each corporate transgression, it became more difficult to ignore the possibility that the pattern of violations signaled a feature, not a bug. Although the Apple miracle contained the seeds of economic reformation, it was poorly understood: a mystery even to itself. Long before the death of its legendary founder, Steve Jobs, its frequent abuses of user expectations raised questions about how well the corporation understood the deep structure and historic potential of its own creations. The dramatic success of Apple's iPod and iTunes instilled internet users with a sense of optimism toward the new digital capitalism, but Apple never did seize the reins on developing the consistent, comprehensive social and institutional processes that would have elevated the iPod's promise to an explicit market form, as Henry Ford and Alfred Sloan had once done.

These developments reflect the simple truth that genuine economic reformation takes time and that the internet world, its investors and shareholders, were and are in a hurry. The credo of digital innovation quickly turned to the language of disruption and an obsession with speed, its campaigns conducted under the flag of "creative destruction." That famous, fateful phrase coined by evolutionary economist Joseph Schumpeter was seized upon as a way to legitimate what Silicon Valley euphemistically calls "permissionless innovation."[77] Destruction rhetoric promoted what I think of as a "boys and their toys" theory of history, as if the winning hand in capitalism is about blowing things up with new technologies. Schumpeter's analysis was, in fact, far more nuanced and complex than modern destruction rhetoric suggests.

Although Schumpeter regarded capitalism as an "evolutionary" process, he also considered that relatively few of its continuous innovations actually rise to the level of evolutionary significance. These rare events are what he called "mutations." These are enduring, sustainable, qualitative shifts in the logic, understanding, and practice of capitalist accumulation, not random, temporary, or opportunistic reactions to circumstances. Schumpeter insisted that this evolutionary mechanism is triggered by new consumer needs, and alignment with those needs is the discipline that drives sustainable mutation: "The capitalist process, not by coincidence but by virtue of its mechanism, progressively raises the standard of life of the masses."[78]

If a mutation is to be reliably sustained, its new aims and practices must be translated into new institutional forms: "The fundamental impulse that sets and keeps the capitalist engine in motion comes from the new consumers' goods, the new methods of production or transportation, the new markets, the new forms of industrial organization that capitalist enterprise creates." Note that Schumpeter says "creates," not "destroys." As an example of mutation, Schumpeter cites "the stages of organizational development from the craft shop to the factory to a complex corporation like U.S. Steel...."[79]

Schumpeter understood creative destruction as one unfortunate by-product of a long and complex process of creative sustainable change. "Capitalism," he wrote, *"creates and destroys."* Schumpeter was adamant on this point: "Creative response shapes the whole course of subsequent events and their 'long-run' outcome.... Creative response changes social and economic situations for good.... This is why creative response is an essential element in the historical process: No deterministic credo avails against this."[80] Finally, and contrary to the rhetoric of Silicon Valley and its worship of speed, Schumpeter argued that genuine mutation demands patience: "We are dealing with a process whose every element takes considerable time in revealing its true features and ultimate effects.... We must judge its performance over time, as it unfolds through decades or centuries."[81]

The significance of a "mutation" in Schumpeter's reckoning implies a high threshold, one that is crossed in time through the serious work of inventing new institutional forms embedded in the new needs of new people. Relatively little destruction is creative, especially in the absence of a robust double movement. This is illustrated in Schumpeter's example of US Steel, founded by some of the Gilded Age's most notorious "robber barons," including Andrew Carnegie and J. P. Morgan. Under pressure from an increasingly insistent double movement, US Steel eventually institutionalized fair labor practices through unions and collective bargaining as well as internal labor markets, career ladders, professional hierarchies, employment security, training, and development, all while implementing its technological advances in mass production.

Mutation is not a fairy tale; it is rational capitalism, bound in reciprocities with its populations through democratic institutions. Mutations fundamentally change the nature of capitalism by shifting it in the direction of those it is supposed to serve. This sort of thinking is not nearly as sexy or exciting as the "boys and their toys" gambit would have us think, but this is what it will take to move the dial of economic history beyond the collision and toward a third modernity.

VI. Surveillance Capitalism Fills the Void

A new breed of economic power swiftly filled the void in which every casual search, like, and click was claimed as an asset to be tracked, parsed, and monetized by some company, all within a decade of the iPod's debut. It was as if a shark had been silently circling the depths all along, just below the surface of the action, only to occasionally leap glistening from the water in pursuit of a fresh bite of flesh. Eventually, companies began to explain these violations as the necessary quid pro quo for "free" internet services. Privacy, they said, was the price one must pay for the abundant rewards of information, connection, and other digital goods when, where, and how you want them. These explanations distracted us from the sea change that would rewrite the rules of capitalism and the digital world.

In retrospect, we can see that the many discordant challenges to users' expectations were actually tiny peepholes into a rapidly emerging institutional form that was learning to exploit second-modernity needs and the established norms of "growth through exclusion" as the means to an utterly novel market project. Over time, the shark revealed itself as a rapidly multiplying, systemic, internally consistent new variant of information capitalism that had set its sights on domination. An unprecedented formulation of capitalism was elbowing its way into history: surveillance capitalism.

This new market form is a unique logic of accumulation in which surveillance is a foundational mechanism in the transformation of investment into profit. Its rapid rise, institutional elaboration, and significant expansion challenged the tentative promise of the inversion and its advocacy-oriented values. More generally, the rise of surveillance capitalism betrayed the hopes and expectations of many "netizens" who cherished the emancipatory promise of the networked milieu.[82]

Surveillance capitalism commandeered the wonders of the digital world to meet our needs for effective life, promising the magic of unlimited information and a thousand ways to anticipate our needs and ease the complexities of our harried lives. We welcomed it into our hearts and homes with our own rituals of hospitality. As we shall explore in detail throughout the coming chapters, thanks to surveillance capitalism the resources for effective life that we seek in the digital realm now come encumbered with a new breed of menace. Under this new regime, the precise moment at which our needs are met is also the precise moment at which our lives are plundered for behavioral data, and all for the sake of others' gain. The result is a perverse amalgam of empowerment inextricably layered with diminishment. In the absence of a decisive societal response that constrains or outlaws this logic of accumulation, surveillance capitalism appears poised to become the dominant form of capitalism in our time.

How did this happen? It is a question that we shall return to throughout this book as we accumulate new insights and answers. For now we can recognize that over the centuries we have imagined threat in the form of state power. This left us wholly unprepared to defend ourselves from new companies with imaginative names run by young geniuses that seemed able to provide us with exactly what we yearn for at little or no cost. This new regime's most poignant harms, now and later, have been difficult to grasp or theorize, blurred by extreme velocity and camouflaged by expensive and illegible machine operations, secretive corporate practices, masterful rhetorical misdirection, and purposeful cultural misappropriation. On this road, terms whose meanings we take to be positive or at least banal—"the open internet," "interoperability," and "connectivity"—have been quietly harnessed to a market process in which individuals are definitively cast as the means to others' market ends.

Surveillance capitalism has taken root so quickly that, with the exception of a courageous cadre of legal scholars and technology-savvy activists, it has cunningly managed to evade our understanding and agreement. As we will discuss in more depth in Chapter 4, surveillance capitalism is inconceivable outside the digital milieu, but neoliberal ideology and policy also provided the habitat in which surveillance capitalism could flourish. This ideology and its practical implementation bends second-modernity individuals to the draconian quid pro quo at the heart of surveillance capitalism's logic of accumulation, in which information and connection are ransomed for the lucrative behavioral data that fund its immense growth and profits. Any effort to interrupt or dismantle surveillance capitalism will have to contend with this larger institutional landscape that protects and sustains its operations.

History offers no control groups, and we cannot say whether with different leadership, more time, or other altered circumstances Apple might have perceived, elaborated, and institutionalized the jewel in its crown as Henry Ford and Alfred Sloan had done in another era. Nor is that opportunity forever lost—far from it. We may yet see the founding of a new synthesis for a third modernity in which a genuine inversion and its social compact are institutionalized as principles of a new rational digital capitalism aligned with a society of individuals and supported by democratic institutions. The fact that Schumpeter reckoned the time line for such institutionalization in decades or even centuries lingers as a critical commentary on our larger story.

These developments are all the more dangerous because they cannot be reduced to known harms—monopoly, privacy—and therefore do not easily yield to known forms of combat. The new harms we face entail challenges to the sanctity of the individual, and chief among these challenges I count the elemental rights that bear on individual sovereignty, including the *right to the future tense* and *the right to sanctuary*. Each of these rights invokes claims to individual agency and personal autonomy as essential prerequisites to freedom of will and to the very concept of democratic order.

Right now, however, the extreme asymmetries of knowledge and power that have accrued to surveillance capitalism abrogate these elemental rights as our lives are unilaterally rendered as data, expropriated, and repurposed in new forms of social control, all of it in the service of others' interests and in the absence of our awareness or means of combat. We have yet to invent the politics and new forms of collaborative action—this century's equivalent of the social movements of the late nineteenth and twentieth centuries that aimed to tether raw capitalism to society—that effectively assert the people's right to a human future. And while the work of these inventions awaits us, this mobilization and the resistance it engenders will define a key battleground upon which the fight for a human future unfolds.

On August 9, 2011, events ricocheted between two wildly different visions of a third modernity. One was based on the digital promise of democratized information in the context of individualized economic and social relations. The other reflected the harsh truths of mass exclusion and elite rule. But the lessons of that day had not yet been fully tallied when fresh answers—or, more modestly, the tenuous glimmers of answers as fragile as a newborn's translucent skin—rose to the surface of the world's attention gliding on scented ribbons of Spanish lavender and vanilla.

VII. For a Human Future

In the wee hours of August 9, 2011, eighteen-year-old Maria Elena Montes sat on the cool marble floor of her family's century-old pastry shop in the El Raval section of Barcelona, nursing her cup of sweet café con leche, lulled by the sunrise scuffling of the pigeons in the plaza as she waited for her trays of rum-soaked gypsy cakes to set.

Pasteleria La Dulce occupied a cramped medieval building tucked into a tiny square on one of the few streets that had escaped both the wrecking ball and the influx of yuppie chic. The Montes family took care that the passing decades had no visible effect on their cherished bakery. Each morning they lovingly filled sparkling glass cases with crispy sugar-studded churros, delicate buñuelos fat with vanilla custard, tiny paper ramekins of strawberry flan, buttery mantecados, coiled ensaimadas drenched in powdered sugar, fluffy magdalenas, crunchy pestiños, and Great-Grandmother Montes's special flaó, a cake made with fresh milk cheese laced with Spanish lavender, fennel, and mint. There were almond and blood-orange tarts prepared, according to Señora Montes, exactly as they had once been served to Queen Isabella. Olive-oil ice cream flavored with anise filled the tubs in the gleaming white freezer along the wall. An old ceiling fan cycled slowly, nudging the perfume of honey and yeast into every corner of the ageless room.

Only one thing had changed. Any other August would have found Maria Elena and her family at their summer cottage nestled into a pine grove near the seaside town of Palafrugell that had been the family's refuge for generations. In 2011, however, neither the Montes nor their customers and friends would take their August holidays. The economic crisis had ripped through the country like the black plague, shrinking consumption and driving unemployment to 21 percent, the highest in the EU, and to an astonishing 46 percent among people under twenty-four years old. In Catalonia, the region that includes Barcelona, 18 percent of its 7.5 million people had fallen below the poverty line.[83] In the summer of 2011, few could afford the simple pleasure of an August spent by the sea or in the mountains.

There was new pressure to sell the building and let the future finally swallow La Dulce. The family could live comfortably on the proceeds of such a sale, even at the bargain rates they would be forced to accept. Business was slow, but Señor Fito Montes refused to lay off any members of a staff that was like an extended family after years of steady employment. Just about everyone they knew said that the end was inevitable and that the Montes should leap at the opportunity for a dignified exit. But the family was determined to make every sacrifice to safeguard Pasteleria La Dulce for the future.

Just three months earlier, Juan Pablo and Maria had made the pilgrimage to Madrid to join thousands of protesters at the Puerta del Sol, where a month-long encampment established Los Indignados, the 15M, as the new voice of a people who had finally been pushed to the breaking point by the economics of contempt. All that was left to say was "*Ya. No mas!*" Enough already! The convergence of so many citizens in Madrid led to a wave of protests across the nation, and eventually those protests would give way to new political parties, including Podemos. Neighborhood assemblies had begun to convene in many cities, and the Montes had attended such a meeting in El Raval just the night before.

With the evening's conversations still fresh, they gathered in the apartment above the shop in the early afternoon of August 9 to share their midday meal and discuss the fate of La Dulce, not quite certain what Papa Montes was thinking.

"The bankers may not know it," Fito Montes reflected, "but the future will need the past. It will need these marble floors and the sweet taste of my gypsy cakes. They treat us like figures in a ledger, like they are reading the number of casualties in a plane crash. They believe the future belongs only to them. But we each have our story. We each have our life. It is up to us to proclaim our right to the future. The future is our home too."

Maria and Juan Pablo breathed a shared sigh of relief as they outlined their plan. Juan Pablo would withdraw temporarily from his university studies, and Maria Elena would postpone her matriculation. They would work on expanding La Dulce's sales with new home-delivery and catering options. Everyone would take a pay cut, but no one would have to leave. Everyone would tighten their belts, except the fat buñuelos and their perfect comrades steadfast in neat, delicious rows.

We know how to challenge the inevitable, they said. We've survived wars; we've survived the Fascists. We'll survive again. For Fito Montes, his family's right to anticipate the future as their home demanded continuity for some things that are elusive, beautiful, surprising, mysterious, inexpressible, and immaterial but without which, they all agreed, life would be mechanical and soulless. He was determined, for example, to ensure that another generation of Spanish children would recognize the bouquet of his blood-orange tarts flecked with rose petals and thus be awakened to the mystery of medieval life in the fragrant gardens of the Alhambra.

On August 9 the heat rose steadily in the shady square, and the sun emptied the avenues where Huns, Moors, Castilians, and Bourbons had each in their turn marched to triumph. Those silent streets bore little evidence of the historic deliberations in Madrid that would be featured in the *New York Times* that very day.[84] But I imagine the two cities linked by invisible ribbons of scent rising from La Dulce high into the bleached Barcelona sky and drifting slowly south and west to settle along the austere facade of the building that housed the Agencia Española de Protección de Datos, where another struggle for the right to the future tense was underway.

The Spanish Data Protection Agency had chosen to champion the claims of ninety ordinary citizens who, like the Montes family, were determined to preserve inherited meaning for a world bent on change at the speed of light.[85] In the name of "the right to be forgotten," the Spaniards had stepped into the bullring brandishing red capes, resolved to master the fiercest bull of all: Google, the juggernaut of surveillance capitalism. When the agency ordered the internet firm to stop indexing the contested links of these ninety individuals, the bull received one of its first and most significant blows.

This official confrontation drew upon the same tenacity, determination, and sentiment that sustained the Montes family and millions of other Spaniards compelled to claw back the future from the self-proclaimed inevitability of indifferent capital. In the assertion of a right to be forgotten, the complexity of human existence, with its thousand million shades of gray, was pitted against surveillance capitalism's economic imperatives that produced the relentless drive to extract and retain information. It was there, in Spain, that the right to the future tense was on the move, insisting that the operations of surveillance capitalism and its digital architecture are not, never were, and never would be inevitable. Instead, the opposition asserted that even Google's capitalism was made by humans to be unmade and remade by democratic processes, not commercial decree. Google's was not to be the last word on the human or the digital future.

Each of the ninety citizens had a unique claim. One had been terrorized by her former husband and didn't want him to find her address online. Informational privacy was essential to her peace of mind and her physical safety. A middle-aged woman was embarrassed by an old arrest from her days as a university student. Informational privacy was essential to her identity and sense of dignity. One was an attorney, Mario Costeja González, who years earlier had suffered the foreclosure of his home. Although the matter had long been resolved, a Google search of his name continued to deliver links to the foreclosure notice, which, he argued, damaged his reputation. While the Spanish Data Protection Agency rejected the idea of requiring newspapers and other originating sites to remove legitimate information—such information, they reasoned, would exist somewhere under any circumstances—it endorsed the notion that Google had responsibility and should be held to account. After all, Google had unilaterally undertaken to change the rules of the information life cycle when it decided to crawl, index, and make accessible personal details across the world wide web without asking anyone's permission. The agency concluded that citizens had the right to request the removal of links and ordered Google to stop indexing the information and to remove existing links to its original sources.

Google's mission to "organize the world's information and make it universally accessible and useful"—starting with the web—changed all of our lives. There have been enormous benefits, to be sure. But for individuals it has meant that information that would normally age and be forgotten now remains forever young, highlighted in the foreground of each person's digital identity. The Spanish Data Protection Agency recognized that not all information is worthy of immortality. Some information should be forgotten because that is only human. Unsurprisingly, Google challenged the agency's order before the Spanish High Court, which selected one of the ninety cases, that of attorney Mario Costeja González, for referral to the Court of Justice of the European Union. There, after lengthy and dramatic deliberations, the Court of Justice announced its decision to assert the right to be forgotten as a fundamental principle of EU law in May of 2014.[86]

The Court of Justice's decision, so often reduced to the legal and technical considerations related to the deletion or de-linking of personal data, was in fact a key inflection point at which democracy began to claw back rights to the future tense from the powerful forces of a new surveillance capitalism determined to claim unilateral authority over the digital future. Instead, the court's analysis claimed the future for the human way, rejecting the inevitability of Google's search-engine technology and recognizing instead that search results are the contingent products of the specific economic interests that drive the action from within the belly of the machine: "The operator of a search engine is liable to affect significantly the fundamental rights to privacy and to the protection of personal data. In the light of the potential seriousness of the interference" with those interests, "it cannot be justified by merely the economic interest which the operator of such an engine has in that processing."[87] As legal scholars Paul M. Schwartz and Karl-Nikolaus Peifer summarized it, "The Luxembourg Court felt that free flow of information matters, but not as much, ultimately, as the safeguarding of dignity, privacy, and data protection in the European rights regime."[88] The court conferred upon EU citizens the right to combat, requiring Google to establish a process for implementing users' de-linking requests and authorizing citizens to seek recourse in democratic institutions, including "the supervisory authority or the judicial authority, so that it carries out the necessary checks and orders the controller to take specific measures accordingly."[89]

In reasserting the right to be forgotten, the court declared that decisive authority over the digital future rests with the people, their laws, and their democratic institutions. It affirmed that individuals and democratic societies can fight for their rights to the future tense and can win, even in the face of a great private power. As the human rights scholar Federico Fabbrini observed, with this vital case the European Court of Justice evolved more assertively into the role of a human rights court, stepping into "the mine-field of human rights in the digital age...."[90]

When the Court of Justice's decision was announced, the "smart money" said that it could never happen in the US, where the internet companies typically seek cover behind the First Amendment as justification for their "permissionless innovation."[91] Some technology observers called the ruling "nuts."[92] Google's leaders sneered at the decision. Reporters characterized Google cofounder Sergey Brin as "joking" and "dismissive." When asked about the ruling during a Q&A at a prominent tech conference, he said, "I wish we could just forget the ruling."[93]

In response to the ruling, Google CEO and cofounder Larry Page recited the catechism of the firm's mission statement, assuring the *Financial Times* that the company "still aims to 'organise the world's information and make it universally accessible and useful.'" Page defended Google's unprecedented information power with an extraordinary statement suggesting that people should trust Google more than democratic institutions: "In general, having the data present in companies like Google is better than having it in the government with no due process to get that data, because we obviously care about our reputation. I'm not sure the government cares about that as much."[94] Speaking to the company's shareholders the day after the court's ruling, Eric Schmidt characterized the decision as a "balance that was struck wrong" in the "collision between a right to be forgotten and a right to know."[95]

The comments of Google's leaders reflected their determination to retain privileged control over the future and their indignation at being challenged. However, there was ample evidence that the American public did not concede the corporation's unilateral power. In fact, the smart money appeared not to be all that smart. In the year following the EU decision, a national poll of US adults found that 88 percent supported a law similar to the right to be forgotten. That year, Pew Research found that 93 percent of Americans believed that it was important to have control of "who can get information about you." A series of polls echoed these findings.[96]

On January 1, 2015, California's "Online Eraser" law took effect, requiring the operator of a website, online service, online application, or mobile application to permit a minor who is a registered user of the operator's service to remove, or to request and obtain removal of, content or information posted by the minor. The California law breached a critical surveillance embattlement, attenuating Google's role as the self-proclaimed champion of an unbounded right to know and suggesting that we are still at the beginning, not the end, of a long and fitful drama.

The Spanish Data Protection Agency and later the European Court of Justice demonstrated the unbearable lightness of the inevitable, as both institutions declared what is at stake for a human future, beginning with the primacy of democratic institutions in shaping a healthy and just digital future. The smart money says that US law will never abandon its allegiance to the surveillance capitalists over the people. But the next decades may once again prove that the smart money can be wrong. As for the Spanish people, their Data Protection Agency, and the European Court of Justice, the passage of time is likely to reveal their achievements as a stirring early chapter in the longer story of our fight for a third modern that is first and foremost a human future, rooted in an inclusive democracy and committed to the individual's right to effective life. Their message is carefully inscribed for our children to ponder: *technological inevitability is as light as democracy is heavy, as temporary as the scent of rose petals and the taste of honey are enduring.*

VIII. Naming and Taming

Taming surveillance capitalism must begin with careful naming, a symbiosis that was vividly illustrated in the recent history of HIV research, and I offer it as an analogy. For three decades, scientists aimed to create a vaccine that followed the logic of earlier cures, training the immune system to produce neutralizing antibodies, but mounting data revealed unanticipated behaviors of the HIV virus that defy the patterns of other infectious diseases.[97]

The tide began to turn at the International AIDS Conference in 2012, when new strategies were presented that rely on a close understanding of the biology of rare HIV carriers whose blood produces natural antibodies. Research began to shift toward methods that reproduce this self-vaccinating response.[98] As a leading researcher announced, "We know the face of the enemy now, and so we have some real clues about how to approach the problem."[99]

The point for us is that every successful vaccine begins with a close understanding of the enemy disease. The mental models, vocabularies, and tools distilled from past catastrophes obstruct progress. We smell smoke and rush to close doors to rooms that are already fated to vanish. The result is like hurling snowballs at a smooth marble wall only to watch them slide down its facade, leaving nothing but a wet smear: a fine paid here, an operational detour there, a new encryption package there.

What is crucial now is that we identify this new form of capitalism on its own terms and in its own words. This pursuit necessarily returns us to Silicon Valley, where things move so fast that few people know what just happened. It is the habitat for progress "at the speed of dreams," as one Google engineer vividly describes it.[100] My aim here is to slow down the action in order to enlarge the space for such debate and unmask the tendencies of these new creations as they amplify inequality, intensify social hierarchy, exacerbate exclusion, usurp rights, and strip personal life of whatever it is that makes it personal for you or for me. If the digital future is to be our home, then it is we who must make it so. We will need to know. We will need to decide. We will need to decide who decides. This is our fight for a human future.

CHAPTER THREE

THE DISCOVERY OF BEHAVIORAL SURPLUS

He watched the stars and noted birds in flight;
A river flooded or a fortress fell:
He made predictions that were sometimes right;
His lucky guesses were rewarded well.

—W. H. AUDEN
SONNETS FROM CHINA, VI

I. Google: The Pioneer of Surveillance Capitalism

Google is to surveillance capitalism what the Ford Motor Company and General Motors were to mass-production–based managerial capitalism. New economic logics and their commercial models are discovered by people in a time and place and then perfected through trial and error. In our time Google became the pioneer, discoverer, elaborator, experimenter, lead practitioner, role model, and diffusion hub of *surveillance capitalism*. GM and Ford's iconic status as pioneers of twentieth-century capitalism made them enduring objects of scholarly research and public fascination because the lessons they had to teach resonated far beyond the individual companies. Google's practices deserve the same kind of examination, not merely as a critique of a single company but rather as the starting point for the codification of a powerful new form of capitalism.

With the triumph of mass production at Ford and for decades thereafter, hundreds of researchers, businesspeople, engineers, journalists, and scholars would excavate the circumstances of its invention, origins, and consequences.[1] Decades later, scholars continued to write extensively about Ford, the man and the company.[2] GM has also been an object of intense scrutiny. It was the site of Peter Drucker's field studies for his seminal *Concept of the Corporation*, the 1946 book that codified the practices of the twentieth-century business organization and established Drucker's reputation as a management sage. In addition to the many works of scholarship and analysis on these two firms, their own leaders enthusiastically articulated their discoveries and practices. Henry Ford and his general manager, James Couzens, and Alfred Sloan and his marketing man, Henry "Buck" Weaver, reflected on, conceptualized, and proselytized their achievements, specifically locating them in the evolutionary drama of American capitalism.[3]

Google is a notoriously secretive company, and one is hard-pressed to imagine a Drucker equivalent freely roaming the scene and scribbling in the hallways. Its executives carefully craft their messages of digital evangelism in books and blog posts, but its operations are not easily accessible to outside researchers or journalists.[4] In 2016 a lawsuit brought against the company by a product manager alleged an internal spying program in which employees are expected to identify coworkers who violate the firm's confidentiality agreement: a broad prohibition against divulging anything about the company to anyone.[5] The closest thing we have to a Buck Weaver or James Couzens codifying Google's practices and objectives is the company's longtime chief economist, Hal Varian, who aids the cause of understanding with scholarly articles that explore important themes. Varian has been described as "the Adam Smith of the discipline of Googlenomics" and the "godfather" of its advertising model.[6] It is in Varian's work that we find hidden-in-plain-sight important clues to the logic of surveillance capitalism and its claims to power.

In two extraordinary articles in scholarly journals, Varian explored the theme of "computer-mediated transactions" and their transformational effects on the modern economy.[7] Both pieces are written in amiable, down-to-earth prose, but Varian's casual understatement stands in counterpoint to his often-startling declarations: "Nowadays there is a computer in the middle of virtually every transaction... now that they are available these computers have several other uses."[8] He then identifies four such new uses: "data extraction and analysis," "new contractual forms due to better monitoring," "personalization and customization," and "continuous experiments."

Varian's discussions of these new "uses" are an unexpected guide to the strange logic of surveillance capitalism, the division of learning that it shapes, and the character of the information civilization toward which it leads. We will return to Varian's observations from time to time in the course of our examination of the foundations of surveillance capitalism, aided by a kind of "reverse engineering" of his assertions, so that we might grasp the worldview and methods of surveillance capitalism through this lens. "Data extraction and analysis," Varian writes, "is what everyone is talking about when they talk about big data." "Data" are the raw material necessary for surveillance capitalism's novel manufacturing processes. "Extraction" describes the social relations and material infrastructure with which the firm asserts authority over those raw materials to achieve economies of scale in its raw-material supply operations.

"Analysis" refers to the complex of highly specialized computational systems that I will generally refer to in these chapters as "machine intelligence." I like this umbrella phrase because it trains us on the forest rather than the trees, helping us decenter from technology to its objectives. But in choosing this phrase I also follow Google's lead. The company describes itself "at the forefront of innovation in machine intelligence," a term in which it includes machine learning as well as "classical" algorithmic production, along with many computational operations that are often referred to with other terms such as "predictive analytics" or "artificial intelligence." Among these operations Google cites its work on language translation, speech recognition, visual processing, ranking, statistical modeling, and prediction: "In all of those tasks and many others, we gather large volumes of direct or indirect evidence of relationships of interest, applying learning algorithms to understand and generalize."[9] These machine intelligence operations convert raw material into the firm's highly profitable algorithmic products designed to predict the behavior of its users. The inscrutability and exclusivity of these techniques and operations are the moat that surrounds the castle and secures the action within.

Google's invention of targeted advertising paved the way to financial success, but it also laid the cornerstone of a more far-reaching development: the discovery and elaboration of surveillance capitalism. Its business is characterized as an advertising model, and much has been written about Google's automated auction methods and other aspects of its inventions in the field of online advertising. With so much verbiage, these developments are both over-described and under-theorized. Our aim in this chapter and those that follow in Part I is to reveal the "laws of motion" that drive surveillance competition, and in order to do this we begin by looking freshly at the point of origin, when the foundational mechanisms of surveillance capitalism were first discovered.

Before we begin, I want to say a word about vocabulary. Any confrontation with the unprecedented requires new language, and I introduce new terms when existing language fails to capture a new phenomenon. Sometimes, however, I intentionally repurpose familiar language because I want to stress certain continuities in the function of an element or process. This is the case with "laws of motion," borrowed from Newton's laws of inertia, force, and equal and opposite reactions.

Over the years historians have adopted this term to describe the "laws" of industrial capitalism. For example, economic historian Ellen Meiksins Wood documents the origins of capitalism in the changing relations between English property owners and tenant farmers, as the owners began to favor productivity over coercion: "The new historical dynamic allows us to speak of 'agrarian capitalism' in early modern England, a social form with distinctive 'laws of motion' that would eventually give rise to capitalism in its mature, industrial form."[10] Wood describes how the new "laws of motion" eventually manifested themselves in industrial production:

> The critical factor in the divergence of capitalism from all other forms of "commercial society" was the development of certain social property relations that generated market imperatives and capitalist "laws of motion"... competitive production and profit-maximization, the compulsion to reinvest surpluses, and the relentless need to improve labour-productivity associated with capitalism.... Those laws of motion required vast social transformations and upheavals to set them in train. They required a transformation in the human metabolism with nature, in the provision of life's basic necessities.[11]

My argument here is that although surveillance capitalism does not abandon established capitalist "laws" such as competitive production, profit maximization, productivity, and growth, these earlier dynamics now operate in the context of a new logic of accumulation that also introduces its own distinctive laws of motion. Here and in following chapters, we will examine these foundational dynamics, including surveillance capitalism's idiosyncratic economic imperatives defined by extraction and prediction, its unique approach to economies of scale and scope in raw-material supply, its necessary construction and elaboration of *means of behavioral modification* that incorporate its machine-intelligence–based "means of production" in a more complex system of action, and the ways in which the requirements of behavioral modification orient all operations toward totalities of information and control, creating the framework for an unprecedented *instrumentarian power* and its societal implications. For now, my aim is to reconstruct our appreciation of familiar ground through new lenses: Google's early days of optimism, crisis, and invention.

II. A Balance of Power

Google was incorporated in 1998, founded by Stanford graduate students Larry Page and Sergey Brin just two years after the Mosaic browser threw open the doors of the world wide web to the computer-using public. From the start, the company embodied the promise of information capitalism as a liberating and democratic social force that galvanized and delighted second-modernity populations around the world.

Thanks to this wide embrace, Google successfully imposed computer mediation on broad new domains of human behavior as people searched online and engaged with the web through a growing roster of Google services. As these new activities were informated for the first time, they produced wholly new data resources. For example, in addition to key words, each Google search query produces a wake of collateral data such as the number and pattern of search terms, how a query is phrased, spelling, punctuation, dwell times, click patterns, and location.

Early on, these behavioral by-products were haphazardly stored and operationally ignored. Amit Patel, a young Stanford graduate student with a special interest in "data mining," is frequently credited with the groundbreaking insight into the significance of Google's accidental data caches. His work with these data logs persuaded him that detailed stories about each user—thoughts, feelings, interests—could be constructed from the wake of unstructured signals that trailed every online action. These data, he concluded, actually provided a "broad sensor of human behavior" and could be put to immediate use in realizing cofounder Larry Page's dream of Search as a comprehensive artificial intelligence.[12]

Google's engineers soon grasped that the continuous flows of collateral behavioral data could turn the search engine into a recursive learning system that constantly improved search results and spurred product innovations such as spell check, translation, and voice recognition. As Kenneth Cukier observed at that time,

> Other search engines in the 1990s had the chance to do the same, but did not pursue it. Around 2000 Yahoo! saw the potential, but nothing came of the idea. It was Google that recognized the gold dust in the detritus of its interactions with its users and took the trouble to collect it up.... Google exploits information that is a by-product of user interactions, or data exhaust, which is automatically recycled to improve the service or create an entirely new product.[13]

What had been regarded as waste material—"data exhaust" spewed into Google's servers during the combustive action of Search—was quickly reimagined as a critical element in the transformation of Google's search engine into a reflexive process of continuous learning and improvement.

At that early stage of Google's development, the feedback loops involved in improving its Search functions produced a balance of power: Search needed people to learn from, and people needed Search to learn from. This symbiosis enabled Google's algorithms to learn and produce ever-more relevant and comprehensive search results. More queries meant more learning; more learning produced more relevance. More relevance meant more searches and more users.[14] By the time the young company held its first press conference in 1999, to announce a $25 million equity investment from two of the most revered Silicon Valley venture capital firms, Sequoia Capital and Kleiner Perkins, Google Search was already fielding seven million requests each day.[15] A few years later, Hal Varian, who joined Google as its chief economist in 2002, would note, "Every action a user performs is considered a signal to be analyzed and fed back into the system."[16] The Page Rank algorithm, named after its founder, had already given Google a significant advantage in identifying the most popular results for queries. Over the course of the next few years it would be the capture, storage, analysis, and learning from the by-products of those search queries that would turn Google into the gold standard of web search.

The key point for us rests on a critical distinction. During this early period, behavioral data were put to work entirely on the user's behalf. User data provided value at no cost, and that value was reinvested in the user experience in the form of improved services: enhancements that were also offered at no cost to users. Users provided the raw material in the form of behavioral data, and those data were harvested to improve speed, accuracy, and relevance and to help build ancillary products such as translation. I call this the *behavioral value reinvestment cycle,* in which all behavioral data are reinvested in the improvement of the product or service (see Figure 1).

The cycle emulates the logic of the iPod; it worked beautifully at Google but with one critical difference: the absence of a sustainable market transaction. In the case of the iPod, the cycle was triggered by the purchase of a high-margin physical product. Subsequent reciprocities improved the iPod product and led to increased sales. Customers were the subjects of the commercial process, which promised alignment with their "what I want, when I want, where I want" demands. At Google, the cycle was similarly oriented toward the individual as its subject, but without a physical product to sell, it floated outside the marketplace, an interaction with "users" rather than a market transaction with customers.

This helps to explain why it is inaccurate to think of Google's users as its customers: there is no economic exchange, no price, and no profit. Nor do users function in the role of workers. When a capitalist hires workers and provides them with wages and means of production, the products that they produce belong to the capitalist to sell at a profit. Not so here. Users are not paid for their labor, nor do they operate the means of production, as we'll discuss in more depth later in this chapter. Finally, people often say that the user is the "product." This is also misleading, and it is a point that we will revisit more than once. For now let's say that users are not products, but rather we are the sources of raw-material supply. As we shall see, surveillance capitalism's unusual products manage to be derived from our behavior while remaining indifferent to our behavior. Its products are about predicting us, without actually caring what we do or what is done to us.

To summarize, at this early stage of Google's development, whatever Search users inadvertently gave up that was of value to the company they also used up in the form of improved services. In this reinvestment cycle, serving users with amazing Search results "consumed" all the value that users created when they provided extra behavioral data. The fact that users needed Search about as much as Search needed users created a balance of power between Google and its populations. People were treated as ends in themselves, the subjects of a nonmarket, self-contained cycle that was perfectly aligned with Google's stated mission "to organize the world's information, making it universally accessible and useful."

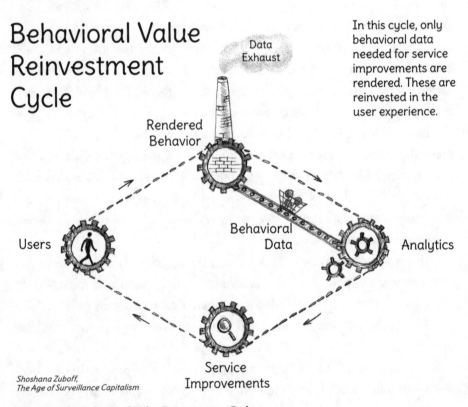

Behavioral Value Reinvestment Cycle

Data Exhaust

In this cycle, only behavioral data needed for service improvements are rendered. These are reinvested in the user experience.

Rendered Behavior

Users

Behavioral Data

Analytics

Service Improvements

Shoshana Zuboff, The Age of Surveillance Capitalism

Figure 1: The Behavioral Value Reinvestment Cycle

III. Search for Capitalism:
Impatient Money and the State of Exception

By 1999, despite the splendor of Google's new world of searchable web pages, its growing computer science capabilities, and its glamorous venture backers, there was no reliable way to turn investors' money into revenue. The behavioral value reinvestment cycle produced a very cool search function, but it was not yet capitalism. The balance of power made it financially risky and possibly counterproductive to charge users a fee for search services. Selling search results would also have set a dangerous precedent for the firm, assigning a price to indexed information that Google's web crawler had already taken from others without payment. Without a device like Apple's iPod or its digital songs, there were no margins, no surplus, nothing left over to sell and turn into revenue.

Google had relegated advertising to steerage class: its AdWords team consisted of seven people, most of whom shared the founders' general antipathy toward ads. The tone had been set in Sergey Brin and Larry Page's milestone paper that unveiled their search engine conception, "The Anatomy of a Large-Scale Hypertextual Web Search Engine," presented at the 1998 World Wide Web Conference: "We expect that advertising funded search engines will be inherently biased towards the advertisers and away from the needs of the consumers. This type of bias is very difficult to detect but could still have a significant effect on the market...we believe the issue of advertising causes enough mixed incentives that it is crucial to have a competitive search engine that is transparent and in the academic realm."[17]

Google's first revenues depended instead on exclusive licensing deals to provide web services to portals such as Yahoo! and Japan's BIGLOBE.[18] It also generated modest revenue from sponsored ads linked to search query keywords.[19] There were other models for consideration. Rival search engines such as Overture, used exclusively by the then-giant portal AOL, or Inktomi, the search engine adopted by Microsoft, collected revenues from the sites whose pages they indexed. Overture was also successful in attracting online ads with its policy of allowing advertisers to pay for high-ranking search listings, the very format that Brin and Page scorned.[20]

Prominent analysts publicly doubted whether Google could compete with its more-established rivals. As the *New York Times* asked, "Can Google create a business model even remotely as good as its technology?"[21] A well-known Forrester Research analyst proclaimed that there were only a few ways for Google to make money with Search: "build a portal [like Yahoo!]... partner with a portal...license the technology...wait for a big company to purchase them."[22]

Despite these general misgivings about Google's viability, the firm's prestigious venture backing gave the founders confidence in their ability to raise money. This changed abruptly in April 2000, when the legendary dot-com economy began its steep plunge into recession, and Silicon Valley's Garden of Eden unexpectedly became the epicenter of a financial earthquake.

By mid-April, Silicon Valley's fast-money culture of privilege was under siege with the implosion of what came to be known as the "dot-com bubble." It is easy to forget exactly how terrifying things were for the valley's ambitious young people and their slightly older investors. Startups with outsized valuations just months earlier were suddenly forced to shutter. Prominent articles such as "Doom Stalks the Dotcoms" noted that the stock prices of Wall Street's most-revered internet "high flyers" were "down for the count," with many of them trading below their initial offering price: "With many dotcoms declining, neither venture capitalists nor Wall Street is eager to give them a dime...."[23] The news brimmed with descriptions of shell-shocked investors. The week of April 10 saw the worst decline in the history of the NASDAQ, where many internet companies had gone public, and there was a growing consensus that the "game" had irreversibly changed.[24]

As the business environment in Silicon Valley unraveled, investors' prospects for cashing out by selling Google to a big company seemed far less likely, and they were not immune to the rising tide of panic. Many Google investors began to express doubts about the company's prospects, and some threatened to withdraw support. Pressure for profit mounted sharply, despite the fact that Google Search was widely considered the best of all the search engines, traffic to its website was surging, and a thousand résumés flooded the firm's Mountain View office each day. Page and Brin were seen to be moving too slowly, and their top venture capitalists, John Doerr from Kleiner Perkins and Michael Moritz from Sequoia, were frustrated.[25] According to Google chronicler Steven Levy,

"The VCs were screaming bloody murder. Tech's salad days were over, and it wasn't certain that Google would avoid becoming another crushed radish."[26]

The specific character of Silicon Valley's venture funding, especially during the years leading up to dangerous levels of startup inflation, also contributed to a growing sense of emergency at Google. As Stanford sociologist Mark Granovetter and his colleague Michel Ferrary found in their study of valley venture firms, "A connection with a high-status VC firm signals the high status of the startup and encourages other agents to link to it."[27] These themes may seem obvious now, but it is useful to mark the anxiety of those months of sudden crisis. Prestigious risk investment functioned as a form of vetting—much like acceptance to a top university sorts and legitimates students, elevating a few against the backdrop of the many—especially in the "uncertain" environment characteristic of high-tech investing. Loss of that high-status signaling power assigned a young company to a long list of also-rans in Silicon Valley's fast-moving saga.

Other research findings point to the consequences of the impatient money that flooded the valley as inflationary hype drew speculators and ratcheted up the volatility of venture funding.[28] Studies of pre-bubble investment patterns showed a "big-score" mentality in which bad results tended to stimulate increased investing as funders chased the belief that some young company would suddenly discover the elusive business model destined to turn all their bets into rivers of gold.[29] Startup mortality rates in Silicon Valley outstripped those for other venture capital centers such as Boston and Washington, DC, with impatient money producing a few big wins and many losses.[30] Impatient money is also reflected in the size of Silicon Valley startups, which during this period were significantly smaller than in other regions, employing an average of 68 employees as compared to an average of 112 in the rest of the country.[31] This reflects an interest in quick returns without spending much time on growing a business or deepening its talent base, let alone developing the institutional capabilities that Joseph Schumpeter would have advised. These propensities were exacerbated by the larger Silicon Valley culture, where net worth was celebrated as the sole measure of success for valley parents and their children.[32]

For all their genius and principled insights, Brin and Page could not ignore the mounting sense of emergency. By December 2000, the *Wall Street*

Journal reported on the new "mantra" emerging from Silicon Valley's investment community: "Simply displaying the ability to make money will not be enough to remain a major player in the years ahead. What will be required will be an ability to show sustained and exponential profits."[33]

IV. The Discovery of Behavioral Surplus

The declaration of a state of exception functions in politics as cover for the suspension of the rule of law and the introduction of new executive powers justified by crisis.[34] At Google in late 2000, it became a rationale for annulling the reciprocal relationship that existed between Google and its users, steeling the founders to abandon their passionate and public opposition to advertising. As a specific response to investors' anxiety, the founders tasked the tiny AdWords team with the objective of looking for ways to make more money.[35] Page demanded that the whole process be simplified for advertisers. In this new approach, he insisted that advertisers "shouldn't even get involved with choosing keywords—*Google would choose them*."[36]

Operationally, this meant that Google would turn its own growing cache of behavioral data and its computational power and expertise toward the single task of matching ads with queries. New rhetoric took hold to legitimate this unusual move. If there was to be advertising, then it had to be "relevant" to users. Ads would no longer be linked to keywords in a search query, but rather a particular ad would be "targeted" to a particular individual. Securing this holy grail of advertising would ensure relevance to users and value to advertisers.

Absent from the new rhetoric was the fact that in pursuit of this new aim, Google would cross into virgin territory by exploiting sensitivities that only its exclusive and detailed collateral behavioral data about millions and later billions of users could reveal. To meet the new objective, the behavioral value reinvestment cycle was rapidly and secretly subordinated to a larger and more complex undertaking. The raw materials that had been solely used to improve the quality of search results would now also be put to use in the service of targeting advertising to individual users. Some data would continue to

be applied to service improvement, but the growing stores of collateral signals would be repurposed to improve the profitability of ads for both Google and its advertisers. These behavioral data available for uses *beyond* service
improvement constituted a surplus, and it was on the strength of this *behavioral surplus* that the young company would find its way to the "sustained
and exponential profits" that would be necessary for survival. Thanks to a
perceived emergency, a new mutation began to gather form and quietly slip
its moorings in the implicit advocacy-oriented social contract of the firm's
original relationship with users.

Google's declared state of exception was the backdrop for 2002, the wa-
tershed year during which surveillance capitalism took root. The firm's appreciation of behavioral surplus crossed another threshold that April, when
the data logs team arrived at their offices one morning to find that a pecu-
liar phrase had surged to the top of the search queries: "Carol Brady's maiden
name." Why the sudden interest in a 1970s television character? It was data
scientist and logs team member Amit Patel who recounted the event to the
New York Times, noting, "You can't interpret it unless you know what else is
going on in the world."[37]

The team went to work to solve the puzzle. First, they discerned that
the pattern of queries had produced five separate spikes, each beginning at
forty-eight minutes after the hour. Then they learned that the query pattern
occurred during the airing of the popular TV show *Who Wants to Be a Mil-
lionaire?* The spikes reflected the successive time zones during which the
show aired, ending in Hawaii. In each time zone, the show's host posed the
question of Carol Brady's maiden name, and in each zone the queries immediately flooded into Google's servers.

As the *New York Times* reported, "The precision of the Carol Brady data
was eye-opening for some." Even Brin was stunned by the clarity of Search's
predictive power, revealing events and trends before they "hit the radar" of
traditional media. As he told the *Times*, "It was like trying an electron microscope for the first time. It was like a moment-by-moment barometer."[38] Google executives were described by the *Times* as reluctant to share their thoughts
about how their massive stores of query data might be commercialized. "There
is tremendous opportunity with this data," one executive confided.[39]

Just a month before the Carol Brady moment, while the AdWords team was already working on new approaches, Brin and Page hired Eric Schmidt, an experienced executive, engineer, and computer science Ph.D., as chairman. By August, they appointed him to the CEO's role. Doerr and Moritz had been pushing the founders to hire a professional manager who would know how to pivot the firm toward profit.[40] Schmidt immediately implemented a "belt-tightening" program, grabbing the budgetary reins and heightening the general sense of financial alarm as fund-raising prospects came under threat. A squeeze on workspace found him unexpectedly sharing his office with none other than Amit Patel.

Schmidt later boasted that as a result of their close quarters over the course of several months, he had instant access to better revenue figures than did his own financial planners.[41] We do not know (and may never know) what other insights Schmidt might have gleaned from Patel about the predictive power of Google's behavioral data stores, but there is no doubt that a deeper grasp of the predictive power of data quickly shaped Google's specific response to financial emergency, triggering the crucial mutation that ultimately turned AdWords, Google, the internet, and the very nature of information capitalism toward an astonishingly lucrative surveillance project.

Google's earliest ads had been considered more effective than most online advertising at the time because they were linked to search queries and Google could track when users actually clicked on an ad, known as the "click-through" rate. Despite this, advertisers were billed in the conventional manner according to how many people viewed an ad. As Search expanded, Google created the self-service system called AdWords, in which a search that used the advertiser's keyword would include that advertiser's text box and a link to its landing page. Ad pricing depended upon the ad's position on the search results page.

Rival search startup Overture had developed an online auction system for web page placement that allowed it to scale online advertising targeted to keywords. Google would produce a transformational enhancement to that model, one that was destined to alter the course of information capitalism. As a Bloomberg journalist explained in 2006, "Google maximizes the revenue it gets from that precious real estate by giving its best position to the advertiser who is likely to pay Google the most in total, based on the price per click *multiplied by Google's estimate of the likelihood that someone will actually click on the ad.*"[42] That pivotal multiplier was the result of Google's advanced computational capabilities trained on its most significant and secret discovery: behavioral surplus. From this point forward, the combination of ever-increasing machine intelligence and ever-more-vast supplies of behavioral surplus would become the foundation of an unprecedented logic of accumulation. Google's reinvestment priorities would shift from merely improving its user offerings to inventing and institutionalizing the most far-reaching and technologically advanced raw-material supply operations that the world had ever seen. Henceforth, revenues and growth would depend upon more behavioral surplus.

Google's many patents filed during those early years illustrate the explosion of discovery, inventiveness, and complexity detonated by the state of exception that led to these crucial innovations and the firm's determination to advance the capture of behavioral surplus.[43] Among these efforts, I focus here on one patent submitted in 2003 by three of the firm's top computer scientists and titled "Generating User Information for Use in Targeted Advertising."[44] The patent is emblematic of the new mutation and the emerging logic of accumulation that would define Google's success. Of even greater interest, it also provides an unusual glimpse into the "economic orientation" baked deep into the technology cake by reflecting the mindset of Google's distinguished scientists as they harnessed their knowledge to the firm's new aims.[45] In this way, the patent stands as a treatise on a new political economics of clicks and its moral universe, before the company learned to disguise this project in a fog of euphemism.

The patent reveals a pivoting of the backstage operation toward Google's new audience of genuine customers. "The present invention concerns advertising," the inventors announce. Despite the enormous quantity of demographic data available to advertisers, the scientists note that much of an ad budget "is simply wasted...it is very difficult to identify and eliminate such waste."[46]

Advertising had always been a guessing game: art, relationships, conventional wisdom, standard practice, but never "science." The idea of being able to deliver a particular message to a particular person at just the moment when it might have a high probability of actually influencing his or her behavior was, and had always been, the holy grail of advertising. The inventors point out that online ad systems had also failed to achieve this elusive goal. The then-predominant approaches used by Google's competitors, in which ads were targeted to keywords or content, were unable to identify relevant ads "for a *particular* user." Now the inventors offered a scientific solution that exceeded the most-ambitious dreams of any advertising executive:

> There is a need to increase the relevancy of ads served for some user request, such as a search query or a document request…to the user that submitted the request.…The present invention may involve novel methods, apparatus, message formats and/or data structures for determining user profile information and using such determined user profile information for ad serving.[47]

In other words, Google would no longer mine behavioral data strictly to improve service for users but rather to read users' minds for the purposes of matching ads to their interests, as those interests are deduced from the collateral traces of online behavior. With Google's unique access to behavioral data, it would now be possible to know what a *particular* individual in a particular time and place was thinking, feeling, and doing. That this no longer seems astonishing to us, or perhaps even worthy of note, is evidence of the profound psychic numbing that has inured us to a bold and unprecedented shift in capitalist methods.

The techniques described in the patent meant that each time a user queries Google's search engine, the system simultaneously presents a specific configuration of a particular ad, all in the fraction of a moment that it takes to fulfill the search query. The data used to perform this instant translation from query to ad, a predictive analysis that was dubbed "matching," went far beyond the mere denotation of search terms. New data sets were compiled that would dramatically enhance the accuracy of these predictions. These data sets were referred to as "user profile information" or "UPI." These new data meant that there would be no more guesswork and far less waste in the advertising budget. Mathematical certainty would replace all of that.

Where would UPI come from? The scientists announce a breakthrough. They first explain that some of the new data can be culled from the firm's existing systems with its continuously accruing caches of behavioral data from Search. Then they stress that even more behavioral data can be hunted and herded from anywhere in the online world. UPI, they write, "may be *inferred*," "*presumed*," and "*deduced*." Their new methods and computational tools could create UPI from integrating and analyzing a user's search patterns, document inquiries, and myriad other signals of online behaviors, even when users do not directly provide that personal information: "User profile information may include any information about an individual user or a group of users. Such information may be provided by the user, provided by a third-party authorized to release user information, *and/or derived from user actions*. Certain user information can be deduced or presumed using other user information of the same user and/or user information of other users. UPI may be associated with various entities."[48]

The inventors explain that UPI can be deduced directly from a user's or group's actions, from any kind of document a user views, or from an ad landing page: "For example, an ad for prostate cancer screening might be limited to user profiles having the attribute 'male' and 'age 45 and over.'"[49] They describe different ways to obtain UPI. One relies on "machine learning classifiers" that predict values on a range of attributes. "Association graphs" are developed to reveal the relationships among users, documents, search queries, and web pages: "user-to-user associations may also be generated."[50] The inventors also note that their methods can be understood only among the priesthood of computer scientists drawn to the analytic challenges of this new online universe: "The following description is presented to enable one skilled in the art to make and use the invention.... Various modifications to the disclosed embodiments will be apparent to those skilled in the art...."[51]

Of critical importance to our story is the scientists' observation that the most challenging sources of friction here are *social,* not technical. Friction arises when users intentionally fail to provide information for no other reason than that they choose not to. "Unfortunately, user profile information is not always available," the scientists warn. Users do not always "voluntarily" provide information, or "the user profile may be incomplete...and hence not comprehensive, *because of privacy considerations,* etc."[52]

A clear aim of the patent is to assure its audience that Google scientists will not be deterred by users' exercise of decision rights over their personal information, despite the fact that such rights were an inherent feature of the original social contract between the company and its users.[53] Even when users do provide UPI, the inventors caution, "it may be *intentionally* or unintentionally inaccurate, it may become stale.... UPI for a user...can be determined (or updated or extended) *even when no explicit information is given to the system....* An initial UPI may include some expressly entered UPI information, *though it doesn't need to.*"[54]

The scientists thus make clear that they are willing—and that their inventions are able—to overcome the friction entailed in users' decision rights. Google's proprietary methods enable it to surveil, capture, expand, construct, and claim behavioral surplus, including data that users intentionally choose not to share. Recalcitrant users will not be obstacles to data expropriation. No moral, legal, or social constraints will stand in the way of finding, claiming, and analyzing others' behavior for commercial purposes.

The inventors provide examples of the kinds of attributes that Google could assess as it compiles its UPI data sets while circumnavigating users' knowledge, intentions, and consent. These include websites visited, psychographics, browsing activity, and information about previous advertisements that the user has been shown, selected, and/or made purchases after viewing.[55] It is a long list that is certainly much longer today.

Finally, the inventors observe another obstacle to effective targeting. Even when user information exists, they say, "Advertisers may not be able to use this information to target ads effectively."[56] On the strength of the invention presented in this patent, and others related to it, the inventors publicly declare Google's unique prowess in hunting, capturing, and transforming surplus into predictions for accurate targeting. No other firm could equal its range of access to behavioral surplus, its bench strength of scientific knowledge and technique, its computational power, or its storage infrastructure. In 2003 only Google could pull surplus from multiple sites of activity and integrate each increment of data into comprehensive "data structures." Google was uniquely positioned with the state-of-the-art knowledge in computer science to convert those data into predictions of who will click on which configuration of what ad as the basis for a final "matching" result, all computed in micro-fractions of a second.

To state all this in plain language, Google's invention revealed new capabilities to infer and deduce the thoughts, feelings, intentions, and interests of individuals and groups with an automated architecture that operates as a one-way mirror irrespective of a person's awareness, knowledge, and consent, thus enabling privileged secret access to behavioral data.

A one-way mirror embodies the specific social relations of surveillance based on asymmetries of knowledge and power. The new mode of accumulation invented at Google would derive, above all, from the firm's willingness and ability to impose these social relations on its users. Its willingness was mobilized by what the founders came to regard as a state of exception; its ability came from its actual success in leveraging privileged access to behavioral surplus in order to predict the behavior of individuals now, soon, and later. The predictive insights thus acquired would constitute a world-historic competitive advantage in a new marketplace where low-risk bets about the behavior of individuals are valued, bought, and sold.

Google would no longer be a passive recipient of accidental data that it could recycle for the benefit of its users. The targeted advertising patent sheds light on the path of discovery that Google traveled from its advocacy-oriented founding toward the elaboration of behavioral surveillance as a full-blown logic of accumulation. The invention itself exposes the reasoning through which the behavioral value reinvestment cycle was subjugated to the service of a new commercial calculation. Behavioral data, whose value had previously been "used up" on improving the quality of Search for users, now became the pivotal—and exclusive to Google—raw material for the construction of a dynamic online advertising marketplace. Google would now secure more behavioral data than it needed to serve its users. That surplus, a behavioral surplus, was the game-changing, zero-cost asset that was diverted from service improvement toward a genuine and highly lucrative market exchange.

These capabilities were and remain inscrutable to all but an exclusive data priesthood among whom Google is the *übermensch*. They operate in obscurity, indifferent to social norms or individual claims to self-determining decision rights. These moves established the foundational mechanisms of surveillance capitalism.

The state of exception declared by Google's founders transformed the youthful Dr. Jekyll into a ruthless, muscular Mr. Hyde determined to hunt his prey anywhere, anytime, irrespective of others' self-determining aims. The new Google ignored claims to self-determination and acknowledged no a priori limits on what it could find and take. It dismissed the moral and legal content of individual decision rights and recast the situation as one of technological opportunism and unilateral power. This new Google assures its actual customers that it will do whatever it takes to transform the natural obscurity of human desire into scientific fact. This Google is the superpower that establishes its own values and pursues its own purposes above and beyond the social contracts to which others are bound.

V. Surplus at Scale

There were other new elements that helped to establish the centrality of behavioral surplus in Google's commercial operations, beginning with its pricing innovations. The first new pricing metric was based on "click-through rates," or how many times a user clicks on an ad through to the advertiser's web page, rather than pricing based on the number of views that an ad receives. The click-through was interpreted as a signal of relevance and therefore a measure of successful targeting, operational results that derive from and reflect the value of behavioral surplus.

This new pricing discipline established an ever-escalating incentive to increase behavioral surplus in order to continuously upgrade the effectiveness of predictions. Better predictions lead directly to more click-throughs and thus to revenue. Google learned new ways to conduct automated auctions for ad targeting that allowed the new invention to scale quickly, accommodating hundreds of thousands of advertisers and billions (later it would be trillions) of auctions simultaneously. Google's unique auction methods and capabilities earned a great deal of attention, which distracted observers from reflecting on exactly what was being auctioned: *derivatives of behavioral surplus*. Click-through metrics institutionalized "customer" demand for these prediction products and thus established the central importance of *economies of scale in surplus supply operations*. Surplus capture would have to become automatic and ubiquitous if the new logic was to succeed, as measured by the successful trading of behavioral futures.

Another key metric called the "quality score" helped determine the price of an ad and its specific position on the page, in addition to advertisers' own auction bids. The quality score was determined in part by click-through rates and in part by the firm's analyses of behavioral surplus. "The clickthrough rate needed to be a *predictive* thing," one top executive insisted, and that would require "all the information we had about the query right then."[57] It would take enormous computing power and leading-edge algorithmic programs to produce powerful predictions of user behavior that became the criteria for estimating the relevance of an ad. Ads that scored high would sell at a lower price than those that scored poorly. Google's customers, its advertisers, complained that the quality score was a black box, and Google was determined to keep it so. Nonetheless, when customers followed its disciplines and produced high-scoring ads, their click-through rates soared.

AdWords quickly became so successful that it inspired significant expansion of the surveillance logic. Advertisers demanded more clicks.[58] The answer was to extend the model beyond Google's search pages and convert the entire internet into a canvas for Google's targeted ads. This required turning Google's newfound skills at "data extraction and analysis," as Hal Varian put it, toward the content of any web page or user action by employing Google's rapidly expanding semantic analysis and artificial intelligence capabilities to efficiently "squeeze" meaning from them. Only then could Google accurately assess the content of a page and how users interact with that content. This "content-targeted advertising" based on Google's patented methods was eventually named AdSense. By 2004, AdSense had achieved a run rate of a million dollars per day, and by 2010, it produced annual revenues of more than $10 billion.

So here was an unprecedented and lucrative brew: behavioral surplus, data science, material infrastructure, computational power, algorithmic systems, and automated platforms. This convergence produced unprecedented "relevance" and billions of auctions. Click-through rates skyrocketed. Work on AdWords and AdSense became just as important as work on Search.

With click-through rates as the measure of relevance accomplished, behavioral surplus was institutionalized as the cornerstone of a new kind of commerce that depended upon online surveillance at scale. Insiders referred to Google's new science of behavioral prediction as the "physics of clicks."[59] Mastery of this new domain required a specialized breed of click physicists who would secure Google's preeminence within the nascent priesthood of behavioral prediction. The firm's substantial revenue flows summoned the greatest minds of our age from fields such as artificial intelligence, statistics, machine learning, data science, and predictive analytics to converge on the prediction of human behavior as measured by click-through rates: computer-mediated fortune-telling and selling. The firm would recruit an authority on information economics, and consultant to Google since 2001, as the patriarch of this auspicious group and the still-young science: Hal Varian was the chosen shepherd of this flock.

Page and Brin had been reluctant to embrace advertising, but as the evidence mounted that ads could save the company from crisis, their attitudes shifted.[60] Saving the company also meant saving themselves from being just another couple of very smart guys who couldn't figure out how to make real money, insignificant players in the intensely material and competitive culture of Silicon Valley. Page was haunted by the example of the brilliant but impoverished scientist Nikola Tesla, who died without ever benefiting financially from his inventions. "You need to do more than just invent things," Page reflected.[61] Brin had his own take: "Honestly, when we were still in the dot-com boom days, I felt like a schmuck. I had an internet startup—so did everybody else. It was unprofitable, like everybody else's."[62] Exceptional threats to their financial and social status appear to have awakened a survival instinct in Page and Brin that required exceptional adaptive measures.[63] The Google founders' response to the fear that stalked their community effectively declared a "state of exception" in which it was judged necessary to suspend the values and principles that had guided Google's founding and early practices.

Later, Sequoia's Moritz recalled the crisis conditions that provoked the firm's "ingenious" self-reinvention, when crisis opened a fork in the road and drew the company in a wholly new direction. He stressed the specificity of Google's inventions, their origins in emergency, and the 180-degree turn from serving users to surveilling them. Most of all, he credited the discovery of behavioral surplus as the game-changing asset that turned Google into a fortune-telling giant, pinpointing Google's breakthrough transformation of the Overture model, when the young company first applied its analytics of behavioral surplus to predict the likelihood of a click:

The first 12 months of Google were not a cakewalk, because the company didn't start off in the business that it eventually tapped. At first it went in a different direction, which was selling its technology—selling licenses for its search engines to larger internet properties and to corporations....Cash was going out of the window at a feral rate during the first six, seven months. And then, very ingeniously, Larry...and Sergey...and others fastened on a model that they had seen this other company, Overture, develop, which was ranked advertisements. They saw how it could be improved and enhanced and made it their own, and that transformed the business.[64]

Moritz's reflections suggest that without the discovery of behavioral surplus and the turn toward surveillance operations, Google's "feral" rate of spending was not sustainable and the firm's survival was imperiled. We will never know what Google might have made of itself without the state of exception fueled by the emergency of impatient money that shaped those crucial years of development. What other pathways to sustainable revenue might have been explored or invented? What alternative futures might have been summoned to keep faith with the founders' principles and with their users' rights to self-determination? Instead, Google loosed a new incarnation of capitalism upon the world, a Pandora's box whose contents we are only beginning to understand.

VI. A Human Invention

Key to our conversation is this fact: surveillance capitalism was invented by a specific group of human beings in a specific time and place. It is not an inherent result of digital technology, nor is it a necessary expression of information capitalism. It was intentionally constructed at a moment in history, in much the same way that the engineers and tinkerers at the Ford Motor Company invented mass production in the Detroit of 1913.

Henry Ford set out to prove that he could maximize profits by driving up volumes, radically decreasing costs, and widening demand. It was an unproven commercial equation for which no economic theory or body of practice existed. Fragments of the formula had surfaced before—in meatpacking plants, flour-milling operations, sewing machine and bicycle factories, armories, canneries, and breweries. There was a growing body of practical knowledge about the interchangeability of parts and absolute standardization, precision machines, and continuous flow production. But no one had achieved the grand symphony that Ford heard in his imagination.

As historian David Hounshell tells it, there was a time, April 1, 1913, and a place, Detroit, when the first moving assembly line seemed to be "just another step in the years of development at Ford yet somehow suddenly dropped out of the sky. Even before the end of the day, some of the engineers sensed that they had made a fundamental breakthrough."[65] Within a year, productivity increases across the plant ranged from 50 percent to as much as ten times the output of the old fixed-assembly methods.[66] The Model T that sold for $825 in 1908 was priced at a record low for a four-cylinder automobile in 1924, just $260.[67]

Much as with Ford, some elements of the economic surveillance logic in the online environment had been operational for years, familiar only to a rarefied group of early computer experts. For example, the software mechanism known as the "cookie"—bits of code that allow information to be passed between a server and a client computer—was developed in 1994 at Netscape, the first commercial web browser company.[68] Similarly, "web bugs"—tiny (often invisible) graphics embedded in web pages and e-mail and designed to monitor user activity and collect personal information—were well-known to experts in the late 1990s.[69]

These experts were deeply concerned about the privacy implications of such monitoring mechanisms, and at least in the case of cookies, there were institutional efforts to design internet policies that would prohibit their invasive capabilities to monitor and profile users.[70] By 1996, the function of cookies had become a contested public policy issue. Federal Trade Commission workshops in 1996 and 1997 discussed proposals that would assign control of all personal information to users by default with a simple automated protocol. Advertisers bitterly contested this scheme, collaborating instead to avert government regulation by forming a "self-regulating" association known as the Network Advertising Initiative. Still, in June 2000 the Clinton administration banned cookies from all federal websites, and by April 2001, three bills before Congress included provisions to regulate cookies.[71]

Google brought new life to these practices. As had occurred at Ford a century earlier, the company's engineers and scientists were the first to conduct the entire commercial surveillance symphony, integrating a wide range of mechanisms from cookies to proprietary analytics and algorithmic software capabilities in a sweeping new logic that enshrined surveillance and the unilateral expropriation of behavioral data as the basis for a new market form. The impact of this invention was just as dramatic as Ford's. In 2001, as Google's new systems to exploit its discovery of behavioral surplus were being tested, net revenues jumped to $86 million (more than a 400 percent increase over 2000), and the company turned its first profit. By 2002, the cash began to flow and has never stopped, definitive evidence that behavioral surplus combined with Google's proprietary analytics were sending arrows to their marks. Revenues leapt to $347 million in 2002, then $1.5 billion in 2003, and $3.2 billion in 2004, the year the company went public.[72] The discovery of behavioral surplus had produced a stunning 3,590 percent increase in revenue in less than four years.

VII. The Secrets of Extraction

It is important to note the vital differences for capitalism in these two moments of originality at Ford and Google. Ford's inventions revolutionized *production*. Google's inventions revolutionized *extraction* and established surveillance capitalism's first economic imperative: the *extraction imperative*. The extraction imperative meant that raw-material supplies must be procured at an ever-expanding scale. Industrial capitalism had demanded economies of scale in production in order to achieve high throughput combined with low unit cost. In contrast, surveillance capitalism demands economies of scale in the extraction of behavioral surplus.

Mass production was aimed at new sources of demand in the early twentieth century's first mass consumers. Ford was clear on this point: "Mass production begins in the perception of a public need."[73] Supply and demand were linked effects of the new "conditions of existence" that defined the lives of my great-grandparents Sophie and Max and other travelers in the first modernity. Ford's invention deepened the reciprocities between capitalism and these populations.

In contrast, Google's inventions destroyed the reciprocities of its original social contract with users. The role of the behavioral value reinvestment cycle that had once aligned Google with its users changed dramatically. Instead of deepening the unity of supply and demand with its populations, Google chose to reinvent its business around the burgeoning demand of advertisers eager to squeeze and scrape online behavior by any available means in the competition for market advantage. In the new operation, *users were no longer ends in themselves but rather became the means to others' ends.*

Reinvestment in user services became the method for attracting behavioral surplus, and users became the unwitting suppliers of raw material for a larger cycle of revenue generation. The scale of surplus expropriation that was possible at Google would soon eliminate all serious competitors to its core search business as the windfall earnings from leveraging behavioral surplus were used to continuously draw more users into its net, thus establishing its de facto monopoly in Search. On the strength of Google's inventions, discoveries, and strategies, it became the mother ship and ideal type of a new economic logic based on fortune-telling and selling—an ancient and eternally lucrative craft that has fed on humanity's confrontation with uncertainty from the beginning of the human story.

It was one thing to proselytize achievements in production, as Henry Ford had done, but quite another to boast about the continuous intensification of hidden processes aimed at the extraction of behavioral data and personal information. The last thing that Google wanted was to reveal the secrets of how it had rewritten its own rules and, in the process, enslaved itself to the extraction imperative. Behavioral surplus was necessary for revenue, and secrecy would be necessary for the sustained accumulation of behavioral surplus.

This is how secrecy came to be institutionalized in the policies and practices that govern every aspect of Google's behavior onstage and offstage. Once Google's leadership understood the commercial power of behavioral surplus, Schmidt instituted what he called the "hiding strategy."[74] Google employees were told not to speak about what the patent had referred to as its "novel methods, apparatus, message formats and/or data structures" or confirm any rumors about flowing cash. Hiding was not a post hoc strategy; it was baked into the cake that would become surveillance capitalism.

Former Google executive Douglas Edwards writes compellingly about this predicament and the culture of secrecy it shaped. According to his account, Page and Brin were "hawks," insisting on aggressive data capture and retention: "Larry opposed any path that would reveal our technological secrets or stir the privacy pot and endanger our ability to gather data." Page wanted to avoid arousing users' curiosity by minimizing their exposure to any clues about the reach of the firm's data operations. He questioned the prudence of the electronic scroll in the reception lobby that displays a continuous stream of search queries, and he "tried to kill" the annual Google Zeitgeist conference that summarizes the year's trends in search terms.[75]

Journalist John Battelle, who chronicled Google during the 2002–2004 period, described the company's "aloofness," "limited information sharing," and "alienating and unnecessary secrecy and isolation."[76] Another early company biographer notes, "What made this information easier to keep is that almost none of the experts tracking the business of the internet believed that Google's secret was even possible."[77] As Schmidt told the *New York Times,* "You need to win, but you are better off winning softly."[78] The scientific and material complexity that supported the capture and analysis of behavioral surplus also enabled the hiding strategy, an invisibility cloak over the whole operation. "Managing search at our scale is a very serious barrier to entry," Schmidt warned would-be competitors.[79]

To be sure, there are always sound business reasons for hiding the location of your gold mine. In Google's case, the hiding strategy accrued to its competitive advantage, but there were other reasons for concealment and obfuscation. What might the response have been back then if the public were told that Google's magic derived from its exclusive capabilities in unilateral surveillance of online behavior and its methods specifically designed to override individual decision rights? Google policies had to enforce secrecy in order to protect operations that were designed to be undetectable because they took things from users without asking and employed those unilaterally claimed resources to work in the service of others' purposes.

That Google had the power to choose secrecy is itself testament to the success of its own claims. This power is a crucial illustration of the difference between "decision rights" and "privacy." Decision rights confer the power to choose whether to keep something secret or to share it. One can choose the degree of privacy or transparency for each situation. US Supreme Court Justice William O. Douglas articulated this view of privacy in 1967: "Privacy involves the choice of the individual to disclose or to reveal what he believes, what he thinks, what he possesses...."[80]

Surveillance capitalism lays claim to these decision rights. The typical complaint is that privacy is eroded, but that is misleading. In the larger societal pattern, privacy is not eroded but redistributed, as decision rights over privacy are claimed for surveillance capital. Instead of people having the rights to decide how and what they will disclose, these rights are concentrated within the domain of surveillance capitalism. Google discovered this necessary element of the new logic of accumulation: it must assert the rights to take the information upon which its success depends.

The corporation's ability to hide this rights grab depends on language as much as it does on technical methods or corporate policies of secrecy. George Orwell once observed that euphemisms are used in politics, war, and business as instruments that "make lies sound truthful and murder respectable."[81] Google has been careful to camouflage the significance of its behavioral surplus operations in industry jargon. Two popular terms—"digital exhaust" and "digital breadcrumbs"—connote worthless waste: leftovers lying around for the taking.[82] Why allow exhaust to drift in the atmosphere when it can be recycled into useful data? Who would think to call such recycling an act of exploitation, expropriation, or plunder? Who would dare to redefine "digital exhaust" as booty or contraband, or imagine that Google had learned how to purposefully construct that so-called "exhaust" with its methods, apparatus, and data structures?

The word "targeted" is another euphemism. It evokes notions of precision, efficiency, and competence. Who would guess that targeting conceals a new political equation in which Google's concentrations of computational power brush aside users' decision rights as easily as King Kong might shoo away an ant, all accomplished offstage where no one can see?

These euphemisms operate in exactly the same way as those found on the earliest maps of the North American continent, in which whole regions were labeled with terms such as "heathens," "infidels," "idolaters," "primitives," "vassals," and "rebels." On the strength of those euphemisms, native peoples—their places and claims—were deleted from the invaders' moral and legal equations, legitimating the acts of taking and breaking that paved the way for church and monarchy.

The intentional work of hiding naked facts in rhetoric, omission, complexity, exclusivity, scale, abusive contracts, design, and euphemism is another factor that helps explain why during Google's breakthrough to profitability, few noticed the foundational mechanisms of its success and their larger significance. In this picture, commercial surveillance is not merely an unfortunate accident or occasional lapse. It is neither a necessary development of information capitalism nor a necessary product of digital technology or the internet. It is a specifically constructed human choice, an unprecedented market form, an original solution to emergency, and the underlying mechanism through which a new asset class is created on the cheap and converted to revenue. Surveillance is the path to profit that overrides "we the people," taking our decision rights without permission and even when we say "no." The discovery of behavioral surplus marks a critical turning point not only in Google's biography but also in the history of capitalism.

In the years following its IPO in 2004, Google's spectacular financial breakthrough first astonished and then magnetized the online world. Silicon Valley investors had doubled down on risk for years, in search of that elusive business model that would make it all worthwhile. When Google's financial results went public, the hunt for mythic treasure was officially over.[83]

The new logic of accumulation spread first to Facebook, which launched the same year that Google went public. CEO Mark Zuckerberg had rejected the strategy of charging users a fee for service as the telephone companies had done in an earlier century. "Our mission is to connect every person in the world. You don't do that by having a service people pay for," he insisted.[84] In May 2007 he introduced the Facebook platform, opening up the social network to everyone, not just people with a college e-mail address. Six months later, in November, he launched his big advertising product, Beacon, which would automatically share transactions from partner websites with all of a user's "friends." These posts would appear even if the user was not currently logged into Facebook, without the user's knowledge or an opt-in function. The howls of protest—from users but also from some of Facebook's partners such as Coca-Cola—forced Zuckerberg to back down swiftly. By December, Beacon became an opt-in program. The twenty-three-year-old CEO understood the potential of surveillance capitalism, but he had not yet mastered Google's facility in obscuring its operations and intent.

The pressing question in Facebook's headquarters—"How do we turn all those Facebook users into money?"—still required an answer.[85] In March 2008, just three months after having to kill his first attempt at emulating Google's logic of accumulation, Zuckerberg hired Google executive Sheryl Sandberg to be Facebook's chief operating officer. The onetime chief of staff to US Treasury Secretary Larry Summers, Sandberg had joined Google in 2001, ultimately rising to be its vice president of global online sales and operations. At Google she led the development of surveillance capitalism through the expansion of AdWords and other aspects of online sales operations.[86] One investor who had observed the company's growth during that period concluded, "Sheryl created AdWords."[87]

In signing on with Facebook, the talented Sandberg became the "Typhoid Mary" of surveillance capitalism as she led Facebook's transformation from a social networking site to an advertising behemoth. Sandberg understood that Facebook's social graph represented an awe-inspiring source of behavioral surplus: the extractor's equivalent of a nineteenth-century prospector stumbling into a valley that sheltered the largest diamond mine *and* the deepest gold mine ever to be discovered. "We have better information than anyone else. We know gender, age, location, and it's real data as opposed to the stuff other people infer," Sandberg said. Facebook would learn to track, scrape, store, and analyze UPI to fabricate its own targeting algorithms, and like Google it would not restrict extraction operations to what people voluntarily shared with the company. Sandberg understood that through the artful manipulation of Facebook's culture of intimacy and sharing, it would be possible to use behavioral surplus not only to satisfy demand but also to *create* demand. For starters, that meant inserting advertisers into the fabric of Facebook's online culture, where they could "invite" users into a "conversation."[88]

VIII. Summarizing the Logic and Operations of Surveillance Capitalism

With Google in the lead, surveillance capitalism rapidly became the default model of information capitalism on the web and, as we shall see in coming chapters, gradually drew competitors from every sector. This new market form declares that serving the genuine needs of people is less lucrative, and therefore less important, than selling predictions of their behavior. Google discovered that *we are less valuable than others' bets on our future behavior.* This changed everything.

Behavioral surplus defines Google's earnings success. In 2016, 89 percent of the revenues of its parent company, Alphabet, derived from Google's targeted advertising programs.[89] The scale of raw-material flows is reflected in Google's domination of the internet, processing over 40,000 search queries every second on average: more than 3.5 billion searches per day and 1.2 trillion searches per year worldwide in 2017.[90]

On the strength of its unprecedented inventions, Google's $400 billion market value edged out ExxonMobil for the number-two spot in market capitalization in 2014, only sixteen years after its founding, making it the second-richest company in the world behind Apple.[91] By 2016, Alphabet/Google occasionally wrested the number-one position from Apple and was ranked number two globally as of September 20, 2017.[92]

It is useful to stand back from this complexity to grasp the overall pattern and how the puzzle pieces fit together:

1. *The logic:* Google and other surveillance platforms are sometimes described as "two-sided" or "multi-sided" markets, but the mechanisms of surveillance capitalism suggest something different.[93] Google had discovered a way to translate its nonmarket interactions with users into surplus raw material for the fabrication of products aimed at genuine market transactions with its real customers: advertisers.[94] The translation of behavioral surplus from outside to inside the market finally enabled Google to convert investment into revenue. The corporation thus created out of thin air and at zero marginal cost an asset class of vital raw materials derived from users' nonmarket online behavior. At first those raw materials were simply "found,"

a by-product of users' search actions. Later those assets were hunted aggressively and procured largely through surveillance. The corporation simultaneously created a new kind of marketplace in which its proprietary "prediction products" manufactured from these raw materials could be bought and sold.

The summary of these developments is that the behavioral surplus upon which Google's fortune rests can be considered as *surveillance assets*. These assets are critical raw materials in the pursuit of *surveillance revenues* and their translation into *surveillance capital*. The entire logic of this capital accumulation is most accurately understood as *surveillance capitalism*, which is the foundational framework for a surveillance-based economic order: a *surveillance economy*. The big pattern here is one of subordination and hierarchy, in which earlier reciprocities between the firm and its users are subordinated to the derivative project of our behavioral surplus captured for others' aims. We are no longer the *subjects* of value realization. Nor are we, as some have insisted, the "product" of Google's sales. Instead, we are the *objects* from which raw materials are extracted and expropriated for Google's prediction factories. Predictions about our behavior are Google's products, and they are sold to its actual customers but not to us. *We are the means to others' ends.*

Industrial capitalism transformed nature's raw materials into commodities, and surveillance capitalism lays its claims to the stuff of human nature for a new commodity invention. Now it is human nature that is scraped, torn, and taken for another century's market project. It is obscene to suppose that this harm can be reduced to the obvious fact that users receive no fee for the raw material they supply. That critique is a feat of misdirection that would use a pricing mechanism to institutionalize and therefore legitimate the extraction of human behavior for manufacturing and sale. It ignores the key point that the essence of the exploitation here is the rendering of our lives as behavioral data for the sake of others' improved control of us. The remarkable questions here concern the facts that our lives are rendered as behavioral data in the first place; that ignorance is a condition of this ubiquitous rendition; that decision rights vanish before one even knows that there is a decision to make; that there are consequences to this diminishment of rights that we can neither see nor foretell; that there is no exit, no voice, and no loyalty, only helplessness, resignation, and psychic numbing; and that encryption is the only positive action left to discuss when we sit around the dinner table and casually ponder how to hide from the forces that hide from us.

2. *The means of production:* Google's internet-age manufacturing process is a critical component of the unprecedented. Its specific technologies and techniques, which I summarize as "machine intelligence," are constantly evolving, and it is easy to be intimidated by their complexity. The same term may mean one thing today and something very different in one year or in five years. For example, Google has been described as developing and deploying "artificial intelligence" since at least 2003, but the term itself is a moving target, as capabilities have evolved from primitive programs that can play tic-tac-toe to systems that can operate whole fleets of driverless cars.

Google's machine intelligence capabilities feed on behavioral surplus, and the more surplus they consume, the more accurate the prediction products that result. *Wired* magazine's founding editor, Kevin Kelly, once suggested that although it seems like Google is committed to developing its artificial intelligence capabilities to improve Search, it's more likely that Google develops Search as a means of continuously training its evolving AI capabilities.[95] This is the essence of the machine intelligence project. As the ultimate tapeworm, the machine's intelligence depends upon how much data it eats. In this important respect the new means of production differs fundamentally from the industrial model, in which there is a tension between quantity and quality. Machine intelligence is the synthesis of this tension, for it reaches its full potential for quality only as it approximates totality.

As more companies chase Google-style surveillance profits, a significant fraction of global genius in data science and related fields is dedicated to the fabrication of prediction products that increase click-through rates for targeted advertising. For example, Chinese researchers employed by Microsoft's Bing's research unit in Beijing published breakthrough findings in 2017. "Accurately estimating the click-through rate (CTR) of ads has a vital impact on the revenue of search businesses; even a 0.1% accuracy improvement in our production would yield hundreds of millions of dollars in additional earnings," they begin. They go on to demonstrate a new application of advanced neural networks that promises 0.9 percent improvement on one measure of identification and "significant click yield gains in online traffic."[96] Similarly, a team of Google researchers introduced a new deep-neural network model, all for the sake of capturing "predictive feature interactions" and delivering "state-of-the-art performance" to improve click-through rates.[97] Thousands of contributions like these, some incremental and some dramatic, equate to an expensive, sophisticated, opaque, and exclusive *twenty-first-century "means of production."*

3. *The products:* Machine intelligence processes behavioral surplus into *prediction products* designed to forecast what we will feel, think, and do: now, soon, and later. These methodologies are among Google's most closely guarded secrets. The nature of its products explains why Google repeatedly claims that it does not sell personal data. What? Never! Google executives like to claim their privacy purity because they do not sell their raw material. Instead, the company sells the predictions that only it can fabricate from its world-historic private hoard of behavioral surplus.

Prediction products reduce risks for customers, advising them where and when to place their bets. The quality and competitiveness of the product are a function of its approximation to certainty: the more predictive the product, the lower the risks for buyers and the greater the volume of sales. Google has learned to be a data-based fortune-teller that replaces intuition with science at scale in order to tell and sell our fortunes for profit to its customers, but not to us. Early on, Google's prediction products were largely aimed at sales of targeted advertising, but as we shall see, advertising was the beginning of the surveillance project, not the end.

4. *The marketplace:* Prediction products are sold into a new kind of market that trades exclusively in future behavior. Surveillance capitalism's profits derive primarily from these *behavioral futures markets.* Although advertisers were the dominant players in the early history of this new kind of marketplace, there is no reason why such markets are limited to this group. The new prediction systems are only incidentally about ads, in the same way that Ford's new system of mass production was only incidentally about automobiles. In both cases the systems can be applied to many other domains. The already visible trend, as we shall see in the coming chapters, is that any actor with an interest in purchasing probabilistic information about our behavior and/or influencing future behavior can pay to play in markets where the behavioral fortunes of individuals, groups, bodies, and things are told and sold (see Figure 2).

The Discovery of Behavioral Surplus

Surveillance capitalism begins with the discovery of behavioral surplus. More behavioral data are rendered than required for service improvements. This surplus feeds machine intelligence - the new means of production - that fabricates predictions of user behavior. These products are sold to business customers in new behavioral futures markets. The Behavioral Value Reinvestment Cycle is subordinated to this new logic.

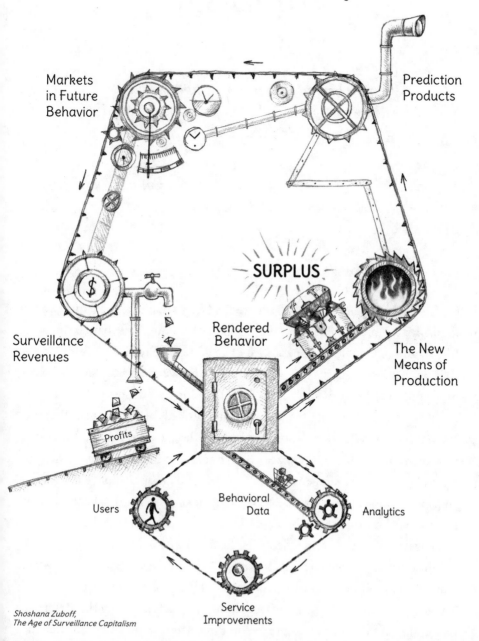

Markets in Future Behavior

Prediction Products

SURPLUS

Surveillance Revenues

Rendered Behavior

The New Means of Production

Profits

Users

Behavioral Data

Analytics

Service Improvements

Shoshana Zuboff,
The Age of Surveillance Capitalism

Figure 2: The Discovery of Behavioral Surplus

CHAPTER FOUR

THE MOAT AROUND
THE CASTLE

The hour of birth their only time in college,
They were content with their precocious knowledge,
To know their station and be right forever.

—W. H. AUDEN

SONNETS FROM CHINA, I

I. Human Natural Resources

Google's former CEO Eric Schmidt credits Hal Varian's early examination of the firm's ad auctions with providing the eureka moment that clarified the true nature of Google's business: "All of a sudden, we realized we were in the auction business."[1] Larry Page is credited with a very different and far more profound answer to the question "What is Google?" Douglas Edwards recounts a 2001 session with the founders that probed their answers to that precise query. It was Page who ruminated, "If we did have a category, it would be *personal information....* The places you've seen. Communications.... Sensors are really cheap.... Storage is cheap. Cameras are cheap. People will generate enormous amounts of data.... Everything you've ever heard or seen or experienced will become searchable. Your whole life will be searchable."[2]

Page's vision perfectly reflects the history of capitalism, marked by taking things that live outside the market sphere and declaring their new life as market commodities. In historian Karl Polanyi's 1944 grand narrative of the "great transformation" to a self-regulating market economy, he described the origins of this translation process in three astonishing and crucial mental inventions that he called "commodity fictions." The first was that human life could be subordinated to market dynamics and reborn as "labor" to be bought and sold. The second was that nature could be translated into the market and reborn as "land" or "real estate." The third was that exchange could be reborn as "money."[3] Nearly eighty years earlier, Karl Marx had described the taking of lands and natural resources as the original "big bang" that ignited modern capital formation, calling it "primitive accumulation."[4]

The philosopher Hannah Arendt complicated both Polanyi's and Marx's notion. She observed that primitive accumulation wasn't just a one-time primal explosion that gave birth to capitalism. Rather, it is a recurring phase in a repeating cycle as more aspects of the social and natural world are subordinated to the market dynamic. Marx's "original sin of simple robbery," she wrote, "had eventually to be repeated lest the motor of capital accumulation suddenly die down."[5]

In our time of pro-market ideology and practice, this cycle has become so pervasive that we eventually fail to notice its audacity or contest its claims. For example, you can now "purchase" human blood and organs, someone to have your baby or stand in line for you or hold a public parking space, a person to comfort you in your grief, and the right to kill an endangered animal. The list grows longer each day.[6]

Social theorist David Harvey builds on Arendt's insight with his notion of "accumulation by dispossession": "What accumulation by dispossession does is to release a set of assets...at very low (and in some instances zero) cost. Over-accumulated capital can seize hold of such assets and immediately turn them to profitable use." He adds that entrepreneurs who are determined to "join the system" and enjoy "the benefits of capital accumulation" are often the ones who drive this process of dispossession into new, undefended territories.[7]

Page grasped that human experience could be Google's virgin wood, that it could be extracted at no extra cost online and at very low cost out in the real world, where "sensors are really cheap." Once extracted, it is rendered as behavioral data, producing a surplus that forms the basis of a wholly new class of market exchange. Surveillance capitalism originates in this act of *digital dispossession,* brought to life by the impatience of over-accumulated investment and two entrepreneurs who wanted to join the system. This is the lever that moved Google's world and shifted it toward profit.

Today's owners of surveillance capital have declared a fourth fictional commodity expropriated from the experiential realities of human beings whose bodies, thoughts, and feelings are as virgin and blameless as nature's once-plentiful meadows and forests before they fell to the market dynamic. In this new logic, *human experience is subjugated to surveillance capitalism's market mechanisms and reborn as "behavior."* These behaviors are rendered into data, ready to take their place in a numberless queue that feeds the machines for fabrication into predictions and eventual exchange in the new behavioral futures markets.

The commodification of behavior under surveillance capitalism pivots us toward a societal future in which market power is protected by moats of secrecy, indecipherability, and expertise. Even when knowledge derived from our behavior is fed back to us as a quid pro quo for participation, as in the case of so-called "personalization," parallel secret operations pursue the conversion of surplus into sales that point far beyond our interests. We have no formal control because we are not essential to this market action.

In this future we are exiles from our own behavior, denied access to or control over knowledge derived from its dispossession by others for others. Knowledge, authority, and power rest with surveillance capital, for which we are merely "human natural resources." We are the native peoples now whose tacit claims to self-determination have vanished from the maps of our own experience.

Digital dispossession is not an episode but a continuous coordination of action, material, and technique, not a wave but the tide itself. Google's leaders understood from the start that their success would require continuous and pervasive fortifications designed to defend their "repetitive sin" from contest and constraint. They did not want to be bound by the disciplines typically imposed by the private market realm of corporate governance or the democratic realm of law. In order for them to assert and exploit their freedom, democracy would have to be kept at bay.

"How did they get away with it?" It is an important question that we will return to throughout this book. One set of answers depends on understanding the conditions of existence that create and sustain demand for surveillance capitalism's services. This theme was summarized in Chapter 2's discussion of the "collision." A second set of answers depends upon a clear grasp of surveillance capitalism's basic mechanisms and laws of motion. This exploration has begun and will continue through Part II.

A third set of answers requires an appreciation of the political and cultural circumstances and strategies that advanced surveillance capitalism's claims and protected them from fatal challenge. It is this third domain that we pursue in the sections that follow. No single element is likely to have done the job, but together a convergence of political circumstances and proactive strategies helped enrich the habitat in which this mutation could root and flourish. These include (1) the relentless pursuit and defense of the founders' "freedom" through corporate control and an insistence on the right to lawless space; (2) the shelter of specific historical circumstances, including the policies and juridical orientation of the neoliberal paradigm and the state's urgent interest in the emerging capabilities of behavioral surplus analysis and prediction in the aftermath of the September 2001 terror attacks; and (3) the intentional construction of fortifications in the worlds of politics and culture, designed to protect the kingdom and deflect any close scrutiny of its practices.

II. The Cry Freedom Strategy

One way that Google's founders institutionalized their freedom was through an unusual structure of corporate governance that gave them absolute control over their company. Page and Brin were the first to introduce a dual-class share structure to the tech sector with Google's 2004 public offering. The two would control the super-class "B" voting stock, shares that each carried ten votes, as compared to the "A" class of shares, which each carried only one vote. This arrangement inoculated Page and Brin from market and investor pressures, as Page wrote in the "Founder's Letter" issued with the IPO: "In the transition to public ownership, we have set up a corporate structure that will make it harder for outside parties to take over or influence Google.... The main effect of this structure is likely to leave our team, especially Sergey and me, with increasingly significant control over the company's decisions and fate, as Google shares change hands."[8]

In the absence of standard checks and balances, the public was asked to simply "trust" the founders. Schmidt would voice this theme on their behalf whenever challenged on the subject. For example, at the Cato Institute in December 2014, Schmidt was asked about the possibility of abuse of power at Google. He simply assured the audience of the continuity of the firm's dynastic line. Page had succeeded Schmidt as CEO in 2011, and the current leaders would handpick future leaders: "We're fine with Larry...same circus, same clowns...it's the same people...all of us who built Google have the same view, and I am sure our successors will have the same view."[9]

By that year, Page and Brin had a 56 percent majority vote, which they used to impose a new tri-class share structure, adding a "C" class of zero-voting-rights stock.[10] As *Bloomberg Businessweek* observed, "The neutered 'C' shares ensure Page and Brin retain control far into the future...."[11] By 2017, Brin and Page controlled 83 percent of the super-voting-class "B" shares, which translated into 51 percent of the voting power.[12]

Many Silicon Valley founders followed Google's lead. By 2015, 15 percent of IPOs were introduced with a dual-class structure, compared to 1 percent in 2005, and more than half of those were for technology companies.[13] Most significantly, Facebook's 2012 IPO featured a two-tiered stock structure that left founder Mark Zuckerberg in control of voting rights. In 2016 the company proposed nonvoting class "C" shares to further solidify Zuckerberg's personal control but later withdrew the proposal under investor pressure.[14]

While financial scholars and investors debated the consequences of these share structures, absolute corporate control enabled the Google and Facebook founders to aggressively pursue acquisitions, establishing an arms race in two critical arenas.[15] State-of-the-art manufacturing depended on machine intelligence, compelling Google and later Facebook to acquire companies and talent representing its disciplines: facial recognition, "deep learning," augmented reality, and more.[16] But machines are only as smart as the volume of their diet allows. Thus, Google and Facebook vied to become the ubiquitous net positioned to capture the swarming schools of behavioral surplus flowing from every computer-mediated direction. To this end the founders paid outsized premiums for the chance to corner behavioral surplus through acquisitions of an ever-expanding roster of key supply routes.

In 2006, for example, just two years after its IPO, Google paid $1.65 billion for a one-and-a-half-year-old startup that had never made any money and was besieged by copyright-infringement lawsuits: YouTube. While the move was called "crazy" and the company was criticized for the outsized price tag, Schmidt went on the offensive, freely admitting that Google had paid a $1 billion premium for the video-sharing site, though saying little about why. By 2009, a canny Forrester Research media analyst had unpacked the mystery: "It actually becomes worth the additional value because Google can tie all of its advertising expertise and search traffic into YouTube... it ensures that these millions and millions of viewers are coming to a Google-owned site rather than someone's else's site.... As a loss leader goes, if it never makes its money back, it's still going to be worth it."[17]

Facebook's Zuckerberg pursued the same strategies, paying "astronomical" prices for a "fast and furious" parade of typically unprofitable startups like virtual reality firm Oculus ($2 billion) and the messaging application WhatsApp ($19 billion), thus ensuring Facebook's ownership of the gargantuan flows of human behavior that would pour through these pipes. Consistent with the extraction imperative, Zuckerberg told investors that he would not consider driving revenue until the service reaches "billions" of users.[18] As one tech journalist put it, "There's no real need for Zuckerberg to chat with the board...there's no way for shareholders to check Zuckerberg's antics...."[19]

It's worth noting that an understanding of this logic of accumulation would have usefully contributed to the EU Commission's deliberations on the WhatsApp acquisition, which was permitted based on assurances that data flows from the two businesses would remain separate. The commission would discover later that the extraction imperative and its necessary of economies of scale in supply operations compel the integration of surplus flows in the quest for better prediction products.[20]

Google's founders constructed a corporate form that gave them absolute control in the market sphere, and they also pursued freedom in the public sphere. A key element of Google's freedom strategy was its ability to discern, construct, and stake its claim to unprecedented social territories that were not yet subject to law. Cyberspace is an important character in this drama, celebrated on the first page of Eric Schmidt and Jared Cohen's book on the digital age: "The online world is not truly bound by terrestrial laws...it's the world's largest ungoverned space."[21] They celebrate their claim to operational spaces beyond the reach of political institutions: the twenty-first-century equivalent of the "dark continents" that drew nineteenth-century European speculators to their shores.

Hannah Arendt's examination of British capitalists' export of over-accumulated capital to Asia and Africa in the mid-nineteenth century helps to develop this analogy: "Here, in backward regions without industries and political organizations, where violence was given more latitude than in any Western country, the so-called laws of capitalism were actually allowed to create realities.... The secret of the new happy fulfillment was precisely that economic laws no longer stood in the way of the greed of the owning classes."[22]

This kind of lawlessness has been a critical success factor in the short history of surveillance capitalism. Schmidt, Brin, and Page have ardently defended their right to freedom from law even as Google grew to become what is arguably the world's most powerful corporation.[23] Their efforts have been marked by a few consistent themes: that technology companies such as Google move faster than the state's ability to understand or follow, that any attempts to intervene or constrain are therefore fated to be ill-conceived and stupid, that regulation is always a negative force that impedes innovation and progress, and that lawlessness is the necessary context for "technological innovation."

Schmidt, Page, and Brin have each been outspoken on these themes. In a 2010 interview with the *Wall Street Journal*, Schmidt insisted that Google needed no regulation because of strong incentives to "treat its users right."[24] In 2011 Schmidt cited former Intel CEO Andy Grove's antidemocratic formula to a *Washington Post* reporter, commenting that Grove's idea "works for me." Google was determined to protect itself from the slow pace of democratic institutions:

> This is an Andy Grove formula.... "High tech runs three-times faster than normal businesses. And the government runs three-times slower than normal businesses. So we have a nine-times gap.... And so what you want to do is you want to make sure that the government does not get in the way and slow things down."[25]

Business Insider covered Schmidt's remarks at the Mobile World Congress that same year, writing, "When asked about government regulation, Schmidt said that technology moves so fast that governments really shouldn't try to regulate it because it will change too fast, and any problem will be solved by technology. 'We'll move much faster than any government.'"[26]

Both Brin and Page are even more candid in their contempt for law and regulation. CEO Page surprised a convocation of developers in 2013 by responding to questions from the audience, commenting on the "negativity" that hampered the firm's freedom to "build really great things" and create "interoperable" technologies with other companies: "Old institutions like the law and so on aren't keeping up with the rate of change that we've caused through technology.... The laws when we went public were 50 years old. A law can't be right if it's 50 years old, like it's before the internet." When asked his thoughts on how to limit "negativity" and increase "positivity," Page reflected, "Maybe we should set aside a small part of the world... as technologists we should have some safe places where we can try out some new things and figure out what is the effect on society, what's the effect on people, without having to deploy kind of into the normal world."[27]

It is important to understand that surveillance capitalists are impelled to pursue lawlessness by the logic of their own creation. Google and Facebook vigorously lobby to kill online privacy protection, limit regulations, weaken or block privacy-enhancing legislation, and thwart every attempt to circumscribe their practices because such laws are existential threats to the frictionless flow of behavioral surplus.[28]

Extraction quarry must be both unprotected and available at zero cost if this logic of accumulation is to succeed. These requirements are also an Achilles heel. Code is law for Google now, but the risk of new laws in its established and anticipated territories remains a persistent danger to surveillance capitalism. If new laws were to outlaw extraction operations, the surveillance model would implode. This market form must either gird itself for perpetual conflict with the democratic process or find new ways to infiltrate, seduce, and bend democracy to its ends if it is to fulfill its own inner logic. The survival and success of surveillance capitalism depend upon engineering collective agreement through all available means while simultaneously ignoring, evading, contesting, reshaping, or otherwise vanquishing laws that threaten free behavioral surplus.

These claims to lawless space are remarkably similar to those of the robber barons of an earlier century. Like the men at Google, the late-nineteenth-century titans claimed undefended territory for their own interests, declared the righteousness of their self-authorizing prerogatives, and defended their new capitalism from democracy at any cost. At least in the US case, we have been here before.

Economic historians describe the dedication to lawlessness among the Gilded Age "robber barons" for whom Herbert Spencer's social Darwinism played the same role that Hayek, Jensen, and even Ayn Rand play for today's digital barons. In the same way that surveillance capitalists excuse their corporations' unprecedented concentrations of information and wealth as the unavoidable result of "network effects" and "winner-take-all" markets, the Gilded Age industrialists cited Spencer's specious, pseudoscientific "survival of the fittest" as proof of a divine plan intended to put society's wealth in the hands of its most aggressively competitive individuals.[29]

The Gilded Age millionaires, like today's surveillance capitalists, stood on the frontier of a vast discontinuity in the means of production with nothing but blank territory in which to invent a new industrial capitalism free from constraints on the use of labor, the nature of working conditions, the extent of environmental destruction, the sourcing of raw materials, or even the quality of their own products. And like their twenty-first-century counterparts, they did not hesitate to exploit the very law that they despised, flying the banner of "private property" and "freedom of contract," much as surveillance capitalists march under the flag of freedom of speech as the justification for unobstructed technological "progress," a topic to which we shall return.

Imbued with the conviction that "the state had neither right nor reason to interfere in the workings of the economy," the Gilded Age millionaires joined forces to defend the "rights of capital" and limit the role of elected representatives in setting policy or developing legislation.[30] There was no need for law, they argued, when one had the "law of evolution," the "laws of capital," and the "laws of industrial society." John Rockefeller insisted that his outsized oil fortune was the result of "the natural law of trade development." Jay Gould, when questioned by Congress on the need for federal regulation of railroad rates, replied that rates were already regulated by "the laws of supply and demand, production and consumption."[31] The millionaires mobilized in 1896 to defeat the populist Democrat William Jennings Bryan, who had vowed to tether economic policy to the political realm, including regulating the railroads and protecting the people from "robbery and oppression."[32]

The bottom line for Gilded Age business elites was that the most effective way to protect the original sin of that economic era was, as David Nasaw put it, "to circumscribe democracy." They did this by lavishly funding their own political candidates as well as through the careful honing and aggressive dissemination of an ideological attack on the very notion of democracy's right to interfere in the economic realm.[33] Their industries were to be "self-regulating": free to follow their own evolutionary laws. "Democracy," they preached, "had its limits, beyond which voters and their elected representatives dared not trespass lest economic calamity befall the nation."[34] In our discussion of "fortifications" we'll see that Google revived all of these strategies and more. But first we explore the unique circumstances that sheltered the young company and protected its discovery of human experience as a limitless resource ripe for the taking.

III. Shelter: The Neoliberal Legacy

Google's leaders were also favored by historical circumstance. Both Google and the wider surveillance capitalist project were the beneficiaries of two developments that contributed to a uniquely sheltering habitat for the surveillance mutation. The first is the neoliberal capture of the governmental machinery for oversight and regulation of the US economy, the framework of which we discussed in Chapter 2.[35]

A fascinating study by University of California law professor Jodi Short empirically illustrates the role of neoliberal ideology as one important explanation for Google's ambitions and successful defense of lawless territory.[36] Short analyzed 1,400 law review articles on the subject of regulation, all published between 1980 and 2005. As the influence of Hayek and Friedman predicts, the dominant theme of this literature was "the coercive nature of administrative government" and the systematic conflation of industry regulation with "tyranny" and "authoritarianism." According to this worldview, all regulation is burdensome, and bureaucracy must be repudiated as a form of human domination. Short observes that during the sample decades these fears were even more influential in shaping regulatory approaches than rational arguments about cost and efficiency, and she identifies two points of origin for these anxieties.

The first source was in the US business community's opposition to New Deal reforms, which, not unlike the propaganda of the Gilded Age millionaires, cast resistance to regulation as a "righteous fight to defend democracy from dictatorship."[37] A second source was the dread of totalitarianism and collectivism incited by World War II and the cold war, a direct inheritance from Hayek. These defensive themes infiltrated and reshaped US political thought, and they gradually transformed policy makers' assumptions about the regulatory role of the state.[38]

Short found several suggested remedies for "coercive" governmental regulation in the literature, but the most salient, especially after 1996—the very years when digital technology and the internet were becoming mainstream—was "self-regulation." The idea here is that firms set their own standards, monitor their own compliance with those standards, and even judge their own conduct in order to "voluntarily report and remediate violations."[39] By the time of Google's public offering in 2004, self-regulation was fully enshrined within government and across the business community as the single most effective tool for regulation without coercion and the antidote to any inclination toward collectivism and the centralization of power.[40]

This neoliberal legacy was a windfall for the new surveillance capitalists. As another legal scholar, Frank Pasquale, observed, it produced a model that treated privacy as a competitive good, assuming that "consumers" engage only with services that offer the level of privacy they seek. Regulatory interference, according to this view, would only undermine competitive diversity. It also credits the "notice and consent" model—click-wrap and its "sadistic" relatives—as accurate signals of individual privacy choices.[41]

The neoliberal zeitgeist also favored Google's leaders, and later their fellow travelers in the surveillance project, as they sought shelter for their inventions beneath claims of First Amendment rights to freedom of expression. This is a complex and contested arena in which constitutional law and political ideology are thoroughly entangled, and I point out just a few elements here in order to better understand the habitat that nurtured the new surveillance market form.[42]

The key dynamic here is that First Amendment jurisprudence, especially over the last two decades, has reflected a "conservative-libertarian" interpretation of First Amendment rights. As constitutional law scholar Steven Heyman suggests, "In recent decades, the First Amendment has become one of the most important means by which judges have sought to advance a conservative-libertarian agenda."[43] This has produced many dramatic judicial decisions, including the US Supreme Court's rejection of any constraints on the role of money in election campaigns, its rejection of restrictions on hate speech and pornography, and its holding that the right to free association takes precedence over state civil rights laws that bar discrimination.

As many legal scholars observe, the ideological orientation of contemporary First Amendment judicial reasoning asserts a close connection between free speech and property rights. The logic that links ownership to an absolute entitlement to freedom of expression has led to a privileging of corporate action as "speech" deserving of constitutional protection.[44] Some scholars regard this as a dangerous reversion to the feudal doctrines from which corporate law evolved in the seventeenth century. Those medieval legal principles limited the sovereign's authority over "the corporations of Aristocracy, Church, guilds, universities, and cities...who asserted the right to rule themselves." One result is that US courts have been "quick to see the possibilities of governmental overreach, but much less willing to see the problems of 'private,' let alone corporate, power."[45]

In this context, surveillance capitalists vigorously developed a "cyberlibertarian" ideology that Frank Pasquale describes as "free speech fundamentalism." Their legal teams aggressively assert First Amendment principles to fend off any form of oversight or externally imposed constraints that either limit the content on their platforms or the "algorithmic orderings of information" produced by their machine operations.[46] As one attorney who has represented many of the leading surveillance capitalists puts it, "The lawyers working for these companies have business reasons for supporting free expression. Indeed, all of these companies talk about their businesses in the language of free speech."[47]

This is one respect in which the surveillance capitalists are not unprecedented. Adam Winkler, a historian of corporate rights, reminds us,

"Throughout American history the nation's most powerful corporations have persistently mobilized to use the Constitution to fight off unwanted government regulations."[48] Although today's mobilizations are not original, Winkler's careful account demonstrates the effects of past mobilizations on the distribution of power and wealth in US society and the strength of democratic values and principles in each era.

The key point for our story in the age of surveillance capitalism is that the expansion of opportunities for free expression associated with the internet has been an emancipatory force in many vital respects, but this fact must not blind us to another condition: free speech fundamentalism has deflected careful scrutiny of the unprecedented operations that constitute the new market form and account for its spectacular success. The Constitution is exploited to shelter a range of novel practices that are antidemocratic in their aims and consequences and fundamentally destructive of the enduring First Amendment values intended to protect the individual from abusive power.

In the US, congressional statutes have played an equally or perhaps even more important role in sheltering surveillance capitalism from scrutiny. The most celebrated of these is a legislative statute known as Section 230 of the Communications Decency Act of 1996, which shields website owners from lawsuits and state prosecution for user-generated content. "No provider or user of an interactive computer service," the statute reads, "shall be treated as the publisher or speaker of any information provided by another information content provider."[49] This is the regulatory framework that enables a site such as TripAdvisor to include negative hotel reviews and permits Twitter's aggressive trolls to roam free without either company being held to the standards of accountability that typically guide news organizations. Section 230 institutionalized the idea that websites are not publishers but rather "intermediaries." As one journalist put it, "To sue an online platform over an obscene blog post would be like suing the New York Public Library for carrying a copy of *Lolita*."[50] As we shall see, this reasoning collapses once surveillance capitalism enters the scene.

Section 230's hands-off stance toward companies perfectly converged with the reigning ideology and practice of "self-regulation," leaving the internet companies, and eventually the surveillance capitalists among them, free to do what they pleased. The statute was crafted in 1995, during the initial phase of the public internet. It aimed to clarify intermediaries' liability for the content on their websites and resolve a controversy created by two contradictory court decisions both involving defamatory posts.[51] In 1991 a court found that CompuServe was not liable for defamation because it had not reviewed the contents of a post before it appeared online. The court reasoned that CompuServe was comparable to a public library, bookstore, or newsstand: a distributor, not a publisher.

Four years later, in 1995, an early provider of web services called Prodigy was sued for a defamatory anonymous posting on one of its message boards. This time a New York state court came to the opposite conclusion. The key problem as the court saw it was that Prodigy had exercised editorial control by moderating its message boards. The company established content guidelines and deleted posts that violated those standards. The court concluded that Prodigy was a publisher, not merely a distributor, because it had taken responsibility for the content on its site. Were the court's ruling to stand, internet companies would face "a paradoxical no-win situation: the more an ISP tried to keep obscene or harmful material away from its users, the more it would be liable for that material."[52] Internet companies faced a binary choice: "free speech savior or shield for scoundrels?"[53]

According to Senator Ron Wyden, Section 230 was intended to resolve that contradiction by encouraging internet companies to exercise some control over content without the risk of legal sanctions. The very first sentence of the statute mentions "protection for 'good samaritan' blocking and screening of offensive material."[54] What Wyden and his colleagues could not have anticipated, and still do not grasp, is that the logic of this early controversy no longer holds. Neither CompuServe nor Prodigy was a surveillance capitalist, but many of today's internet intermediaries are committed to the pursuit of surveillance revenues.

This fact fundamentally changes the relationship between the company and the content on its platforms, and it explains why surveillance capitalists cannot be compared to the New York Public Library as the neutral caretaker of Nabokov's venerated book. Far from it. Under the regime of surveillance capitalism, content is a source of behavioral surplus, as is the behavior of the people who provide the content, as are their patterns of connection, communication, and mobility, their thoughts and feelings, and the meta-data expressed in their emoticons, exclamation points, lists, contractions, and salutations. That book on the bookshelf—along with the records of anyone who may have touched it and when, their location, behavior, networks, and so on—is now the diamond mine ready for excavation and plunder, to be rendered into behavioral data and fed to the machines on their way to product fabrication and sales. Section 230's protection of the "intermediaries" now functions as another bulwark that shelters this extractive surveillance capitalist operation from critical examination.

There is nothing neutral about the surveillance intermediary now, as the extraction imperative and its demand for economies of scale in surplus supply mean that the surveillance capitalists must use every means to attract a never-ending tide of content to their shores. They no longer merely host content but aggressively, secretly, and unilaterally extract value from that content. As we shall see in Chapter 18, economic imperatives require them to forgo as few of these raw materials as possible. That means moderating only those extremes that threaten the volume and velocity of surplus by repelling users or attracting regulatory scrutiny. This is the reason that firms such as Facebook, Google, and Twitter have been reluctant to remove even the most egregious content from their landscapes, and it helps to explain why "lawyers for tech companies litigate ferociously to prevent even a sliver of erosion" in Section 230.[55] A statute once crafted to nurture an important new technological milieu is now the legal bulwark that protects the asymmetric wealth, knowledge, and power of a rogue capitalism.

IV. Shelter: Surveillance Exceptionalism

In his book *Surveillance After September 11,* surveillance scholar David Lyon writes that in the aftermath of the attacks that day, existing surveillance practices were intensified and previous limits were lifted: "After several decades in which data-protection officials, privacy watchdogs, civil rights groups, and others have tried to mitigate negative social effects of surveillance, we are witnessing a sharp tilt toward more exclusionary and intrusive surveillance practices."[56] This abrupt refocusing of governmental power and policy after the 9/11 attacks in New York City and Washington, DC, is a second historical condition that lent shelter to the fledgling market form.

Lyon's characterization is accurate.[57] In the years before 9/11, the Federal Trade Commission emerged as the key actor defining the debate on internet privacy in the US. For reasons that we have already reviewed, the FTC favored self-regulation, and it cajoled internet companies to establish codes of conduct, privacy policies, and methods of enforcement.[58] But the FTC eventually concluded that self-regulation would not be sufficient to protect individual consumers' privacy on the web. In 2000, still one year before the 9/11 attacks, Google's discovery of behavioral surplus, or the success of AdWords, a majority of FTC commissioners issued a report in which they recommended legislation to regulate online privacy: "Because self-regulatory initiatives to date fall far short of broad-based implementation of self-regulatory programs," they wrote, "the Commission has concluded that such efforts alone cannot ensure that the online marketplace as a whole will follow the standards adopted by industry leaders...notwithstanding several years of industry and governmental effort." The reported noted that a mere 8 percent of popular websites featured a seal of approval from one of the industry privacy watchdogs.[59]

The commissioners proceeded to outline federal legislation that would have protected consumers online despite the dominant bias against regulation and in favor of treating internet operations as free speech. The recommendations demanded "clear and conspicuous" notice of information practices; consumer choice over how personal information is used; access to all personal information, including rights to correct or delete; and enhanced security of personal information.[60] Had these been translated into law, it is quite possible that many of the foundational elements of surveillance capitalism would have been either plainly illegal or at least subject to public examination and contest.

The FTC effort was short-lived. According to Peter Swire, chief counselor for Privacy in the Clinton Administration and later a member of President Obama's Review Group on Intelligence and Communication Technologies, "With the attacks of September 11, 2001, everything changed. The new focus was overwhelmingly on security rather than privacy."[61] The privacy provisions debated just months earlier vanished from the conversation more or less overnight. In both the US Congress and across the EU, legislation was quickly put in place that decisively expanded surveillance activities. The US Congress passed the Patriot Act, created the Terrorist Screening Program, and instituted a host of other measures that dramatically increased the warrantless collection of personal information. The events of 9/11 also triggered a steady stream of legislation that expanded the powers of intelligence and law-enforcement agencies across Europe, including Germany (a country that had been highly sensitized to surveillance under the hammer of both Nazi and Stalinist totalitarianism), the UK, and France.[62]

In the US the failure to "connect the dots" on the terrorist attack was a source of shame and dismay that overwhelmed other concerns. Policy guidelines shifted from "need to know" to "need to share" as agencies were urged to tear down walls and blend databases for comprehensive information and analysis.[63] In a parallel development, privacy scholar Chris Jay Hoofnagle observes that the threat of comprehensive privacy legislation had also mobilized the business community and its lobbyists to either "shape or stop" any potential bill. In the post–9/11 political environment, the two forces converged for an easy victory.[64]

The internet was the critical target. CIA Director Michael Hayden conceded as much in 2013 when he told an audience that in the years following 9/11, the CIA "could be fairly charged with the militarization of the world wide web."[65] Legislation to regulate online privacy was an immediate casualty. Marc Rotenberg, the director of the Electronic Privacy Information Center (EPIC), testified to the 9/11 Commission on the sudden reversal of privacy concerns, observing that before 9/11, "There was hardly any positive discussion about the development of techniques that would enable massive surveillance while attempting to safeguard privacy."[66] Swire concurred, noting that as a result of the new emphasis on information sharing, "Congress lost interest in regulating information usage in the private sector.... Without the threat of legislation, the energy went out of many of the self-regulatory efforts that industry had created."[67] At the FTC, the focus shifted from the broader concerns of privacy rights to a more politically palatable "harms-based" strategy, pursuing cases in which concrete physical harms or economic injuries could be defined, such as identify theft or database security.[68]

With legislation off the table, other forces shaped the political environment in which surveillance capitalism would root and grow. The 9/11 terrorist attacks thrust the intelligence community into an unfamiliar demand curve that insisted on exponential increases in velocity. For all its secrecy, even the NSA was subject to the temporalities and legal restrictions of a democratic state. The tempos of democracy are slow by design, weighted by redundancies, checks and balances, laws and rules. The agencies sought methods of deployment that could rapidly bypass legal and bureaucratic restrictions.

In this environment of trauma and anxiety, a "state of exception" was invoked to legitimate a new imperative: speed at any cost. As Lyon put it, "What 9/11 did was to produce socially negative consequences that hitherto were the stuff of repressive regimes and dystopian novels.... The suspension of normal conditions is justified with reference to the 'war on terrorism.'"[69] Critical to our story is the fact that this state of exception favored Google's growth and the successful elaboration of its surveillance-based logic of accumulation.

Google's mission was to "organize and make accessible the world's information," and by late 2001 the intelligence community established "information dominance" in the public's house, quickly institutionalizing it in hundreds of billions of dollars' worth of state-sponsored global technology infrastructure, personnel, and practice. The contours of a new interdependency between public and private agents of information dominance began to emerge, one that is best understood through the lens of what the sociologist Max Weber once called an "elective affinity" born of a mutual magnetism that originates in shared meanings, interests, and reciprocities.[70]

The elective affinity between public intelligence agencies and the fledgling surveillance capitalist Google blossomed in the heat of emergency to produce a unique historical deformity: *surveillance exceptionalism*. The 9/11 attacks transformed the government's interest in Google, as practices that just hours earlier were careening toward legislative action were quickly recast as mission-critical necessities. Both institutions craved certainty and were determined to fulfill that craving in their respective domains at any price. These elective affinities sustained surveillance exceptionalism and contributed to the fertile habitat in which the surveillance capitalism mutation would be nurtured to prosperity.

The elective affinity between public and private missions was evident as early as 2002, when former NSA Chief Admiral John Poindexter proposed his Total Information Awareness (TIA) program with a vision that reads like an early guide to the foundational mechanisms of behavioral surplus capture and analysis:

> If terrorist organizations are going to plan and execute attacks against the United States, their people must engage in transactions and they will leave signatures in this information space.... We must be able to pick this signal out of the noise...the relevant information extracted from this data must be made available in large-scale repositories with enhanced semantic content for analysis to accomplish this task.[71]

As CIA Director George Tenet had declared in 1997, "The CIA needs to swim in the Valley," referring to the need to master the new technologies flowing from Silicon Valley.[72] In 1999 it opened a CIA-funded venture firm in the valley, In-Q-Tel, as a conduit for cutting-edge technologies. The operation was meant to be an agency experiment, but after 9/11 it became a critical source of new capabilities and relationships, including with Google. As Silicon Valley's *Mercury News* reported, "There's a new urgency with the CIA to find technology that makes sense of all the unstructured data floating around on the internet and elsewhere. The agency can't train analysts quickly enough." In-Q-Tel's CEO described the government's agencies as "scrambling" and noted that "we're in a state of hyperactivity now."[73]

Surveillance exceptionalism thrived in that hyperactivity. Poindexter's Total Information Awareness program did not obtain congressional support, but an analysis in the *MIT Technology Review* showed that many of the TIA initiatives were quietly reassigned to the Pentagon's Advanced Research and Development Activity (ARDA), which in 2002 received $64 million to fund a research program in "novel intelligence from massive data." In 2004 the US General Accounting Office surveyed 199 data-mining projects across dozens of federal agencies and more than 120 programs developed to collect and analyze personal data to predict individual behavior.[74] The *New York Times* reported in 2006 that the intelligence agencies, backed by a $40 billion annual budget, regularly fielded secretive shopping expeditions to Silicon Valley in search of new data-mining and analysis technologies.[75]

State security agencies sought ways to avail themselves of Google's rapidly developing capabilities and simultaneously use Google to further develop, commercialize, and diffuse security and surveillance technologies with proven intelligence value. If TIA could not be fully developed and integrated in Washington, parts of the job could be delegated to Silicon Valley and its standout in information dominance: Google. By late summer 2003, Google was awarded a $2.07 million contract to outfit the agency with Google search technology. According to documents obtained by Consumer Watchdog under the Freedom of Information Act, the NSA paid Google for a "search appliance capable of searching 15 million documents in twenty-four languages." Google extended its services for another year at no cost in April 2004.[76]

In 2003 Google also began customizing its search engine under special contract with the CIA for its Intelink Management Office, "overseeing top-secret, secret and sensitive but unclassified intranets for CIA and other IC agencies."[77] Key agencies used Google systems to support an internal wiki called Intellipedia that allowed agents to share information across organizations as quickly as it was vacuumed up by the new systems.[78] In 2004 Google acquired Keyhole, a satellite mapping company founded by John Hanke, whose key venture backer was the CIA venture firm, In-Q-Tel. Keyhole would become the backbone for Google Earth, and Hanke would go on to lead Google Maps, including the controversial Street View Project. In 2009 Google Ventures and In-Q-Tel both invested in a Boston-based startup, Recorded Future, that monitors every aspect of the web in real time in order to predict future events. *Wired* reported that it was the first time the CIA-backed venture firm and Google had funded the same startup and that both firms had seats on Recorded Future's board of directors.[79]

In the decade that followed 9/11, surveillance exceptionalism was also expressed in the flattery of imitation, as the NSA tried to become more like Google, emulating and internalizing Google's capabilities in a variety of domains. In 2006 General Keith Alexander outlined his vision for a search tool called ICREACH that "would allow unprecedented volumes of...metadata to be shared and analyzed across the many agencies in the Intelligence

Community." By late 2007, the program was piloted, boosting the number of communications events it shared from 50 billion to more than 850 billion. The system was designed with a "Google-like" search interface that enabled analysts to run searches against meta-data "selectors" and to extract vital behavioral surplus for analyses that could reveal "social networks," "patterns of life," and "habits," and in general "predict future behavior."[80] In 2007 two NSA analysts wrote an internal training manual on how to find information on the internet. It expressed the agency's keen interest in all things Google with a detailed chapter devoted to a deconstruction of Google Search and the Google "hacks" that can uncover information not intended for public distribution.[81]

That year, the elective affinities that infused the intelligence community's interest in Google were also highlighted when Google research director and AI expert Peter Norvig presented at a Pentagon Highlands Forum meeting: an exclusive networking event where military and intelligence officials commune with members of the high-tech industry, elected officials, elite academics, top corporate executives, and defense contractors. In 2001 the forum's director, Richard O'Neill, described its work to a Harvard audience as "an idea engine, so the ideas that emerge from meetings are available for use by decision makers as well as by people from the think tanks."[82] It was to be a bridge between the government and commercial leaders, especially in Silicon Valley.[83] According to one highly detailed account by investigative journalist Nafeez Ahmed and cited by legal scholar Mary Anne Franks, the forum was both a support system and an incubator of Google's growth, as well as a connecting and convening force for the Pentagon, intelligence agencies, and the young company: "The US intelligence community's incubation of Google from inception occurred through a combination of direct sponsorship and informal networks of financial influence, themselves closely aligned with Pentagon interests."[84] Another legal scholar described the "collaboration" between Google and the intelligence community, especially the NSA, as "unprecedented."[85]

During these years, scholars noted the growing interdependencies between the intelligence agencies, resentful of constitutional constraints on their prerogatives, and the Silicon Valley firms.[86] The agencies craved the lawlessness that a firm such as Google enjoyed. In his 2008 essay "The

Constitution in the National Surveillance State," law professor Jack Balkin observed that the Constitution inhibits government actors from high-velocity pursuit of their surveillance agenda, and this creates incentives for the government "to rely on private enterprise to collect and generate information for it."[87] Balkin noted that the Supreme Court has imposed few privacy restrictions on business records and information that people give to third parties. E-mail is typically held in private servers, making its protection "limited if not nonexistent." This absence of law made private companies attractive partners for government actors bound to democratic constraints.

The government's need to evade constitutional oversight, argues legal scholar Jon Michaels, leads to secret public-private intelligence collaborations that tend to be "orchestrated around handshakes rather than legal formalities, such as search warrants, and may be arranged this way to evade oversight and, at times, to defy the law."[88] He observed that intelligence agencies are irresistibly drawn to "and in some respects dependent upon" firms' privately held data resources.[89]

Both scholars' observations were confirmed in 2010, when former NSA Director Mike McConnell offered another glimpse into the elective affinities between Google and the intelligence community. Writing in the *Washington Post,* McConnell made clear that Google's surveillance-based operations in data capture, extraction, and analysis were both taken for granted and coveted. Here the boundaries of private and public melt in the intense heat of new threats and their high-velocity demands that must be met in "milliseconds." In McConnell's future there is one "seamless" surveillance empire in which the requirements of self-preservation leave no opportunity for the amenities of democracy, with its time-wasting practices of due process, evidence, warrants, and law. As McConnell insisted,

> An effective partnership with the private sector must be formed so information can move quickly back and forth from public to private and classified to unclassified...to protect the nation's critical infrastructure. Recent reports of possible partnership between Google and the government point to the kind of joint efforts—and shared challenges—that we are likely to see in the future...such arrangements will muddy the waters between the traditional roles of the

government and the private sector....Cyberspace knows no borders, and our defensive efforts must be similarly seamless.[90]

In the final months of the Obama administration, then Secretary of Defense Ash Carter toured Silicon Valley, where he announced a new Defense Innovation Advisory Board, meant to formalize a channel between the tech executives and the DOD. Carter appointed Schmidt to the new board and tasked him with selecting its members. As *Wired* concluded, "The government needs Silicon Valley more than ever as it seeks to defend from security threats in cyberspace."[91] These facts are amply illustrated in a comprehensive treatment of "bulk collection" by an international group of scholars and edited by Indiana University's Fred Cate and Berkeley's James Dempsey. Cate and Dempsey note the "expansive aggregation" of personal data in the hands of private companies: "Governments understandably want access to this data....Essentially every government in the world claims the power to compel disclosure of this data by the companies that hold it."[92] Had it not been for surveillance exceptionalism, it is possible that these data would not even exist, at least not in their current volume and detail.

Surveillance exceptionalism helped to shape the evolutionary course of information capitalism by creating an environment in which Google's budding surveillance practices were coveted rather than contested. Once again, history offers us no control groups, and we cannot know with certainty whether information capitalism might have developed in a different direction had it not been for the sudden new interest in surveillance capabilities. For now, it appears that one unanticipated consequence of this public-private "elective affinity" was that the fledgling practices of surveillance capitalism were allowed to root and grow with little regulatory or legislative challenge, emboldening Google's young leaders to insist on lawlessness as a natural right and, in ways that are even more opaque, emboldening the state to grant them that freedom.

Powerful elective affinities favored the acquisition of certainty at any price, and part of that price appears to have been the shelter of surveillance capitalism. In the fullness of time, historians will no doubt discover the specifics of these relationships and the ways in which Google's discoveries in the capture and use of behavioral surplus were sheltered from scrutiny, at least in part, because of this new habitat of militarized demand.

In the context of new military purpose, the digital capabilities that were aimed toward the advocacy-oriented values of the behavioral value reinvestment cycle flowed toward surveillance without impediment. Surveillance assets thrived without risk of sanction and attracted surveillance capital. Revenue followed. The situation recalls the auto, steel, and machine tool industries at mid-century, when military orders kept plants operating at full capacity. In the end, however, this turned out to be more curse than blessing. Military demand distorted and suppressed the innovation process and drove a wedge between these industries and their civilian customers, leaving them vulnerable to foreign competitors in the globalizing markets of the late 1970s and early 1980s.[93]

Similarly, under the conditions of surveillance exceptionalism, Google's leaders were not compelled to undertake the arduous and risky work of inventing an exchange-based advocacy-oriented market form when the surveillance model was so lucrative. Why risk experimentation with more-organic paths to monetization when surveillance and extraction operations were safe from law and hugely profitable? Eventually, it wasn't just Google asking these questions; every other internet business faced the same choices. Once surveillance revenues set the bar for venture capitalists and Wall Street analysts, it became that much easier for internet companies to go with the flow. Then it became onerous not to.

V. Fortifications

Why is it that so many years after the events that triggered the mania for information dominance, surveillance capitalism still runs with relatively little impediment, especially in the US? The intervening years have seen the proliferation of thousands of institutional facts that normalized surveillance capitalism's practices and made them appear necessary and inevitable: the discovery of behavioral surplus and the massive accumulations of capital and material that followed, the proliferation of devices and services, the integration of data flows, and the institutionalization of futures markets in human behavior.

This does not mean that we should succumb to the natural fallacy and interpret this flourishing as a signal of surveillance capitalism's inherent worthiness or inevitability. In the coming chapters we will uncover many additional factors that have contributed to this success, but here I want to focus on Google's proactive efforts to build fortifications around its supply chains in order to protect surplus flows from challenge.

Although many elements of this fortification strategy have been well publicized, their importance for our story lies in the fact that each is one aspect of a multipronged effort that deflects scrutiny from core operations in order to maintain the flow of free, unregulated behavioral surplus. Fortifications have been erected in four key arenas to protect Google, and eventually other surveillance capitalists, from political interference and critique: (1) the demonstration of Google's unique capabilities as a source of competitive advantage in electoral politics; (2) a deliberate blurring of public and private interests through relationships and aggressive lobbying activities; (3) a revolving door of personnel who migrated between Google and the Obama administration, united by elective affinities during Google's crucial growth years of 2009–2016; and (4) Google's intentional campaign of influence over academic work and the larger cultural conversation so vital to policy formation, public opinion, and political perception. The results of these four arenas of defense contribute to an understanding of how surveillance capitalism's facts came to stand and why they continue to thrive.

First, Google demonstrated that the same predictive knowledge derived from behavioral surplus that had made the surveillance capitalists wealthy could also help candidates win elections. To make the point, Google was ready to apply its magic to the red-hot core of twenty-first-century campaigning, beginning with the 2008 Obama presidential campaign. Schmidt had a leading role in organizing teams and guiding the implementation of cutting-edge data strategies that would eclipse the traditional political arts with the science of behavioral prediction.[94] Indeed, "At Obama's Chicago headquarters...they remodeled the electorate in every battleground state each weekend...field staff could see the events' impact on the projected behaviors and beliefs of every voter nationwide."[95]

Research by media scholars Daniel Kreiss and Philip Howard indicates that the 2008 Obama campaign compiled significant data on more than 250 million Americans, including "a vast array of online behavioral and relational data collected from use of the campaign's web site and third-party social media sites such as Facebook...."[96] Journalist Sasha Issenberg, who documented these developments in his book *The Victory Lab*, quotes one of Obama's 2008 political consultants who likened predictive modeling to the tools of a fortune-teller: "We knew who...people were going to vote for before they decided."[97]

Obama used his proximity to Schmidt to cement his own identity as the innovation candidate poised to disrupt business as usual in Washington.[98] Once elected, Schmidt joined the Transition Economic Advisory Board and appeared next to Obama at his first postelection press conference.[99] According to *Politico*, "The image alone of Schmidt standing elbow-to-elbow with Obama's top economic thinkers was enough to send shivers up the spine of Google's competitors. 'This terrifies Microsoft,' said a Democratic lobbyist familiar with the industry. 'There's a reason why people are scared to death of Google.'"[100]

Schmidt's role in President Obama's election was but one chapter in a long, and by now fabled, relationship that some have described as a "love affair."[101] Not surprisingly, Schmidt took on an even more prominent role in the 2012 reelection campaign. He led in fundraising and in breaking new technical ground, and he "personally oversaw the voter-turnout system on election night."[102]

Political correspondent Jim Rutenberg's *New York Times* account of the data scientists' seminal role in the 2012 Obama victory offers a vivid picture of the capture and analysis of behavioral surplus as a political methodology. The campaign knew "every single wavering voter in the country that it needed to persuade to vote for Obama, by name, address, race, sex, and income," and it had figured out how to target its television ads to these individuals. One breakthrough was the "persuasion score" that identified how easily each undecided voter could be persuaded to vote for the Democratic candidate.[103]

The facts of behavioral surplus and its predictive power were kept top secret in the Obama campaigns, just as they are in Google, Facebook, and other domains of information dominance. As Rutenberg observed, "The extent to which the campaign used the newest tech tools to look into people's lives and the sheer amount of personal data its vast servers were crunching remained largely shrouded. The secrecy...was partly...to maintain their competitive edge. But it was also no doubt because they worried that practices like 'data mining' and 'analytics' could make voters uncomfortable."[104]

Second, with the 2012 election in sight, an interview with the *Washington Post* in 2011 found Schmidt boasting about another fortification strategy: "The staffers are young—the staffers get it....So that's what we depend on. And of course we've hired ex-staffers as well. They all know each other. So that's how it really works."[105] Google's political utilities paved the way for the unusually crowded and fast-spinning revolving door between the East Coast and West Coast centers of power. The Google Transparency Project analyzed the movement of staff between the Googlesphere (the company plus its affiliates and its law and lobbying firms) and the government (including the White House, Congress, government agencies, federal commissions, and national political campaigns) during the Obama years. It found that by April 2016, 197 individuals had migrated from the government into the Googlesphere, and 61 had moved in the other direction. Among these, 22 White House officials went to work for Google, and 31 Googlesphere executives joined the White House or federal advisory boards with direct relevance to Google's business.[106]

Third, just to be on the safe side, Google shared its largesse throughout the political system. In Schmidt's 2014 book, coauthored with longtime Google executive Jonathan Rosenberg, the CEO aggressively developed the theme of government as the shill of incumbents colluding to inhibit change, with Google on the outside: an upstart and a disrupter. The authors voiced their disdain for politicians and lobbyists, writing, "This is the natural path of politicians since incumbents tend to have a lot more money than disrupters and are quite expert in using it to bend the political will of any democratic government."[107]

That same year, while Schmidt disparaged incumbents and their political sway, Google spent more on lobbying than any other corporation—over $17 million and nearly twice as much as surveillance rival Facebook. In the next few years, as the White House changed hands, Google maintained its pace, outspending every other company with a more than $18 million lobbying outlay in 2018 as the company fended off privacy legislation and other initiatives that might impede its freedom to capture and process behavioral surplus. Google was also among the wealthiest of all registered lobbyists in the EU, second only to a lobbying group that represents a confederation of European corporations.[108]

The firm also learned to engineer sophisticated lobbying efforts at the state level, primarily geared to fight back any proposed legislation that would augment privacy and curtail behavioral surplus operations. For example, Google won the right to put its self-driving cars on the road—anticipated as important supply chains—after enlisting Obama officials to lobby state regulators for key legislation.[109] Both Google and Facebook currently lead aggressive state-level lobbying campaigns aimed at repelling or weakening statutes to regulate biometric data and protect privacy. As one report put it, "They want your body."[110]

In the fourth arena of fortifications the corporation learned to infiltrate and influence academic research and civil society advocacy in ways that softened or in some cases thwarted the examination of its practices. The *Washington Post* describes Google as a "master of Washington influence" and notes the subtlety with which the firm grasps and directs its own narrative. Schmidt was hands-on in this work as well. Already a board member of the New America Foundation, a public-policy think tank that played an influential role in shaping the Obama administration's approach to economic issues, he assumed the chairmanship in 2013 with a personal donation of $1 million, a significant percentage of its $12.9 million annual budget that year. Between 1999 and 2016, when Schmidt left the board, the foundation received $21 million from a combination of Google, Schmidt, and Schmidt's family foundation.[111]

The *Washington Post* published an elaborate exposé of Google's meticulous work in this fourth domain, illustrated in the backstage intrigues that accompanied a three-part series on internet search competition held at George Mason University's Law and Economics Center, a "free-market–oriented" academic center that had received significant funding from Google.[112] The meetings occurred in May 2012, just as the FTC was investigating the Google antitrust case. Reporters found that company staffers worked closely with the center, choosing pro-Google speakers and participants, many of whom were Google employees. Their efforts included "sending the center's staff a detailed spreadsheet listing members of Congress, FTC commissioners, and senior officials with the

Justice Department and state attorney general's office." Reporters noted that the conference's panels were dominated by "leading technology and legal experts" who forcefully rejected the need for government action against Google, "making their arguments before some of the very regulators who would help determine its fate." Many participants had no idea that Google was involved in crafting the meetings because Google and center staffers had agreed to conceal the corporation's backstage involvement.[113]

The FTC antitrust investigation appears to have heightened Google's fears of a regulatory threat to surveillance capitalism. That year, Google's grant-making operation aimed at civil society organizations took an aggressive turn. According to the Center for Media and Democracy's investigatory research report, "The Googlization of the Far Right," the corporation's 2012 list of grantees featured a new group of antigovernment groups known for their opposition to regulation and taxes and their support for climate-change denial, including Grover Norquist's Americans for Tax Reform, the Koch brothers–funded Heritage Action, and other antiregulatory groups such as the Federalist Society and the Cato Institute.[114] The corporation also quietly acknowledged its membership in the corporate lobbying group ALEC, known for its opposition to gun control and emissions curbs, and for its support for voter-suppression schemes, tobacco industry tax breaks, and other far-right causes.[115] Meanwhile, a list of Google Policy Fellows for 2014 included individuals from a range of nonprofit organizations whom one would expect to be leading the fight against that corporation's concentrations of information and power, including the Center for Democracy and Technology, the Electronic Frontier Foundation, the Future of Privacy Forum, the National Consumers League, the Citizen Lab, and the Asociación por los Derechos Civiles.[116]

In July 2017 the *Wall Street Journal* reported that since 2009, Google had actively sought out and provided funding to university professors for research and policy papers that support Google's positions on matters related to law, regulation, competition, patents, and so forth.[117] In many cases, Google weighed in on the papers before publication, and some of the authors did not disclose Google as a source of funding. Although Google publicly claimed that "the check came with no requirements," another case in 2017 belied that notion. That summer, one of the New America Foundation's most highly regarded scholars and a specialist in digital monopolies, Barry Lynn, posted a statement praising the EU's historic decision to levy a $2.7 billion fine on Google as the result of a multiyear antitrust investigation. According to the *New York Times* and Lynn's own account, New America's director bent to pressure from Schmidt, firing Lynn and his Open Markets team of ten researchers. "Google is very aggressive in throwing its money around Washington and Brussels, and then pulling strings," Lynn told the *New York Times.* "People are so afraid of Google now." The reporters cite Google's "muscular and sophisticated" influence operation as surpassing any other US company's.[118]

With Google in the lead, surveillance capitalism vastly expanded the market dynamic as it learned to expropriate human experience and translate it into coveted behavioral predictions. Google and this larger surveillance project have been birthed, sheltered, and nurtured to success by the historical conditions of their era—second-modernity needs, the neoliberal inheritance, and the realpolitik of surveillance exceptionalism—as well as by their own purpose-built fortifications designed to protect supply chain operations from scrutiny through political and cultural capture.

Surveillance capitalism's ability to keep democracy at bay produced these stark facts. Two men at Google who do not enjoy the legitimacy of the vote, democratic oversight, or the demands of shareholder governance exercise control over the organization and presentation of the world's information. One man at Facebook who does not enjoy the legitimacy of the vote, democratic oversight, or the demands of shareholder governance exercises control over an increasingly universal means of social connection along with the information concealed in its networks.

CHAPTER FIVE

THE ELABORATION OF SURVEILLANCE CAPITALISM: KIDNAP, CORNER, COMPETE

All words like Peace and Love,
all sane affirmative speech,
had been soiled, profaned, debased
to a horrid mechanical screech.

—W. H. AUDEN
"WE TOO HAD KNOWN GOLDEN HOURS"

I. The Extraction Imperative

"Our ultimate ambition is to transform the overall Google experience, making it beautifully simple," Larry Page said, "*almost automagical because we understand what you want* and can deliver it instantly."[1] In the drive to fulfill this ambition, the extraction imperative produces a relentless push for scale in supply operations. There can be no boundaries that limit scale in the hunt for behavioral surplus, no territory exempted from plunder. The assertion of decision rights over the expropriation of human experience, its translation into data, and the uses of those data are collateral to this process, inseparable as a shadow. This explains why Google's supply chains began with Search but steadily expanded to encompass new and even-more-ambitious territories far from clicks and queries. Google's stores of behavioral surplus now embrace everything in the online milieu: searches, e-mails, texts, photos, songs, messages, videos, locations, communication patterns, attitudes, preferences, interests, faces, emotions, illnesses, social networks, purchases, and so on. A new continent of behavioral surplus is spun each moment from the many virtual threads of our everyday lives as they collide with Google, Facebook, and, more generally, every aspect of the internet's computer-mediated architecture. Indeed, under the direction of surveillance capitalism the global reach of computer mediation is repurposed as an *extraction architecture.*

This process originated online but has spread to the real world as well, a fact that we will examine more closely in Part II. If Google is a search company, why is it investing in smart-home devices, wearables, and self-driving cars? If Facebook is a social network, why is it developing drones and augmented reality? This diversity sometimes confounds observers but is generally applauded as visionary investment: far-out bets on the future. In fact, activities that appear to be varied and even scattershot across a random selection of industries and projects are actually all the same activity guided by the same aim: behavioral surplus capture. Each is a slightly different configuration of hardware, software, algorithms, sensors, and connectivity designed to mimic a car, shirt, cell phone, book, video, robot, chip, drone, camera, cornea, tree, television, watch, nanobot, intestinal flora, or any online service, but they all share the same purpose: behavioral surplus capture.

Google is a shape-shifter, but each shape harbors the same aim: to hunt and capture raw material. *Baby, won't you ride my car? Talk to my phone? Wear my shirt? Use my map?* In all these cases the varied torrent of creative shapes is the sideshow to the main event: the continuous expansion of the extraction architecture to acquire raw material at scale to feed an expensive production process that makes prediction products that attract and retain more customers. When confronted in 2008 with a question about why Google had 150 "products," its CEO, Eric Schmidt, responded: "That can be stated as criticism, but it can also be stated as strategy. The goal of the company is customer satisfaction. You should think of Google as one product: customer satisfaction."[2] Those customers are the world's advertisers and others who pay for its predictions. "Customer satisfaction" therefore equates to Google's dominant market share in lucrative new behavioral futures markets, fed by its ever-expanding extraction architecture.

New supply routes are continuously constructed and tested, and only some go operational. Routes that reliably produce scale, such as the Android smartphone operating system or Gmail, are elaborated and institutionalized.

Those that fail are shuttered or modified. If one route is blocked, another is found. Successful supply routes double as canvases for targeted advertising, expanding the reach of behavioral futures markets and simultaneously engaging users in ways that yield yet more behavioral surplus. There will always be a changing roster of supply routes, but all variations share the same operational mandate: the capture of behavioral surplus and the acquisition of decision rights. Like a river running to the sea, if one route is blocked, another is found.

In this chapter we follow the consequences of the extraction imperative as it drives the elaboration of the new market form and its competitive dynamics. The imperative elevates surplus supply operations to a defining role in every aspect of the surveillance capitalist enterprise. This begins with a continuous parade of innovations aimed at cornering raw-material supplies. Cornering is not simply a technological achievement. Sustainable dispossession requires a highly orchestrated and carefully phased amalgam of political, communicative, administrative, legal, and material strategies that audaciously asserts and tirelessly defends claims to new ground. The success of these strategies, first at Google and later at Facebook, established both their feasibility and their rewards, drawing new competitors into an increasingly ruthless cycle of kidnapping human experience, cornering surplus supplies, and competing in new behavioral futures markets.

II. Cornered

The discovery of behavioral surplus in 2001–2002 meant that Google Search would be the first Google "service" to be re-crafted as a supply route. The resulting shifts in the mechanisms of Search were nearly impossible for people to imagine, let alone detect. When Harvard Business School's Benjamin Edelman researched these hidden mechanisms in 2010, he found that the "enhanced features" option of a product called Google Toolbar—a plug-in for Microsoft's Internet Explorer web browser that lets users search without having to go to google.com—transmits to the company "the full URL of every page view, including searches at competing search engines." Edelman discovered that it was "strikingly easy" to activate this option but impossible to disable it. Even when a user specifically instructed that the toolbar be disabled, and even when it appeared to be disabled because it had disappeared from view, the toolbar continued to track browsing behavior.[3] As Google now hosts "trillions" of searches annually, its varied search-related tracking mechanisms combined with its robust and nearly inescapable cookies (bits of tracking code inserted in your computer) ensure immense economies of scale that constitute the bedrock of Google's supply operations.[4]

In 2015 internet legal scholar Tim Wu joined Harvard Business School's Michael Luca and a team of data scientists from Yelp to research hidden mechanisms in Google Search that function, from our perspective, to expand crucial supply functions. They discovered that Google was systematically corrupting Search results to favor its own content and "downstream products":

> Google has begun to develop its own content over time, such as its own price results for shopping and its own reviews for local businesses…. Google is acting both as a search engine and a content provider. To use its search dominance to promote this content, Google has developed a feature called "universal search," through which it intentionally excludes content competitors and only shows Google's content.[5]

The ceaseless requirement for surplus at scale predicts corporate behavior that favors exclusivity. Because Search is the foundation of Google's supply operations, the company has every incentive to entice users to its Search platform, content, and ancillary services and then to use its backstage "methods, apparatus, and data structures" for efficient extraction. The tilt toward exclusivity produces a range of practices considered "monopolistic" in the perspective of twentieth-century regulatory frameworks. These characterizations, while valid, omit the most-salient elements of the new order. The extraction imperative demands that everything be possessed. In this new context, goods and services are merely surveillance-bound supply routes. It's not the car; it's the behavioral data from driving the car. It's not the map; it's the behavioral data from interacting with the map. The ideal here is continuously expanding borders that eventually describe the world and everything in it, all the time.

Traditionally, monopolies on goods and services disfigure markets by unfairly eliminating competition in order to raise prices at will. Under surveillance capitalism, however, many of the practices defined as monopolistic actually function as means of *cornering* user-derived raw-material supplies. There is no monetary price for the user to pay, only an opportunity for the company to extract data. Cornering practices are not designed to protect product niches but rather to protect critical supply routes for the unregulated commodity that is behavioral surplus. In another time, rogue market actors might corner markets in copper or magnesium, but in our time it is behavioral surplus. The corporation unfairly impedes competitors in Search in order to protect the dominance of its most important supply route, not primarily to fix prices.

These cornering operations are not abstractions, with distant effects on minerals or crops that eventually wind their way toward the price of goods. In this picture it is we who are "cornered." We are the source of the coveted commodity; our experience is the target of extraction. As surveillance capitalism migrates from Silicon Valley to a range of other firms and sectors, we gradually find ourselves in a world of no escape, "cornered" by converging, overlapping, and relentlessly expanding dispossession operations. It is important to say—and we will revisit this theme more than once—that regulatory interventions designed to constrain Google's monopoly practices are likely to have little effect on the fundamental operations of this market form. New supply routes are continuously discovered, opened, and secured. Dispossession activities are compelled to circumvent every obstacle and will continue to do so, short of a genuine existential threat.

Google's Android mobile platform offers an example of the governing role of surplus capture and defense. Internet use went mobile with the rise of the smartphone and the tablet, and Google was forced to find new ways to defend and expand its primary supply chain in Search. Android quickly became the corporation's second critical supply route for behavioral surplus. In 2008 Google led an alliance of technology manufacturers and wireless operators to develop an "open and comprehensive platform for mobile devices." Some observers thought that an Android phone was Google's opportunity to compete with Apple for the lucrative margins on smartphones, but Google insiders had grasped the even greater potential for growth and profit through behavioral surplus and its fabrication into prediction products.

Google licensed Android to mobile handset makers for free because it was intended to draw users into Google Search and other Google services, establishing a ubiquitous mobile supply apparatus to sustain known terrains of behavioral surplus and open up new ones, including geolocation and mobile payment systems that are highly coveted by advertisers.[6] As Google's chief financial officer told financial analysts in 2009, "If we move forward the adoption of these mobile phones by lowering the cost because it is open source, think of how many searches [that will produce]."[7] A prominent Silicon Valley venture capitalist described Android devices in 2011 as

> not "products" in the classic business sense…they are not trying to make a profit on Android.…They want to take any layer that lives between themselves and the consumer and make it free (or even less than free).…In essence, they are not just building a moat; Google is also scorching the earth for 250 miles around the outside of the castle to ensure no one can approach it.[8]

Supply operations were the protected treasure within the fortified castle, and Android's development policies were key to the success of this supply strategy. Unlike the iPhone, the Android platform was "open source," which made it easy for applications developers around the world to create apps for Android users. Eventually, Google bundled this valuable new universe of apps into its Google Play store. Manufacturers who wanted to preinstall Google Play on their devices were required to license and install Google's mobile services as exclusive or default capabilities: Search, Gmail, Google Pay, YouTube, Google Maps, Google Photos, and whatever other supply routes happen to be in ascendance at the time.

In 2016 Google's Android practices became the focus of a European Union antitrust investigation, and its complaints were a mirror image of Google's purposeful construction and protection of vital supply routes in Search and Mobile. Once again, governmental opposition to Google's monopolistic activities emphasized traditional competitive harms at the expense of surveillance capitalism's new harms. By April of 2013, Eric Schmidt told a conference devoted to "all things" digital that "our goal with Android is to reach everyone. We'll cross one billion Android devices in six to nine months. In a year or two, we'll hit two billion.... *A relatively inexpensive smartphone with a browser is all you need to get the world's information.*" That final sentence presumably was intended to describe benefits to Android's users. However, it is an even-more-effective summary of Google's own ambitions and an insight into the vital economies of scale associated with this mobile supply route.[9]

Google fiercely defends threatened supply routes. Any disruption of its extraction operations and its exclusive claims to raw material is the line that cannot be breached. In 2009 Android manufacturer Motorola chose to replace Google's free location services with those of Skyhook Wireless, which, Motorola believed, produced more-reliable results. A Google product manager admitted Skyhook's superiority but expressed his concerns in an e-mail to a Google executive, noting that if other manufacturers switched to Skyhook, it "would be awful for Google, because it will cut off our ability to continue collecting data" for the company's Wi-Fi location database. Court documents from Skyhook's eventual lawsuit against Motorola (and Samsung) include an e-mail from Google's senior vice president of Mobile to Motorola's CEO, insisting that the interruption of Google's data collection was a "stop-ship issue."[10]

Another legal fracas further illustrates how products such as Android are valued more for supply than for sales. Disconnect, Inc., founded in 2011 by two former Google engineers and a privacy-rights attorney, developed desktop and mobile applications "to protect the privacy and security of internet users by blocking invisible, unsolicited network connections between a user's browser or mobile device and sites/services that engage in invisible tracking or are known or suspected distributors of malware...not only when he browses the web, but also when he uses other third-party mobile applications."[11] Disconnect specifically took aim at the "invisible, unsolicited and frequently undisclosed" network connections from third-party sites and services that occur as soon as you visit a website or open a mobile application.

Unfortunately for Disconnect, the very process that it aimed to impede had been established as a significant supply route for Google and other surveillance capitalists.[12] Several studies explain the extent of Google's extraction architecture, including the Web Privacy Census, which primarily measured cookies. The census analyzed the top 100, 1,000, and 25,000 websites in 2011, 2012, and 2015, years of feverish discovery and elaboration for surveillance capitalists. The comparisons between 2012 and 2015 revealed more than twice as many sites with 100 or more cookies and more than three times the number of sites with 150 or more cookies. In 2015 the team found that anyone who simply visited the 100 most popular websites would collect over 6,000 cookies in his or her computer, 83 percent of which were from third parties unrelated to the website that was visited. The census found "Google tracking infrastructure" on 92 of the top 100 sites and 923 of the top 1,000 sites, concluding that "Google's ability to track users on popular websites is unparalleled, and it approaches the level of surveillance that only an Internet Service Provider can achieve."[13]

Another 2015 analysis, this one of the top one million websites, by Timothy Libert of the University of Pennsylvania, found that 90 percent leak data to an average of nine external domains that track, capture, and expropriate user data for commercial purposes. Among these websites, 78 percent initiate third-party transfers to a domain owned by one company: Google. Another 34 percent transfer to a Facebook-owned domain.[14] Steven Englehardt and Arvind Narayanan from Princeton University reported in 2016 on the results of their measurement and analysis of tracking data from one million websites.[15] They identified 81,000 third parties, but only 123 of those are present on more than 1 percent of sites. Of that group, the top five third parties and twelve of the top twenty are Google-owned domains. "In fact," they conclude, "Google, Facebook, and Twitter are the only third-party entities present on more than 10% of sites." Chinese researchers investigated 10,000 apps from the top third-party app markets in 2017. They found a "covert" process in which an app autonomously launches other apps in the background of your phone, and they concluded that this "app collusion" was most prevalent in third-party Android markets. Of the 1,000 top apps on one of China's popular platforms, 822 launched an average of 76 other apps, and of these launches, 77 percent were triggered by cloud-based "push services" that are intended to update apps but are obviously doing a lot more. In the Android environment, the researchers note, Google provides the push service.[16]

Finally, extraordinary research from the French nonprofit Exodus Privacy and the Yale Privacy Lab in 2017 documented the exponential proliferation of tracking software. Exodus identified 44 trackers in more than 300 apps for Google's Android platform, some of which are also produced for Apple's operating system. Altogether, these apps have been downloaded billions of times. Two themes stand out in the research report: ubiquity and intensification. First, there is hardly an innocent app; if it's not tracking you now, it may be doing so in the next week or month: "There is an entire industry based upon these trackers, and apps identified as 'clean' today may contain trackers that have not yet been identified. Tracker code may also be added by developers to new versions of apps in the future." Second is that even the most innocent-seeming applications such as weather, flashlights, ride sharing, and dating apps are "infested" with dozens of tracking programs that rely on increasingly bizarre, aggressive, and illegible tactics to collect massive amounts of behavioral surplus ultimately directed at ad targeting. For example, the ad tracker FidZup developed "communication between a sonic emitter and a mobile phone...." It can detect the presence of mobile phones and therefore their owners by diffusing a tone, inaudible to the human ear, inside a building: "Users installing 'Bottin Gourmand,' a guide to restaurants and hotels in France, would thus have their physical location tracked via retail outlet speakers as they move around Paris. Their experience would be shared by readers of a car magazine app 'Auto Journal' and the TV guide app 'TeleStar.'" In a pattern foreshadowed by the Google patent that we examined in Chapter 3 and that we shall see repeatedly in the coming chapters, the research findings emphasize that the always-on tracking is impervious to the Android "permissions system," despite its promises of user control.[17]

Given the hostility and intensity of these supply operations, it is not too surprising that Disconnect software was banned from Google Play's vast catalog of mobile apps, leading to Disconnect's lawsuit against Google in 2015. The startup's complaint explains that "advertising companies including Google use these invisible connections to 'track' the user as he/she browses the web or opens other mobile applications, in order to collect personal information about the user, create a 'profile' of the user, and make money targeting advertising to the user."[18] It went on to argue that the privacy protections offered by Google "invariably permit the company to continue to gather private information...."[19] Google's ban of the Disconnect app is exceptionally revealing in light of the fact that, unlike Apple, Google is notoriously "libertarian" when it comes to the millions of apps sold or downloaded for "free" from its app store. Its loose guidelines attempt to identify and exclude malicious applications but little else.[20]

Disconnect's founders attempted to challenge the extraction imperative, but they could not do so alone. After trying—without success—to negotiate with Google, they eventually joined other organizations in filing a complaint against Google in the EU, helping to precipitate an Android-focused antitrust probe.[21] As Disconnect argued,

> Google is under enormous pressure from the financial community to increase the "effectiveness" of its tracking, so that it can increase revenues and profits. Giving a user the ability to control his own privacy information (and to protect himself from malware) by blocking invisible connections to problematic sites constitutes an existential threat to Google.[22]

As ex-Googlers, the founders of Disconnect thought that they knew their adversary well, but they underestimated the progress of surveillance capitalism's institutionalization and the ferocity with which the corporation was prepared to fend off "existential threats" to its supply routes.

III. The Dispossession Cycle

Long before Disconnect, Google had discovered that successful dispossession is not a single action but rather an intricate convergence of political, social, administrative, and technical operations that requires cunning management over a substantial period of time. Its dispossession operations reveal *a predictable sequence of stages* that must be crafted and orchestrated in great detail in order to achieve their ultimate destination as a system of facts through which surplus extraction is normalized.

The four stages of the cycle are *incursion, habituation, adaptation,* and *redirection.* Taken together, these stages constitute a "theory of change" that describes and predicts dispossession as a political and cultural operation supported by an elaborate range of administrative, technical, and material capabilities. There are many vivid examples of this cycle, including Google's Gmail; Google's efforts to establish supply routes in social networks, first with Buzz and then with Google+; and the company's development of Google Glass. In this chapter we focus on the Street View narrative for a close look at the dispossession cycle and its management challenges.

The first stage of successful dispossession is initiated by unilateral *incursion* into undefended space: your laptop, your phone, a web page, the street where you live, an e-mail to your friend, your walk in the park, browsing online for a birthday gift, sharing photos of your kids, your interests and tastes, your digestion, your tears, your attention, your feelings, your face. The incursion is when dispossession operations rely on their virtual capabilities to kidnap behavioral surplus from the nonmarket spaces of everyday life where it lives. The incursion initiates Google's most basic and prolific form of dispossession: Arendt's repeated "original sin of simple robbery." Incursion moves down the road without looking left or right, continuously laying claim to decision rights over whatever is in its path. "I'm taking this," it says. "These are mine now."

The company has learned to launch incursions and proceed until resistance is encountered. It then seduces, ignores, overwhelms, or simply exhausts its adversaries. Seduction means a cascade of golden enticements: unprecedented storage, access to new qualities of information, new conveniences. When necessary, the company can just as easily pivot toward harsher tactics that aim to deplete its adversaries' time, money, and grit. There are hundreds of cases launched against Google by countries, states, groups, and individuals, and there are many more cases that never become public. According to Marc Rotenberg, executive director of the Electronic Privacy Information Center (EPIC), no one knows precisely how many lawsuits there are around the world.[23] The legal challenges are varied, but they nearly always come back to the same thing: unilateral incursion met by resistance.

Legal opposition and social protest have surfaced in relation to the digitalization of books,[24] the collection of personal information through Street

View's Wi-Fi and camera capabilities,[25] the capture of voice communications,[26] the bypassing of privacy settings,[27] the manipulation of search results,[28] the extensive retention of search data,[29] the tracking of smartphone location data,[30] wearable technologies and facial-recognition capabilities,[31] the secret collection of student data for commercial purposes,[32] and the consolidation of user profiles across all Google's services and devices,[33] just to name several instances. Expect to see drones, body sensors, neurotransmitters, "digital assistants," and other sensored devices on this list in the years to come. Meanwhile, Google is consistently stunning in its sense of entitlement, resolve, and audacity. The extraction imperative compels it to push new boundaries into undefended space.

In a second stage the aim is *habituation*. Whereas lawsuits and investigations unwind at the tedious pace of democratic institutions, Google continues the development of its contested practices at high velocity. During the elapsed time of FTC and FCC inquiries, court cases, judicial reviews, and EU Commission investigations, the new contested practices become more firmly established as institutional facts, rapidly bolstered by growing ecosystems of stakeholders. People habituate to the incursion with some combination of agreement, helplessness, and resignation. The sense of astonishment and outrage dissipates. The incursion itself, once unthinkable, slowly worms its way into the ordinary. Worse still, it gradually comes to seem inevitable. New dependencies develop. As populations grow numb, it becomes more difficult for individuals and groups to complain.

In a third stage of the cycle, when Google is occasionally forced to alter its practices, its executives and engineers produce superficial but tactically effective *adaptations* that satisfy the immediate demands of government authorities, court rulings, and public opinion. Meanwhile, in a final stage the corporation regroups to cultivate new rhetoric, methods, and design elements that *redirect* contested supply operations just enough so that they appear to be compliant with social and legal demands. The creativity, financial resources, and determination brought to the task of managing this staged process are flexible and dynamic. In contrast, the operational necessity for economies of scale in the capture of behavioral surplus is a perpetual-motion machine whose implacable rhythms offer no room for divergence.

The theory and practice of dispossession were developed and refined as the company learned how to counter and transform public resistance as an essential condition for the protection and expansion of its behavioral surplus franchise. Google's launch of Gmail on April Fool's Day, 2004, provided an early occasion to climb this learning curve as the corporation faced down public outrage over the automated scanning of e-mail content intended as a fresh source of surplus for targeted ads. Eventually, the dispossession cycle was refined as an explicit theory of change that framed a tactical game plan, which is by now regularly evoked as the surveillance capitalist corporation's battle-tested response to societal resistance.

The dispossession cycle at Google was so successful in facing down the threats to Gmail that it was replicated and further elaborated in the battles over Google Street View, the street-mapping operation launched in 2007. Once again, the company did not ask permission. It simply repeated the "original sin of simple robbery" and took what it wanted, waiting for resistance to run its course as it devoured and datafied the world's public spaces, streets, buildings, and homes.

Stage One: Incursion

Street View first entered public awareness with an apparently benign blog post. Peter Fleischer, Google's "privacy counsel," helped launch the new "service" by writing a paean celebrating America's "noble tradition" of public spaces, where, he claimed, "people don't have the same expectations of privacy as they do in their homes." As a lawyer, Fleischer knows the work that words do in establishing contracts and setting precedents, so it pays to read his 2007 words with care. His casually declarative prose undertakes some extraordinary work as he asserts that all public spaces are fair game for Google's taking. In his account, any public space is a fitting subject for the firm's new breed of incursion without authorization, knowledge, or agreement. Homes, streets, neighborhoods, villages, towns, cities: they are no longer local scenes where neighbors live and walk, where residents meet and talk. Google Street View, we are informed, claims every place as just another object among objects in an infinite grid of GPS coordinates and camera angles.

In his declaration, Fleischer intends to establish Google's prerogative to empty every place of the subjective meanings that unite the human beings who gather there. Yes, when we leave our home we know we will be seen, but we expect to be seen by one another in spaces that we choose. Instead, it's all impersonal spectacle now. My house, my street, my neighborhood, my favorite café: each is redefined as a living tourist brochure, surveillance target, and strip mine, an object for universal inspection and commercial expropriation.

Google had already taken everything on the web, but Street View and Google's other mapping operations, Google Maps and Google Earth (the company's 3-D view of the world using satellite and aerial imagery), announced an even-more-ambitious vision. Everything in the world was to be known and rendered by Google, accessed through Google, and indexed by Google in its infinite appetite for behavioral surplus. The presumption is that nothing is beyond Google's borders. The world is vanquished now, on its knees, and brought to you by Google.

The blog post that accompanied Street View is a precise replica of the invaders who once landed on that blameless Caribbean beach. Those *adelantados* concealed the bare facts of invasion in elaborate gestures of friendship and humility that made it impossible to discern the clear and present danger implicit in their arrival. Fleischer similarly assures his audience of friendly terms. Street View, which used cartoonishly wrapped cars with a large 360-degree camera mount on the roof to capture the imagery it sought, was designed to "respect the privacy of people who happen to be walking down a public street," Fleischer wrote. "That's why we designed a simple process for anyone to contact us and have their image removed," and he promised that it would respect laws and customs "in other parts of the world."[34]

Resistance came swiftly and often. By January 2009, Street View was facing opposition in Germany and Japan. John Hanke, by then the vice president for Google Maps–related products, dismissed the uproar. (You will recall that Hanke had founded the CIA-funded satellite mapping company, Keyhole, and after Google's purchase he led its transformation into Google Earth.) He told a reporter that it was all simply part of a "cycle of people understanding exactly what it is and what it isn't and what they shouldn't really be concerned about"—in other words, the dispossession cycle. Google Earth was also under fire, blamed for aiding a deadly terrorist attack in

Mumbai, but Hanke insisted that the debate over Google Earth or Street View had "mostly died off" in "the West." He cleverly equated any resistance to Google's incursions with the anti-freedom-of-expression interests of authoritarian governments and their "closed information societies."[35] This would become a standard rhetorical device for Google and its allies as they executed their offense.

Was Hanke surprised, then, when in April 2009 residents of the quiet English village of Broughton blocked a Street View car that tried to breach the village perimeter, calling it an unwelcome intrusion? This was "the West," after all, but the debate over privacy, self-determination, and decision rights was anything but dead. Privacy International submitted a formal complaint to the UK privacy authority, citing more than 200 reports from people who were identifiable on Street View images and demanded that the service be suspended.

Google's executives had apparently missed Fleischer's memo on respecting privacy claims. Instead, Hanke dismissed the protesters out of hand. He told the *London Times* that the company was undeterred and planned to complete coverage of the UK by the end of that year. He declared that Street View's information was "good for the economy and good for us as individuals....It is about giving people powerful information so that they can make better choices."[36]

Hanke's remarks were wishful thinking, of course, but they were consistent with Google's wider practice: it's great to empower people, but not *too* much, lest they notice the pilfering of their decision rights and try to reclaim them. The firm wants to enable people to make better choices, but not if those choices impede Google's own imperatives. Google's ideal society is a population of distant users, not a citizenry. It idealizes people who are informed, but only in the ways that the corporation chooses. It means for us to be docile, harmonious, and, above all, grateful.

In 2010 the German Federal Commission for Data Protection announced that Google's Street View operation actually camouflaged a covert data sweep; Street View cars were secretly collecting personal data from private Wi-Fi networks.[37] Google denied the charge, insisting that it was gathering only publicly broadcast Wi-Fi network names and the identifying addresses of Wi-Fi routers, but not personal information sent over the network.[38]

Within days, an independent analysis by German security experts proved decisively that Street View's cars were extracting unencrypted personal information from homes. Google was forced to concede that it had intercepted and stored "payload data," personal information grabbed from unencrypted Wi-Fi transmissions. As its apologetic blog post noted, "In some instances entire emails and URLs were captured, as well as passwords." Technical experts in Canada, France, and the Netherlands discovered that the payload data included names, telephone numbers, credit information, passwords, messages, e-mails, and chat transcripts, as well as records of online dating, pornography, browsing behavior, medical information, location data, photos, and video and audio files. They concluded that such data packets could be stitched together for a detailed profile of an identifiable person.[39]

Google's "Spy-Fi" scandal filled headlines around the world. Many believed that the Street View revelations would inflict irreparable damage to Google. In Germany, where the firm's actions were in clear violation of privacy and data-protection laws, officials reacted angrily and warned that Google would face EU investigations and consequences in the German courts. A bill was introduced into the German Parliament that proposed to fine Google for displaying personal property without owners' consent. Google faced fresh litigation in Switzerland, Canada, France, and the Netherlands. By 2012, there were multiple investigations in twelve countries, including most of Europe, the North Atlantic, and Australia, and Google had been found guilty of violating laws in at least nine countries.[40]

In the US, attorneys general from thirty-eight states launched a probe into Google's Street View practices. Private citizens filed numerous class-action suits, eight of which were consolidated in the Northern California US District Court. The head of Privacy International said that Google was becoming "Big Brother."[41] The Electronic Privacy Information Center championed substantial legal resistance in the US against Google's efforts to avoid repercussions in the wake of the Spy-Fi scandal, and it maintained a detailed and continuously updated online overview of the worldwide outrage, protests, investigations, litigation, and settlements in response to Google Street View and its extraction tactics.[42]

Google characterized Street View's "privacy violations" as a "mistake" made by a single engineer working on an "experimental" project, whose code had inadvertently made it into Street View's software. The firm refused to release the identity of the mystery engineer and insisted that the project's leaders were unaware of the data capture and "had no intention" of using those data. As Eric Schmidt told the *Financial Times,* "We screwed up," noting that the engineer in question would face an internal investigation for his clear "violation" of Google's policies. Unbowed, Schmidt insisted on the validity of Google's mission to index all the world's information.[43]

A 2012 investigation by the Federal Communications Commission described the case as "a deliberate software-design decision by one of the Google employees working on the Street View project."[44] The engineer had been selected for the team because of his unique expertise in Wi-Fi "wardriving," the practice of driving around using equipment to locate wireless networks.[45] His design notes indicated that user traffic and location data would be logged along with "information about what they are doing" that would "be analyzed offline for use in other initiatives." The notes identified but then dismissed "privacy considerations."[46]

The FCC found evidence that contradicted Google's scapegoating narrative. The records showed that the engineer had e-mailed links to his software documentation to project leaders, who then shared them with the entire Street View team. It also found evidence that on at least two occasions, the engineer told his colleagues that Street View was collecting personal data. Despite these facts along with evidence of the company's exhaustive internal software reviews and testing procedures and the regular transfer of payload data from Street View's hard disks to Google's Oregon data center, Google's engineers denied any knowledge of personal data collection.[47]

Stage Two: Habituation

Hanke's bet that the "cycle" would eventually wear down resistance reflects a key operational component of the extraction imperative, discovered in Search, refined with Gmail, and elaborated with Street View. The messages that come through are "Don't look back. Wait them out. Step on them, if necessary."

The April 2012 FCC report is heart wrenching in its way, a melancholic depiction of democracy's vulnerability in the face-off with a wealthy, determined, and audacious surveillance capitalist opponent. In November 2010 the FCC sent Google a letter of inquiry requesting necessary information. Little was forthcoming. By March of the next year, a second "supplemental" letter was sent. Google's response was incomplete information and lack of cooperation, which produced another "demand letter" in August. Google's continued lack of engagement required yet another letter in late October. The FCC staff was burdened with following up and chasing down evasive corporate executives and their representatives for an entire year.

The document is a revelation of negative space and a saga of democracy rebuffed. The FCC's detailed initial request produced "only five documents" and no e-mails. The corporation said that it had no time to undertake a comprehensive review, calling it "burdensome." Google "failed" to identify relevant individuals. It "redacted" names. It asserted that the information requested "serves no useful purpose." It "failed" to verify information. When asked for specific submissions, "Google did not do so." Google "argued" that it should "not be required" to provide access to the payload data it had illicitly collected. "Google waited...." The phrases "failed to respond" and "failed to provide" are repeated throughout the account. "Google violated Commission orders...delaying...." Affidavits were requested five times, but the company did not provide any of these until September 2011, after the FCC threatened a subpoena. The mystery engineer simply refused to speak with investigators, citing his Fifth Amendment right to avoid self-incrimination. As the report concludes, "There is evidence that Google's failure to cooperate with the Bureau was in many or all cases deliberate." It might have said "imperative."

Ultimately, the corporation's lawyers prevailed, defending Google's data sweeps with a single obscure passage in a decades-old wiretap law. Perhaps the most telling element of the entire episode is that the same democratic system of laws and rules that the corporation openly treated with scorn was invoked to protect it from accountability. In the end the FCC fined Google only $25,000 for obstructing its investigation. Google did not evade legal consequences because society agreed with its practices, but because there was not enough relevant law to protect citizens from its incursions.

The thirty-eight attorneys general did not fare much better. When the group's leader, Connecticut's Richard Blumenthal, issued a civil investigative demand (equivalent to a subpoena) to get access to the infamous private data, "Google ignored it."[48] The company finally agreed to settle with the states in 2013, accepting a mere $7 million fine and a series of agreements regarding "aggressive" self-policing. The *New York Times* announced that Google had finally admitted that "it had violated people's privacy during its Street View mapping project, when it casually scooped up...personal information," as if this scandal had been the only contested element in the whole affair. State officials crowed that "the industry giant...is committing to change its corporate culture to encourage sensitivity to issues of personal data privacy."[49] In light of the fact that the extraction imperative is what makes this giant a giant, one doesn't know whether to laugh or cry at the confidence of the attorneys general in Google's commitment to privacy self-regulation.

There are two key elements here that illuminate habituation tactics. The first is the simple fact of the elapsed time between Street View's initial incursion in 2007, the 2010 scandal, the 2012 conclusion of the FCC inquiry, and the 2013 conclusion of the states' investigation. The German investigation also closed in late 2012, with little to show for its trouble. Other contests and lawsuits lumbered on. Despite all the sound and fury, Google continued to operate Street View during those years. Between 2008 and 2010, 600 billion bytes of personal information were collected "illegitimately" around the world, 200 billion of those in the US.[50] The corporation said that it had discontinued personal data collection. Had it? Can anyone say with certainty? Even if it had, the original incursion that was Street View continued unscathed.

The second point is that in retrospect, one sees that the very idea of a single rogue engineer was designed and elaborated as a brilliant piece of misdirection, a classic scapegoating ploy. It directed attention away from the ambitious and controversial agenda of the extraction imperative toward a different narrative of a single infected cell excised from the flesh of an enormous but innocent organism. All that was left was to excise the infected flesh and let the organism declare itself cured of its privacy kleptomania. Then—a return to the streets, born again.

Google achieved exactly what Hanke had predicted. Street View's fundamental audacity, the unprecedented astonishing incursion that drew English villagers into the streets to block a Google camera car, enjoyed six more years of rooting in global consciousness. The corporation's strategic discipline when it comes to stonewalling, snubbing, and exploiting democracy brought six more years of people using Street View data and six more years to build the tacit case for Google's inevitability and our helplessness. There were six more years for this simple robbery of decision rights to shade into normalcy and even to be reckoned as "convenient," "useful," or "marvelous."

Stage Three: Adaptation

In October 2010, just before the corporation received the FCC's first letter of inquiry, Google's senior vice president of Engineering and Research announced "stronger privacy controls" in an official Google blog post. "We've failed badly here," he said. The Street View scandal was framed as an inadvertent error, a single blemish on a company that works hard "to earn your trust." The post assured the public that the corporation was talking to external regulators "about possible improvements to our policies" and promised that it would make changes to ensure user privacy. Alma Whitten, a Google executive with credentials in computer security and privacy controls, was appointed as director of Privacy across engineering and product management. The blog also introduced a new internal training emphasis on "responsible collection, use and handling of user data." Finally, the post pledged new internal controls to oversee how data are handled. "We are mortified by what happened," it read, "but confident that these changes to our processes and structure will significantly improve our internal privacy and security practices for the benefit of all our users."[51]

Though pledging reform to the public, the corporation was simultaneously forced to adapt to governmental demands in a range of countries—including Australia, Belgium, Canada, France, Holland, Hong Kong, Ireland, Israel, Italy, New Zealand, Poland, Spain, South Korea, the UK, and the US—where Street View was subjected to litigation, fines, and/or regulation. In Japan, homeowners complained of Street View cameras that peered above privacy fencing to record private homes. Google agreed to government demands to lower its cameras, reshoot all images, and blur identifiable facial images and license plates. In Germany, Google allowed residents to request that their homes be blurred in any Street View images. Nearly 250,000 households made opt-out requests in 2009–2010, requiring Google to hire 200 temporary programmers to meet the demand.[52] Google was fined 145,000 euros by the Hamburg data-protection supervisor who had first discovered the Street View illicit data gathering, just short of the 150,000 euro fee he could have imposed.[53] It was the largest fine ever to have been levied by European regulators for privacy concerns. The discount reflected Google's assurances that it would swiftly and thoroughly delete the payload data. In 2011 Google ended its Street View program in Germany, continuing to support but no longer update the images it had already collected.[54]

Other countries imposed bans on Street View operations. Switzerland initially banned the service in 2009, insisting that Google remove all imagery it had posted of Swiss towns and cities. Eventually, the ban was lifted, but the Swiss Federal Administrative Court imposed a series of strict guidelines, including blurring faces, instituting opt-out measures, and reducing camera heights. By 2016, Google's service remained confined to outdoor tourist sites.[55] The corporation also faced Street View bans in Austria, the Czech Republic, Greece, India, and Lithuania. By the summer of 2017, though, Street View data were available from at least some regions of each of these countries.[56]

Stage Four: Redirection

What Google did not say in its mea culpa blog post, the one thing that it could not say, was that it would abandon its fundamental logic of accumulation: the principles of surveillance capitalism that had brought the behemoth into existence and sustained its growth. The message of Street View's redirection campaign was that Google would not exempt anything from the grid. Everything must be corralled for conversion into raw material. Short of institutional suicide, there is little that Google *can* say or do to ensure "user privacy." This helps to explain why, as one 2015 article celebrating the history of Google Maps observes, "Google Maps was attracting all sorts of privacy controversies...people were freaked out....But that doesn't mean Street View has been tamped down as a project. It's now available in 65 of Google Maps' 200-some countries."[57]

Alma Whitten's job was to repair Google's privacy reputation, not to dismantle the extraction imperative and its relentless demand for economies of scale in the supply function. This is to say that her job was a logical impossibility. That she may have nevertheless taken it seriously is suggested by the fact that just two-and-a-half years after her appointment as privacy czar, she announced her retirement from Google in April 2013. Indeed, it is painful to watch Whitten testify about Google's practices to an early-2013 congressional hearing. She was under questioning from Congress, and one sees the effort required as she hunts for the words to convey an answer without conveying the truth.[58] The time had come to regroup and redirect the global mapping project, not to end it.

That nothing much had changed or would change was immediately suggested by the fate of Google's mystery engineer in the two years that followed the scandal. Within days of the FCC report in April 2012, a former state investigator who had been assigned to the Street View inquiry identified Google's "rogue" actor as Marius Milner, a celebrated hacker and wardriving specialist. It had been two years since the supposedly irreparable damage that he inflicted on Google and his "clear violation" of policy, yet he continued to be employed at the firm in its YouTube operations. Later that year, he would be one of six inventors on a team led by John Hanke to patent "A System and Method for Transporting Virtual Objects in a Parallel Reality Game."[59]

The invention in which Milner participated was related to a virtual reality game called Ingress, also developed by Hanke and his team at Google. (Hanke would eventually establish his own shop, Niantic Labs, within Google's new Alphabet holding company.) Ingress became a test bed for many of the foundational concepts that reappeared in another "game," Pokémon Go, a prototype of a second phase of surveillance capitalist expansion that we examine closely in Part II. In this next phase, Google's maps are a critical resource for the expansion of digital dispossession from the virtual world to the one that we call "real." In light of those plans, Street View could not be allowed to die or even to be constrained. The corporation's senior product manager for Google Maps framed it succinctly in September 2012, just four months after the FCC investigation: "If you look at the offline world, the real world in which we live, that information is not entirely online. Increasingly as we go about our lives, we are trying to bridge that gap between what we see in the real world and [the online world], and Maps really plays that part."[60]

Google's closely guarded "Ground Truth" project, initiated in 2008 but only publicly revealed just four months after the FCC report in 2012, exemplifies the point. Ground Truth is the "deep map" that contains the detailed "logic of places": walking paths, goldfish ponds, highway on-ramps, traffic conditions, ferry lines, parks, campuses, neighborhoods, buildings, and more.[61] Getting these details right is a source of competitive advantage in the contest for behavioral surplus accrued from mobile devices. The construction of the deep map draws on public resources such as geographic databases from the US Census Bureau and the US Geological Survey,[62] but what distinguishes Google's maps from all others is the integration of its exclusive proprietary data from Street View. In other words, data compiled through public investments are augmented with data taken from a unilateral transfer of surplus behavior and decision rights. The composite results are then reclassified as private assets.

One of the first journalists invited to see demonstrations of Ground Truth in 2012, Alexis Madrigal, observed that "the Maps team, largely driven by Street View, is publishing more imagery data every two weeks than Google possessed in total in 2006. . . . Google is up to five million miles driven now." Street View cars are likened to Google Search's early web crawlers, which quietly commandeered web pages for indexing and access in the corporation's original act of dispossession. By 2012, Street View data also provided street signs and addresses. Soon, Madrigal wrote, "any word that is visible from a road will become a part of Google's index of the physical world" thanks to Street View. Madrigal's look at the Ground Truth operation concludes, "The geographic data Google has assembled is not likely to be matched by any other company. . . . *They've built this whole playground as an elaborate lure for you.*"[63]

As one project leader put it, "The challenge of deciding you're going to map the world is that you can't ever stop."[64] So it was that by 2016, Google's Street View website celebrated its successful evolution by stating, "We've come a long way since our initial U.S. launch in 2007; today we've expanded our 360-degree panoramic views to include locations on all seven continents." Street View's fleet of surveillance-gathering tools had been augmented to include a wearable backpack, a three-wheeled pedicab, a snow-mobile, and a trolley, all of which were designed to capture places that Street View cars could not traverse. Tourist boards and nonprofits were offered the use of the company's Trekker equipment (the backpack camera) to "collect views of remote and unique places" that were, literally and figuratively, "off the grid."[65]

What Google couldn't build, it bought. In 2013 the corporation won a re-ported bidding war with Facebook for Israeli social mapping startup Waze, a firm that pioneered community-sourced real-time traffic information. In 2014 it acquired real-time satellite imaging startup Skybox just as the US Department of Commerce lifted restrictions on high-resolution satellite imagery. As an expert explained,

> If you imagine a satellite above your office then the old resolution could probably make out your desk. The new imagery—where each pixel measures around 31 cm—can now make out what's on your desk. When you reach this sort of frequency you can begin to add in what we call "pattern of life" analysis. This means looking at activity in terms of movement—not just identification.[66]

In this context one appreciates the significance of another aspect of Google's redirection campaign: a 2011 announcement that the corporation had breached "a new frontier" with the introduction of an "indoor positioning system" that enabled it to locate and follow people "when you're inside an airport, shopping mall, or retail store." Eventually, these new capabilities would include sensors and embedded cameras that let users map and navigate interior spaces.[67] In a September 2014 blog, Google Maps' dynamic new capabilities were showcased to the public as your new "co-pilot for deciding everything from turn-by-turn directions, to discovering new restaurants, to deciding which hiking trails to climb." The post credits Street View with providing these wondrous new capabilities and announces the expansion of the whole incursion with the introduction of a mobile mapping tool dubbed "Cartographer," worn as a backpack and able to map the interior of buildings.[68] Cartographer's information could be added to the growing navigational database of interior spaces, amplifying Google's ability to locate people and devices as they moved between outdoor and indoor spaces.

Building interiors had eluded Street View and the extraction imperative; few homeowners were likely to invite those cameras indoors. Instead, Cartographer's capabilities were bundled into the larger Street View redirection campaign and pitched to businesses as a way to enhance consumer trust, allay anxiety, and substantially increase revenues. Google exhorted consumer-facing businesses to "invite customers inside." With its "Business View," consumers would be able to see inside thousands of hotels, restaurants, and other destinations. Search listings would feature the new Street View content. Hotel listings would offer a virtual tour of the properties. "Give them the confidence they're seeking," Google told its business marketplace, by allowing consumers "to experience your location before they arrive." Google asserted that virtual tours "double bookings," and it instated a certification program that enabled businesses to hire a Google-approved freelance photographer to produce images for Street View. These extraordinary new redirection tactics aimed to flip the old pattern. They reframed Street View from an edgy incursion circumventing resistance through stealth to an opulent VIP tent where businesses scrambled for an entry pass.

Street View's redirection and elaboration announced a critical shift in the orientation and ambition of the surveillance program: *it would no longer be only about routes, but about routing.* We will examine this new episode of dispossession in the chapters that follow. For now, suffice to say that Street View and the larger project of Google Maps illustrate the new and even more ambitious goals toward which this cycle of dispossession would soon point: the migration from an online data source to a real-world monitor to an advisor to an active shepherd—from knowledge to influence to control. Ultimately, Street View's elaborate data would become the basis for another complex of spectacular Google incursions: the self-driving car and "Google City," which we learn more about in Chapter 7. Those programs aim to take surplus capture to new levels while opening up substantial new frontiers for the establishment of behavioral futures markets in the real world of goods and services. It is important to understand that each level of innovation builds on the one before and that all are united in one aim, the extraction of behavioral surplus at scale.

In this progression, Google perceives an opportunity that it hopes its customers will come to appreciate: its ability to influence actual behavior as it occurs in the real spaces of everyday life. In 2016, for example, the corporation introduced a new Maps app feature called "Driving Mode" that suggests destinations and travel times without users even selecting where they want to go. If you searched for a hammer online, then "Driving Mode" can send you to a hardware store when you buckle up your seat belt. "Google is integrating this 'push' technology into its main mobile search app," reported the *Wall Street Journal.*[69]

With this app, Google the "copilot" prompts an individual to turn left and right on a path defined by its continuously accruing knowledge of the person *and* the context. Predictions about where and why a person might spend money are derived from Google's exclusive access to behavior surplus and its equally exclusive analytic capabilities: *"Eat here." "Buy this."* Google's surplus analysis can predict that you are likely to buy an expensive woolen suit, and its real-time location data can trigger the proprietor or advertiser's real-time prompt, matched to your profile and delivered at the very moment that you are within sight of the flannels, tweeds, and cashmeres. Push and pull, suggest, nudge, cajole, shame, seduce: Google wants to be your copilot for life itself. Each human response to each commercial prompt yields more data to refine into better prediction products. The prompts themselves are bought and paid for in a novel iteration of Google's online ad markets: *real-time, real-world trading in behavioral futures.* Your future.

The stakes are high in this market frontier, where *unpredictable behavior is the equivalent of lost revenue.* Google cannot leave anything to chance.[70] In September 2016 the tech newsletter the *Register* revealed that the Google Play app preinstalled in the latest Android phone continuously checks a user's location, sending that information to your third-party apps as well as to Google's own servers. One security researcher was shocked when his Android phone prompted him to download the McDonald's app at the very moment that he crossed the threshold of the fast-food restaurant. He later discovered that Google Play had monitored his location thousands of times. Similarly, Google Maps "doesn't give you a decent option of turning it off." If you do, the operating system warns, "basic features of your device may no longer function as intended."[71] Google's insistence reflects the authoritarian politics of the extraction imperative as well as the corporation's own enslavement to the implacable demands of its economics.

The historic point for us to consider here is that the once-spurned Street View found new life in its contribution to the decisive expansion of behavioral futures markets both online and in the real world. Once dedicated to targeted online advertising, these markets now grow to encompass predictions about what human beings will do now, soon, and later, whether they make their way online, on sidewalks and roads, or through rooms, halls, shops, lobbies, and corridors. These ambitious goals foreshadow fresh incursions and dispossessions as resistance is neutralized and populations fall into dulled submission.

Google discovered by chance or intention the source of every mapmaker's power. The great historian of cartography, John B. Harley, said it succinctly: "Maps created empire." They are essential for the effective "pacification, civilization, and exploitation" of territories imagined or claimed but not yet seized in practice. Places and people must be known in order to be controlled. "The very lines on the map," wrote Harley, were a language of conquest in which "the invaders parcel the continent among themselves in designs reflective of their own complex rivalries and relative power." The first US rectangular land survey captured this language perfectly in its slogan: "Order upon the Land."[72] The cartographer is the instrument of power as the author of that order, reducing reality to only two conditions: the map and oblivion. The cartographer's truth crystallizes the message that Google and all surveillance capitalists must impress upon all humans: *if you are not on our map, you do not exist.*

IV. The Dogs of Audacity

Projects such as Street View taught Google that it could assume the role of arbiter of the future and get away with it. It learned to sustain even the most-contested dispossession efforts when they are necessary to secure vital new supply lines. For example, while Street View protests erupted around the world and just months before Germany's announcement that Street View was secretly capturing personal information from unprotected Wi-Fi networks, Google introduced Buzz—a platform intended to float Google's nets in the path of the coveted behavioral surplus that streamed from social networks. The invasive practices introduced with Buzz—it commandeered users' private information to establish their social networks by fiat—set off a fresh round of the dispossession cycle and its dramatic contests.

As Google learned to successfully redirect supply routes, evading and nullifying opposition, it became even more emboldened to let slip the dogs of audacity and direct them toward havoc. Among many examples, Google Glass neatly illustrates the tenacity of the extraction imperative and its translation into commercial practice. Google Glass combined computation, communication, photography, GPS tracking, data retrieval, and audio and video recording capabilities in a wearable format patterned on eyeglasses. The data it gathered—location, audio, video, photos, and other personal information—moved from the device to Google's servers, merging with other supply routes to join the titanic one-way flow of behavioral surplus.

The project was seen as a precursor to more flexible and less overt forms of wearable computation and surplus capture. John Hanke described its familiar shape in the form of eyewear as suitable for "the early adoption phases" of wearable technology in much the same way that the first automobiles resembled horse-drawn buggies. In other words, the "glasses" were intended to disguise what was in fact unprecedented: "Ultimately we will want these technologies, wherever they are on your body, to be totally optimized based on the job they're doing, not on what is more socially acceptable at that first moment of creation, just because it reminds people of something they've seen in the past."[73] Introduced with great flair in the spring of 2012 as fashion-forward futurism, it wasn't long before the public registered fresh horror at this bizarre invasion. Those who wore the device were recast as "glassholes," and some businesses banned the glasses from their premises.[74]

Privacy advocates protested the "always on" but "undetectable" recording of people and places that eliminates a person's reasonable expectation of privacy and/or anonymity. They warned of new risks as facial-recognition software is applied to these new data streams and predicted that technologies like Glass would fundamentally alter how people behave in public. By May 2013, a congressional privacy caucus asked CEO Larry Page for assurances on privacy safeguards for Glass, even as a Google conference was held to coach developers on creating apps for the new device. In April 2014 Pew Research announced that 53 percent of Americans thought that smart wearables were "a change for the worse," including 59 percent of American women.[75]

Google continued to tough it out, waiting for habituation to kick in. That June it announced that Glass would offer the Livestream video-sharing app, enabling Glass users to stream everything around them to the internet in real time. When asked about these controversial and intrusive capabilities in the hands of any owner of the device, Livestream's CEO reckoned, "Google is ultimately in charge of...setting the rules."[76] Sergey Brin made it clear that any resistance would be categorically rejected when he told the *Wall Street Journal*, "People always have a natural aversion to innovation."[77]

Adaptation began in 2015 with the announcement that Glass would no longer be available. The company said nothing to acknowledge the public's revulsion or the social issues that Glass had raised. A short blog post announced, "Now we're ready to put on our big kid shoes and learn how to run...you'll start to see future versions of Glass when they're ready."[78] An eyewear designer was tasked with transforming the look from a futuristic device to something more beautiful.

Redirection began quietly. In June 2015 the FCC's Office of Engineering and Technology received new design plans for Glass, and September brought fresh headlines announcing that Glass "is getting a new name, and a new lease on life."[79] A year later, Eric Schmidt, now Google's chairman, put the situation into perspective: "It is a big and very fundamental platform for Google." He explained that Glass was withdrawn from public scrutiny only to "make it ready for users...these things take time."[80] As more information trickled out of the corporation, it became clear that there was no intention of ceding potential new supply routes in wearable technologies, no matter the public reaction. Glass was the harbinger of a new "wearables" platform that would help support the migration of behavioral surplus operations from the online to the offline world.[81]

In July 2017 the redirection phase went public with a blog post introducing a new iteration of Google Glass to the world, now as "Glass Enterprise Edition."[82] This time there would be no frontal attack on public space. Instead, it was to be a tactical retreat to the workplace—the gold standard of habituation contexts, where invasive technologies are normalized among captive populations of employees. "Workers in many fields, like manufacturing, logistics, field services, and healthcare find it useful to consult a wearable device for information and other resources while their hands are busy," wrote the project's leader, and most press accounts lauded the move, citing productivity and efficiency increases in factories that deployed the new Glass.[83] There was little acknowledgment that habituation to Glass at work was most certainly a back door to Glass in our streets or that the intrusive surveillance properties of the device would, with equal certainty, be imposed on the women and men required to use them as a condition of their employment.

The lesson of Glass is that when one route to a supply source encounters obstacles, others are constructed to take up the slack and drive expansion. The corporation has begrudgingly learned to pay more attention to the public relations of these developments, but the unconditional demands of the extraction imperative mean that the dispossession cycle must proceed at full throttle, continuously claiming new territory.

Dispossession may be an act of "simple robbery" in theory, but in fact it is a complex, highly orchestrated political and material process that exhibits discernible stages and predictable dynamics. The theory of change exhibited here systematically transfers knowledge and rights from the many to the few in a glorious fog of Page's "automagic." It catalogues public contest as the unfortunate but predictable outcry of foolish populations who exhibit a knee-jerk "resistance to change," wistfully clinging to an irretrievable past while denying an inevitable future: Google's future, surveillance capitalism's future. The theory indicates that opposition must simply be weathered as the signature of the first difficult phases of incursion. It assumes that opposition is fleeting, like the sharp yelp of pain when a Novocain needle first pierces the flesh, before numbness sets in.

V. Dispossession Competition

Google's spectacular success in constructing the mechanisms and principles of surveillance capitalism and attracting surveillance revenues ignited competition in an escalating war of extraction. Google began in a blank space, but it would soon contend with other firms drawn to surveillance revenues. Facebook was the first and has remained the most aggressive competitor for behavioral surplus supplies, initiating a wave of *incursions* at high speed, establishing a presence on the free and lawless surplus frontier while denying its actions, repelling criticism, and thoroughly confusing the public. The "Like" button, introduced widely in April 2010 as a communications tool among friends, presented an early opportunity for Facebook's Zuckerberg to master the dispossession cycle. By November of that year, a study of the *incursion* already underway was published by Dutch doctoral candidate and privacy researcher Arnold Roosendaal, who demonstrated that the button was a powerful supply mechanism from which behavioral surplus is continuously captured and transmitted, installing cookies in users' computers whether or not they click the button. Presciently describing the operation as an "alternative business model," Roosendaal discovered that the button also tracks non-Facebook members and concluded that Facebook was potentially able to connect with, and therefore surveil, "all web users."[84] Only two months earlier, Zuckerberg had characterized Facebook's growing catalogue of privacy violations as "missteps."[85] Now he stuck to the script, eventually calling Roosendaal's discovery a "bug."[86]

By 2011 the *habituation* stage of the cycle was in full swing. A May *Wall Street Journal* report confirmed Facebook's tracking, even when users don't click the button, and noted that the button was already installed on one-third of the world's one thousand most-visited websites. Meanwhile, Facebook's Chief Technology Officer said of the button, "We don't use them for tracking and they're not intended for tracking."[87] On September 25, Australian hacker Nik Cubrilovic published findings showing that Facebook continued to track users even after they logged out of the site.[88] Facebook announced that it would fix "the glitch," explaining that certain cookies were tracking users in error, and noting that it could not cease the practice entirely due to "safety" and "performance" considerations.[89] Journalists discovered that just three days before Cubrilovic's revelations, the corporation received a patent on specialized techniques for tracking users across web domains. The new data methods enabled Facebook to track users, create personal profiles on individuals and their social networks, receive reports from third parties on each action of a Facebook user, and log those actions in the Facebook system in order to correlate them with specific ads served to specific individuals.[90] The company immediately denied the relevance and importance of the patent.[91]

With Facebook's unflinching assertions that it did not track users, even in the face of many robust facts, specialists grew more frustrated and the public more confused. This appears to have been the point. By denying every accusation and pledging its commitment to user well-being, Facebook secured a solid year and a half with which to habituate the world to its "Like" button, institutionalizing that iconic thumb turned toward the sky as an indispensable prosthetic of virtual communication.[92]

This solid achievement paved the way for the *adaptation* stage of the dispossession cycle, when in late November 2011, Facebook consented to a settlement with the FTC over charges that it had systematically "deceived consumers by telling them that they could keep their Facebook information private, and then repeatedly allowing it to be shared and made public."[93] The complaint brought by EPIC and a coalition of privacy advocates in 2009 initiated an FTC investigation that yielded plenty of evidence of the corporation's broken promises.[94] These included website changes that made private information public, third-party access to users' personal data, leakage of personal data to third-party apps, a "verified apps" program in which nothing was verified, enabling advertisers to access personal information, allowing access to personal data after accounts were deleted, and violations of the Safe Harbor Framework, which governs data transfers between the United States and the EU. In the parallel universe of surveillance capitalism, each one of these violations was worthy of a five-star rating from the extraction imperative. The FTC order barred the company from making further privacy misrepresentations, required users' affirmative consent to new privacy policies, and mandated a comprehensive privacy program to be audited every two years for twenty years. FTC Chairman Jon Leibowitz insisted that "Facebook's innovation does not have to come at the expense of consumer privacy."[95] But Leibowitz was not up against a company; he was up against a new market form with distinct and intractable imperatives whose mandates can be fulfilled *only* at the expense of user privacy.

Redirection came swiftly. In 2012 the company announced it would target ads based on mobile app use, as it worked with Datalogix to determine when online ads result in a real-world purchase. This gambit required mining personal information, including e-mail addresses, from user accounts. In 2012 Facebook also gave advertisers access to targeting data that included users' e-mail addresses, phone numbers, and website visits, and it admitted that its system scans personal messages for links to third-party websites and automatically registers a "like" on the linked web page.[96] By 2014, the corporation announced that it would be tracking users across the internet using, among its other digital widgets, the "Like" button, in order to build detailed profiles for personalized ad pitches. Its "comprehensive privacy program" advised users of this new tracking policy, reversing every assertion since April 2010 with a few lines inserted into a dense and lengthy terms-of-service agreement. No opt-out privacy option was offered.[97] The truth was finally out: the bug was a feature.

Meanwhile, Google maintained the pledge that had been critical to the FTC's approval of its 2007 acquisition of the ad-tracking behemoth Double-Click when it agreed not to combine data from the tracking network with other personally identifiable information in the absence of a user's opt-in consent. In this case, Google appears to have waited for Facebook to extend the surveillance capitalist frontier and bear the brunt of incursion and habituation. Later, in the summer of 2016, Google crossed that frontier with an announcement that a user's DoubleClick browsing history "may be" combined with personally identifiable information from Gmail and other Google services. Its promised opt-in function for this new level of tracking was presented with the headline "Some new features for your Google account." One privacy scholar characterized the move as the final blow to the last "tiny semblance" of privacy on the web. A coalition of privacy groups presented a new complaint to the FTC, implicitly recognizing the logic of the dispossession cycle: "Google has done incrementally and furtively what would plainly be illegal if done all at once."[98]

Facebook's IPO in 2012 was notoriously botched when last-minute downward revisions of its sales projections, precipitated by the rapid shift to mobile devices, led to some unsavory dealings among its investment bankers and their clients. But Zuckerberg, Sheryl Sandberg, and their team quickly mastered the nuances of the dispossession cycle, this time to steer the company toward mobile ads. They learned to be skilled and ruthless hunters of behavioral surplus, capturing supplies at scale, evading and resisting law, and upgrading the means of production to improve prediction products.

Surveillance revenues flowed fast and furiously, and the market lavishly rewarded the corporation's shareholders. By 2017, the *Financial Times* hailed the company's 71 percent earnings surge with the headline "Facebook: The Mark of Greatness" as Facebook's market capitalization rose to just under $500 billion, with 2 billion average monthly active users. Facebook ranked seventh in one important tally of the top 100 companies in the first quarter of 2017, when just a year earlier it hadn't figured anywhere in the top 100. Advertising, primarily mobile, accounted for nearly every dollar of the company's revenue in the second quarter of 2017: $9.2 billion of a total $9.3 billion and a 47 percent increase over the prior year.[99]

The *Guardian* reported that Google and Facebook accounted for one-fifth of global ad spending in 2016, nearly double the figure of 2012, and by one accounting Google and Facebook owned almost 90 percent of the growth in advertising expenditures in 2016.[100] Surveillance capitalism had propelled these corporations to a seemingly impregnable position.

Among the remaining three of the largest internet companies, Microsoft, Apple, and Amazon, it was Microsoft that first and most decisively turned toward surveillance capitalism as the means to restore its leadership in the tech sector, with the appointment of Satya Nadella to the role of CEO in February 2014. Microsoft had notoriously missed several key opportunities to compete with Google in the search business and develop its targeted advertising capabilities. As early as 2009, when Nadella was a senior vice president and manager of Microsoft's search business, he publicly criticized the company's failure to recognize the financial opportunities associated with that early phase of surveillance capitalism. "In retrospect," he lamented, "it was a terrible decision" to end the search-ad service: "None of us saw the paid-search model in all its glory." Nadella recognized then that Microsoft's Bing search engine could not compete with Google because it lacked scale in behavioral surplus capture, the critical factor in the fabrication of high-quality prediction products: "When you look at search... it's a game of scale. Clearly we don't have sufficient scale today and that hinders... the quality of the ad relevance which is perhaps the bigger issue we have today."[101]

Less than three months after assuming his new role, Nadella announced his intention to redirect the Microsoft ship straight into this game of scale with the April release of a study that the company had commissioned from market intelligence firm IDC.[102] It concluded that "companies taking advantage of their data have the potential to raise an additional $1.6 trillion in revenue over companies that don't," and Nadella was determined to make landfall on the far shores of this rich new space. Microsoft would reap the advantages of its own data, and it would specialize in "empowering" its clients to do the same. Nadella composed a blog to signal the new direction, writing, "The opportunity we have in this new world is to find a way of catalyzing this data exhaust from ubiquitous computing and converting it into fuel for ambient intelligence."[103] As a video outlining the new "data vision" explains, "Data that was once untapped is now an asset."

Many of Nadella's initiatives aim to make up for lost time in establishing robust supply routes to surplus behavior and upgrading the company's means of production. Bing's search engineering team built its own model of the digital and physical world with a technology it calls Satori: a self-learning system that is adding 28,000 DVDs of content every day.[104] According to the project's senior director, "It's mind-blowing how much data we have captured over the last couple of years. The line would extend to Venus and you would still have 7 trillion pixels left over."[105] All those pixels were being put to good use. In its October 2015 earnings call, the company announced that Bing had become profitable for the first time, thanks to around $1 billion in search ad revenue from the previous quarter.

Another strategy to enhance Bing's access to behavioral surplus was the corporation's "digital assistant," Cortana, to which users addressed more than one billion questions in the three months after its 2015 launch.[106] As one Microsoft executive explains, "Four out of five queries go to Google in the browser. In the [Windows 10] task bar [where Cortana is accessed,] five out of five queries go to Bing.... We're all in on search. Search is a key component to our monetization strategy."[107]

Cortana generates more than search traffic. As Microsoft's privacy policy explains, "Cortana works best when you sign in and let her use data from your device, your personal Microsoft account, other Microsoft services, and third-party services you choose to connect."[108] Like Page's automagic, Cortana is intended to inspire awestruck and grateful surrender. One Microsoft executive characterizes Cortana's message: "'I know so much about you. I can help you in ways you don't quite expect. I can see patterns that you can't see.' That's the magic."[109]

Nevertheless, the company made a canny decision not to disclose the true extent of Cortana's knowledge to its users. It wants to know everything about you, but it does not want you to know how much it knows or that its operations are entirely geared to continuously learning more. Instead, the "bot" is programmed to ask for permission and confirmation. The idea is to avoid spooking the public by presenting Cortana's intelligence as "progressive" rather than "autonomous," according to the project's group program manager, who noted that people do not want to be surprised by how much their phones are starting to take over: "We made an explicit decision to be a little less 'magical' and little more transparent."[110]

Nadella envisions a new platform of "conversations" in which users interact with bots that induce them to eagerly disclose the details of their daily lives.[111] The platform promises to deliver experiences such as "conversational commerce,"[112] where, for example, a bot

> knows what shoes you bought last week, it knows your preferences from your past purchases, it knows your profile and can call a recommendations model to determine what products you have the most affinity to buy.... Using the power of data and analytics, the bot can respond back with recommendations that it determines are most relevant for you. It can also invite people from your social network to help you make a choice. Once you make the selection, it will use your size information, shipping address, payment information to ship the selected dress to you.[113]

The release of Microsoft's new operating system, Windows 10, in July 2015 drove home the seriousness of purpose and urgency that the corporation now assigned to establishing and securing supply routes to behavioral surplus.[114] One software engineer writing in *Slate* described it as "a privacy morass in dire need of reform" as he detailed how the system "gives itself the right to pass loads of your data to Microsoft's servers, use your bandwidth for Microsoft's own purposes, and profile your Windows usage."[115]

As many analysts quickly discovered, the system pushed users toward the "express install" function, in which every default setting enabled the maximum flow of personal information to the corporation's servers. An investigation by tech website *Ars Technica* revealed that even when those default settings were reversed and key services such as Cortana disabled, the system continued to access the internet and transmit information to Microsoft. In some instances those transmissions appeared to contain personal information, including a machine ID, user content, and location data.[116]

According to an analysis by the Electronic Frontier Foundation (EFF), even users who opted out of Cortana were subject to an "unprecedented" amount of information capture, including text, voice, and touch input; web tracking; and telemetry data on their general use, programs, session duration, and more. The EFF also found that the company chose to hold security functions hostage to personal data flows, claiming that security updates for the operating system would not function properly if users chose to limit location reporting.[117]

In 2016 Microsoft acquired LinkedIn, the professional social network, for $26.2 billion. The aim here is to establish reliable supply routes to the social network dimension of surplus behavior known as the "social graph." These powerful new flows of social surplus from 450 million users can substantially enhance Microsoft prediction products, a key fact noted by Nadella in his announcement of the acquisition to investors: "This can drive targeting and relevance to the next level."[118] Of the three key opportunities that Nadella cited to investors upon the announcement of the acquisition, one was "Accelerate monetization through individual and organization subscriptions and targeted advertising." Among the key factors here would be unified professional profiles across all services, devices, and channels and Microsoft's comprehensive knowledge of each individual user: "Today Cortana knows about you, your organization and about the world. In the future, Cortana will also know your entire professional network to connect dots on your behalf and stay one step ahead."[119]

Once again, the market richly rewarded Microsoft, and Nadella, for the pivot toward surveillance revenues. When Nadella climbed into the CEO's chair in February 2014, the company's shares were trading at around $34, and its market value was roughly $315 billion. Three years later, in January 2017, the corporation's market capitalization topped $500 billion for the first time since 2000, and its shares rose to an all-time high of $65.64.[120]

VI. The Siren Song of Surveillance Revenues

The unprecedented successes of Google, Facebook, and then Microsoft exerted a palpable magnetism on the global economy, especially in the US, where the politics of lawlessness were most firmly entrenched. It did not take long before companies from established sectors with roots far from Silicon Valley demonstrated their determination to compete for surveillance revenues. Among the first in this second wave were the telecom and cable companies that provide broadband service to millions of individuals and households. Although there is some debate about whether these companies can effectively compete with the established internet giants, the facts on the ground suggest that the ISPs are nonetheless determined to try. "Armed with their expansive view over the entire web, internet providers may even be in a position to out-Facebook Facebook, or out-Google Google," observed the *Washington Post*.[121]

The largest of these corporations—Verizon, AT&T, and Comcast—made strategic acquisitions that signaled a shift away from their long-standing models of fees for service in favor of monetizing behavioral surplus. Their tactical maneuvers demonstrate the generalizability of surveillance capitalism's foundational mechanisms and operational requirements, and are evidence that this new logic of accumulation defines a wholly new territory of broad-based market endeavor.

Verizon—the largest telecom company in the US and the largest in the world as measured by market capitalization[122]—publicly introduced its shift toward surveillance revenues in the spring of 2014, when an article in *Advertising Age* announced the company's move into mobile advertising. Verizon's VP of data marketing argued that such advertising had been limited by "addressability...the growing difficulty of tracking consumers as they move between devices." As one marketing expert complained, "There isn't a pervasive identity that tracks users from mobile applications and your mobile browser." The article explained that Verizon had developed "a cookie alternative for a marketing space vexed by the absence of cookies." Verizon aimed to solve advertisers' tracking needs by assigning a hidden and undeletable tracking number, called a PrecisionID, to each Verizon user.[123]

In fact, Verizon's *incursion* had launched two years earlier in 2012 but had been carefully hidden from the public. That was probably because the ID enables the corporation to identify and monitor individuals' habits on their smartphones and tablets, generating behavioral surplus while bypassing customers' awareness. The tracker can neither be turned off nor evaded with private browsing or other privacy tools and controls. Whenever a Verizon subscriber visits a website or mobile app, the corporation and its partners use this hidden ID to aggregate and package behavioral data, all without customers' knowledge.

Verizon's indelible tracking capabilities provided a distinct advantage in the growing competition for behavioral surplus. Advertisers hungry to redefine your walk in the park as their "marketing space" could now reliably target ads to your phone on the strength of the corporation's indelible personal identifier. Verizon also entered into partnership with Turn, an advertising technology firm already notorious for the invention of an unusual "zombie cookie" or "perma-cookie" that immediately "respawns" when a user chooses to opt out of ad tracking or deletes tracking cookies. As a Verizon partner, the Turn zombie cookie attached itself to Verizon's secret tracking number, adding even more protection from discovery and scrutiny. Turn's chief "privacy officer" defended the arrangement, saying, "We are trying to use the most persistent identifier that we can in order to do what we do."[124]

By the fall of 2014, Verizon's stealthy new claim on free raw material was outed by Jacob Hoffman-Andrews, a technologist with the Electronic Frontier Foundation. An article in *Wired* called attention to Hoffman-Andrews' analysis of Verizon's surveillance program and his additional discovery that AT&T was using a similar tracking ID. The article quoted a Verizon spokesperson admitting, "There's no way to turn it off."[125] Hoffman-Andrews observed that even when customers opt out of Verizon's targeted ads, its tracking ID persists, as the corporation bypasses or overrides all signals of a user's intentions, including the Do Not Track setting, Incognito and other private browsing modes, and cookie deletion. The ID is then broadcast to every "unencrypted website a Verizon customer visits from a mobile device. It allows third-party advertisers and websites to assemble a deep, permanent profile of visitors' web browsing habits without their consent."[126] Alarmed by the threat of fresh competition, Google, posing as a privacy advocate, launched a campaign for a new internet protocol that would prevent "header injections" such as Verizon's PrecisionID.[127]

Privacy expert and journalist Julia Angwin and her colleagues at *ProPublica* reported that similar tracking IDs were becoming standard throughout the telecom industry. As one ad executive put it, "What we're excited about is the carrier-level ID, a higher-level recognition point that lets us track with certainty...." Hoffman-Andrews would eventually call the telecom's tactics "a spectacular violation of Verizon users' privacy."[128] True as this may be, the corporation's tactical operations suggest an even more far-ranging development.

Verizon would not retreat from the territory already claimed with its incursion. The hidden ID would stay, and the company assured customers that "it is unlikely that sites and ad entities will attempt to build customer profiles."[129] However, it didn't take long for experts to discover that Twitter's mobile advertising arm already relied on the Verizon ID to track Twitter users' behavior.[130] Then computer scientist and legal scholar Jonathan Mayer found that Turn's zombie cookie sent and received data from more than thirty businesses, including Google, Facebook, Yahoo!, Twitter, Walmart, and WebMD. Mayer investigated both Verizon and Turn's opt-out policies and found both to be deceptive, concluding that every one of Verizon's public statements on the privacy and security of its tracking ID was false. "For an ordinary user," he wrote, "there simply is no defense."[131]

Verizon's substantial entry into surveillance capitalism necessarily tethered the corporation's interests to the extraction imperative. We can see this in the way that Verizon discovered and implemented the dispossession cycle, moving rapidly through its sequence of tactical phases from incursion to redirection. Verizon's initial incursion bought it three years of internal experimentation and discovery. During that time it crossed the threshold of public awareness, beginning the gradual process of public *habituation* to its new practices. Once its strategies were public, it endured a barrage of critical news articles and the scrutiny of privacy experts, but it also bought more time to explore revenue opportunities and supply route expansion. Public reaction to its incursion forced the corporation to map the next phases of the cycle.

Public pressure triggered the shift toward *adaptation* in early 2015. An FCC investigation into Verizon's illicit tracking practices had been launched a few months earlier. The Electronic Privacy Information Center circulated a petition in January 2015 demanding that the FCC penalize the company. By the end of that month, the Senate Committee on Commerce, Science, and Transportation published a letter to Verizon expressing "deep concern" over its new practices.[132] The committee chastised Verizon and Turn for their "seemingly" deliberate "violation of consumer privacy" and "circumvention of customer choice."[133] Within a day of the letter's publication, Verizon announced, "We have begun working to expand the opt-out to include the identifier referred to as the UIDH [unique identifier header], and expect that to be available soon." The *New York Times* called Verizon's announcement "a major revision of its mobile ad-targeting program."[134]

The *Times* could not have known that the *redirection* phase of the dispossession cycle was already in motion. In May 2015 Verizon agreed to purchase AOL for $4.4 billion. As many analysts quickly appreciated, the real attraction of AOL was its CEO, Tim Armstrong, the first head of advertising sales at Google and the man who oversaw its transition from Madison Avenue–style advertising to AdWords' breakthrough discoveries. He was president of Google's Americas sales division when, like Sheryl Sandberg before him, Armstrong left Google for AOL in 2009 with a profound grasp of AdWords' surveillance DNA and the determination to rescue AOL's balance sheet with surveillance capitalism gene therapy. As Verizon's president of Operations told investors, "For us, the principal interest was around the ad tech platform that Tim Armstrong and his team have done a really terrific job building." *Forbes* observed that Armstrong needed Verizon's resources "to challenge the duopoly of Google and Facebook."[135]

Any serious challenge to the giant surveillance capitalists must begin with economies of scale in behavioral surplus capture. To that end, Verizon immediately redirected its supply routes through AOL's advertising platforms. Within a few months of the acquisition, Verizon quietly posted a new privacy notice on its website that few of its 135 million wireless customers would ever read. A few lines slipped into the final paragraphs of the post told the story: PrecisionID is on the move again. Verizon and AOL would now work together "to deliver services that are more personalized and useful to you… we will combine Verizon's existing advertising programs… into the AOL Advertising Network. The combination will help make the ads you see more valuable across the different devices and services you use." The new notice asserted that "the privacy of our customers is important to us," though not important enough to compromise the extraction imperative and allow raw-material providers to challenge the corporation's dispossession program. Opt-out procedures were available but, as usual, complex, difficult to ascertain, and time-consuming. "Please note," the post concluded, "that using browser controls such as clearing cookies on your devices or clearing your browser history is not an effective way to opt out of the Verizon or AOL advertising programs."[136]

The FCC settlement with Verizon was another gloomy example of a public institution outmatched by the velocity and resources of a determined surveillance capitalist. In March 2016, long after the announcement of Verizon's tactical redirection, the FCC reached a $1.35 million settlement with Verizon over its hidden ID privacy violations. Although Verizon agreed to relaunch its cookie on an opt-in basis, the settlement did not extend to AOL's advertising network, which is where the action had moved. Verizon's burgeoning new supply routes would remain unchallenged.[137] Later that month, Armstrong would meet with ad buyers, a rendezvous described by the *Wall Street Journal* as "his first real chance to pitch that AOL—fresh off its sale to Verizon Communications Inc.—intended to become a credible threat to Facebook Inc. and Google.…"[138]

On March 31, 2016, the FCC issued a Notice of Proposed Rulemaking that would establish privacy guidelines for ISPs. The companies would be allowed to continue to collect behavioral data that enhanced the security and effectiveness of their own services, but all other uses of "consumer data" would require opt-in consent. "Once we subscribe to an ISP," FCC Chairman Tom Wheeler wrote, "most of us have little flexibility to change our mind or avoid that network rapidly."[139] The proposal was aimed exclusively at ISPs, considered under the jurisdiction of the FCC, but did not include internet companies, which the Federal Trade Commission is charged with regulating.

In light of the high-stakes dispossession competition already under way among key ISPs, it is not surprising that the proposal quickly became a political lightning rod. ISPs, their lobbyists, policy advisors, and political allies lined up to kill the effort, stating that ISPs' competitive prospects would be unfairly impeded: "Telecom companies are against this proposal, arguing it puts them on an unequal footing with other internet companies that collect data on users, like Google...."[140] In October 27, 2016, FCC commissioners in a 3–2 vote delivered a landmark ruling in favor of, in this case, consumer protection on the internet. It was a historic day not only in the young life of surveillance capitalism but also in the venerable and long life of the FCC, an agency that had never before passed such online protections.[141]

Neither the original FCC proposals nor the final vote chilled Verizon's bid for economies of scale in behavioral surplus. If law was coming to its town, it would simply buy a new town without a sheriff. In June 2017 Verizon closed on the purchase of Yahoo!'s core business, thus acquiring the former internet giant's one billion active monthly users, including its 600 million monthly active mobile users, for a mere $4.48 billion.[142] "Scale is imperative," Armstrong had told journalists a year earlier.[143] "If you want to play in the Olympics you have to compete against Google and Facebook."[144] Armstrong touted Verizon's advantages: its complete view of users' behavior and download activity twenty-four hours per day and its continuous tracking of their locations.

By 2017, the elements of Verizon's new ambitions were finally in place. The new internet company headed by Armstrong and dubbed Oath would combine Yahoo! and AOL for a total of 1.3 billion monthly users. As the *New York Times* summarized, "Verizon hopes to use its range of content and new forms of advertising to attract more viewers and marketers as it competes against Google and Facebook."[145]

In a chilling epilogue to this chapter in the history of surveillance capitalism, on March 28, 2017, a newly elected Republican Congress voted in favor of a resolution to overturn the broadband privacy regulations over which the FCC had struggled just months earlier. The rules had required cable and phone companies to obtain meaningful consent before using personal information for ads and profiling. The companies understood, and they persuaded Republican senators, that the principle of *consent* would strike a serious blow to the foundational mechanisms of the new capitalism: the legitimacy of uni-lateral surplus dispossession, ownership rights to surplus, decision rights over surplus, and the right to lawless space for the prosecution of these activities.[146] To this end the resolution also prevented the FCC from seeking to establish similar protections in the future. Writing in the *New York Times,* Democratic FCC appointee Wheeler went to the heart of the problem:

> To my Democratic colleagues and me, the digital tracks that a con-sumer leaves when using a network are the property of that con-sumer. They contain private information about personal preferences, health problems and financial matters. Our Republican colleagues on the commission argued the data should be available for the net-work to sell.[147]

The reversal meant that although federal laws protected the privacy of a telephone call, the same information transmitted by internet immediately en-ters the ISPs' surplus supply chains. This roust finally signaled the end of the myth of "free." The Faustian pact that had been sold to the world's internet users posed surveillance as the bitter price of free services such as Google's Search and Facebook's social network. This obfuscation is no longer tena-ble, as every consumer who pays his or her monthly telecom bill now also purchases the privilege of a remote and abstract but nevertheless rapacious digital strip search.[148]

New and established companies from every sector—including re-tail, finance, fitness, insurance, automotive, travel, hospitality, health, and education—are joining the migratory path to surveillance revenues, lured by the magnetism of outsized growth, profit, and the promise of the lavish rewards that only the financial markets can confer. We will explore many examples drawn from these sectors in the coming chapters.

In another trend, surveillance in the interest of behavioral surplus cap-ture and sale has become a service in its own right. Such companies are often referred to as "software-as-a-service" or SaaS, but they are more accurately termed "surveillance as a service," or "SVaaS." For example, a new app-based approach to lending instantly establishes creditworthiness based on detailed mining of an individual's smartphone and other online behaviors, including texts, e-mails, GPS coordinates, social media posts, Facebook profiles, retail transactions, and communication patterns.[149] Data sources can include intimate details such as the frequency with which you charge your phone battery, the number of incoming messages you receive, if and when you return phone calls, how many contacts you have listed in your phone, how you fill out online forms, or how many miles you travel each day. These behavioral data yield nuanced patterns that predict the likelihood of loan default or repayment and thus enable continuous algorithmic development and refinement. Two economists who researched this approach discovered that these qualities of surplus produce a predictive model comparable to traditional credit scoring, observing that "the method quantifies rich aspects of behavior typically considered 'soft' information, making it legible to formal institutions."[150] "You're able to get in and really understand the daily life of these customers," explained the CEO of one lending company that analyzes 10,000 signals per customer.[151]

Such methods were originally developed for markets in Africa to help the "unbanked"—people without established credit—to qualify for loans. One lending group interviewed potential customers in low-income countries and concluded that it would be easy to exploit the already beleaguered poor: "Most said they had no problem sharing personal details in exchange for much-needed funds." But these app-based lending startups are typically developed and funded in Silicon Valley, and it is therefore not surprising that the same techniques have become part of a wider trend of exploiting American families that have been economically hollowed out by the financial crisis and neoliberalism's austerity medicine. As the *Wall Street Journal* reports, new startups such as Affirm, LendUp, and ZestFinance "use data from sources such as social media, online behavior and data brokers to determine the creditworthiness of tens of thousands of U.S. consumers who don't have access to loans," more evidence that decision rights and the privacy they enable have become luxuries that too many people cannot afford.[152]

Another example of surveillance-as-a-service is a firm that sells deep vetting of potential employees and tenants to employers and landlords. For instance, a prospective tenant receives a demand from her potential landlord that requires her to grant full access to all social media profiles. The service then "scrapes your site activity," including entire conversation threads and private messages, runs it through natural language processing and other analytic software, and finally spits out a report that catalogues everything from your personality to your "financial stress level," including exposing protected status information such as pregnancy and age. There is no opportunity for affected individuals to view or contest information. As in the case of digital lenders, although a prospective tenant must formally "opt in" to the service, it is those who have less money and fewer options who are trapped in this Faustian bargain in which privacy is forfeit to social participation. "People will give up their privacy to get something they want," celebrates the CEO of this service firm.[153]

Another genre of SVaaS firms employs data science and machine learning to scour the internet for behavioral surplus about individuals, either to sell it or to analyze and fabricate it into lucrative prediction products. Legal scholar Frank Pasquale describes this as "the dark market for personal data."[154] For example, hiQ markets its prediction products to corporate human resources professionals. It scrapes the web for information related to a client's employees, including social media and public available data; then its "data science engine extracts strong signals from that noise that indicate someone may be a flight risk." Machine learning models assign risk scores to each employee, enabling clients "to pinpoint with laser-like accuracy the employees that are highest risk...." The company claims that it provides "a crystal ball" and that its predictions are "virtually identical" to observed turnover. With hiQ's information, companies can preemptively intervene. They might make an effort to retain an employee, or they may choose to preemptively terminate someone who is predicted to be a "flight risk."[155]

Another example is Safegraph, a company that partners with all those apps that are tracking your behavior to amass "high precision/low false positive" data collected "in background from large populations." According to the *Washington Post*, the company collected 17 trillion location markers from 10 million smartphones in November 2016 alone, data that were sold to two university researchers, among others, for a detailed study of political influences on patterns of family behavior on Thanksgiving Day that year.[156] Despite the widely employed euphemisms of "anonymization" and "deidentification," Safegraph tracks individual devices and the movement of their owners throughout the day, producing data that are sufficiently granular to be able to identify individuals' home locations.

Surveillance capitalism was born digital, but as we shall see in following chapters, it is no longer confined to born-digital companies. This logic for translating investment into revenue is highly adaptive and exceptionally lucrative as long as raw-material supplies are free and law is kept at bay. The rapid migration to surveillance revenues that is now underway recalls the late-twentieth-century shift from revenues derived from goods and services to revenues derived from mastering the speculative and shareholder-value-maximizing strategies of financial capitalism. Back then, every company was forced to obey the same commandments: shrink head count, offshore manufacturing and service facilities, reduce expenditures on product and service quality, diminish commitments to employees and consumers, and automate the customer interface, all radical cost-reduction strategies designed to support the firm's share price, which was held hostage to an increasingly narrow and exclusionary view of the firm and its role in society.

As competition for surveillance assets heats up, new laws of motion rise to salience. Eventually, these will shape an even-more-merciless imperative to predict future behavior with greater certainty and detail, forcing the whole project to break loose from the virtual world in favor of the one that we call "real." In Part II we follow this migration to the real world, as competitive dynamics force the expansion of supply operations and an ever-more-complex extraction architecture reaches both further and deeper into new territories of human experience.

Before we undertake that project, it is time to stop and check our bearings. I have suggested that the dangers of surveillance capitalism cannot be fully grasped through either the lens of privacy or of monopoly. In Chapter 6 I offer a new way of thinking about danger. The threats we face are even more fundamental as surveillance capitalists take command of the essential questions that define knowledge, authority, and power in our time: *Who knows? Who decides? Who decides who decides?*

CHAPTER SIX

HIJACKED: THE DIVISION OF LEARNING IN SOCIETY

They wondered why the fruit had been forbidden:
It taught them nothing new. They hid their pride,
But did not listen much when they were chidden:
They knew exactly what to do outside.

—W. H. AUDEN

SONNETS FROM CHINA, I

I. The Google Declarations

On December 4, 1492, Columbus escaped the onshore winds that had prevented his departure from the island that we now call Cuba. Within a day he dropped anchor off the coast of a larger island known to its people as Quisqueya or Bohio, setting into motion what historians call the "conquest pattern." It's a design that unfolds in three phases: the invention of legalistic measures to provide the invasion with a gloss of justification, a declaration of territorial claims, and the founding of a town to legitimate and institutionalize the conquest.[1] The sailors could not have imagined that their actions that day would write the first draft of a pattern whose muscle and genius would echo across space and time to a digital twenty-first century.

On Bohio, Columbus finally found a thriving material culture worthy of his dreams and the appetites of the Spanish monarchs. He saw gold, elaborate stone and woodwork, "ceremonial spaces...stone-lined ball courts...stone collars, pendants, and stylized statues...richly carved wooden thrones... elaborate personal jewelry...." Convinced that the island was "his best find so far, with the most promising environment and the most ingenious inhabitants," he declared to Queen Isabella, "it only remains to establish a Spanish presence and order them to perform your will. For...they are yours to command and make them work, sow seed, and do whatever else is necessary, and build a town, and teach them to wear clothes and adopt our customs."[2]

According to the philosopher of language John Searle, a declaration is a particular way of speaking and acting that establishes facts out of thin air, creating a new reality where there was nothing. Here is how it works: sometimes we speak to simply describe the world—"you have brown eyes"—or to change it—"Shut the door." A declaration combines both, asserting a new reality by describing the world as if a desired change were already true: "All humans are created equal." "They are yours to command." As Searle writes, "We make something the case by representing it as being the case."[3]

Not every declaration is a spoken statement. Sometimes we just describe, refer to, talk about, think about, or even act in relation to a situation in ways that "create a reality by representing that reality as created." For example, let's say the waiter brings my friend and me two identical bowls of soup, placing one bowl in front of each of us. Without saying anything, he has declared that the bowls are not the same: one bowl is my friend's, and the other bowl is mine. We strengthen the facts of his declaration when I take soup only from "my" bowl and my friend takes soup from his. When "his" bowl is empty, my friend is still hungry, and he asks permission to take a spoonful of soup from the bowl in front of me, further establishing the fact that it is *my* bowl of soup. In this way declarations rise or fall on the strength of others' acceptance of the new facts. As Searle concludes, "All of institutional reality, and therefore...all of human civilization is created by...declarations."[4]

Declarations are inherently invasive because they impose new facts on the social world while their declarers devise ways to get others to agree to those facts. Columbus's declaration reflects this "conquest pattern," as historian Matthew Restall writes:

> Sixteenth-century Spaniards consistently presented their deeds and those of their compatriots in terms that prematurely anticipated the completion of Conquest campaigns and imbued Conquest chronicles with an air of inevitability. The phrase "Spanish Conquest" and

all it implies has come down through history because the Spaniards were so concerned to depict their endeavors as conquests and pacifications, as contracts fulfilled, as providential intention, as *faits accomplis.*[5]

The Spanish conquerors and their monarchs were eager to justify their invasion as one way to induce agreement, especially among their European audience. They developed measures intended to impart "a legalistic veneer by citing and following approved precedents."[6] To this end the soldiers were tasked with reading the Monarchical Edict of 1513 known as the *Requirimiento* to indigenous villagers before attacking them.[7] The edict declared that the authority of God, the pope, and the king was embodied in the conquistadors and then declared the native peoples as vassals subordinate to that authority: "You Cacics and Indians of this Continent.... We declare or be it known to you all, that there is but one God, one hope, and one King of Castile, who is Lord of these Countries; appear forth without delay, and take the oath of Allegiance to the Spanish King, as his Vassals."[8]

The edict went on to enumerate the sufferings that would befall the villagers if they failed to comply. In this world-shattering confrontation with the unprecedented, the native people were summoned, advised, and forewarned in a language they could not fathom to surrender without resistance in recognition of authorities they could not conceive. The exercise was so cynical and cruel that the approaching invaders often dispatched their obligation by mumbling the edict's long paragraphs into their beards in the dead of night as they hid among the thick vegetation waiting to pounce: "Once the Europeans had discharged their duty to inform, the way was clear for pillage and enslavement." The friar Bartolomé de las Casas, whose account bears witness to this history of Spanish atrocities, wrote that the *Requirimiento* promised the native people fair treatment upon surrender but also spelled out the consequences of defiance. Every act of indigenous resistance was framed as "revolt," thereby legitimizing brutal "retaliation" that exceeded military norms, including grotesque torture, the burning of whole villages in the dark of night, and hanging women in public view: "I will do to you all the evil and damages that a lord may do to vassals who do not obey or receive him. And I solemnly declare that the deaths and damages received from such will be your fault and not that of His Majesty, nor mine, nor of the gentlemen who came with me."[9]

Conquest by declaration should sound familiar because the facts of surveillance capitalism have been carried into the world on the strength of six critical *declarations* pulled from thin air when Google first asserted them. That the facts they proclaimed have been allowed to stand is evident in the dispossession strategies of Verizon and other new entrants to the surveillance capitalist firmament. In the rapture of the young firm's achievements, Google's founders, fans, and adoring press passed over in silence the startling vision of invasion and conquest concealed in these assertions.[10]

The six declarations laid the foundation for the wider project of surveillance capitalism and its original sin of dispossession. They must be defended at any cost because each declaration builds on the one before it. If one falls, they all fall:

- We claim human experience as raw material free for the taking. On the basis of this claim, we can ignore considerations of individuals' rights, interests, awareness, or comprehension.
- On the basis of our claim, we assert the right to take an individual's experience for translation into behavioral data.
- Our right to take, based on our claim of free raw material, confers the right to own the behavioral data derived from human experience.
- Our rights to take and to own confer the right to know what the data disclose.
- Our rights to take, to own, and to know confer the right to decide how we use our knowledge.
- Our rights to take, to own, to know, and to decide confer our rights to the conditions that preserve our rights to take, to own, to know, and to decide.

Thus, the age of surveillance capitalism was inaugurated with six declarations that defined it as an age of conquest. Surveillance capitalism succeeded by way of aggressive declaration, and its success stands as a powerful illustration of the invasive character of declarative words and deeds, which aim to conquer by imposing a new reality. These twenty-first-century invaders do not ask permission; they forge ahead, papering the scorched earth with faux-legitimation practices. Instead of cynically conveyed monarchical edicts, they offer cynically conveyed terms-of-service agreements whose stipulations are just as obscured and incomprehensible. They build their fortifications, fiercely defending their claimed territories, while gathering strength for the next incursion. Eventually, they build their towns in intricate ecosystems of commerce, politics, and culture that declare the legitimacy and inevitability of all that they have accomplished.

Eric Schmidt asked for trust, but Google's "declarations" ensured that it did not require our trust to succeed. Its declarative victories have been the means through which it amassed world-historic concentrations of knowledge and power. These are the bulwarks that enable its continued progress. Schmidt has occasionally revealed something like this point. When describing "modern technology platforms," he writes that "almost nothing, short of a biological virus, can scale as quickly, efficiently, or aggressively as these technology platforms, and this makes the people who build, control, and use them powerful too."[11]

On the strength of its unprecedented concentrations of knowledge and power, surveillance capitalism achieves dominance over *the division of learning in society*—the axial principle of social order in an information civilization. This development is all the more dangerous because it is unprecedented. It cannot be reduced to known harms and therefore does not easily yield to known forms of combat. What is this new principle of social order, and how do surveillance capitalists take command of it? These are the questions that we pursue in the sections that follow. The answers help us reflect on what we have learned and prepare for what lies ahead.

II. Who Knows?

This book began by recalling an urgent question posed to me by a young pulp mill manager in a small southern town: *"Are we all going to be working for a smart machine, or will we have smart people around the machine?"* In the years that followed that rainy evening, I closely observed the digitalization of work in the pulp mill. As I described it in *In the Age of the Smart Machine*, the shift to information technology transformed the mill into an "electronic text" that became the primary focus of every worker's attention. Instead of the hands-on tasks associated with raw materials and equipment, doing "a good job" came to mean monitoring data on screens and mastering the skills to understand, learn from, and act through the medium of this electronic text. What seems ordinary today was extraordinary then.

These obvious changes, I argued, signaled a deep and significant transformation. The ordering principle of the workplace had shifted from a division of labor to a *division of learning*. I wrote about the many women and men who surprised themselves and their managers as they conquered new intellectual skills and learned to thrive in the information-rich environment, but I also documented the bitter conflicts that attended those achievements, summarized as dilemmas of *knowledge, authority,* and *power*.

Any consideration of the division of learning must resolve these dilemmas expressed in three essential questions. The first question is *"Who knows?"* This is a question about the distribution of knowledge and whether one is included or excluded from the opportunity to learn. The second question is *"Who decides?"* This is a question about authority: which people, institutions, or processes determine who is included in learning, what they are able to learn, and how they are able to act on their knowledge. What is the legitimate basis of that authority? The third question is *"Who decides who decides?"* This is a question about power. What is the source of power that undergirds the authority to share or withhold knowledge?

The young manager would ultimately find his answers, but they were not what either of us had hoped for. Even as the pulp mill workers struggled and often triumphed, Hayek's worldview was taking hold at the highest policy levels and Jensen's operational disciplines were finding an eager welcome on Wall Street, which quickly learned to impose them on every public company. The result was a cost-down business model oriented to its Wall Street audience, which insisted on automating and exporting jobs rather than investing in the digital skills and capabilities of the US worker. The answer to the question *Who knows?* was that the machine knows, along with an elite cadre able to wield the analytic tools to troubleshoot and extract value from information. The answer to *Who decides?* was a narrow market form and its business models that decide. Finally, in the absence of a meaningful double movement, the answer to *Who decides who decides?* defaults entirely to financial capital bound to the disciplines of shareholder-value maximization.

It is not surprising that nearly forty years later, a Brookings Institution report laments that millions of US workers are "shut out of decent middle-skill opportunities" in the face of "rapid digitalization." The report exhorts companies to "invest urgently in IT upskilling strategies for incumbent workers, knowing that digital skills represent a key channel of productivity gains."[12] How different might our society be if US businesses had chosen to invest in people as well as in machines?

Most companies opted for the smart machine over smart people, producing a well-documented pattern that favors substituting machines and their algorithms for human contributors in a wide range of jobs. By now, these include many occupations far from the factory floor.[13] This results in what economists call "job polarization," which features some high-skill jobs and other low-skill jobs, with automation displacing most of the jobs that were once "in the middle."[14] And although some business leaders, economists, and technologists describe these developments as necessary and inevitable consequences of computer-based technologies, research shows that the division of learning in the economic domain reflects the strength of neoliberal ideology, politics, culture, and institutional patterns. For example, in continental and northern Europe, where key elements of the double movement have survived in some form, job polarization is moderated by substantial investments in workforce education that produce a more inclusive division of learning as well as high-quality innovative products and services.[15]

Most critical to our story is that we now face a second historical phase of this conflict. The division of learning in the economic domain of production and employment is critical, but it is only the beginning of a new struggle over the even larger question of the division of learning in society. The dilemmas of knowledge, authority, and power have burst through the walls of the workplace to overwhelm our daily lives. As people, processes, and things are reinvented as information, the division of learning in society becomes the ascendant principle of social ordering in our time.

A wholly new electronic text now extends far beyond the confines of the factory or office. Thanks to our computers, credit cards, and phones, and the cameras and sensors that proliferate in public and private spaces, just about everything we now do is mediated by computers that record and codify the details of our daily lives at a scale that would have been unimaginable only a few years ago. We have reached the point at which there is little that is omitted from the continuous accretion of this new electronic text. In later chapters we review many illustrations of the new electronic text as it spreads silently but relentlessly, like a colossal oil slick engulfing everything in its path: your breakfast conversation, the streets in your neighborhood, the dimensions of your living room, your run in the park.

The result is that both the world and our lives are pervasively rendered as information. Whether you are complaining about your acne or engaging in political debate on Facebook, searching for a recipe or sensitive health information on Google, ordering laundry soap or taking photos of your nine-year-old, smiling or thinking angry thoughts, watching TV or doing wheelies in the parking lot, all of it is raw material for this burgeoning text. Information scholar Martin Hilbert and his colleagues observe that even the foundational elements of civilization, including "language, cultural assets, traditions, institutions, rules, and laws...are currently being digitized, and for the first time, explicitly put into visible code," then returned to society through the filter of "intelligent algorithms" deployed to govern a rapidly multiplying range of commercial, governmental, and social functions.[16] The essential questions confront us at every turn: *Who knows? Who decides? Who decides who decides?*

III. Surveillance Capital and the Two Texts

There are important parallels with the late nineteenth and early twentieth centuries, when the division of labor first emerged as the foremost principle of social organization in the nascent industrial societies of Europe and North America. These experiences can offer guidance and alert us to what is at stake. For example, when the young Emile Durkheim wrote *The Division of Labor in Society*, the title itself was controversial. The division of labor had been understood as a critical means of achieving labor productivity through the specialization of tasks. Adam Smith memorably wrote about this new principle of industrial organization in his description of a pin factory, and the division of labor remained a topic of economic discourse and controversy throughout the nineteenth century. Durkheim recognized labor productivity as an economic imperative of industrial capitalism that would drive the division of labor to its most extreme application, but that was not what held his fascination.

Instead, Durkheim trained his sights on the social transformation already gathering around him, observing that "specialization" was gaining "influence" in politics, administration, the judiciary, science, and the arts. He concluded that the division of labor was no longer quarantined in the industrial *workplace*. Instead, it had burst through those factory walls to becoming the critical organizing principle of industrial *society*. This is also an example of Edison's insight: that the principles of capitalism initially aimed at production eventually shape the wider social and moral milieu. "Whatever opinion one has about the division of labor," Durkheim wrote, "everyone knows that it exists, and is more and more becoming one of the fundamental bases of the social order."[17]

Economic imperatives predictably mandated the division of labor in production, but what was the purpose of the division of labor in society? This was the question that motivated Durkheim's analysis, and his century-old conclusions are still relevant for us now. He argued that the division of labor accounts for the interdependencies and reciprocities that link the many diverse members of a modern industrial society in a larger prospect of solidarity. Reciprocities breed mutual need, engagement, and respect, all of which imbue this new ordering principle with moral force.

In other words, the division of labor was summoned into society at the beginning of the twentieth century by the rapidly changing circumstances of the first modernity's new individuals, discussed in Chapter 2. It was an essential response to their new "conditions of existence." As people like my great-grandparents joined the migration to a modern world, the old sources of meaning that had bonded communities across space and time melted away. What would hold society together in the absence of the rules and rituals of clan and kin? Durkheim's answer was the division of labor. People's needs for a coherent new source of meaning and structure were the cause, and the effect was an ordering principle that enabled and sustained a healthy modern community. As the young sociologist explained,

> The most remarkable effect of the division of labor is not that it increases output of functions divided, but that it renders them solidary. Its role...is not simply to embellish or ameliorate existing societies, but to render societies possible which, without it, would not exist....It passes far beyond purely economic interests, for it consists in the establishment of a social and moral order sui generis.[18]

Durkheim's vision was neither sterile nor naive. He recognized that things can take a dark turn and often do, resulting in what he called an "abnormal" (sometimes translated as "pathological") division of labor that produces social distance, injustice, and discord in place of reciprocity and interdependency. In this context, Durkheim singled out the destructive effects of social inequality on the division of labor in society, especially what he viewed as the most dangerous form of inequality: *extreme asymmetries of power* that make "conflict itself impossible" by "refusing to admit the right of combat." Such pathologies can be cured only by a politics that asserts the people's right to contest, confront, and prevail in the face of unequal and illegitimate power over society. In the late nineteenth century and most of the twentieth century, that contest was led by labor and other social movements that asserted social equality through institutions such as collective bargaining and public education.

The transformation that we witness in our time echoes these historical observations as the division of learning follows the same migratory path from the economic to the social domain once traveled by the division of labor. Now the division of learning "passes far beyond purely economic interests," for it establishes the basis for our social order and its moral content.

The division of learning is to us, members of the second modernity, what the division of labor was to our grandparents and great-grandparents, pioneers of the first modernity. In our time the division of learning emerges from the economic sphere as a new principle of social order and reflects the primacy of learning, information, and knowledge in today's quest for effective life. And just as Durkheim warned his society a century ago, today our societies are threatened as the division of learning drifts into pathology and injustice at the hands of the unprecedented asymmetries of knowledge and power that surveillance capitalism has achieved.

Surveillance capitalism's command of the division of learning in society begins with what I call *the problem of the two texts*. The specific mechanisms of surveillance capitalism compel the production of two "electronic texts," not just one. When it comes to the first text, we are its authors and readers. This public-facing text is familiar and celebrated for the universe of information and connection it brings to our fingertips. Google Search codifies the informational content of the world wide web. Facebook's News Feed binds the network. Much of this public-facing text is composed of what we inscribe on its pages: our posts, blogs, videos, photos, conversations, music, stories, observations, "likes," tweets, and all the great massing hubbub of our lives captured and communicated.

Under the regime of surveillance capitalism, however, the first text does not stand alone; it trails a shadow close behind. The first text, full of promise, actually functions as the supply operation for the second text: the *shadow text*. Everything that we contribute to the first text, no matter how trivial or fleeting, becomes a target for surplus extraction. That surplus fills the pages of the second text. This one is hidden from our view: "read only" for surveillance capitalists.[19] In this text our experience is dragooned as raw material to be accumulated and analyzed as means to others' market ends. The shadow text is a burgeoning accumulation of behavioral surplus and its analyses, and it says more about us than we can know about ourselves. Worse still, it becomes increasingly difficult, and perhaps impossible, to refrain from contributing to the shadow text. It automatically feeds on our experience as we engage in the normal and necessary routines of social participation.

More mystifying still are the ways in which surveillance capitalists apply what they learn from their exclusive shadow text to shape the public text to their interests. There have been myriad revelations of Google and Facebook's manipulations of the information that we see. For now I'll simply point out that Google's algorithms, derived from surplus, select and order search results, and Facebook's algorithms, derived from surplus, select and order the content of its News Feed. In both cases, researchers have shown that these manipulations reflect each corporation's commercial objectives. As legal scholar Frank Pasquale describes it, "The decisions at the Googleplex are made behind closed doors...the power to include, exclude, and rank is the power to ensure which public impressions become permanent and which remain fleeting.... Despite their claims of objectivity and neutrality, they are constantly making value-laden, controversial decisions. They help create the world they claim to merely 'show' us."[20] When it comes to the shadow text, surveillance capitalism's laws of motion compel both its secrecy and its continuous growth. We are the objects of its narratives, from whose lessons we are excluded. As the source from which all the treasure flows, this second text is *about* us, but it is not *for* us. Instead, it is created, maintained, and exploited outside our awareness for others' benefit.

The result is that the division of learning is both the ascendant principle of social ordering in our information civilization and already a hostage to surveillance capitalism's privileged position as the dominant composer, owner, and guardian of the texts. Surveillance capitalism's ability to corrupt and control these texts produces unprecedented asymmetries of knowledge and power that operate precisely as Durkheim had feared: the relatively free rein accorded to this market form and the innately illegible character of its action have enabled it to impose substantial control over the division of learning outside of our awareness and without means of combat. When it comes to the essential questions, surveillance capital has gathered the power and asserted the authority to supply all the answers. However, even authority is not enough. Only surveillance capital commands the material infrastructure and expert brainpower to rule the division of learning in society.

IV. The New Priesthood

Scientists warn that the world's capacity to produce information has substantially exceeded its ability to process and store information. Consider that our technological memory has roughly doubled about every three years. In 1986 only 1 percent of the world's information was digitized and 25 percent in 2000. By 2013, the progress of digitalization and datafication (the application of software that allows computers and algorithms to process and analyze raw data) combined with new and cheaper storage technologies had translated 98 percent of the world's information into a digital format.[21]

Information is digital, but its volume exceeds our ability to discern its meaning. As the solution to this problem, information scholar Martin Hilbert counsels, "The only option we have left to make sense of all the data is to fight fire with fire," using "artificially intelligent computers" to "sift through the vast amounts of information.... Facebook, Amazon, and Google have promised to...create value out of vast amounts of data through intelligent computational analysis."[22] The rise of surveillance capitalism necessarily turns Hilbert's advice into a dangerous proposition. Although he does not mean to, Hilbert merely confirms the privileged position of the surveillance capitalists and the asymmetrical power that enables them to bend the division of learning to their interests.

Google's asymmetrical power draws on all the social sources that we have considered: its declarations, its defensive fortifications, its exploitation of law, the legacy of surveillance exceptionalism, the burdens of second-modernity individuals, and so on. But its power would not be operational without the gargantuan material infrastructure that surveillance revenues have bought. Google is the pioneer of "hyperscale," considered to be "the largest computer network on Earth."[23] Hyperscale operations are found in high-volume information businesses such as telecoms and global payments firms, where data centers require millions of "virtual servers" that exponentially increase computing capabilities without requiring substantial expansion of physical space, cooling, or electrical power demands.[24] The machine intelligence at the heart of Google's formidable dominance is described as "80 percent infrastructure," a system that comprises custom-built, warehouse-sized data centers spanning 15 locations and, in 2016, an estimated 2.5 million servers in four continents.[25]

Investors deem Google "harder to catch than ever" because it is unmatched in its combination of infrastructure scale *and* science. Google is known as a "full stack AI company" that uses its own data stores "to train its own algorithms running on its own chips deployed on its own cloud." Its dominance is further strengthened by the fact that machine learning is only as intelligent as the amount of data it has to train on, and Google has the most data.[26] By 2013, the company understood that its shift into the "neural networks" that define the current frontier of artificial intelligence would substantially increase computational demands and require a doubling of its data centers. As Urs Hölzle, Google's senior vice president of technical infrastructure, put it, "The dirty secret behind [AI] is that they require an insane number of computations to just actually train the network." If the company had tried to process the growing computational workload with traditional CPUs, he explained, "We would have had to double the entire footprint of Google—data centers and servers—just to do three minutes or two minutes of speech recognition per Android user per day."[27]

With data center construction as the company's largest line item and power as its highest operating cost, Google invented its way through the infrastructure crisis. In 2016 it announced the development of a new chip for "deep learning inference" called the tensor processing unit (TPU). The TPU would dramatically expand Google's machine intelligence capabilities, consume only a fraction of the power required by existing processors, and reduce both capital expenditure and the operational budget, all while learning more and faster.[28]

Global revenue for AI products and services is expected to increase 56-fold, from $644 million in 2016 to $36 billion in 2025.[29] The science required to exploit this vast opportunity and the material infrastructure that makes it possible have ignited an arms race among tech companies for the 10,000 or so professionals on the planet who know how to wield the technologies of machine intelligence to coax knowledge from an otherwise cacophonous data continent. Google/Alphabet is the most aggressive acquirer of AI technology and talent. In 2014–2016 it purchased nine AI companies, twice as many as its nearest rival, Apple.[30]

The concentration of AI talent at Google reflects a larger trend. In 2017, US companies are estimated to have allocated more than $650 million to fuel the AI talent race, with more than 10,000 available positions at top employers across the country. The top five tech companies have the capital to crowd out competitors: startups, universities, municipalities, established corporations in other industries, and less wealthy countries.[31] In Britain, university administrators are already talking about a "missing generation" of data scientists. The huge salaries of the tech firms have lured so many professionals that there is no one left to teach the next generation of students. As one scholar described it, "The real problem is these people are not dispersed through society. The intellect and expertise is concentrated in a small number of companies."[32]

On the strength of its lavish recruitment efforts, Google tripled its number of machine intelligence scientists in just the last few years and has become the top contributor to the most prestigious scientific journals—four to five times the world average in 2016. Under the regime of surveillance capitalism, the corporation's scientists are not recruited to solve world hunger or eliminate carbon-based fuels. Instead, their genius is meant to storm the gates of human experience, transforming it into data and translating it into a new market colossus that creates wealth by predicting, influencing, and controlling human behavior.

More than six hundred years ago, the printing press put the written word into the hands of ordinary people, rescuing the prayers, bypassing the priesthood, and delivering the opportunity for spiritual communion directly into the hands of the prayerful. We have come to take for granted that the internet enables an unparalleled diffusion of information, promising more knowledge for more people: a mighty democratizing force that exponentially realizes Gutenberg's revolution in the lives of billions of individuals. But this grand achievement has blinded us to a different historical development, one that moves out of range and out of sight, designed to exclude, confuse, and obscure. In this hidden movement the competitive struggle over surveillance revenues reverts to the pre-Gutenberg order as the division of learning in society shades toward the pathological, captured by a narrow priesthood of privately employed computational specialists, their privately owned machines, and the economic interests for whose sake they learn.

V. The Privatization of the Division of Learning in Society

The division of learning in society has been hijacked by surveillance capitalism. In the absence of a robust double movement in which democratic institutions and civil society tether raw information capitalism to the people's interests—however imperfectly—we are thrown back on the market form of the surveillance capitalist companies in this most decisive of contests over the division of learning in society. Experts in the disciplines associated with machine intelligence know this, but they have little grasp of its wider implications. As data scientist Pedro Domingos writes, "Whoever has the best algorithms and the most data wins.... Google with its head start and larger market share, knows better what you want...whoever learns fastest wins...." The *New York Times* reports that Google CEO Sundar Pichai now shares a floor with the company's AI research lab and notes it as a trend among many CEOs: a literal take on the concentration of power.[33]

Just over thirty years ago, legal scholar Spiros Simitis published a seminal essay on the theme of privacy in an information society. Simitis grasped early on that the already visible trends in public and private "information processing" harbored threats to society that transcended narrow conceptions of privacy and data ownership: "Personal information is increasingly used to enforce standards of behavior. Information processing is developing, therefore, into an essential element of long-term strategies of manipulation intended to mold and adjust individual conduct."[34] Simitis argued that these trends were incompatible not only with privacy but with the very possibility of democracy, which depends upon a reservoir of individual capabilities associated with autonomous moral judgment and self-determination.

Building on Simitis's work, Berkeley's Paul M. Schwartz warned in 1989 that computerization would transform the delicate balance of rights and obligations upon which privacy law depends: "Today the enormous amounts of personal data available in computers threaten the individual in a way that renders obsolete much of the previous legal protection." Most important, Schwartz foresaw that the scale of the still-emerging crisis would impose risks that exceed the scope of privacy law: "The danger that the computer poses is to human autonomy. The more that is known about a person, the easier it is to control him. Insuring the liberty that nourishes democracy requires a structuring of societal use of information and even permitting some concealment of information."[35]

Both Simitis and Schwartz sensed the ascent of the division of learning as the axial principle of a new computational societal milieu, but they could not have anticipated the rise of surveillance capitalism and its consequences. Although the explosive growth of the information continent shifts a crucial axis of the social order from a twentieth-century division of labor to a twenty-first-century division of learning, it is surveillance capitalists who command the field and unilaterally lay claim to a disproportionate share of the decision rights that shape the division of learning in society.

Surveillance capitalists' acts of digital dispossession impose a new kind of control upon individuals, populations, and whole societies. Individual privacy is a casualty of this control, and its defense requires a reframing of privacy discourse, law, and judicial reasoning. The "invasion of privacy" is now a predictable dimension of social inequality, but it does not stand alone. It is the systematic result of a "pathological" division of learning in society in which surveillance capitalism knows, decides, and decides who decides. Demanding privacy from surveillance capitalists or lobbying for an end to commercial surveillance on the internet is like asking Henry Ford to make each Model T by hand or asking a giraffe to shorten its neck. Such demands are existential threats. They violate the basic mechanisms and laws of motion that produce this market leviathan's concentrations of knowledge, power, and wealth.

So here is what is at stake: surveillance capitalism is profoundly antidemocratic, but its remarkable power does not originate in the state, as has historically been the case. Its effects cannot be reduced to or explained by technology or the bad intentions of bad people; they are the consistent and predictable consequences of an internally consistent and successful logic of accumulation. Surveillance capitalism rose to dominance in the US under conditions of relative lawlessness. From there it spread to Europe, and it continues to make inroads in every region of the world. Surveillance capitalist firms, beginning with Google, dominate the accumulation and processing of information, especially information about human behavior. They know a great deal about us, but our access to their knowledge is sparse: hidden in the shadow text and read only by the new priests, their bosses, and their machines.

This unprecedented concentration of knowledge produces an equally unprecedented concentration of power: asymmetries that must be understood as the *unauthorized privatization of the division of learning in society.* This means that powerful private interests are in control of the definitive principle of social ordering in our time, just as Durkheim warned of the subversion of the division of labor by the powerful forces of industrial capital a century ago. As things currently stand, it is the surveillance capitalist corporations that *know.* It is the market form that *decides.* It is the competitive struggle among surveillance capitalists that *decides who decides.*

VI. *The Power of the Unprecedented: A Review*

The titanic power struggles of the twentieth century were between industrial capital and labor, but the twenty-first century finds surveillance capital pitted against the entirety of our societies, right down to each individual member. The competition for surveillance revenues bears down on our bodies, our homes, and our cities in a battle for power and profit as violent as any the world has seen. Surveillance capitalism cannot be imagined as something "out there" in factories and offices. Its aims and effects are *here...* are *us*.

Ours is not simply a case of being ambushed and outgunned. We were caught off guard because there was no way that we could have imagined these acts of invasion and dispossession, any more than the first unsuspecting Taíno cacique could have foreseen the rivers of blood that would flow from his inaugural gesture of hospitality toward the hairy, grunting, sweating men, the *adelantados* who appeared out of thin air waving the banner of the Spanish monarchs and their pope as they trudged across the beach. Why have we been slow to recognize the "original sin of simple robbery" at the heart of this new capitalism? Like the Taínos, we faced something altogether new to our story: the unprecedented. And, like them, we risk catastrophe when we assess new threats through the lens of old experience.

On the "supply side," surveillance capitalists deftly employed the entire arsenal of the declaration to assert their authority and legitimacy in a new and undefended digital world. They used declarations to take without asking. They camouflaged their purpose with illegible machine operations, moved at extreme velocities, sheltered secretive corporate practices, mastered rhetorical misdirection, taught helplessness, purposefully misappropriated cultural signs and symbols associated with the themes of the second modernity—empowerment, participation, voice, individualization, collaboration—and baldly appealed to the frustrations of second-modernity individuals thwarted in the collision between psychological yearning and institutional indifference.

In this process the pioneer surveillance capitalists at Google and Facebook evaded the disciplines of corporate governance and rejected the disciplines of democracy, protecting their claims with financial influence and political relationships. Finally, they benefited from history, born in a time when regulation was equated with tyranny and the state of exception precipitated by the terrorist attacks of 9/11 produced surveillance exceptionalism, further enabling the new market to root and flourish. Surveillance capitalists' purposeful strategies and accidental gifts produced a form that can romance and beguile but is also ruthlessly efficient at extinguishing space for democratic deliberation, social debate, individual self-determination, and the right to combat as it forecloses every path to exit.

On the "demand side," second-modernity populations starved for enabling resources were so enraptured by the plentiful bags of rice and powdered milk thrown from the back of the digital truck that little attention was paid to the drivers or their destination. We needed them; we even believed that we couldn't live without them. But under scrutiny, those long-awaited delivery trucks look more like automated vehicles of invasion and conquest: more Mad Max than Red Cross, more Black Sails than Carnival Cruise. The wizards behind their steering wheels careen across every hill and hollow, learning how to scrape and stockpile our behavior over which they unabashedly assert their rights as conquerors' plunder.

In the absence of a clear-minded appreciation of this new logic of accumulation, every attempt at understanding, predicting, regulating, or prohibiting the activities of surveillance capitalists will fall short. The primary frameworks through which our societies have sought to assert control over surveillance capitalism's audacity are those of "privacy rights" and "monopoly." Neither the pursuit of privacy regulations nor the imposition of constraints on traditional monopoly practices has so far interrupted the key mechanisms of accumulation, from supply routes to behavioral futures markets. On the contrary, surveillance capitalists have extended and elaborated their extraction architectures across every human domain as they master the practical and political requirements of the dispossession cycle. This success now threatens the deepest principles of social order in an information civilization as surveillance capitalism takes unauthorized command over the division of learning in society.

If there is to be a fight, let it be a fight over capitalism. Let it be an insistence that raw surveillance capitalism is as much a threat to society as it is to capitalism itself. This is not a technical undertaking, not a program for advanced encryption, improved data anonymity, or data ownership. Such strategies only acknowledge the inevitability of commercial surveillance. They leave us hiding in our own lives as we cede control to those who feast on our behavior for their own purposes. Surveillance capitalism depends on the social, and it is only in and through collective social action that the larger promise of an information capitalism aligned with a flourishing third modernity can be reclaimed.

In Part I we have seen how Google built its extraction architecture in the online world. As competition for surveillance revenues intensified, a second economic imperative rose to prominence driving an expansion of that architecture into another world, the one that we call "real."

Now the story of surveillance capitalism moves in this new direction. In Part II, I invite you to rekindle your sense of astonishment as we follow the trail of this second economic imperative defined by the prediction of human behavior. The *prediction imperative* enlarges the complexity of surplus operations as economies of scale are joined by *economies of scope* and *economies of action*. These new disciplines drive surveillance capitalism far into the intimate reaches of our daily lives and deep into our personalities and our emotions. Ultimately, they compel the development of highly inventive but resolutely secret new means to interrupt and modify our behavior for the sake of surveillance revenues. These operations challenge our elemental *right to the future tense*, which is the right to act free of the influence of illegitimate forces that operate outside our awareness to influence, modify, and condition our behavior. We grow numb to these incursions and the ways in which they deform our lives. We succumb to the drumbeat of inevitability, but nothing here is inevitable. Astonishment is lost but can be found again.

PART II

THE ADVANCE OF SURVEILLANCE CAPITALISM

CHAPTER SEVEN

THE REALITY BUSINESS

Falling in love with Truth before he knew Her,
He rode into imaginary lands,
By solitude and fasting hoped to woo Her,
And mocked at those who served Her with their hands.

—W. H. AUDEN

SONNETS FROM CHINA, VI

I. The Prediction Imperative

There could not have been a more fitting setting for Eric Schmidt to share his opinion on the future of the web than the World Economic Forum in Davos, Switzerland. In 2015, during a session at the winter playground for neoliberals—and increasingly surveillance capitalists—Schmidt was asked for his thoughts about the future of the internet. Sitting alongside his former Google colleagues Sheryl Sandberg and Marissa Mayer, he did not hesitate to share his belief that "The internet will disappear. There will be so many IP addresses... so many devices, sensors, things that you are wearing, things that you are interacting with, that you won't even sense it. It will be part of your presence all the time. Imagine you walk into a room and the room is dynamic."[1] The audience gasped in astonishment, and shortly thereafter, headlines around the world exploded in shock at the former Google CEO's pronouncement that the end of the internet was at hand.

Schmidt was, in fact, merely paraphrasing computer scientist Mark Weiser's seminal 1991 article, "The Computer for the 21st Century," which has framed Silicon Valley's technology objectives for nearly three decades. Weiser introduced what he called "ubiquitous computing" with two legendary sentences: "The most profound technologies are those that disappear. They weave themselves into the fabric of everyday life until they are indistinguishable from it." He described a new way of thinking "that allows the computers themselves to vanish into the background.... Machines that fit the human environment instead of forcing humans to enter theirs will make using a computer as refreshing as taking a walk in the woods."[2]

Weiser understood that the virtual world could never be more than a shadow land no matter how much data it absorbs: "Virtual reality is only a map, not a territory. It excludes desks, offices, other people... weather, trees, walks, chance encounters and, in general, the infinite richness of the universe." He wrote that virtual reality "simulates" the world rather than "invisibly enhancing the world that already exists." In contrast, ubiquitous computing would infuse that *real* world with a universally networked apparatus of silent, "calm," and voracious computing. Weiser refers to this apparatus as the new "computing environment" and delights in the possibilities of its limitless knowledge, such as knowing "the suit you looked at for a long time last week because it knows both of your locations, and it can retroactively find the designer's name even though that information did not interest you at the time."[3]

Schmidt was not describing the end of the internet but rather its successful unshackling from dedicated devices such as the personal computer and the smartphone. For surveillance capitalists, this transition is not a choice. Surveillance profits awakened intense competition over the revenues that flow from new markets for future behavior. Even the most sophisticated process of converting behavioral surplus into products that accurately forecast the future is only as good as the raw material available for processing. Surveillance capitalists therefore must ask this: what forms of surplus enable the fabrication of prediction products that most reliably foretell the future? This question marks a critical turning point in the trial-and-error elaboration of surveillance capitalism. It crystallizes a second economic imperative—the *prediction imperative*—and reveals the intense pressure that it exerts on surveillance capitalist revenues.

The first wave of prediction products enabled targeted online advertising. These products depended upon surplus derived at scale from the internet. I have summarized the competitive forces that drive the need for surplus at scale as the "extraction imperative." Competition for surveillance revenues eventually reached a point at which the volume of surplus became a necessary but insufficient condition for success. The next threshold was defined by the quality of prediction products. In the race for higher degrees of certainty, it became clear that the best predictions would have to approximate observation. The prediction imperative is the expression of these competitive forces (see Figure 3).

Google/Alphabet, Facebook, Microsoft, and many more companies now drawn to surveillance revenues have staked their claims on the internet's "disappearance" because they must. Compelled to improve predictions, surveillance capitalists such as Google understood that they had to widen and diversify their extraction architectures to accommodate new sources of surplus and new supply operations. Economies of scale would still be vital, of course, but in this new phase, supply operations were enlarged and intensified to accommodate *economies of scope* and *economies of action*. What does this entail?

The shift toward economies of scope defines a new set of aims: behavioral surplus must be vast, but it must also be varied. These variations are developed along two dimensions. The first is the *extension* of extraction operations from the virtual world into the "real" world, where we actually live our actual lives. Surveillance capitalists understood that their future wealth would depend upon new supply routes that extend to real life on the roads, among the trees, throughout the cities. Extension wants your bloodstream and your bed, your breakfast conversation, your commute, your run, your refrigerator, your parking space, your living room.

Economies of scope also proceed along a second dimension: *depth*. The drive for economies of scope in the depth dimension is even more audacious. The idea here is that highly predictive, and therefore highly lucrative, behavioral surplus would be plumbed from intimate patterns of the self. These supply operations are aimed at your personality, moods, and emotions, your lies and vulnerabilities. Every level of intimacy would have to be automatically captured and flattened into a tidal flow of data points for the factory conveyor belts that proceed toward manufactured certainty.

Just as scale became necessary but insufficient for higher-quality predictions, it was also clear that economies of scope would be necessary but insufficient for the highest quality of prediction products able to sustain competitive advantage in the new markets for future behavior. Behavioral surplus must be vast and varied, but the surest way to predict behavior is to intervene at its source and shape it. The processes invented to achieve this goal are what I call *economies of action.* In order to achieve these economies, machine processes are configured to intervene in the state of play in the real world among real people and things. These interventions are designed to enhance certainty by doing things: they nudge, tune, herd, manipulate, and modify behavior in specific directions by executing actions as subtle as inserting a specific phrase into your Facebook news feed, timing the appearance of a BUY button on your phone, or shutting down your car engine when an insurance payment is late.

This new level of competitive intensity characterized by scope and action ratchets up the invasive character of supply operations and initiates a new era of surveillance commerce that I call the *reality business.* Economies of scale were implemented by machine-based extraction architectures in the online world. Now the reality business requires machine-based architectures in the real world. These finally fulfill Weiser's vision of ubiquitous automated computational processes that "weave themselves into the fabric of everyday life until they are indistinguishable from it," but with a twist. Now they operate in the interests of surveillance capitalists.

There are many buzzwords that gloss over these operations and their economic origins: "ambient computing," "ubiquitous computing," and the "internet of things" are but a few examples. For now I will refer to this whole complex more generally as the "apparatus." Although the labels differ, they share a consistent vision: the everywhere, always-on instrumentation, datafication, connection, communication, and computation of all things, animate and inanimate, and all processes—natural, human, physiological, chemical, machine, administrative, vehicular, financial. Real-world activity is continuously rendered from phones, cars, streets, homes, shops, bodies, trees, buildings, airports, and cities back to the digital realm, where it finds new life as data ready for transformation into predictions, all of it filling the ever-expanding pages of the shadow text.[4]

As the prediction imperative gathers force, it gradually becomes clear that extraction was the first phase of a far-more-ambitious project. Economies of action mean that real-world machine architectures must be able *to know* as well as *to do*. Extraction is not enough; now it must be twinned with execution. The extraction architecture is combined with a new *execution architecture*, through which hidden economic objectives are imposed upon the vast and varied field of behavior.[5]

Gradually, as surveillance capitalism's imperatives and the material infrastructures that perform extraction-and-execution operations begin to function as a coherent whole, they produce a twenty-first-century "means of behavioral modification." The aim of this undertaking is not to impose behavioral norms, such as conformity or obedience, but rather to produce behavior that reliably, definitively, and certainly leads to desired commercial results. The research director of Gartner, the well-respected business advisory and research firm, makes the point unambiguously when he observes that mastery of the "internet of things" will serve as "a key enabler in the transformation of business models from 'guaranteed levels of performance' to '*guaranteed outcomes*.'"[6]

This is an extraordinary statement because there can be no such guarantees in the absence of the power to make it so. This wider complex that we refer to as the "means of behavioral modification" is the expression of this gathering power. The prospect of guaranteed outcomes alerts us to the force of the prediction imperative, which demands that surveillance capitalists make the future for the sake of predicting it. Under this regime, ubiquitous computing is not just a knowing machine; it is an actuating machine designed to produce more certainty *about* us and *for* them.

This gradually accruing, smart, *and* muscular apparatus is gradually being assembled around us. No one knows what the real magnitude is or will be. It is a domain plagued by hyperbole, where projections frequently outrun actual results. Despite this, the planning, investment, and invention necessary to draw this vision of ubiquity into reality are well underway. The visions and aims of its architects, the work that has already been accomplished, and the programs that are currently in development constitute a turning point in the evolution of surveillance capitalism.

Finally, I want to underscore that although it may be possible to imagine something like the "internet of things" without surveillance capitalism, it is impossible to imagine surveillance capitalism without something like the "internet of things." Every command arising from the prediction imperative requires this pervasive real-world material "knowing and doing" presence. The new apparatus is the material expression of the prediction imperative, and it represents a new kind of power animated by the economic compulsion toward certainty. Two vectors converge in this fact: the early ideals of ubiquitous computing and the economic imperatives of surveillance capitalism. This convergence signals the metamorphosis of the digital infrastructure *from a thing that we have to a thing that has us.*

Futuristic as this may sound, the vision of individuals and groups as so many objects to be continuously tracked, wholly known, and shunted this way or that for some purpose of which they are unaware has a history. It was coaxed to life nearly sixty years ago under the warm equatorial sun of the Galapagos Islands, when a giant tortoise stirred from her torpor to swallow a succulent chunk of cactus into which a dedicated scientist had wedged a small machine.

It was a time when scientists reckoned with the obstinacy of free-roaming animals and concluded that surveillance was the necessary price of knowledge. Locking these creatures in a zoo would only eliminate the very behavior that scientists wanted to study, but how were they to be surveilled? The solutions once concocted by scholars of elk herds, sea turtles, and geese have been refurbished by surveillance capitalists and presented as an inevitable feature of twenty-first-century life on Earth. All that has changed is that now *we are the animals.*

II. The Tender Conquest of Unrestrained Animals

It was a 1964 international expedition to the Galapagos Islands that presented a unique opportunity to explore telemetry, a frontier technology based on the long-distance transmission of computer data. A new breed of scientists who combined biology, physics, engineering, and electronics championed this new tech, and chief among these was R. Stuart MacKay, a physicist cum

The Dynamic of Behavioral Surplus Accumulation

Surveillance capitalism's master motion is the accumulation of new sources of behavioral surplus with more predictive power. The goal is predictions comparable to guaranteed outcomes in real-life behavior. Extraction begins online, but the prediction imperative increases the momentum, driving extraction toward new sources in the real world.

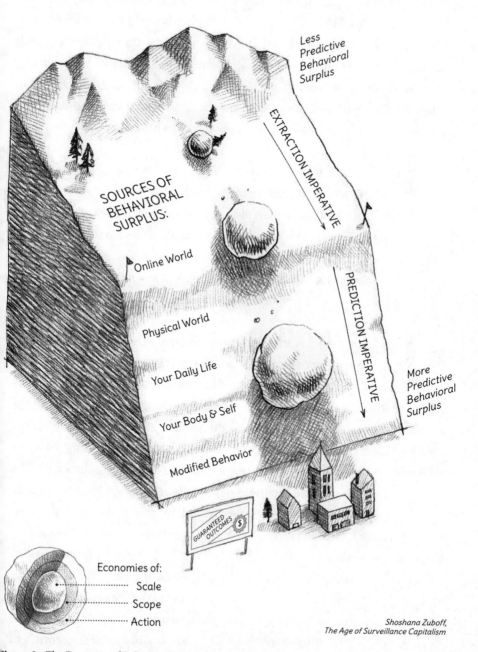

Less Predictive Behavioral Surplus

EXTRACTION IMPERATIVE

SOURCES OF BEHAVIORAL SURPLUS:

Online World

PREDICTION IMPERATIVE

Physical World

Your Daily Life

More Predictive Behavioral Surplus

Your Body & Self

Modified Behavior

GUARANTEED OUTCOMES $

Economies of:
- Scale
- Scope
- Action

Shoshana Zuboff,
The Age of Surveillance Capitalism

Figure 3: The Dynamic of Behavioral Surplus Accumulation

electrical engineer, biologist, and surgeon who was known among his scientific peers as the experts' expert.[7]

MacKay viewed telemetry as a means of enhancing and protecting the well-being of animal populations. A photo of MacKay from the Galapagos expedition shows him poised tenderly beside a giant tortoise that had swallowed his tiny machine; in another he gently holds a rare marine iguana with a sensor attached to its torso, all of it to measure the animals' internal body temperatures. He emphasized the key element that distinguished telemetry from other forms of monitoring: the possibility of capturing behavior in its natural habitat with sensors of such compactness that they could disappear into the body without triggering the animal's awareness:

> The use of a radio signal from a transmitter, in or on a subject, to carry information to a remote receiver for recording allows flexibility of movement and permits disturbance-free exploration of otherwise inaccessible parts of the body without the subject even being aware of the measuring process...the methods leave the subject in a relatively normal psychological and physiological state, and do not interfere with the continuation of normal activities.[8]

MacKay's published work focused primarily on the technical aspects of his studies, although occasionally there is a glimpse of larger purpose. Telemetry created the possibility of huge data sets and the opportunity for correlational studies at the scale of entire animal populations. He noted that the same techniques could be applied to the static world: forest canopies, the curing of concrete, chemical reaction vessels, and food processing. MacKay envisioned whole populations of connected data-emitting individuals. His first-generation "wearable technologies" made it possible to study "unrestrained animals" among every species, including people. Biomedical telemetry, he stressed, was uniquely suited to gather information that would be otherwise impossible to collect "in the wild." The key principle was that his telematics operated outside an animal's awareness. This was especially useful in solving problems such as the difficulty of measuring "uncooperative animals" and the need to gather data even when herds roamed through "inaccessible regions." In other words, MacKay's inventions enabled scientists to render animals as information even when they believed themselves to be free, wandering and resting, unaware of the incursion into their once-mysterious landscapes.

MacKay stressed that the transmission and monitoring of sensor data were only part of the story. The route was not enough; it had to be the routing. He argued for a "reverse process" of telestimulation that would not only monitor behavior but also reveal how it could be modified and optimized, providing what he regarded as "a remote dialogue between the subject and the experimenter."[9]

MacKay's broad vision has come to fruition in the digital age. Satellite acuity combined with the explosive growth of computational power fitted onto tiny slivers of silicon, advanced sensors, internet-enabled networks, and "big data" predictive analytics have produced extraordinary systems that reveal the meanings and movements of whole animal populations and their individuals: anywhere, anytime. The same wearables traveling on and in the bodies of animals have also become broad sensors of the planet's climate, geography, and ecology, enabling "a quorum sensing of our planet, using a variety of species to tap into the diversity of senses that have evolved across animal groups," producing a "sixth sense of the global animal collective."[10] As you will already have guessed, there is little reason to suppose that these capabilities will remain trained on nonhuman species.

Indeed, the threshold has already been crossed.[11] In 2014 a team of University of Washington researchers led by Jenq-Neng Hwang announced a "super GPS" assembled from municipal surveillance cameras "to enable the dynamic visualization of the realistic situation of humans walking on the road and sidewalks, so eventually people can see the animated version of the real-time dynamics of city streets on a platform like Google Earth."[12] If this were a novel, then Professor MacKay's brilliant work, along with that of the many dedicated scientists who followed in his footsteps, would serve as foreshadowing.

In a metamorphosis that MacKay did not foresee, the science of animal tracking that grew from his pathbreaking vision became the template for surveillance capitalism's next phase of evolution as telematics now applied to human behavior succumbed to the thrall of a new and lucrative logic of accumulation. The requirements of prediction that would later merge into an economic imperative were already evident in MacKay's work. The need for economies of scope, both in extension and depth, is reflected in his foundational framework that aimed to render information about populations *and* the details of individuals, reaching into the distant corners of previously inaccessible regions. Today those regions include the front seat of your car, your kitchen, and your kidneys. His "reverse process" of telestimulation is resurrected in the economies of action that automatically stimulate behavior, not to save the human herd from catastrophe but rather to heighten the predictability of its behavior.

MacKay yearned for discovery, but today's "experimenters" yearn for certainty as they translate our lives into calculations. MacKay's animals were unrestrained and innately uncooperative because they felt themselves to be free, sheltering and roaming in unknown terrain. Now, the un-self-conscious, easy freedom enjoyed by the human animal—the sense of being unrestrained that thrives in the mystery of distant places and intimate spaces—is simply friction on the path toward surveillance revenues.

III. Human Herds

MacKay's legacy is reimagined for our time in the work of Professor Joseph Paradiso of the MIT Media Lab, where some of surveillance capitalism's most valuable capabilities and applications, from data mining to wearable technologies, were invented.

Paradiso's brilliant group of data scientists, engineers, musicians, and artists reconceives the world through the lens of Google Search by applying the same disciplines that mastered the web—datafication, indexing, browsing, and searching—to master reality itself. Without "ubiquitous sensate environments," Paradiso writes, "the cognitive engines of this everywhere-enabled world are deaf, dumb, and blind, and can't respond relevantly to the real-world events that they aim to augment."[13] In other words, ubiquitous computing is meaningless without the ubiquitous sensing that conveys the experience for computation.

To this end, Paradiso's students invented a "ListenTree," which emits streaming sound that "invites attention" and "points to a future where digital information might become a seamless part of the physical world." He and his colleagues populated a 250-acre marsh with hundreds of sensors that measure and record temperature, humidity, moisture, light motion, wind, sound, tree sap flow, chemical levels, and more. They developed "inertial sensors" that track and compute complex movements and "flexible sensate fibers" to create "radically new functional substrates that can impact medicine, fashion, and apparel...and bring electronics into all things stretchable or malleable." There are electronics that attach directly to skin in the form of tattoos and makeup, while fingernails and wrists are transformed into computational interfaces that can read finger gestures, even in the absence of hand movements. "Sensor tape" and "stickers" can adhere "to inaccessible surfaces and building materials," where they can be "wirelessly interrogated...."[14]

Paradiso and his colleagues wrestled with the paradox of, on the one hand, proliferating sensor data in nearly every environment—from smartphones to home devices to streets to cameras to cars—and, on the other hand, the difficulties involved in integrating sensor-generated data flows and producing meaningful analyses. Their answer was "DoppelLab," a digital platform for combining and visually representing sensor data.[15] The idea is to transform any physical space, from the interior of an office building to an entire city, into a "browse-able environment" where you can see and hear everything going on in that space as it flows from thousands or billions or trillions of sensors. Just as browsers like Netscape first "gave us access to the mass of data contained on the internet, so will software browsers enable us to make sense of the flood of sensor data that is on the way."[16]

The aim here is a grand synthesis: the collation and fusion of every sort of sensor data from every channel and device to develop a "virtual sensor environment" in which "crawlers will constantly traverse data...calculating state and estimating other parameters derived from the data" collected from everywhere from office interiors to entire cities.

Paradiso is confident that "a proper interface to this artificial sensoria promises to produce...a *digital omniscience*...a pervasive everywhere augmented reality environment...that can be intuitively browsed" just as web browsers opened up the data contained on the internet. He insists that ubiquitous sensor information and computing will be "an extension of ourselves rather than an embodiment of an 'other.'" Information will stream "directly into our eyes and ears once we enter the age of wearables... the boundaries of the individual will be very blurry in this future."[17]

According to Paradiso and his coauthor, Gershon Dublon, the next great technological challenge is "context aggregation," which means the ability to assemble rapidly expanding sensor information into new "applications." The idea is that every physical space and every trace of behavior within that space—bees buzzing, your smile, the temperature fluctuations in my closet, their breakfast conversation, the swoosh of the trees—will be "informated" (translated into information). Spaces can be aggregated into a seamless flow of searchable information, sights, and sounds in much the same way that Google once aggregated web pages for indexing and searching: "This shift will create a seamless nervous system that covers the planet—and one of the main challenges for the computing community now is how to merge the rapidly evolving 'omniscient' electronic sensoria onto human perception."[18]

For all their brilliance, these creative scientists appear to be unaware of the restless economic order eager to commandeer their achievements under the flag of surveillance revenues. Paradiso does not reckon with the translation of his paradise of omniscience into the realpolitik of surveillance capitalism as the prediction imperative insists on surplus culled from these new flows and surveillance capitalists fill the front seats of the classroom of digital omniscience.

IV. Surveillance Capitalism's Realpolitik

Waning levels of government leadership and funding for "ubiquitous computing" leave the technology companies to lead in basic research and applications, each vying to become "the Google" of the new apparatus and its architectures of extraction and execution.[19] Despite the radical prospects of the ubiquitous connected sensate computational apparatus and the often-repeated claim "It will change everything," technology firms in the US have, thus far, continued their run of relative lawlessness, unimpeded by any comprehensive social or regulatory vision. As Intel's chief strategist for the "internet of things" commented in response to concerns over privacy implications, "One thing that we absolutely believe is that though we hear the conversation around policy, we don't want policy to get in the way of technological innovation...."[20]

In place of "policy" or a "social contract," it is capitalism, and increasingly surveillance capitalism, that shapes the action. New behavioral futures markets and "targeted applications" are eagerly awaited. As Microsoft's director of the machine intelligence platform for integrating and analyzing data from the "internet of things" says, "The part that's equally cool and creepy is what happens after everybody and their competitor gets on board with smart devices: a big secondary market for data...a secondary revenue source." These markets, he explains, are "just like Google or Facebook's" markets for targeted advertising."[21] An IBM report concurs: "Thanks to the internet of things, physical assets are turning into participants in real-time global digital markets. The countless types of assets around us will become as easily indexed, searched and traded as any online commodity.... We call this the 'liquification of the physical world.'"[22]

In an ominous parallel to the rhetoric of "data exhaust" as the prelude to dispossession, this second phase of expropriation also requires new rhetoric that can simultaneously legitimate and distract from the real action unleashed by the prediction imperative. A new euphemism, "dark data," plays this role. For example, Harriet Green directed IBM's $3 billion investment in the "internet of things," a resource commitment that aimed to make the company a serious contender to become "the Google" of ubiquitous computing. Green says that digital omniscience is impeded by the fact that most of the data collected by companies are "unstructured," making them difficult to "datafy" and code.[23] IBM's customers are plagued by the question "What can we do with this [unstructured] data to make us more efficient or to create new products and services that we can sell to optimize what we're doing or create new things for clients?"[24]

Unstructured data cannot merge and flow in the new circuits of liquefied assets bought and sold. They are friction. Green fixes on the declarative term that simultaneously names the problem and justifies its solution: *dark data*. The message we saw honed in the online world—"If you're not in the system, you don't exist"—is refined for this new phase of dispossession. Because the apparatus of connected things is intended to be *everything*, any behavior of human or thing absent from this push for universal inclusion is *dark*:

menacing, untamed, rebellious, rogue, out of control. The stubborn expanse of dark data is framed as the enemy of IBM's and its customers' ambitions. Note the echoes of MacKay here, with his determination to penetrate the secrets of *unrestrained animals* and *inaccessible regions*. The tension is that no thing counts until it is *rendered* as behavior, translated into electronic data flows, and channeled into the light as observable data. *Everything* must be illuminated for counting and herding.

In this way the notion of "dark data" handily becomes the "data exhaust" of ubiquitous computing. It provides the moral, technical, commercial, and legal rationale for powerful systems of machine intelligence that can capture and analyze behaviors and conditions never intended for a public life. For those who seek surveillance revenues, dark data represent lucrative and necessary territories in the dynamic universal jigsaw constituted by surveillance capitalism's urge toward scale, scope, and action. Thus, the technology community casts dark data as the intolerable "unknown unknown" that threatens the financial promise of the "internet of things."[25]

It is therefore understandable that Green portrays machine intelligence—and specifically IBM's anthropomorphized artificial intelligence system called "Watson"—as the authoritative savior of an apparatus threatened by waste and incomprehensibility. Machine intelligence is referred to as "cognitive computing" at IBM, presumably to avoid the uneasy connotations of inscrutable power associated with words like *machine* and *artificial*.

Under the leadership of CEO Ginni Rometty, the corporation invested heavily in "Watson," heralded by the company as "the brains of the 'internet of things.'" Rometty wants IBM to dominate the machine learning functions that will translate ubiquitous data into ubiquitous knowledge and action. "The first discussion is around how much dark data you have that only Watson and cognitive can really interrogate," Green says. "You know the amount of data being created on a daily basis—much of which will go to waste unless it is utilized. This so-called dark data represents a phenomenal opportunity...the ability to use sensors for everything in the world to basically be a computer, whether it's your contact lens, your hospital bed, or a railway track."[26] The message is that surveillance capitalism's new instruments will render the entire world's actions and conditions as behavioral flows. Each rendered bit is liberated from its life in the social, no longer inconveniently encumbered by moral reasoning, politics, social norms, rights, values, relationships, feelings, contexts, and situations. In the flatness of this flow, data are data, and behavior is behavior. The body is simply a set of coordinates in time and space where sensation and action are translated as data. All things animate and inanimate share the same existential status in this blended confection, each reborn as an objective and measurable, indexable, browsable, searchable "it."

From the vantage point of surveillance capitalism and its economic imperatives, world, self, and body are reduced to the permanent status of *objects* as they disappear into the bloodstream of a titanic new conception of markets. His washing machine, her car's accelerator, and your intestinal flora are collapsed into a single dimension of equivalency as information assets that can be disaggregated, reconstituted, indexed, browsed, manipulated, analyzed, reaggregated, predicted, productized, bought, and sold: anywhere, anytime.

The worldview elaborated by scientists such as Paradiso and business leaders such as Green has been swept into action on many fronts where digital omniscience is eagerly welcomed as the recipe for certainty in the service of certain profits. The next section is an opportunity to see this worldview in action, in a business sector far from the pioneers of surveillance capitalism: automobile insurance. Extraction and prediction become the hallmarks of a new logic of accumulation as insurers and their consultants plot their approach to surveillance revenues. In the plans and practices of these new actors, we witness both the determination to institutionalize economies of scope and action and the drift toward a dark new world in which the automatic and closely targeted means of behavioral modification are understood as the path to profit.

V. Certainty for Profit

In Chapter 3 we met Google's Hal Varian, and now once again he lights the way, exposing the significance and specific requirements of the prediction imperative. Recall that Varian identified four new "uses" of the computer mediation of transactions.[27] The first of these was "data extraction and analysis," from which we deduced the extraction imperative as one of the foundational mechanisms of surveillance capitalism. Varian says that the other three new uses—"new contractual forms due to better monitoring," "personalization and customization," and "continuous experiments"—"will, in time, become even more important than the first."[28] That time has come.

"Because transactions are now computer-mediated we can observe behavior that was previously unobservable and write contracts on it," Varian says. "This enables transactions that were simply not feasible before." He gravitates to the example of "vehicular monitoring systems," recognizing their paradigmatic power. Varian says that if someone stops making monthly car payments, "Nowadays it's a lot easier just to instruct the vehicular monitoring system not to allow the car to be started and to signal the location where it can be picked up."[29] Insurance companies, he notes, can also rely on these monitoring systems to check if customers are driving safely and thus determine whether to maintain the insurance policy, vary the cost of premiums, and decide whether to pay a claim.

Varian's new uses of computer mediation in this insurance realm are entirely dependent upon internet-enabled devices that know and do. In fact, they are impossible to imagine without the material means of extraction and execution architectures planted in and permeating the real world. The vehicular monitoring system that he prescribes, for example, provides economies of scope *and* action. It knows *and* intervenes in the state of play, monitoring data *and* acting on programmed instructions to shut off the car's engine, thus allowing the repo man to locate the disabled automobile and its vanquished driver.

As the prediction imperative pulls supply operations into the real world, product or service providers in established sectors far from Silicon Valley are enthralled by the prospects of surveillance revenues. For example, the CEO of Allstate Insurance wants to be like Google: "There are lots of people who are monetizing data today. You get on Google, and it seems like it's free. It's not free. You're giving them information; they sell your information. Could we, should we, sell this information we get from people driving around to various people and capture some additional profit source…? It's a long-term game."[30] Automobile insurers appear to be especially eager to implement Varian's vision and MacKay's telematics. The fight for your car, it turns out, is an illustration of the intensity of purpose with which companies great and small now pursue behavioral surplus.

Auto insurers have long known that risk is highly correlated with driver behavior and personality, but there was little that they could do about it.[31] Now the remote sensate monitoring systems of modern telematics can provide a continuous stream of data about where we are, where we're going, the details of our driving behavior, and the conditions of our vehicle. App-based telematics can also calculate how we are feeling and what we are saying by integrating dashboard and even smartphone information.

Auto insurers are besieged by consultants and would-be technology partners who proffer surveillance capitalistic strategies that promise a new chapter of commercial success. "Uncertainty will be strongly reduced," intones a McKinsey report on the future of the insurance industry. "This leads to demutualization and a focus on predicting and managing individual risks rather than communities."[32] A report by Deloitte's Center for Financial Services counsels "risk minimization"—a euphemism for guaranteed outcomes—through monitoring and enforcing policyholder behavior in real time, an approach called "behavioral underwriting." "Insurers can monitor policyholder behavior directly," Deloitte advises, by "recording the times, locations, and road conditions when they drive, whether they rapidly accelerate or drive at high or even excessive speeds, how hard they brake, as well as how rapidly they make turns and whether they use their turn signals."[33] Telematics produce continuous data flows, so real-time behavioral surplus can replace the traditional "proxy factors," such as demographic information, that had previously been used to calculate risk. This means that surplus must be both plentiful (economies of scale) and varied (economies of scope) in both range and depth.

Even smaller underwriters that cannot afford extensive capital outlays for telematics are advised that they can accomplish most of these aims with a smartphone application, eliminating costly hardware and data-transmission expenses: "These insurers may also benefit because a mobile app gathers first-hand data on the behavior and performance of the driver carrying the smartphone...yielding a 360-degree view of the total exposure being underwritten...."[34]

As certainty replaces uncertainty, premiums that once reflected the necessary unknowns of everyday life can now rise and fall from millisecond to millisecond, informed by the precise knowledge of how fast you drive to work after an unexpectedly hectic early morning caring for a sick child or if you perform wheelies in the parking lot behind the supermarket. "We know that 16-year old drivers have a whole lot of accidents...but not every 16-year-old is a lousy driver," observes one insurance industry telematics expert. Rates based on actual behavior are "a big advantage in being able to price appropriately."[35] This kind of certainty means that insurance contracts designed to mitigate risk now give way to machine processes that respond "almost immediately" to nuanced infractions of prescribed behavioral parameters and thus substantially decrease risk or eliminate it entirely.[36]

Telematics are not intended merely to know but also to do (economies of action). They are hammers; they are muscular; they enforce. Behavioral underwriting promises to reduce risk through machine processes designed to modify behavior in the direction of maximum profitability. Behavioral surplus is used to trigger punishments, such as real-time rate hikes, financial penalties, curfews, and engine lockdowns, or rewards, such as rate discounts, coupons, and gold stars to redeem for future benefits. The consultancy firm AT Kearney anticipates "IoT enriched relationships" to connect "more holistically" with customers "to influence their behaviors."[37]

Varian's blithe statement that "it's a lot easier" to instruct a vehicular monitoring system to shut off a car when a payment is late is not hyperbole. For example, Spireon, which describes itself as "the largest aftermarket vehicle telematics company" and specializes in tracking and monitoring vehicles and drivers for a variety of customers such as lenders, insurers, and fleet owners, offers a system akin to Varian's ideal.[38] Its "Loan-Plus Collateral Management System" pushes alerts to drivers when they have fallen behind in their payments, remotely disables the vehicle when delinquency exceeds a predetermined period, and locates the vehicle for the repo man to recover.

Telematics announce a new day of behavioral control. Now the insurance company can set specific parameters for driving behavior. These can include anything from fastening the seat belt to rate of speed, idling times, braking and cornering, aggressive acceleration, harsh braking, excessive hours on the road, driving out of state, and entering a restricted area.[39] These parameters are translated into algorithms that continuously monitor, evaluate, and rank the driver, calculations that translate into real-time rate adjustments.

According to a patent held by Spireon's top strategist, insurers can eliminate uncertainty by shaping behavior.[40] The idea is to continuously optimize the insurance rate based on monitoring the driver's adherence to behavioral parameters defined by the insurer. The system translates its behavioral knowledge into power, assigning credits or imposing punishments on drivers. Surplus is also translated into prediction products for sale to advertisers. The system calculates "behavioral traits" for advertisers to target, sending ads directly to the driver's phone. A second patent is even more explicit about triggers for punitive measures.[41] It identifies a range of algorithms that activate consequences when the system's parameters are breached: "a violation algorithm," "a curfew algorithm," "a monitoring algorithm," "an adherence algorithm," "a credit algorithm."

The consultancy firms are aligned in advising all their insurance clients to get into the surveillance game. AT Kearney acknowledges that the "connected car" is a proving ground for what is to come: "Ultimately, IoT's true value depends on customers adjusting their behaviors and risk profiles based on feedback from their 'things.'"[42] Health insurers are another target: "Wearable accelerometers" could "improve traceability of their compliance" with prescribed exercise regimes, and "digestible sensors" could track compliance with dietary and medication schedules, "providing higher truth and better granularity than a monthly refill."[43]

Deloitte acknowledges that according to its own survey data, most consumers reject telematics on the basis of privacy concerns and mistrust companies that want to monitor their behavior. This reluctance can be overcome, the consultants advise, by offering cost savings "significant enough" that people are willing "to make the [privacy] trade-off," in spite of "lingering concerns...." If price inducements don't work, insurers are counseled to present behavioral monitoring as "fun," "interactive," "competitive," and "gratifying," rewarding drivers for improvements on their past record and "relative to the broader policy holder pool."[44] In this approach, known as "gamification," drivers can be engaged to participate in "performance based contests" and "incentive based challenges."[45]

If all else fails, insurers are advised to induce a sense of inevitability and helplessness in their customers. Deloitte counsels companies to emphasize "the multitude of other technologies already in play to monitor driving" and that "enhanced surveillance and/or geo-location capabilities are part of the world we live in now, for better or worse."[46]

Behavioral underwriting offers auto insurers cost savings and efficiencies, but it is not the endgame for a revitalized insurance industry. The analytics that produce targeted advertising in the online world are repurposed for the real world, laying the foundation for new behavioral futures markets that trade in predictions of customer behavior. This is where the real drive for surveillance revenues is focused. For example, an executive of cloud services provider Covisint advises clients aiming to "cash in" on automotive telematics to move beyond targeted ads to "targeted applications." These are not ads on a screen but real-life experiences shaped by the same capabilities as targeted ads and designed to lure you into real places for the sake of others' profit. That means selling driver data to third parties that will figure out where you are, where you're going, and what you want: "They know what restaurants you like because you drive your car there, so they can recommend restaurants as you're driving and the restaurants will pay...."[47]

Behavioral surplus is understood as the raw material for products that establish "co-marketing" with other services such as "towing, auto repair, car washes, restaurants, retail outlets...."[48] The consultants at McKinsey make a similar recommendation, advising insurers that the "internet of things" enables their expansion into "completely new areas" such as "data marketplaces." Health surplus can be "monetized," says Deloitte, by providing "relevant referrals." The firm advises its clients, especially those unlikely to reach scale in telematics, to establish partnerships with "digital players."[49] The model is a 2016 deal between IBM and General Motors that announced the creation of "OnStar Go," the car industry's "first cognitive mobility platform." Dell and Microsoft have launched "internet of things" insurance "accelerators." Dell provides insurers with hardware, software, analytics, and services to "more accurately predict risk and take preventative measures," and Microsoft has linked up with American Family Insurance to develop startups focused on home automation.[50]

The data companies were once regarded as mere "suppliers," but it is more likely that the auto companies will become suppliers to the data behemoths. "Google tries to accompany people throughout their day, to generate data and then use that data for economic gain," acknowledges

Daimler's CEO. "It's at that point where a conflict with Google seems pre-programmed."[51] Google and Amazon are already locked in competition for the dashboard of your car, where their systems will control all communication and applications. From there it is a short step to telemetry and related data. Google already offers applications developers a cloud-based "scaleable geolocation telemetry system" using Google Maps. In 2016 Google France announced its interest in partnerships with insurance companies "to develop bundles of products which blend technology and hardware with insurance." That same year a report from Cap Gemini consultants found that 40 percent of insurers see Google "as a potential rival and threat because of its strong brand and ability to manage customer data."[52]

VI. Executing the Uncontract

These examples drawn from the ordinary world of automobile insurance teach some extraordinary lessons. Drivers are persuaded, induced, incentivized, or coerced into a quid pro quo that links pricing to the expansion of a real-world extraction/execution architecture aimed at new behavioral surplus flows (economies of scope). Behavioral data drawn from their experience are processed, and the results flow in two directions. First, they return to the drivers, executing procedures to interrupt and shape behavior in order to enhance the certainty, and therefore profitability, of predictions (economies of action). Second, prediction products that rank and sort driver behavior flow into newly convened behavioral futures markets in which third parties lay bets on what drivers will do now, soon, and later: Will he maintain a high safety rating? Will she act in compliance with our rules? Will he drive like a girl? These bets translate into pricing, incentive structures, and monitoring and compliance regimes. In both operations, *surplus drawn from the driver's experience is repurposed as the means to shape and compel the driver's experience for the sake of guaranteed outcomes.* Most of this occurs, as MacKay advised, outside the driver's awareness while she still thinks that she is free.

The Google declarations underwrite all the action here. As Varian writes, "Because transactions are now computer-mediated *we* can observe behavior that was *previously unobservable* and *write* contracts on it. This enables transactions that were *simply not feasible before.*"[53] Varian's "we" refers to those with privileged access to the shadow text into which behavioral data flow. Our behavior, once unobservable, is declared as free for the taking, theirs to own, and theirs to decide how to use and how to profit from. This includes the production of "new contractual forms" that compel us in ways that would not have been possible but for surveillance capitalism's original declarations of dispossession.

Varian recognized that the subregions of automotive telematics exemplify this new economic frontier when he wrote, "Nowadays it's a lot easier just to instruct the vehicular monitoring system not to allow the car to be started and to signal the location where it can be picked up."[54] Yawn. But wait. "A lot easier" for whom? He means, of course, a lot easier for the "we" that now observes what was, until surveillance capitalism, unobservable and executes actions that were, until surveillance capitalism, not feasible. Varian's laid-back, simple prose is a kind a lullaby that makes his observations seem banal, so ordinary as to barely warrant comment. But in Varian's scenario, what happens to the driver? What if there is a child in the car? Or a blizzard? Or a train to catch? Or a day-care center drop-off on the way to work? A mother on life support in the hospital still miles away? A son waiting to be picked up at school?

It was not long ago that Varian's prosaic proclamations were regarded as the stuff of nightmares. In his 1967 book *The Year 2000* the hyper-rational wunderkind futurist Herman Kahn anticipated many of the capabilities that Varian now assigns to the new extraction/execution architecture.[55] Kahn was no shrinking violet. He was rumored to be director Stanley Kubrick's model for the title character in *Dr. Strangelove,* and he was well-known for arguing that nuclear war is both "winnable" and "survivable." Yet it was Kahn who foresaw innovations such as Varian's vehicular monitoring system and characterized them as "a twenty-first century nightmare." Among his many technology-oriented insights, Kahn foresaw automated computer systems that track all vehicular movements and also listen to and record conversations with all the capability available for high-speed scan and search. He imagined computer systems able to detect and respond to individual behavior—a raised voice, a threatening tone: "Such computers may also be able to apply a great deal of inferential logic on their own—they may become a sort of transistorized Sherlock Holmes making hypotheses and investigating leads in a more or less autonomous or self-motivated manner...."[56] Anyone who wields this kind of knowledge, he concluded, is, like Faust, "less immoral than amoral...indifferent to the fate of those who stand in his way rather than brutal."[57]

Contemporary reviewers of Kahn's book invariably seized upon the dark "nightmare scenarios" of the computerized surveillance theme, the science-fiction–like forms of control that, they assumed, "will be actively feared and resented by many."[58] Despite the wide range of scenarios that Kahn presented in his book on the distant year 2000, Kahn's voyage into the "unthinkable" was viewed by the public as a way to prepare for "the worst possible outcome" in a terrifying "nightmare of social controls."[59] Yet now that same nightmare is rendered as an enthusiastic progress report on surveillance capitalism's latest triumphs. Varian's update is delivered without self-consciousness or a hint of controversy, rather than the astonishment and revulsion that were predicted just decades ago. How has the nightmare become banal? Where is our sense of astonishment and outrage?

Political scientist Langdon Winner grappled with this question in his seminal book *Autonomous Technology*, published in 1977. His answer? "What we lack is our bearings," he wrote. Winner painstakingly described the ways in which our experience of "things technological" confounds "our vision, our expectations, and our capacity to make intelligent judgments. Categories, arguments, conclusions, and choices that would have been entirely obvious in earlier times are obvious no longer."[60]

So let us establish our bearings. What Varian celebrates here is not a new form of contract but rather a final solution to the enduring uncertainty that is the raison d'être of "contract" as a means of "private ordering." In fact, the use of the word *contract* in Varian's formulation is a perfect example of the horseless-carriage syndrome. Varian's invention is unprecedented and cannot be understood as simply another kind of contract. It is, in fact, the *annihilation* of contract; this invention is better understood as the *uncontract*.

The uncontract is a feature of the larger complex that is the means of behavioral modification, and it is therefore an essential modality of surveillance capitalism. It contributes to economies of action by leveraging proprietary behavioral surplus to preempt and foreclose action alternatives, thus replacing the indeterminacy of social processes with the determinism of programmed machine processes. This is not the automation of society, as some might think, but rather the replacement of society with machine action dictated by economic imperatives.

The uncontract is not a space of contractual relations but rather a unilateral execution that makes those relations unnecessary. The uncontract desocializes the contract, manufacturing certainty through the substitution of automated procedures for promises, dialogue, shared meaning, problem solving, dispute resolution, and trust: the expressions of solidarity and human agency that have been gradually institutionalized in the notion of "contract" over the course of millennia. The uncontract bypasses all that social work in favor of compulsion, and it does so for the sake of more-lucrative prediction products that approximate observation and therefore guarantee outcomes.

This substitution of machine work for social work is possible thanks to the success of Google's declarations and the road that Google paved for surveillance capitalists' dominance of the division of learning. Sitting in the catbird seat, Google can observe what was previously unobservable and know what was previously unknowable. As a result, the company can do what was previously undoable: bypass social relations in favor of automated machine processes that compel the behaviors that advance commercial objectives. When we celebrate the uncontract, as Varian and others do, we celebrate the asymmetries of knowledge and power that produce these new possibilities. The uncontract is a signpost that reminds us of our bearings as we follow the next sections of this chapter toward a clearer picture of surveillance capitalism's growing ambitions in the annexation of "reality" to its kingdom of conquered human experience.

VII. Inevitabilism

It is difficult to keep your bearings when everyone around you is losing theirs. The transition to ubiquitous computing, "when sensors are everywhere," Paradiso writes, won't be "incremental" but rather "a revolutionary phase shift much like the arrival of the world wide web."[61] The same "phase shift" that is understood by its architects as the universal antidote to uncertainty is anticipated with absolute certainty. Paradiso is not alone here. On the contrary, the rhetoric of inevitability is so "ubiquitous" that within the tech community it can be considered a full-blown ideology of *inevitabilism*.

The sense of incontestable certainty that infuses Paradiso's vision has long been recognized as a key feature of utopianism. In their definitive history of utopian thought, Frank and Fritzie Manuel wrote that "since the end of the eighteenth century the predictive utopia has become a major form of imaginative thought and has preempted certain scientific techniques of forecasting... the contemporary utopia... binds past, present, and future together as though fated. The state they depict appears virtually ordained either by god or by history; there is a carry-over of millenarian certainty...."[62]

The Manuels, along with many other historians, consider Marxism to be the last great modern utopia.[63] There are hundreds of passages in Karl Marx's writing that convey his inevitabilism. In the very first section of *The Communist Manifesto*, published in 1848, Marx wrote the following: "What the bourgeoisie, therefore, produces, above all, is its own grave-diggers. Its fall and the victory of the proletariat are equally inevitable."[64]

Before the rise of the modern utopia, the genre was largely composed of fantastical narratives in which isolated pockets of human perfection were discovered in exotic mountain aeries, hidden valleys, or faraway islands. Modern utopias such as Marxism diverge from those fairy tales, addressing "the reformation of the entire species" with a rational systemic vision "whose province was the whole world." No longer content as mere weavers of dreams, modern utopianists shifted toward totalistic and universal visions, prophecies of "the ineluctable end toward which mankind was moving."[65]

Now the proselytizers of ubiquitous computing join Marx and other modern utopianists in postulating a new phase of history, like Paradiso's "revolutionary phase shift," in which all of society is reassembled in a novel and superior pattern. Despite the fact that inevitability is the opposite of politics and history, apostles of the apparatus routinely hijack historical metaphors that lend a veneer of gravitas to their claims. The rise of the apparatus is alternatively cast as the inauguration of a new "age," "era," "wave," "phase," or "stage." This kind of historical framing conveys the futility of opposition to the categorical inevitability of the march toward ubiquity.

Silicon Valley is the *axis mundi* of inevitabilism. Among high-tech leaders, within the specialist literature, and among expert professionals there appears to be universal agreement on the idea that *everything* will be connected, knowable, and actionable in the near future: ubiquity and its consequences in total information are an article of faith.

Not surprisingly, Google's leaders are passionate inevitabilists. The very first sentences of Schmidt and Cohen's 2013 book, *The New Digital Age*, exemplify this thrust: "Soon everyone on Earth will be connected," they write. So-called predictive "laws" such as "Moore's Law" and "photonics" are called upon to signal this new iron law of necessity that will produce exponential growth in connectivity and computational power.[66] And later, "The collective benefit of sharing human knowledge and creativity grows at an exponential rate. In the future, information technology will be everywhere, like electricity. It will be a given."[67] When the book's assertions garnered some criticism, the authors confronted their critics in an afterword to the paperback edition: "But bemoaning the inevitable increase in the size and reach of the technology sector distracts us from the real question.... Many of the changes we discuss are inevitable. They're coming."

Despite its pervasiveness both in Silicon Valley and in the wider culture of data scientists and technology developers, inevitabilism is rarely discussed or critically evaluated. Paradiso's conception of a "digital omniscience" is taken for granted, with little discussion of politics, power, markets, or governments. As in most accounts of the apparatus, questions of individual autonomy, moral reasoning, social norms and values, privacy, decision rights, politics, and law take the form of afterthoughts and genuflections that can be solved with the correct protocols or addressed with still more technology solutions. If information will stream "directly into our eyes and ears" and "the boundaries of the individual will be very blurry," then who can access that information? What if I don't want my life streaming through your senses? *Who knows? Who decides? Who decides who decides?* The answers to such questions are drowned in the thrum of all things continuously illuminated, registered, counted, controlled, and judged.

The best that Paradiso can offer up is a suggestion that "the law could give a person ownership or control of data generated in his or her vicinity; a person could then choose to encrypt or restrict those data from entering the network."[68] Paradiso imagines a society in which it falls to each individual to protect herself from the omniscient ubiquitous sensate computational systems of the new apparatus. Rather than paradise, it seems a recipe for a new breed of madness. Yet this is precisely the world that is now under construction around us, and this madness appears to be a happy feature of the plan.

Between 2012 and 2015, I interviewed 52 data scientists and specialists in the "internet of things." They came from 19 different companies with a combined 586 years of experience in high-technology corporations and startups, primarily in Silicon Valley. I spoke with them about the prominence of inevitability rhetoric among the purveyors of the new apparatus, and I posed the same question to each one: why do so many people say that ubiquitous computing is inevitable? The agreement among their responses was striking. Although they did not have the language of surveillance capitalism, nearly every interviewee regarded inevitability rhetoric as a Trojan horse for powerful economic imperatives, and each one of them lamented the lack of any critical discussion of these assumptions.

As the marketing director of a Silicon Valley firm that sells software to link smart devices told me, "There's all that dumb real estate out there and we've got to turn it into revenue. The 'internet of things' is all push, not pull. Most consumers do not feel a need for these devices. You can say 'exponential' and 'inevitable' as much as you want. The bottom line is that the Valley has decided that this has to be the next big thing so that firms here can grow."

I spoke with a senior engineer from a large tech company that invests heavily in the "internet of things." The response:

> Imagine you have a hammer. That's machine learning. It helped you climb a grueling mountain to reach the summit. That's machine learning's dominance of online data. On the mountaintop you find a vast pile of nails, cheaper than anything previously imaginable. That's the new smart sensor tech. An unbroken vista of virgin board stretches before you as far as you can see. That's the whole dumb

world. Then you learn that any time you plant a nail in a board with your machine learning hammer, you can extract value from that formerly dumb plank. That's data monetization. What do you do? You start hammering like crazy and you never stop, unless somebody makes you stop. But there is nobody up here to make us stop. This is why the "internet of everything" is inevitable.

A senior systems architect laid out the imperative in the clearest terms: "The IoT is inevitable like getting to the Pacific Ocean was inevitable. It's manifest destiny. Ninety-eight percent of the things in the world are not connected. So we're gonna connect them. It could be a moisture temperature that sits in the ground. It could be your liver. That's *your* IoT. The next step is what we do with the data. We'll visualize it, make sense of it, and monetize it. That's *our* IoT."

VIII. Men Made It

The relentless drumbeat of inevitabilist messages presents the new apparatus of ubiquity as the product of technological forces that operate beyond human agency and the choices of communities, an implacable movement that originates outside history and exerts a momentum that in some vague way drives toward the perfection of the species and the planet. The image of technology as an autonomous force with unavoidable actions and consequences has been employed across the centuries to erase the fingerprints of power and absolve it of responsibility. The monster did it, not Victor Frankenstein. However, the ankle bracelet does not monitor the prisoner; the criminal justice system does that.

Every doctrine of inevitability carries a weaponized virus of moral nihilism programmed to target human agency and delete resistance and creativity from the text of human possibility. Inevitability rhetoric is a cunning fraud designed to render us helpless and passive in the face of implacable forces that are and must always be indifferent to the merely human. This is the world of the robotized interface, where technologies work their will, resolutely protecting power from challenge.

No one has expressed this with more insight and economy than John Steinbeck in the opening chapters of his masterwork, *The Grapes of Wrath*, which describes the dustbowl farmers who are thrown out of their Oklahoma homes during the Great Depression and then head west to California. The families are forced off the land that they have tended for generations. They plaintively argue their case to the bank agents sent to impress upon them the facts of their helplessness. But the agents respond with "The bank is something else than men. It happens that every man in a bank hates what the bank does, and yet the bank does it. The bank is something more than men, I tell you. It's the monster. Men made it, but they can't control it."[69]

This theme of supposed technological autonomy is a venerable one among technology scholars. Langdon Winner again proves to be a worthy guide when he reminds us that an unquestioning acceptance of technology has become a feature of modern life: "The changes and disruptions that an evolving technology repeatedly caused in modern life were accepted as given or inevitable simply because no one bothered to ask whether there were other possibilities."[70]

Winner observes that we have allowed ourselves to become "committed" to a pattern of technological "drift," defined as "accumulated unanticipated consequences." We accept the idea that technology must not be impeded if society is to prosper, and in this way we surrender to technological determinism. Rational consideration of social values is considered "retrograde," Winner writes, "not the ticket that scientific technology gives to civilization.... To this day, any suggestions that the forward flow of technological innovation be in any way limited...violate a fundamental taboo....Instead we accept the change, later looking back upon what we have done to ourselves as a topic of curiosity."[71] To Winner's "curiosity" I add another theme: remorse.

Surveillance capitalist leaders assume that we will succumb to the naturalistic fallacy as Steinbeck's farmers were meant to do. Because Google is successful—because surveillance capitalism is successful—its rules must obviously be right and good. Like the bank agents, Google wants us to accept that its rules simply reflect the requirements of autonomous processes, something that people cannot control. However, our grasp of the inner logic of surveillance capitalism suggests otherwise. Men and women made it, and they can control it. They merely choose not to do so.

Inevitabilism enshrines the apparatus of ubiquity as progress but conceals the realpolitik of surveillance capitalism at work behind the scenes. We know that there can be alternative paths to a robust information capitalism that produces genuine solutions for a third modernity. We have seen that surveillance capitalism was discovered and honed in history, handcrafted by men and women to serve the interests of impatient capital. It is this same logic that now demands ubiquity, ready to colonize technical developments for the sake of its imperatives and growth. Inevitabilism operates in the service of these imperatives as it distracts attention from the ambitions of a rising economic order and the competitive anxieties that drive the surveillance project toward certainty, thus necessitating its ever-more-voracious claims on our behavior.

Inevitabilism precludes choice and voluntary participation. It leaves no room for human will as the author of the future. This raises questions: At what point does inevitabilism's claim to ubiquitous extraction and execution shade into abuse? Will inevitabilism's utopian declarations summon new forms of coercion designed to quiet restless populations unable to quell their hankering for a future of their choice?[72]

IX. To the Ground Campaign

Google's declarations; surveillance capitalism's dominance over the division of learning in society and its laws of motion; ubiquitous architectures of extraction and execution; MacKay's penetration of inaccessible regions while observing unrestrained animals with methods that elude their awareness; the uncontract and its displacement of society; Paradiso's ubiquitous sensate environment; dark data; the inevitabilism evangelists: there is one place where all these elements come together and transform a shared public space built for human engagement into a petri dish for the reality business of surveillance capitalism. That place is the city.

Cisco has 120 "smart cities" globally, some of which have embraced Cisco Kinetic, which as Jahangir Mohammed, the company's vice president and general manager of IoT, explains in a blog post, "is a cloud-based platform that helps customers *extract,* compute, and move data from connected things to IoT applications to deliver *better outcomes*.... Cisco Kinetic gets the right data to the right applications at the right time... while *executing policies to enforce data ownership, privacy, security and even data sovereignty laws.*"[73] But, as is so often the case, the most audacious effort to transform the urban commons into the surveillance capitalist's equivalent of Paradiso's 250-acre marsh comes from Google, which has introduced and legitimated the concept of the "for-profit city." Just as MacKay had counseled and Weiser proselytized, the computer would be operational everywhere and detectable nowhere, always beyond the edge of individual awareness.

In 2015, shortly after Google reorganized itself into a holding company called Alphabet, Sidewalk Labs became one of nine "confirmed companies" under the Alphabet corporate umbrella. Whether what even Sidewalk CEO Dan Doctoroff, a former private equity financier, CEO of Bloomberg, and deputy mayor of New York City in the Bloomberg administration, refers to as a "Google city" succeeds, the company has interested the public by recasting our central gathering place as a commercial operation in which once-public assets and functions are reborn as the cornered raw materials earmarked for a new marketplace. In this vision, MacKay and Paradiso's conceptions come to fruition under the auspices of surveillance capitalism in a grand scheme of vertically integrated supply, production, and sales.

Sidewalk Labs' first public undertaking was the installation of several hundred free internet-enabled kiosks in New York City, ostensibly to combat the problem of "digital inequality." As we saw with Google Street View, the company can siphon a lot of valuable information about people from a Wi-Fi network, even if they don't use the kiosks.[74] Doctoroff has characterized the Sidewalk Labs' kiosks as "fountains of data" that will be equipped with environmental sensors and also collect "other data, all of which can create very hyperlocal information about conditions in the city."

In 2016 the US Department of Transportation (DOT) announced a partnership with Sidewalk Labs "to funnel transit data to city officials." The DOT worked to draw cities into Google's orbit with a competition for $40 million in grants. Winners would work with Sidewalk Labs to integrate technology into municipal operations, but Sidewalk Labs was eager to work with finalists in order to develop its own traffic-management system, Flow.[75] Flow relies on

Google Maps, Street View vehicles, and machine intelligence to capture and analyze data from drivers and public spaces.[76] These analyses produce prediction products described as "inferences about where people are coming from or going," enabling administrators "to run virtual experiments" and improve traffic flow.[77]

Doctoroff postulates a city presided over by digital omniscience: "We're taking everything from anonymized smartphone data from billions of miles, trips, sensor data, and bringing that into a platform."[78] Sidewalk refers to its high-tech services as "new superpowers to extend access and mobility." Algorithms designed to maintain critical behaviors within a prescribed zone of action would manage these data flows: "In a world in which we can monitor things like noise or vibrations, why do we need to have these very prescriptive building codes?" As an alternative, Doctoroff suggests "performance-based zoning" administered by the ubiquitous apparatus through the medium of algorithms. These processes, like Varian's vehicular monitoring systems, are indifferent to why you behave as long as they can monitor and control the behavior you produce. As Doctoroff explains it, "I don't care what you put here as long as you don't exceed performance standards like noise levels...." This is preferable, he says, because it enhances "the free flow of property...that is a logical extension of...these technologies."[79] Why should citizens have any say over their communities and the long-term implications of how luxury high-rises, hotels, or a residential building going commercial could affect rents and local businesses as long as an algorithm is satisfied with noise thresholds?

When Columbus, Ohio, was named winner of the DOT competition, it began a three-year demonstration project with Sidewalk, including a hundred kiosks and free access to the Flow software. Documents and correspondence from this collaboration eventually obtained by the *Guardian* describe innovations such as "dynamic parking," "optimized parking enforcement," and a "shared mobility marketplace" that reveal a more troubling pattern than the rhetoric suggests. Sidewalk's data flows combine public and private assets for sale in dynamic, real-time virtual markets that extract maximum fees from citizens and leave municipal governments dependent upon Sidewalk's proprietary information. For example, public and private parking spaces are combined in online markets and rented "on demand" as the cost of parking varies in real time, substantially increasing parking income. Optimized parking enforcement depends on Sidewalk's algorithms "to calculate the most lucrative routes for parking cops," earning cities millions of extra dollars that they desperately need but that arrive at the expense of their citizens.

Cities are required to invest substantial public monies in Sidewalk's technology platform, including channeling municipal funds earmarked for low-cost public bus service into "mobility markets" that rely on private ridesharing companies such as Uber. The company insists that cities "share public transport data with ride-sharing companies, allowing Uber to direct cars to overcrowded bus stops." The Flow Transit system integrates information and payment for nearly every kind of transport into Google Maps, and cities are obligated to "upgrade" to Sidewalk's mobile payment system "for all existing transit and parking services." Just as it requires public-transit data, Sidewalk also insists that cities share all parking and ridership information with Sidewalk Labs in real time.[80] When asked, Doctoroff has emphasized the novel blending of public functions and private gain, assuring his listeners on both counts that "our mission is to use technology to change cities... to bring technology to solve big urban problems.... We expect to make a lot of money from this."[81]

In April 2016 a "curated group of leaders" in tech, media, and finance met at the Yale Club in Manhattan to hear Sidewalk CEO Dan Doctoroff's talk: "Google City: How the Tech Juggernaut Is Reimagining Cities—Faster Than You Realize."[82] His remarks provide a candid assessment of the "Google city" as a market operation shaped by the prediction imperative. He could not have been more direct in articulating Sidewalk Labs' approach as a translation of Google's online world to the reality of city life:

> In effect, what we're doing is replicating the digital experience in physical space.... So ubiquitous connectivity; incredible computing power including artificial intelligence and machine learning; the ability to display data; sensing, including cameras and location data as well as other kinds of specialized sensors.... We fund it all... through a very novel advertising model.... We can actually then target ads to people in proximity, and then obviously over time track

them through things like beacons and location services as well as their browsing activity.[83]

Later that year, Sidewalk announced collaborations with sixteen additional cities, noting that achieving scale would enable it to improve its Flow software products. Doctoroff referred to these collaborations as "inevitable."[84]

The vast and varied ground campaign already underway turns the prediction imperative into concrete activity. In pursuit of economies of scope, a wave of novel machine processes are honed for extraction, rendering people and things as behavioral data. For the sake of economies of action, the apparatus learns to interrupt the flow of personal experience in order to influence, modify, and direct our behavior, guided by the plans and interests of self-authorizing commercial actors and the buzzing market cosmos in which they participate. In nearly every case the agents of institutionalization present their novel practices as if they are one thing, when they are, in fact, something altogether different. The realpolitik of commercial surveillance operations is concealed offstage while the chorus of actors singing and dancing under the spotlights holds our attention and sometimes even our enthusiasm. They sweat under the stage lights for the sake of one aim: that we fail to notice the answers or, better yet, forget to ask the questions: *Who knows? Who decides? Who decides who decides?*

In light of these ambitions, it is not surprising that Doctoroff, like Page, prefers lawless space. Press reports confirmed that Alphabet/Google was actively considering a proposal for a new city and that more than a hundred urban planners, researchers, technologists, building experts, economists, and consultants were involved in the project.[85] The *Wall Street Journal* reported that although it was unclear how the company would fund the tens of billions of dollars necessary for such a large-scale undertaking, "One key element is that Sidewalk would be seeking autonomy from many city regulations, so it could build without constraints...."[86]

In October 2017, Doctoroff appeared with Alphabet Executive Chairman Eric Schmidt and Canadian Prime Minister Justin Trudeau to reveal that Toronto would be the site of Sidewalk's planned development. Its intent is to develop the right mix of technology that it can then license to cities around the world. "The genesis of the thinking for Sidewalk Labs came from Google's founders getting excited thinking of 'all the things you could do if someone would just give us a city and put us in charge,'" Toronto's *Globe and Mail* reported Schmidt as saying, noting that "he joked he knew there were good reasons that doesn't happen." Then, just as quickly, the paper related Schmidt's reaction when he first learned that Sidewalk, and by extension Alphabet, had secured this opportunity in Toronto: "Oh my God! We've been selected. Now, it's our turn."[87]

CHAPTER EIGHT

RENDITION: FROM EXPERIENCE TO DATA

You take a picture of 'em, they'll kill you.
They think you're takin' somethin' away from 'em.
That you only got so much...stuff!...and if other
People are takin' it all, then there ain't none left for yourself.

—ROBERT GARLAND, *THE ELECTRIC HORSEMAN*

To photograph is to appropriate the thing photographed.
It means putting oneself into a certain relation to the world
that feels like knowledge—and, therefore, like power.

—SUSAN SONTAG, *ON PHOTOGRAPHY*

I. Terms of Sur-Render

We worry about companies that amass our personal data, and we wonder why they should profit. "Who owns the data?" we ask. But every discussion of data protection or data ownership omits the most important question of all: why is our experience rendered as behavioral data in the first place? It has been far too easy to overlook this important step in the chain of events that produces behavioral surplus. This chapter and the next draw our attention to the gap between experience and data, as well as to the specific operations that target this gap on a mission to transform the one into the other. I call these operations *rendition*. We have seen that the dispossession of human experience is the original sin of surveillance capitalism, but this dispossession is not mere abstraction. *Rendition* describes the concrete operational practices through which dispossession is accomplished, as human experience is claimed as raw material for datafication and all that follows, from manufacturing to sales. A focus on these intermediate practices illustrates that the apparatus of ubiquity is not a passive one-way mirror. Rather, it actively creates its own stores of knowledge through rendition.

The noun *rendition* derives from the verb *render,* a most unusual word whose double meanings describe a two-sided equation that perfectly captures what happens in the gap between human experience and behavioral data. On one side of the equation, the verb describes a process in which something is formed out of something else that is originally given. It designates the causal action of turning one thing into another, such as rendering oil from fat (extraction) or rendering an English text from the original Latin (translation). These meanings have also found their way into the vocabulary of digital technology. For example, a "rendering engine" converts the coded content of an HTML page for display and printing.

On the other side of the equation, *render* also describes the way in which the thing that is changed gives itself over to this process: it *sur-renders.* The verb *rendre* first appears in tenth-century French, meaning "to give back, present, yield," as in "rendering an account" or "the tree renders its fruit." By the fourteenth century, the word also incorporated the idea of handing over, delivering, or acknowledging dependency or obligation, as in "Render unto Caesar." These meanings are active today when we say "rendering a verdict," "rendering service," or "rendering property."

Surveillance capitalism must work both sides of the equation. On one side, its technologies are designed to render our experience into data, as in rendering oil from fat. This typically occurs outside of our awareness, let alone our consent. On the other side of the equation, every time we encounter a digital interface we make our experience available to "datafication," thus "rendering unto surveillance capitalism" its continuous tithe of raw-material supplies.

This two-sided equation is a novel arrangement. As we saw in Chapter 1, the Aware Home project developed at Georgia Tech just a year before the invention of surveillance capitalism employed different practices that embodied very different assumptions: (1) that it must be the individual alone who decides what experience is rendered as data, (2) that the purpose of the data is to enrich the individual's life, and (3) that the individual is the sole arbiter of how the data are shared or put to use. Nearly two decades later, the Aware Home is barely more than an archeological fragment reminding us of the road not taken toward an empowering digital future and a more-just division of learning in society. Down that road, it is the individual who knows, decides, and decides who decides: an end in herself, not a means to others' ends. The lesson of the Aware Home is that there can be rendition without surveillance capitalism. However, the lesson of this chapter and the next is that *there can be no surveillance capitalism without rendition.*

Nothing is exempt, as products and services from every sector join devices like the Nest thermostat in the competition for surveillance revenues. For example, in July 2017 iRobot's autonomous vacuum cleaner, Roomba, made headlines when the company's CEO, Colin Angle, told Reuters about its data-based business strategy for the smart home, starting with a new revenue stream derived from selling floor plans of customers' homes scraped from the machine's new mapping capabilities. Angle indicated that iRobot could reach a deal to sell its maps to Google, Amazon, or Apple within the next two years. In preparation for this entry into surveillance competition, a camera, new sensors, and software had already been added to Roomba's premier line, enabling new functions, including the ability to build a map while tracking its own location. The market had rewarded iRobot's growth vision, sending the company's stock price to $102 in June 2017 from just $35 a year earlier, translating into a market capitalization of $2.5 billion on revenues of $660 million.[1]

Privacy experts raised alarms, knowing that such data streams have virtually no legal or security protection. But Angle assured the public that iRobot would not sell data without its customers' permission and expressed confidence that "most would give their consent in order to access the smart home functions."[2] Why was Angle so confident?

According to the company's privacy policy, it is true that owners of the Roomba can control or stop the collection of usage data by "disconnecting your WiFi or Bluetooth from the app, for example, by changing your WiFi password." However, as Angle told tech site *Mashable* in July 2017, even when customers do not opt-in to the mapping service, the Roomba captures mapping and usage data, but only usage data "is sent to the cloud so it can be shown on your mobile device."[3] What Angle neglected to say was that a customer who refuses to share his or her home's interior mapping data with iRobot also loses most of the smart functionality of the "autonomous" vacuum cleaner, including the ability to use one's phone to start or pause a cleaning, schedule cleanings, review "Clean Map reports," receive automatic software updates, or start a "SPOT Clean to focus on a particularly dirty area."[4]

Angle's confidence-enhancing strategy goes to the heart of the larger rendition project, for which surveillance capitalist purveyors of "smart" home products have developed a singular approach. On the one hand, they stress that customers can opt in to data sharing. On the other hand, customers who refuse to opt in face limited product functionality and data security. In these *Requirimiento*-style relationships, instead of the *adelantados'* message, "Bend the knee or we destroy you," the message here is "Bend the knee or we degrade your purchase."

Under this new regime, something as simple as buying a mattress now requires careful legal scrutiny of the "abusive contracts" that nearly everyone ignores. Consider the Sleep Number bed, with its "smart bed technology and sleep tracking."[5] The company's website features a beautiful couple snuggled in bed happily glued to their smartphones as they delight in data from their SleepIQ app. The bed's base and mattress are "customizable" with features that raise or lower the angle of the bed and sensors that soften or firm up the mattress. Other sensors measure heart rate, breathing, and movement: "Every morning you'll get your SleepIQ® score, representing your individual quality and length of sleep... your restful sleep, restless sleep and time out of bed... and what adjustments you can make." The company suggests that you connect your sleep app to your fitness tracker and your thermostat in order to see how your workout or bedroom temperature affects your sleep.

A dense, twelve-page privacy policy accompanies the bed. Customers are advised that providing information is an affirmation of consent to use that information in-line with the policy, which employs the usual onerous terms: third-party sharing, Google analytics, targeted advertising, and much more. In addition, if customers create a user profile to maximize the effectiveness of the app, the company also collects "biometric and sleep-related data about how You, a Child, and any person that uses the Bed slept, such as that person's movement, positions, respiration, and heart rate while sleeping." It also collects all the audio signals in your bedroom. As with most such policies, customers are advised that the company can "share" or "exploit" personal information even "after You deactivate or cancel the Services and/or your Sleep Number account or User Profile(s)." Customers are warned that no data transmission or storage "can be guaranteed to be 100% secure" and that it does not honor "Do Not Track" notifications. Finally, on page 8 of the document, the policy addresses a customer's choices regarding the use of personal information: "Whether You submit Information to Us is entirely up to You. If you decide to not submit Information, We may not be able to provide certain features, products, and/or services to You."[6]

This same coercive *Requirimiento* twist can be found in the lengthy, dense legal compacts associated with Alphabet-owned Nest thermostats. The terms-of-service and end-user licensing agreements reveal oppressive privacy and security consequences in which sensitive information is shared with other devices, unnamed personnel, and third parties for the purposes of analysis and ultimately for trading in behavioral futures markets, action that ricochets back to the owner in the form of targeted ads and messages designed to push more products and services. Despite this, courts have generally upheld companies' claims that they bear no liability without a clear demonstration of economic harm to the consumer.

Nest takes little responsibility for the security of that information and none for how other companies will put it to use. In fact, University of London legal scholars Guido Noto La Diega and Ian Walden, who analyzed these documents, reckon that were one to enter into the Nest ecosystem of connected devices and apps, each with their own equally burdensome terms, the purchase of a single home thermostat entails the need to review nearly a thousand "contracts."[7]

This absurdity is compounded by the fact that virtually no one reads even one such "contract." A valuable empirical study of 543 participants familiar with surveillance and privacy law issues found that when asked to join a new online service, 74 percent opted for the "quick join" procedure, bypassing the terms-of-service agreement and the privacy policy. Among those who did scroll through the abusive contracts, most went directly to the "accept" button. The researchers calculated that the documents required at least forty-five minutes for adequate comprehension, but for those who looked at the agreements, the median time they spent was fourteen seconds.[8]

Should the customer refuse to agree to Nest's stipulations, the terms of service indicate that the functionality and security of the thermostat itself will be deeply compromised, no longer supported by the necessary updates meant to ensure its reliability and safety. The consequences can range from frozen pipes to failed smoke alarms to an easily hacked internal home system. In short, the effectiveness and safety of the product are brazenly held hostage to its owners' submission to rendition as conquest, by and for others' interests.

One can easily choose not to purchase a Roomba, a SleepNumber bed, or a Nest thermostat, but each of these is merely emblematic of the immense project of rendition as the first and vital step in the construction of the apparatus of ubiquity. Thousands of "internet of things" objects are becoming available. As La Diega and Walden conclude, in this new product regime the simple product functions that we seek are now hopelessly enmeshed in a tangled mixture of software, services, and networks.[9]

The very idea of a functional, effective, affordable product or service as a sufficient basis for economic exchange is dying. Where you might least expect it, products of every sort are remade by the new economic requirements of connection and rendition. Each is reimagined as a gateway to the new apparatus, praised for being "smart" while traditional alternatives are reviled for remaining "dumb." It is important to acknowledge that in this context, "smart" is a euphemism for rendition: intelligence that is designed to render some tiny corner of lived experience as behavioral data. Each smart object is a kind of marionette; for all its "smartness," it remains a hapless puppet dancing to the puppet master's hidden economic imperatives. Products, services, and applications march to the drumbeat of inevitabilism toward the promise of surveillance revenues hacked from the still-wild spaces that we call "my reality," "my home," "my life," and "my body." Every smart product repeats our essential questions: What does a smart product know, and whom does it tell? *Who knows? Who decides? Who decides who decides?*

Examples of products determined to render, monitor, record, and communicate behavioral data proliferate, from smart vodka bottles to internet-enabled rectal thermometers, and quite literally everything in between.[10] The business developer for a spirits company thus cites his plan for a "connected bottle": "The more we learn about consumers and their behaviors, the better services we can connect to them."[11] Many brands are determined "to give packaging a speaking role in an increasingly interactive marketplace." Global spirits distributor Diageo promises "smart sensor-equipped bottles" that can track purchases and sales data, and, most importantly, "communicate with consumers' devices and switch gears—recipes versus sales promos—once the bottle is opened." A producer of bar equipment states the case plainly enough: "It really is all about… allowing these [bar] owners to see stuff that they couldn't see before and maximize their profits."[12]

Today our homes are in surveillance capitalism's crosshairs, as competitors chased a $14.7 billion market for smart-home devices in 2017, up from $6.8 billion just a year earlier and expected to reach more than $101 billion by 2021.[13] You may have already encountered some of the early absurdities: smart toothbrushes, smart lightbulbs, smart coffee mugs, smart ovens, smart juicers, and smart utensils said to improve your digestion. Others are often more grim: a home security camera with facial recognition; an alarm system that monitors unusual vibrations before a break-in occurs; indoor GPS locators; sensors that attach to any object to analyze movement, temperature, and other variables; every kind of connected appliance; cyborg cockroaches designed to detect sound. Even the baby's nursery is reconceived as a source of fresh behavioral surplus.[14]

An appreciation of the surveillance logic of accumulation that drives this action suggests that this network of things is already evolving into a network of coercion, in which mundane functions are ransomed for behavioral surplus.[15] A December 2013 letter from Google's finance director to the US Securities and Exchange Commission's Division of Corporate Finance provides a vivid glimpse of these facts. The letter was composed in response to an SEC query on the segmentation of Google's revenues between its desktop and mobile platforms.[16] Google answered by stating that users would be "viewing our ads on an increasingly wide diversity of devices in the future" and that its advertising systems were therefore moving toward "device agnostic" design that made segmentation irrelevant and impractical. "A few years from now," the letter stated, "we and other companies could be serving ads and other content on refrigerators, car dashboards, thermostats, glasses, and watches, to name just a few possibilities."

Here is at least one endgame: the "smart home" and its "internet of things" are the canvas upon which the new markets in future behavior inscribe their presence and assert their demands in our most-intimate spaces. Key to the story is that all of this action is prosecuted in support of a larger market process that bets on the future of our behavior and over which we have no knowledge or control. Each node in the network—the vacuum cleaner, the mattress, the thermostat—must play its part, beginning with the frictionless rendition of behavior, as the whole team of seething insistent "smart" things joins the migration to surveillance revenues. As we are shorn of alternatives, we are forced to purchase products that we can never own while our payments fund our own surveillance and coercion. Adding insult to injury, data rendered by this wave of things are notoriously insecure and easily subject to breaches. Moreover, manufacturers have no legal responsibility to notify device owners when data are stolen or hacked.

There are other, even more grandiose ambitions for the rendition of all solitary things. Companies such as Qualcomm, Intel, and ARM are developing tiny, always-on, low-power computer vision modules that can be added to any device, such as your phone or refrigerator, or any surface. A Qualcomm executive says that appliances and toys can know what's going on around them: "A doll could detect when a child's face turns toward it."[17]

Consider "smart skin," developed by brilliant university scientists and now poised for commercial elaboration. Initially valued for its ability to monitor and diagnosis health conditions from Parkinson's disease to sleep disorders, smart skin is now hailed for its promise of ultra-unobtrusive ubiquity. Researchers at Georgia Tech developed a version of "smart skin" that sucks energy from radio waves and other energy sources, eliminating the need for batteries. Smart skin, described as "the ultimate sensing tool that could potentially allow for the mass implementation of perpetual wireless networks,"[18] can cognize, sense, analyze, wirelessly communicate, and "modify parameters" using simple radio frequency (RFID) technology.[19] As in the case of Paradiso's "sensor tape," the researchers stress that it can also "be applied everywhere" to "monitor, sense, and interact with the world around us in a perpetual way, thus significantly enhancing ambient intelligence," all of it as inconspicuous as a "decal sticker." They suggest, for example, the shelves of grocery stores, where revenue opportunities are plentiful.[20]

Rendition has become a surveillance capitalist project shaped by its imperatives and directed toward its objectives. In the composition of the shadow text, rendition is Step One: the concrete operationalization of the "original sin of simple robbery" that defined this market project from the start. Google rendered the Earth, its streets, and its dwelling places, bypassing our consent and defying our protests. Facebook rendered the social network and its limitless details for the sake of the company's behavioral futures markets. Now the ubiquitous apparatus is the means to the ubiquitous rendition of human experience. We have seen the urgency with which surveillance capitalists pursue the elimination of "friction" as a critical success factor in supply operations. The prediction imperative makes boundaries and borders intolerable, and surveillance capitalists will do almost anything to eliminate them. This pursuit transforms "connection" into a commercial imperative and transforms individual autonomy into a threat to surveillance revenues.

Surveillance capitalism's rendition practices overwhelm any sensible discussion of "opt in" and "opt out." There are no more fig leaves. The euphemisms of consent can no longer divert attention from the bare facts: under surveillance capitalism, rendition is typically unauthorized, unilateral, gluttonous, secret, and brazen. These characteristics summarize the asymmetries of power that put the "surveillance" in surveillance capitalism. They also highlight a harsh truth: it is difficult to be where rendition is not. As industries far beyond the technology sector are lured by surveillance profits, the ferocity of the race to find and render experience as data has turned rendition into a global project of surveillance capital.

This chapter and the next survey a range of rendition activities in the pursuit of economies of scope. The remainder of this chapter concentrates on *extension, the first dimension of scope*, as rendition operations move into the real world, seizing fresh unexpected chunks of human experience. Extension wants every corner and crevice, every utterance and gesture on the path to dispossession. All that is moist and alive must hand over its facts. There can be no shadow, no darkness. The unknown is intolerable. The solitary is forbidden. Later, in Chapter 9, we move into the *depth* dimension. The net is cast wide over the waters of daily life, but there are also submarines exploring the depths in search of new sources of surplus prized for their rare predictive powers: your personality, emotions, and endorphins. The examples in these chapters are not intended to be exhaustive but rather to illustrate the seriousness of purpose, tenacity, and subterfuge with which surveillance capitalists pursue their search for new aspects of human experience that can be monetized as certainty.

In this pursuit, we are necessarily involved in citing specific actors, products, and techniques, knowing that the details of persons and companies are constantly churning. Firms will be bought and sold, fail or succeed; people will come and go. Specific technologies, products, and techniques will be abandoned, refined, and surpassed. When they fall, new ones will take their place, as long as surveillance capitalism is allowed to flourish. Velocity and churn have been critical to surveillance capitalism's success, and we cannot allow the constant movement to inhibit our determination to grasp the "laws of motion" that command this roiling landscape. It is the pattern and its purpose that we want to grasp.

II. Body Rendition

The rendition of your body begins quite simply with your phone. Even when your city is not "smart" or owned and operated by Google, market players with an interest in your behavior know how to find your body.[21] For all of the elaborate ways in which surveillance capitalists labor to render reality as behavior for surplus, the simplest and most profound is their ability to know exactly where you are all the time. Your body is reimagined as a behaving object to be tracked and calculated for indexing and search. Most smartphone apps demand access to your location even when it's not necessary for the service they provide, simply because the answer to this question is so lucrative.

Location data can be extracted from "geotags" created when your smartphone automatically embeds your identity and location in photos and videos. Retailers use "geofencing" to demarcate a geographical area and send alerts to smartphones within those parameters: "Come here now!" "Buy this here!" "An offer, just for you!"[22] Download the Starbucks app, and then leave your house if you want to see this in action. As one marketing consultancy advises, "Mobile advertising, the ultimate form of geo-targeting, is the holy grail of advertising."[23] "Tips and tricks" for location-based marketing are generously offered by a firm that specializes in mobile advertising: "It allows you to tap into people's compulsive nature by encouraging impulse buys with the notifications you send out.... It also allows you to gain insight on your current customers by reading what they're saying on Yelp and Facebook...."[24]

Another mobile marketing firm recommends "life pattern marketing" based on techniques derived from military intelligence known as "patterns of life analysis." These involve gathering location and other data from phones, satellites, vehicles, and sensors to assemble intelligence on the daily behavioral patterns of a "person of interest" in order to predict future behavior. Marketers are exhorted to "map the daily patterns" of a "target audience" in order to "intercept people in their daily routines with brand and promotional messages." As the firm emphasizes, "The psychological power of the perception of ubiquity is profound. Life Pattern Marketing forms a powerful psychological imprint on consumers."[25]

You can shut down the GPS locator on your phone, but most people do not, both because they rely on its utilities and because they are ignorant of its operations. According to Pew Research, 74 percent of US smartphone owners in 2013 used apps that required location data, and 90 percent did so in 2015—that's about 153 million people, more than those who listen to music or watch video on their phones.[26] Surveillance capitalism's reliance on secret operations means that most of us simply do not and cannot know the extent to which our phone doubles as a tracking device for corporate surveillance.

A study by researchers at Carnegie Mellon University proves the point efficiently.[27] During a three-week period, twenty-three participants were continuously informed of the number of apps accessing their location information and the total number of accesses in a given period. They were flabbergasted by the sheer volume of the onslaught as they each variously learned that their locations were accessed 4,182 times, 5,398 times, 356 times, and so on, over a 14-day period—all for the sake of advertisers, insurers, retailers, marketing firms, mortgage companies, and anyone else who pays to play in these behavioral markets.[28] As one participant summed it up, "It felt like I'm being followed by my own phone. It's scary."[29] Fifty-eight percent of participants subsequently restricted the permissions granted to their mobile apps.

Unsurprisingly, Google represents the vanguard of location-based tracking. A 2016 affidavit from law-enforcement officials seeking a search warrant for a California bank robber made plain why Google location data are unparalleled: "Google collects and retains location data from Android-enabled mobile devices. Google collects this data whenever one of their services is activated and/or whenever there is an event on the mobile device such as a phone call, text messages, internet access, or email access." The officials on the case requested location information from Google because it offers far more detail than even the phone companies can provide. The location systems in Android combine cell-tower data with GPS, Wi-Fi networks, and other information culled from photos, videos, and other sources: "That lets Android pinpoint users to a single building, rather than a city block."[30] In November 2017 *Quartz* investigative reporters discovered that since early 2017, Android phones had been collecting location information by triangulating the nearest cell towers, even when location services were disabled, no apps were running, and no carrier SIM card was installed in the phone. The information was used to manage Google's "push" notifications and messages sent to users on their Android phones, enabling the company to track "whether an individual with an Android phone or running Google apps has set foot in a specific store, and use that to target the advertising a user subsequently sees."[31]

Google's location history system is a product of the corporation's global mapping operations. Though active for over a decade, it was unveiled to the public only in 2015 as "Your Timeline," a feature that "allows you to visualize your real-world routines."[32] The corporation calculated that any negative reaction to the volume and persistence of tracking revealed by Timeline would be mitigated by the value of users' active contributions to their own stocks of behavioral surplus as they fine-tune the information, add relevant photos, insert comments, and so forth. This was represented as an individual's investment in personalized services such as Google Now so that it might more effectively comb your e-mail and apps in order to push relevant traffic and weather updates, notifications, suggestions, and reminders. Location data is the quid pro quo for these services.

This transaction is greased by the usual promises of privacy and control: "Your Timeline is private and visible only to you; and you control the locations you choose to keep." But Google uses your location data to target ads; indeed, these are among the most significant sources of surplus in Google's advertising markets with a direct impact on click rates. The standard account from Google and other surveillance capitalists is that behavioral surplus is retained only as meta-data, which are then aggregated across large numbers of individual users. We are told that it's not possible to identify individuals from these large-scale amalgamations. However, with as little as three bits of data easily culled from the public record—birth date, zip code, and sex—reidentification science has demonstrated its ability to de-anonymize meta-data with "disturbing ease."[33] In a summary of this research, legal scholar Paul Ohm writes that "re-identification makes all our secrets fundamentally easier to discover and reveal. Our enemies will find it easier to connect us to facts that they can use to blackmail, harass, defame, frame or discriminate against us.... This mistake pervades nearly every information privacy law." Regarding the massive caches of supposedly anonymous behavior surplus, Ohm calls them "databases of ruin."[34]

When it comes to location data, the situation is just as bad. In 2013 a group of MIT and Harvard computer scientists demonstrated that because individuals tend to have idiosyncratic mobility signatures, any analyst with the right tools can easily extract the mobility pattern of a specific individual within a large anonymized data set of location meta-data. Another research team demonstrated that data collected by seemingly "innocuous" smartphone-embedded sensors, such as accelerometers, gyroscopes, and magnetometers, can be used to infer "an ever-growing range of human activities and even moods." Their work also shows that these sensor data can be used "to obtain sensitive information about specific users from anonymized datasets."[35]

Companies are putting these surveillance capabilities to work. Broadcom produced a "global navigation satellite system" in a chip that combines satellite communications with the sensors in your cell phone to create a "positioning engine" that can find your location even if you are not connected to a network, including your location in a building, how many steps you've taken, in what direction, at what altitude. All this depends solely on one factor, says a company vice president: "the device in your hand."[36] Princeton computer scientists Arvind Narayanan and Edward Felten summed it up this way: "There is no known effective method to anonymize location data, and no evidence that it's meaningfully achievable."[37]

Even without "de-anonymization," location meta-data constitute an unrivaled concentration of knowledge within private firms and an extraordinary advantage in the division of learning. In 2016 Chinese search engine Baidu, often referred to as the Google of China, announced that its "Big Data Lab" uses location data from its 600 million users to track and predict the dynamics of the Chinese economy. The company built an "employment index" for the national economy as well as a "consumption index." It also touted its ability to generate quite-specific predictions such as Apple's second-quarter earnings in China that year. "To the best of our knowledge," Baidu researchers wrote, "we are the first to measure the second-largest economy by mining such unprecedentedly large scale and fine granular spatial-temporal data."[38]

As powerful as location data are, wearable technologies and their applications are another significant proving ground in the act of body rendition.[39] One 2017 report describes a new generation of wearables "armed with more sensors and smarter algorithms... focused on biometric monitoring... and... body parts as conduits for data collection...." These complex sensors can access "environmental context... smells... emotional state...."[40] Google has developed internet-enabled fabrics, claiming that it aims to bring inductive yarns to every garment and fabric on Earth. "If you can weave the sensor into the textile, as a material," explains project leader Ivan Poupyrev, "you're moving away from the electronics. You're making the basic materials of the world around us interactive." A partnership with Levi Strauss has already yielded "interactive denim," including a jacket first brought to market in September 2017. The material is described as able to "infer behavior" in order to be "interactive yet authentic."[41] The jacket contains sensors that can "see" through the fabric to detect and decipher gestures as subtle as the twitch of your finger.

There is a numbing repetition of MacKay's themes throughout the literature on wearables. Just as he insisted that telemetric devices must operate "outside the awareness" of "unrestrained animals," today's developers stress that wearables must be "unobtrusive" to avoid raising alarm. They are to be "continuous," "pervasive," and, crucially, "low cost" in order to achieve economies of scope.[42] The digital marketing firm Ovum forecasts 650 million wearables by 2020, nearly double the number used in 2016, and its research suggests that growth is largely driven by the lure of surveillance revenues. Mobile advertisers, they report, see wearables as "a source of very granular data insights and also new types of behavioral and usage data. Wearables of the future will have the ability to capture a wide array of data related to a user's contextual activity, health and emotional state. This information can be used to enhance and tailor both products and marketing messages to a very high degree...."[43]

Health care is an especially active proving ground for wearable sensored technologies, a development that's particularly pernicious given the more-innocent origins of this idea. When telemetry first shifted from MacKay's gaggles, flocks, and herds to the human animal, one of its first applications was as a means to oversee the vulnerable in the form of push-button pendants for elderly people alone in their homes. In 2002, the year that a still-secret surveillance capitalism achieved its first breakthroughs, a review of "wireless telemedicine" stressed the value of home monitoring for the elderly and for the expansion of health services in remote areas. As in the case of the Aware Home, a diagram of the proposed digital architecture for such home-monitoring services features only three parties: a closed loop that exclusively links a patient at home, her hospital's servers, and her physician.[44] There are no extra parties imagined in either of these designs, no companies capturing your behavior, no behemoth tech firms with their porous platforms and proprietary servers transforming your life into surplus so they can make book on what you will want next and enable their customers to sell it to you first.

Before the birth and spread of surveillance capitalism, it was possible to imagine digital representations of your body as an enrichment of the intimate relationships between a patient and a trusted doctor, a mother and her child, elderly parents and their adult children. As surveillance capitalism overwhelms the digital milieu, that vision has been made ridiculous. Both the Aware Home and the telemedicine design assume that all behavioral data are reinvested in service to the human being who is the subject of these arrangements, providing serenity, trust, and dignity: a chance for real knowledge and empowerment.

Many articles on health monitoring continue to emphasize its utility for the elderly, but the conversation has decisively moved on from this earlier state of grace. Some researchers anticipate the fusion of "smart cities" and what's now called "m-health" to produce "smart health," defined as "the provision of health services by using the context-aware network and sensing infrastructure of smart cities."[45] Toward that end, there are now reliable sensors for rendering an increasing range of physiological processes as behavioral data, including body temperature, heart rate, brain activity, muscle motion, blood pressure, sweat rate, energy expenditure, and body and limb motion. There are sensors that can render audio, visual, and physiological data during postsurgical patient recovery and rehabilitation. A flexible, sensored textile patch has been developed that can render breathing, hand movements, swallowing, and walking as behavioral data. In other applications, "wearable micromachined sensors" provide "accurate biomechanical analysis" as you walk or run, and a "body area network" records and analyzes walking and running "under extreme conditions."[46]

It is eloquent testimony to the health care system's failure to serve the needs of second-modernity individuals that we now access health data and advice from our phones while these pocket computers aggressively access us. M-health has triggered an explosion of rendition and behavioral surplus capture as individuals turn in record numbers to their fitness bands and diet apps for support and guidance.[47] By 2016, there were more than 100,000 mobile health apps available on the Google Android and Apple iOS platforms, double the number in 2014.[48] These rich data can no longer be imagined as cloistered within the intimate closed loops between a patient and her doctor or between an application and its dieters or runners. That bucolic vision has its holdouts, to be sure, but for surveillance capitalists this vision is but a faded daguerreotype.

In the US, most health and fitness applications are not subject to health privacy laws, and the laws that do exist do not adequately take into account either new digital capabilities or the ferocity of surveillance capitalist operations. Companies are expected to self-regulate by following guidelines suggested by the Federal Trade Commission (FTC) and other government agencies. For example, in 2016 the FTC issued a list of best practices for developers of mobile health applications aimed at increasing transparency, privacy, and security. Among these suggestions, developers are encouraged to "make sure your app doesn't access consumer information it doesn't need," "let consumers select particular contacts, rather than having your app request access to all user contacts through the standard API," and let users "choose privacy-protective default settings." That year the Food and Drug Administration announced that it would also *not* seek to regulate health and fitness apps, citing their "low-level risk." Instead, the agency released its own set of voluntary guidelines for software developers.[49]

The agencies' well-meaning guidelines overlook the inconvenient truth that transparency and privacy represent friction for surveillance capitalists in much the same way that improving working conditions, rejecting child labor, or shortening the working day represented friction for the early industrial capitalists. It took targeted laws to change working conditions back then, not suggestions. Then as now, the problems to which these pleas for self-restraint are addressed cannot be understood as excesses, mistakes, oversights, or lapses of judgment. They are necessitated by the reigning logic of accumulation and its relentless economic imperatives.

A legal review of mobile health apps concludes that most of them "take the consumers' private information and data without the consumers' permission and...do not generally disclose to the user that this information will be sent to advertising companies." These conclusions are borne out by a long queue of studies,[50] but let's focus on a 2016 in-depth investigation by scholars from the Munk School of Global Affairs at the University of Toronto in association with Open Effect, a nonprofit focused on digital privacy and security. This study looked at the collection, processing, and usage activities associated with nine fitness trackers.[51] Seven were chosen for their popularity, one was made by a Canadian company, and the ninth was an app that specialized in women's health. All but two apps transmitted every logged fitness event to the company's servers, which enabled backup and sharing with one's friends but also "data analytics" and distribution to third parties. Some of the trackers transmitted device identification numbers; others passively and continuously transmitted the user's precise longitude and latitude coordinates. These identifiers "could link fitness and biographical data to a single mobile phone hardware, or single specific fitness wearable...." None of this sensitive information was necessary for the tracker to operate effectively, and most of the privacy policies were opaque at best and allowed data to be "sold or exchanged with third parties." As we know, once a third party captures your surplus, it is shared with other third parties, who share with other third parties, and so on.

The team also examined the trackers' transmission of the Bluetooth Media Access Controller or "MAC" address that is unique to each phone. When this address is publicly discoverable, any third party with an interest in your movements—retailers who want to know your mall activity, insurers concerned about your compliance with an exercise regime—can "persistently" track your phone. Multiple data sets logged over time can be combined to form a fine-grained picture of your movements, enabling targeted applications and heightening the probability of guaranteed outcomes. The only real protection is when an app randomly but regularly generates a new MAC address for your phone, but of the nine trackers, only Apple's performed this operation.

The report also identifies a general pattern of careless security as well as the ability to generate false data. The researchers observed that consumers are likely to be misled and confused, overestimating the extent of security measures and underestimating "the breadth of personal data collected by fitness tracking companies." As they concluded, "We discovered severe security vulnerabilities, incredibly sensitive geolocation transmissions that serve no apparent benefit to the end user, and…policies leaving the door open for the sale of users' fitness data to third parties without express consent of the users."

If you are inclined to dismiss this report because fitness trackers can be written off as toys, let's consider a look at an incisive investigation into Android-based diabetes apps in a 2016 *Journal of American Medicine* research report and, with it, ample illustration of the frenzy of body rendition. The researchers note that although the FDA approved the prescription of a range of apps that transmit sensitive health data, the behind-the-scenes practices of these apps are "*understudied.*" They examined 211 diabetes apps and randomly sampled 65 of them for close analysis of data-transmission practices.[52]

Among these apps, merely downloading the software automatically "authorized collection and modification of sensitive information." The researchers identified a great deal of backstage action, including apps that modify or delete your information (64 percent), read your phone status and identity (31 percent), gather location data (27 percent), view your Wi-Fi connections (12 percent), and activate your camera in order to access your photos and videos (11 percent). Between 4 percent and 6 percent of the apps went even further: reading your contact lists, calling phone numbers found in your device, modifying your contacts, reading your call log, and activating your microphone to record your speech.

Finally, the research team unearthed an even darker secret: privacy policies do not matter. Of the 211 apps in the group, 81 percent did not have privacy policies, but for those that did, "not all of the provisions actually protected privacy." Of those apps *without* privacy policies, 76 percent shared sensitive information with third parties, and of those *with* privacy policies, 79 percent shared data while only about half admitted doing so in their published disclosures. In other words, privacy policies are more aptly referred to as *surveillance policies,* and that is what I suggest we call them.

There are many new territories of body rendition: organs, blood, eyes, brain waves, faces, gait, posture. Each of these expresses the same patterns and purpose that we have seen here. The surveillance capitalists relentlessly fight any attempts to constrain rendition. The ferocity with which they claim their "right to rendition" out of thin air is ample evidence of its foundational importance in the pursuit of surveillance revenues.

This ferocity is well illustrated in surveillance capitalists' determination to discourage, eliminate, or weaken any laws governing the rendition of biometric information, especially facial recognition. Because there is no federal law in the US that regulates facial recognition, these battles occur at the state level. Currently, the Illinois Biometric Privacy Act offers the most comprehensive legal protections, requiring companies to obtain written consent before collecting biometric information from any individual and, among other stipulations, granting individuals the right to sue a company for unauthorized rendition.[53]

The Center for Public Integrity, along with journalists, privacy advocates, and legal scholars, has documented the active opposition of surveillance capitalists to the Illinois law and similar legislative proposals in other states. With its unique competitive advantages in facial recognition, Facebook is considered the most uncompromising of all the tech companies when it comes to biometric data, described as "working feverishly to prevent other states from enacting a law like the one in Illinois."[54]

Facebook's considerable political muscle had been cultivated in just a few years as it learned to emulate Google's playbook of political and cultural fortifications. The company's founder, Mark Zuckerberg, demonstrated an iron determination to preserve his freedom in lawless space, pushing the boundaries of existing regulations and vigorously opposing even the whisper of new law. Between 2009 and 2017, the company increased its lobbying spend by a factor of fifty, building "a massive lobbying entourage of Washington power brokers." Facebook's $4.6 million in donations during the 2016 election cycle complemented its lobbying budget of $11.5 million in 2017.[55]

Zuckerberg's advantages in biometrics are significant. In 2017 Facebook boasted two billion monthly users uploading 350 million photos every day, a supply operation that the corporation's own researchers refer to as "practically infinite."[56] In 2018 a Facebook research team announced that it had "closed the gap" and was now able to recognize faces "in the wild" with 97.35 percent accuracy, "closely approaching human-level performance." The report highlights the corporation's supply and manufacturing advantages, especially the use of "deep learning" based on "large training sets."[57] Facebook announced its eagerness to use facial recognition as a means to more powerful ad targeting, but even more of the uplift would come from the immense machine training opportunities represented by so many photos. By 2018, its machines were learning to discern activities, interests, mood, gaze, clothing, gait, hair, body type, and posture.[58] The marketing possibilities are infinite.

It should not surprise any student of the prediction imperative that with these advantages in hand, Facebook is unwilling to accept anything less than total conquest in its bid to render faces for the sake of more-lucrative prediction products. So far, Facebook and its brethren have been successful, turning back legislative proposals in Montana, New Hampshire, Connecticut, and Alaska, and fatally weakening a bill that was passed in Washington state. Among the tech companies, only Facebook continued to oppose even the diminished terms of the Washington legislation.[59]

If rendition is interrupted, surveillance capitalism cannot stand, for the whole enterprise rests on this original sin. This fact is amply displayed in the public drama surrounding the ill-fated 2015 attempt to produce public guidelines on the creation and use of biometric information through a voluntary "privacy multi-stakeholder" process convened by the National Telecommunications and Information Association (NTIA) under the auspices of the US Department of Commerce. After weeks of negotiations, consumer advocates walked out in protest over the hard-line position of the tech companies and their lobbyists on the single most pivotal issue: consent.

The companies insisted on their right to use facial-recognition systems to identify a "stranger on the street" without first obtaining the individual's consent. As one lobbyist in the talks told the press, "Everyone has the right to take photographs in public...if someone wants to apply facial recognition, should they really need to get consent in advance?" Privacy scholars were quick to respond that there is no lawfully established right to such actions, let alone a First Amendment right.[60] Nobody reckoned with the fact that the prediction imperative makes individual ignorance the preferred condition for rendition operations, just as Arendt had observed and MacKay had prescribed for animals in the wild. Original sin prefers the dark.

The talks continued without the advocates, and in 2016 the NTIA issued its "Privacy Best Practice Recommendations for Commercial Facial Recognition Use." The guidelines should be understood as the "best" for the surveillance capitalists but as the "worst" for everyone else. In the language of these guidelines, the tech companies, retailers, and others determined to chase surveillance revenues are simply "encouraged" to make their policies on facial recognition "available to consumers, in a reasonable manner...." Where companies impose facial recognition on a physical location, they are "encouraged" to provide "notice" to consumers.[61] Rendition operations are tacitly accorded legitimacy, not only in the lack of contest, but because they stand as the immovable facts draped in the cheap garlands of toothless "best practices." Georgetown University legal scholar Alvaro Bedoya, a member of the advocacy group that quit the deliberations, blasted the recommendations as "a mockery of the Fair Information Practice Principles on which they claim to be grounded"; they offer "no real protection for individuals" and "cannot be taken seriously."[62]

Under the regime of surveillance capitalism, individuals do not render their experience out of choice or obligation but rather out of ignorance and the dictatorship of no alternatives. The ubiquitous apparatus operates through coercion and stealth. Our advance into life necessarily takes us through the digital, where involuntary rendition has become an inescapable fact. We are left with few rights to know, or to decide who knows, or to decide who decides. This abnormal division of learning is created and sustained by secret fiat, implemented by invisible methods, and directed by companies bent to the economic imperatives of a strange new market form. Surveillance capitalists impose their will backstage, while the actors perform the stylized lullabies of disclosure and agreement for the public.

The prediction imperative transforms the things that we have into things that have us in order that it might render the range and richness of our world, our homes, and our bodies as behaving objects for its calculations and fabrications on the path to profit. The chronicles of rendition do not end here, however. Act II requires a journey from our living rooms and streets to another world below the surface, where inner life unfolds.

CHAPTER NINE

RENDITION FROM
THE DEPTHS

I couldn't feel, so I tried to touch...

—LEONARD COHEN
"HALLELUJAH"

I. Personalization as Conquest

Microsoft CEO Satya Nadella introduced Cortana, the corporation's "personal digital assistant," at the firm's annual Ignite conference in 2016:

> This new category of the personal digital assistant is a runtime, a new interface. It can take text input. It can take speech input. It knows you deeply. It knows your context, your family, your work. It knows the world. It is unbounded. In other words, it's about you; it's not about any one device. It goes wherever you go. It's available on any phone—iOS, Android, Windows—doesn't matter. It is available across all of the applications that you will use in your life.[1]

This is a new frontier of behavioral surplus where the dark data continent of your inner life—your intentions and motives, meanings and needs, preferences and desires, moods and emotions, personality and disposition, truth telling or deceit—is summoned into the light for others' profit. The point is not to cure but to render all of it as immeasurably tiny bits of behavior available for calculation so that each can take its place on the assembly line that moves from raw materials to product development, manufacturing, and sales.

The machine invasion of human depth is prosecuted under the banner of "personalization," a slogan that betrays the zest and cynicism brought to the grimy challenge of exploiting second-modernity needs and insecurities for outsize gain. From the point of view of the prediction imperative, personalization is a means of "individualizing" supply operations in order to secure a continuous flow of behavioral surplus from the depths. This process can be accomplished successfully only in the presence of our unrelenting hunger for recognition, appreciation, and most of all, support.

Recall that Hal Varian, Google's chief economist, helped chart this course. "Personalization and customization" are the third "new use" of computer-mediated transactions. Instead of having to ask Google questions, it should "know what you want and tell you before you ask the question." Google Now, the corporation's first digital assistant, was charged with this task. Varian warned that people would have to give Google even more of themselves in order to reap the value of the application: "Google Now has to know a lot about you and your environment to provide these services. This worries some people." He rationalizes any concern, arguing that rendering personal information to Google is no different from sharing intimacies with doctors, lawyers, and accountants. "Why am I willing to share all this private information?" he asks. "Because I get something in return.... These digital assistants will be so useful that everyone will want one." Varian is confident that the needs of second-modernity individuals will subvert any resistance to the rendition of personal experience as the quid pro quo for the promise of a less stressful and more effective life.[2]

In fact, Varian's notion of personalization is the precise opposite of the relationships with trusted professionals to which he refers. Doctors, accountants, and attorneys are held to account by mutual dependencies and reciprocities dictated by the extensive institutionalization of professional education, codes of conduct, and procedures for evaluation and review. Violation of these rules risks punishment in the form of professional sanction and public law. Google and its brethren in surveillance capitalism bear no such risks.

Varian's remarks constitute one of those rare occasions in which the fog of technology rhetoric parts just enough to discern the utilities of social and economic inequality to surveillance capitalism's larger mission. Varian reasons that inequality offers an opportunity to raise the ante on Google's quid pro quo for effective life. He counsels that the way to predict the future is to observe what rich people have because that's also what the middle class and the poor will want. "What do rich people have now?" he asks rhetorically. "Personal assistants."

That the luxuries of one generation or class become the necessities of the next has been fundamental to the evolution of capitalism during the last five hundred years. Historians describe the "consumer boom" that ignited the first industrial revolution in late-eighteenth-century Britain, when, thanks to visionaries like Josiah Wedgewood and the innovations of the early modern factory, families new to the middle class began to buy the china, furniture, and textiles that only the rich had previously enjoyed. This new "propensity to consume" is considered "unprecedented in the depth to which it penetrated the lower reaches of society...."[3] In 1767 the political economist Nathaniel Forster worried that "fashionable luxury" was spreading "like a contagion," and he complained of the "perpetual restless ambition in each of the inferior ranks to raise themselves to the level of those immediately above them."[4] Adam Smith wrote insightfully on this social process, noting that upper-class luxuries can in time be recast as "necessaries." This occurs as "the established rules of decency" change to reflect new customs introduced by elites, triggering lower-cost production methods that transform what was once unattainable into newly affordable goods and services.[5] Ford's Model T is the outstanding twentieth-century example of this progression.

Varian casts personalization as a twenty-first-century equivalent of these historical dynamics, the new "necessaries" for the harried masses bent under the weight of stagnant wages, dual-career obligations, indifferent corporations, and austerity's hollowed-out public institutions. Varian's bet is that the digital assistant will be so vital a resource in the struggle for effective life that ordinary people will accede to its substantial forfeitures. "There is no putting the genie back in the bottle," Varian the inevitabilist insists. "Everyone will expect to be tracked and monitored, since the advantages, in terms of convenience, safety, and services, will be so great...continuous monitoring will be the norm."[6] Everyone, that is, except those wealthy or stubborn enough to achieve effective life without Google's assistance and thus escape the worst excesses of rendition. As decision rights and self-determination become privileges of the wealthy, what will Varian offer to answer those who clamor for the same?

Historically, breakthroughs in lower-cost goods and services unleashed expansions of production and employment, higher wages, and an improved standard of living for many. Varian has no such reciprocities in mind. Instead, he pokes his finger into the open wound of second-modernity insecurities and bends our pain to the objectives of the surveillance project. With Varian, the hunger for new necessities is construed as an opportunity to dispossess, even as it conveniently provides the justification for that dispossession right down to the depths.

Google Now was a first step, although later it would look more like a stalking horse and habituation exercise paving the way for what was to come. Christened "predictive search," it combined every system that Google had ever built, including the corporation's achievements in voice search and neural networking, its knowledge of the world represented in its one-billion-entity "knowledge graph," and its unparalleled machine intelligence capabilities. All of this firepower was amassed in order to learn from your content, context, and behavior not only through search, e-mail, and calendar activity but also from the data in your phone, including movement, location, activities, voice, and apps. This time, the aim was not only to sell ads but rather "to guess the information you'll need at any given moment" as you move through the real world.[7]

As a promotional video crows, "Google Now is always one step ahead so you can feel more confident as you navigate your day...with the predictive power of Now, you get just what you need to know, right when you need it." One writer described the new service as "having the search engine come to you."[8] The app's information cards swim into view on your phone's home screen in anticipation of your needs: notification of a changed flight time, impending weather and traffic, nearby restaurants and shops, that museum you've been wanting to visit. One Google executive reasoned that Google already knows all of this about you, so it might as well turn it into a service that can provide the company with access to even more information: "Google's going to know when my flight is, whether my package has gotten here yet, and where my wife is and how long it's going to take her to get home this afternoon.... Of course Google knows that stuff."[9] Google Now's predictive capabilities follow the pattern we have seen throughout: they derive from machine processes trained on unceasing flows of virtual and real-world behavior. Why did Google devote so much machine power and valuable surplus in order to thoughtfully assist you through your day? The reason is that Google Now signaled a new breed of prediction products.

Google's breakthrough crawler enabled the lightning-fast indexing of the world wide web, the apparatus of ubiquity then enabled new operations to crawl reality, and now in this third phase, distinct supply operations are required to crawl our lives. In Google Now one sees an initial foray into this new space, in which the web crawler's ability to find information combines with new *life-crawling* operations intended to render, anticipate, and, as we shall see, ultimately modify your behavior. Online and offline behavioral surplus—your e-mail content, where you went this afternoon, what you said, what you did, how you felt—are combined into prediction products that can serve an emerging marketplace in which every aspect of your daily reality is up for bid.

Facebook's "M," launched in 2015 as part of its Messenger application, is another example of this new phase. It was introduced as a "personal digital assistant...that completes tasks and finds information on your behalf... powered by artificial intelligence that's trained and supervised by people."[10] Facebook's vice president in charge of messaging products described the company's goals for M by saying, "We start capturing all of your intent from the things you want to do. Intent often leads to buying something, or to a transaction and that's an opportunity for us to [make money] over time." Most importantly, the VP stressed, "M learns from human behaviors."[11] The corporation's machines would be trained on surplus from Messenger's 700 million daily users. Eventually, it was hoped, M's operations would be fully automated and would not need human trainers.

By 2017, Facebook had scaled back its machine intelligence ambitions and focused its personal assistant on the core mission: commerce. "The team in there now is finding ways to activate commercial intent inside Messenger," a Facebook executive reported.[12] The idea is to "prioritize commerce-driven experiences" and design new ways for users to "quickly buy things" without the tedium of entering credit card information, flipping pages, or opening applications. Pop-up buttons appear during your conversations with friends whenever the system detects a possible "commercial intention." Just tap to order, buy, or book, and let the system do the rest.[13]

In this way the "personal digital assistant" is revealed as a market avatar, another Trojan horse in which the determination to render and monetize your life is secreted under the veil of "assistance" and embellished with the poetry of "personalization." Its friendly recommendations, advice, and eagerness to act on your behalf barely conceal an aggressive new market cosmos hovering over any and every aspect of your daily life. It may be composed of restaurants, banks, plumbers, merchants, ticket sellers, airlines, and a limitless queue of possible strangers summoned by their interests in your behavior: now, soon, and later. They are standing by to cash in on your walk to work, your conversation with your teenager, or your aging running shoes. A digital assistant may derive its character from your inclinations and preferences, but it will be skewed and disfigured in unknown measure by the hidden market methods and contests that it conceals.

Google joined other tech companies determined to establish "conversation" as the medium through which humans engage with the apparatus. In time, the obsession with voice may be surpassed or joined by others so that merely thinking a thought or waving a finger can translate into and initiate action. For now, there are compelling reasons for the race to the spoken word. The first is obvious: reliable voice recognition can translate a sprawling landscape of service interactions into low-cost automated processes of theoretically unlimited scale and scope, a fact that has been noted by labor economists for some time.[14] The competitive race between a new crop of "personal digital assistants" is best understood from this point of view. The voice that rises to dominance, the One Voice, will be the colossus of behavioral surplus pipelines with a potentially insurmountable competitive advantage in its ability to corner and kidnap the dominant share of human experience.

"Conversation" stands alone in its promise to dominate raw-material supply, and the rewards to the One Voice would be astronomical. Casual talk helps to blur the boundaries between "it"—the apparatus saturated with commercial agents—and us. In conversation we imagine friendship. The more we fancy the apparatus as our confidante, nanny, governess, and support system—a disembodied, pervasive "Mrs. Doubtfire" for each person—the more experience we allow it to render, and the richer its supply operations grow. Communication is the first human joy, and a conversational interface is prized for the frictionless ease in which a mere utterance can trigger action, especially market action: "Let there be light." "Let there be new running shoes." What could be dreamier than to speak and have it be so? An Amazon senior vice president comments on the company's voice-activated home devices: "The nice thing about the Amazon device business is that when we sell a device, generally people buy more blue jeans. And little black dresses. And shoes. And so that's good." "Voice shopping," he concludes, is good for business and good for predicting business.[15]

In a conversation addressed to a digital thing, as opposed to a conversation in a shop, words can occur on the fly with less friction and effort; less inhibition, fretting, and comparing; less concern about the limits of one's bank account or where a product or service is sourced; less doubt and hesitation; less memory and remorse. The speaker feels herself at the center of a seamlessly flowing universe. The seams are all backstage, where the machines confront and conquer stubborn sources of friction such as distinct apps and entities; recalcitrant administrative service, distribution, payments, and delivery systems; and boundaries and borders that threaten the flows of desire and satisfaction. Spontaneous and fluid, universally burbling "conversation" turns the new personal digital assistant into a voice that sits between your life and the new markets for your life, between your experience and the auctioning of your experience: "a runtime, a new interface" that creates the sensation of mastery while, in fact, giving it away.

In this commercial dreamscape, words that were once conceived of as "behind closed doors" are eagerly rendered as surplus. These new supply operations convert your talk into behavior for surplus in two ways. The first derives from *what* you say, the second from *how* you say it. Smart-home devices such as Amazon's Echo or Google Home render rivers of casual talk from which sophisticated content analyses produce enhanced predictions that "anticipate" your needs. Google used its 2016 developers conference to introduce its conversational reimagining of Google Now, rechristened "Assistant" and integrated across the company's devices, services, tools, and applications. "We want users to have an ongoing, two-way dialogue with Google. We want to help you get things done in your real world and we want to do it for you," explained Google CEO Sundar Pichai. "For example, you can be in front of this structure in Chicago and ask Google, 'Who Designed This?' You don't need to say 'the bean' or 'the cloud gate.' We understand your context and we answer that the designer is Anish Kapoor."[16]

Google's Assistant arrived already integrated into Google's new messaging app, Allo, where it can search for information, execute tasks, or even compose routine messages on your behalf. Most importantly, Assistant animates the firm's home device, Google Home.[17] The idea is that in time, the device (or its successor) will claim for rendition a theoretically limitless scope of animate and inanimate domestic activities: conversations, lightbulbs, queries, schedules, movement, travel planning, heating systems, purchases, home security, health concerns, music, communication functions, and more.

There was a time when you searched Google, but now Google searches you. Advertisements for Google Home feature loving families leading busy, intricate lives but visibly relieved to return home and fall into the arms of this omniscient, efficient caretaker. This second-modernity dream come true extracts an unusually high tax for its promise of a more effective life. For each user to have his or her own individual Google, as Pichai envisions, Google must have each individual.[18]

The caretaker's ability to effectively serve you depends entirely upon the degree to which your life is rendered, knowingly or unknowingly, to its ministrations. The breadth and depth of rendered life correspond to the scale of market action that can be triggered and mediated by Assistant. There are differences among the various incarnations of "personalization" and "assistance" offered by the tech giants, but these are trivial compared with the collective urge toward total knowledge—about your inner states, real-world context, and specific daily life activities—all in the service of successfully training the machines that they might better target market operations to each moment of life.

All the potential market action associated with what you say depends upon voice activation, recognition, and response. These, in turn, are the product of highly sophisticated machine systems trained on vast global stockpiles of spoken words. The more structural insights the machines glean from spoken surplus, the more commerce flows from its content. This means that the value of what you say cannot be realized without machines that can learn from precisely *how* you say it. This form of surplus derives from the *structure* of your speech: vocabulary, pronunciation, intonation, cadence, inflection, dialect.

The competition for supplies of talk turns your phrases into this second form of surplus as companies determined to develop and perfect voice capabilities scour the world for speech. "Amazon, Apple, Microsoft and China's Baidu have embarked on a worldwide hunt for terabytes of human speech," reports *Bloomberg Businessweek.* "Microsoft has set up mock apartments in cities around the globe to record volunteers speaking in a home setting." The tech firms capture flows of talk from their smart devices and phones as they record and retain your words. Chinese search firm Baidu collects speech in every dialect: "Then they take all that data and use it to teach their computers how to parse, understand, and respond to commands and queries."[19]

Pieces of your talk are regularly farmed out in bulk to third-party firms that conduct "audio review processes" in which virtual scorers, tasked to evaluate the degree of match between the machine's text and the original chunk of human speech, review audio recordings retained from smartphones, messaging apps, and digital assistants. Companies such as Amazon, Google, and Microsoft use these audio analyses to improve the algorithms of their voice systems. The tech companies insist that such recordings are anonymous, nothing more than voices without identities. "Partners do not have the ability to tie the voice samples back to specific people," one Microsoft executive asserted. But one journalist who signed on to a virtual job as an audio recording analyst concluded just the opposite, as she listened to recordings full of pathos, intimacy, and easily identifiable personal information:

> Within the recordings themselves, users willingly surrender personal information—information that is especially valuable in these review processes because they are so specific. Uncommon names, difficult-to-pronounce cities and towns, hyperlocal oddities.... I heard people share their full names to initiate a call or offer up location-sensitive information while scheduling a doctor's appointment... the recordings capture people saying things they'd never want heard, regardless of anonymity.... There isn't much to keep people who are listening to these recordings from sharing them.[20]

There is also substantial capital investment directed at talk, and Samsung's Smart TV illustrates some of this behind-the-scenes action. Business forecasts routinely predict strong growth in the market for internet-enabled appliances, and Samsung is among a small group of market leaders. Its appliances use the Android operating system platform, and early on the firm established alliances with both the Alphabet/Google subsidiary Nest and with Cisco. "Our first mission is to bring your home to your connected life," a top executive explained in 2014.[21] In 2015 privacy advocates discovered that the corporation's smart TVs were actually too smart, recording everything said in the vicinity of the TV—*please pass the salt; we're out of laundry detergent; I'm pregnant; let's buy a new car; we're going to the movies now; I have a rare disease; she wants a divorce; he needs a new lunch box; do you love me?*—and sending all that talk to be transcribed by another market leader in voice-recognition systems, Nuance Communications.[22]

The TV's "surveillance policy"—yes, even a TV has a *surveillance policy* now—reveals the layers of surveillance effort and commercial interest that operate outside of awareness in our homes. Samsung acknowledges that the voice commands aimed at triggering the TV's voice-recognition capabilities are sent to a third party and adds, "Please be aware that if your spoken words include personal or other sensitive information, that information will be among the data captured and transmitted to a third party through your use of Voice Recognition."[23] Samsung disclaims responsibility for the policies of third-party firms, as nearly all surveillance policies do, including the one that actually collects and translates the talk of its unsuspecting customers. Samsung advises that "You should exercise caution and review the privacy statements applicable to the third-party websites and services you use."[24] The intrepid consumer determined to study these documents will find no succor in Nuance's privacy policy, only the same catechism found in Samsung's and that of nearly every company. It also encourages you to read the privacy policies of the companies to whom it's selling your conversations, and so it goes: a forced march toward madness or surrender.[25]

In California, at least, the legislature passed a law prohibiting connected TVs from collecting voice data without "prominently informing" customers and further outlawed the use of such data for third-party advertising.[26] As we know from our examination of the dispossession cycle, however, the economic imperatives that drive surveillance capitalists to capture behavioral surplus are not easily deterred. By 2016, Samsung had doubled down on its smart-TV–based secret rendition and behavioral surplus supply chains, positioning its new models as the hub of a new "Samsung SmartThings smart-home ecosystem in an open platform that supports thousands of devices," including fans, lights, thermostats, security cameras, and locks—and all with the aid of a single universal remote able to capture your every spoken command.[27]

In 2017 the FTC reached a $2.2 million settlement of a complaint initiated by the Office of the New Jersey Attorney General against Vizio, one of the world's largest manufacturers and distributors of internet-enabled smart televisions. Vizio's supply operations appear to be even more aggressive than Samsung's. Investigators discovered that "on a second-by-second basis, Vizio collected a selection of pixels on the screen that it matched to a database of TV, movie, and commercial content." The company then identified additional viewing data "from cable or broadband service providers, set-top boxes, streaming devices, DVD players, and over-the-air broadcasts." All of this amounted to as much as 100 billion data points each day just from the 11 million TVs identified in the settlement.[28] Vizio disguised its supply operations behind a setting called "Smart Interactivity," described to consumers as a feature that "enables program offers and suggestions" without any indication of its actual functions.

In an unusually vivid blog post, the FTC describes Vizio's direct sales of this behavioral surplus:

> Vizio then turned that mountain of data into cash by selling consumers' viewing histories to advertisers and others. And let's be clear: We're not talking about summary information about national viewing trends. According to the complaint, Vizio got personal. The company provided consumers' IP addresses to data aggregators, who then matched the address with an individual consumer or household. Vizio's contracts with third parties prohibited the re-identification of consumers and households by name, but allowed a host of other personal details—for example, sex, age, income, marital status, household size, education, and home ownership. And Vizio permitted these companies to track and target its consumers across devices.[29]

A concurring statement from FTC Acting Chairwoman Maureen K. Ohlhausen emphasized that the settlement broke new ground in its allegation

that "individualized television viewing activity falls within the definition of sensitive information" that merits protection by the FTC.[30] This
finger in the dike would not hold back the tidal wave of similar incursions as the prediction imperative cracks the whip to drive the hunt for
unexplored pieces of talk from daily life. Rendition takes command of even
the most benign supply sources, such as toys, which have now become
"toys that spy." A new breed of interactive dolls and toy robots, including
a girl doll called "My Friend Cayla," turn out to be supply hubs for under-
age behavioral surplus, subjecting young children and their parents' smartphones "to ongoing surveillance... without any meaningful data protection
standards."[31]

The popular playthings, marketed by Genesis Toys, are bundled with a
mobile application that, once downloaded to a smartphone, "provides the
data processing" to enable the toy's ability to capture and understand whatever the child says.[32] Along the way, the app accesses most of the phone's
functions, including many that are irrelevant to the toy's operations such as
contact lists and the camera. The app establishes a Bluetooth connection that
links the toy to the internet, and it records and uploads conversations as the
toy actively engages the child in discussion. One doll targeted in the complaint
systematically prompts children to submit a range of personal information, including where they live.

The child's conversations are translated into text by third-party voice-
recognition software from, once again, Nuance Communications, and that
information is used to retrieve answers to the child's questions using Google Search and other web sources. Researchers discovered that the audio files
of children's talk (Nuance calls them "dialogue chunks") are uploaded to the
company's servers, where they are analyzed and stored.[33] As you might expect, those dialogue chunks continue their journey as behavioral surplus, in
much the same way as Samsung's TV-captured audio, to be sold again and
again "for other services and products," as the Genesis terms-of-service
agreement indicates.

Meanwhile, Mattel, one of the world's largest toy companies, was gaining ground with its innovations in interactive, internet-enabled, machine-
intelligence–powered toys, led by the new conversational Barbie Doll and its
Barbie Dream House.[34] The voice-activated smart dollhouse could respond to

more than one hundred commands, such as "lower the elevator" and "turn on the disco ball," a new kind of habituation exercise aimed at normalizing ubiquity in intimate spaces. "Barbie's New Smart Home Is Crushing It So Hard," exulted *Wired*. "Barbie's ultimate crib is voice controlled.... It's what a real smart home should be: Straight crushing it with universal voice control instead of a hodgepodge of disparate appliances hogging up app space on your phone.... The future is served."[35]

In this future, children learn the principles of the One Voice—a run time, a new interface. It is available everywhere to execute their commands, anticipate their desires, and shape their possibilities. The omnipresence of the One Voice, with its fractious, eager marketplace-of-you concealed under its skirts, changes many things. Intimacy as we have known it is compromised, if not eliminated. Solitude is deleted. The children will learn first that there are no boundaries between self and market. Later they will wonder how it could ever have been different.

When Mattel hired a new CEO in January 2017, it's no surprise that the company looked to Google, selecting its president for the Americas with responsibility for Google's commercial and advertising sales operations.[36] Most analysts agreed that the appointment heralded Mattel's commitment to its innovations in internet-enabled toys and virtual reality, but the appointment underscores the shift in focus from making great products *for* you to collecting great data *about* you.

The doll that was once a beloved mirror of a child's unfettered imagination, along with all the other toys in the toy box—and the box, and the room that hosts the box, and the house that hosts the room—are all earmarked for rendition, calculation, connection, and profit. No longer mere things, they are reinvented as vehicles for a horde of commercial opportunities fabricated from our dialogue chunks and assorted gold dust.

In 2017 Germany's Federal Network Agency banned the Cayla doll as an illegal surveillance device and urged parents to destroy any such dolls in their possession. In the US the FTC had yet to take any action against the doll or Genesis Toys. Meanwhile, the connected dollhouse prepares our children and families for the connected room (a project that Mattel announced in January 2017 and then nine months later shelved amid an uproar from parents and privacy advocates), which paves the way for the connected home, whose pur-veyors hope will numb us to the connected world as we travel the path of ubiquity's manifest destiny and its promise of surveillance revenues.[37]

In pursuit of the *what* and the *how* of voice surplus, the logic of competi-tion is to corner as much supply as possible. The urge toward totality gener-ates competitive pressures to become *the* run time and *the* new interface: the dominant, if not exclusive, medium through which we access and engage the apparatus as it engages us. It's a race to corner all the talk as a prerequisite for achieving the privileged status of the One Voice, which will bestow upon the winner the ability to anticipate and monetize all the moments of all the peo-ple during all the days.

The messianic urge toward totality and supremacy is evident in the rheto-ric and strategies of key competitors in this race. Although Google, Microsoft, Amazon, and Samsung each have aspirations to dominate voice capture, it is Amazon, its machine learning assistant Alexa, and its expanding line of Echo hubs and Dot speakers that offer the most telling case here. Alexa appears to be a threshold event that will define Amazon not only as an aggressive capi-talist but also as a surveillance capitalist.[38]

Amazon aggressively opened Alexa to third-party developers in order to expand the assistant's range of "skills," such as reading a recipe or ordering a pizza. It also opened its platform to smart-home device makers from lighting systems to dishwashers, turning Alexa into a single voice for controlling your home systems and appliances. In 2015 Amazon announced that Alexa would be sold as a service, known as "Amazon Lex," enabling any company to inte-grate Alexa's brain into its products. Amazon Lex is described as "a service for building conversational interfaces into any application using voice and text.... Lex enables you to define entirely new categories of products."[39] As Alexa's se-nior vice president explained, "Our goal is to try to create a kind of open, neu-tral ecosystem for Alexa...and make it as pervasive as we possibly can."[40]

By 2018, Amazon had inked deals with home builders, installing its Dot speakers directly into ceilings throughout the house as well as Echo devices and Alexa-powered door locks, light switches, security systems, door bells, and thermostats. As one report put it, "Amazon can acquire more com-prehensive data on people's living habits...." The company wants to sell real-world services such as house cleaning, plumbing, and restaurant delivery, but according to some insiders, the vision is more far-reaching: an omniscient voice that knows all experience and anticipates all action.[41] Already, forward-looking Amazon patents include the development of a "voice-sniffer algorithm" integrated into any device and able to respond to hot words such as "bought," "dislike," or "love" with product and service offers.[42]

Amazon is on the hunt for behavioral surplus.[43] This explains why the company joined Apple and Google in the contest for your car's dashboard, forging alliances with Ford and BMW. "Shopping from the steering wheel" means hosting behavioral futures markets in your front seat. Alexa is ready with restaurant recommendations or advice on where to get your tires checked. "As pervasive as possible" explains why Amazon wants its Echo/Alexa device to also function as a home phone, able to make and receive calls; why it inked an agreement to install Echo in the nearly 5,000 rooms of the Wynn resort in Las Vegas; and why it is selling Alexa to call centers to automate the process of responding to live questions from customers by phone and text.[44] Each expansion of Alexa's territory increases the volume of voice surplus accrued to Amazon's servers and fed to Alexa.

The path to the coronation of the One Voice is not an easy one, and there are other travelers determined to elbow their way to the finish line. Google also wants its "personal assistant," Google Home, to double as your home phone. Samsung reappears as another contender with its acquisition of "Viv," a powerful new voice system designed by the original developers of Apple's personal assistant Siri, who were frustrated with the constraints of Apple's approach. Viv's lead developer explained that "you can get things done by talking to things…a marketplace that will become the next big area…."[45]

If life is a wild horse, then the digital assistant is one more means by which that horse is to be broken by rendition. Unruly life is brought to heel, rendered as behavioral data and reimagined as a territory for browsing, searching, knowing, and modifying. Just as surveillance capitalism transformed the web into a market onslaught fueled by the capture and analysis of behavioral surplus, so everyday life is set to become a mere canvas for the explosion of a new always-on market cosmos dedicated to our behavior and from which there is no escape.

II. *Rendition of the Self*

"We are used to face-to-face interaction where words disappear....I assumed that keyboard communication was like a letter or a phone call, but now I understand that it doesn't disappear. The myth is that electronic communication is invisible...."[46] He was a brilliant research scientist at a large pharmaceuticals company that I called "Drug Corp" in my 1988 book, *In the Age of the Smart Machine*.[47] I had spent several years visiting the research group there as they shifted much of their daily communications from face-to-face meetings to DIALOG, one of the world's first "computer conferencing" systems. DIALOG was a precursor to a technology that we now call "social media." The DIALOG platform created a new social space in which the scientific community at Drug Corp elaborated and extended "their networks of relationships, access to information, thoughtful dialogue, and social banter," as I described it then. They embraced DIALOG with great enthusiasm, but it did not end well. "In time," I wrote, "it became clear that they had also unwittingly exposed once evanescent and intangible aspects of their social exchange to an unprecedented degree of hierarchical scrutiny." The interviews that stretched across those years documented the scientists' gradual awakening to new dangers as dimensions of personal experience that were implicit and private suddenly became explicit and public in ways that they did not anticipate and with consequences that they lamented deeply.

Thanks to the new computer-mediated milieu, the scientists' social and professional conversations now appeared as an electronic text: visible, knowable, shareable. It enriched their work in many ways, but it also created unexpected vulnerabilities as dispositions, values, attitudes, and social interactions were recast as objects of scrutiny. In a series of conflicts that unfolded over the years, I watched as the managers and executives at Drug Corp were simply unable to quell their inclination to use the new social text as a means to evaluate, critique, and punish. On more than one occasion I learned of managers who had printed out pages of DIALOG conversations in order to spread them out on the floor and analyze opinions on a particular subject, sometimes actually taking scissors to the pages and organizing the entries by theme or person. In many cases these investigations were pure fact gathering, but in other instances managers wanted to identify the individuals who agreed with or opposed their directives.

The enduring witness of the text was adapted as a medium through which managers attempted "to control and channel what had always been the most ephemeral aspects of subordinates' behavior."[48] All the original excitement and promise melted into cynicism and anxiety as the scientists gradually withdrew from DIALOG, opting instead for a routine e-mail application and a preference for more-perfunctory, impersonal messages.

Decades later, the children and grandchildren of those scientists, along with most of us, communicate freely through our smartphones and social media, unaware that we are reliving the bitter lessons of Drug Corp but now at an entirely new level of rendition. The scientists were rattled to see their casual talk reified and converted into an object of hierarchical inspection. Now, the interiors of our lives—simplistically summarized as "personality" or "emotions"—are recast as raw material available to anyone who can make or buy a new generation of supply-chain accessories and the means of production to analyze this new genus of behavioral surplus and fabricate extremely lucrative prediction products.

"Personalization" is once again the euphemism that spearheads this generation of prediction products manufactured from the raw materials of the self. These innovations extend the logic of earlier iterations of dispossession: from web crawling to reality crawling to life crawling to *self crawling*. As has been the case in each iteration, insights and techniques once intended to illuminate and enrich quickly disappear into the magnetic field of the commercial surveillance project, only to reappear later as ever more cunning methods of supply, manufacture, and sales.

The two billion and counting Facebook users are the most poignant descendants of Drug Corp's scientists. Many of them joined Facebook to escape the pervasive hierarchical oversight of workplace communications that had become routine in the years since Drug Corp's first experiments. Facebook, they once thought, was "our place," as benign and taken for granted and as the old Ma Bell, a necessary utility for association, communication, and participation. Instead, Facebook became one of the most authoritative and threatening sources of predictive behavioral surplus from the depths. With a new generation of research tools it learned to plunder your "self" right through to your most intimate core. New supply operations can render as measurable behavior everything from the nuances of your personality to your sense of time, sexual orientation, intelligence, and scores of other personal characteristics. The corporation's immense machine intelligence capabilities transform these data into vivid prediction products.

The stage was set for the discovery of your Facebook profile's easy pillage in 2010, when a collaboration of German and US scholars came to the unexpected conclusion that Facebook profiles are not idealized self-portraits, as many had assumed. Instead, they found that the information on Facebook reflects the user's actual personality, as independently assessed by the well-validated protocols of the five-factor personality model and as compared to study participants' own descriptions of their "ideal self."[49]

There is compelling evidence to suggest that the unique dynamics of the Facebook milieu eventually complicated this picture of "real personality," as we shall explore in Chapter 16, but in 2011 these early findings encouraged three University of Maryland researchers to take the next logical step. They developed a method that relies on sophisticated analytics and machine intelligence to accurately predict a user's personality from publicly available information in his or her Facebook profile.[50]

In the course of this research, the team came to appreciate the magic of behavioral surplus, discovering, for example, that a person's disclosure of specific personal information such as religion or political affiliation contributes less to a robust personality analysis than the fact that the individual is willing to share such information in the first place. This insight alerted the team to a new genre of powerful behavioral metrics. Instead of analyzing the content of user lists, such as favorite TV shows, activities, and music, they learned that simple "meta-data"—such as the *amount* of information shared—"turned out to be much more useful and predictive than the original raw data." The computations produced on the strength of these behavioral metrics, when combined with automated linguistic analysis and internal Facebook statistics, led the research team to conclude that "we can predict a user's score for a personality trait to within just more than one-tenth of its actual value."[51] The University of Maryland team began what would become a multiyear journey toward the instrumentalization of data from the depths for the purposes of a highly intentional program of manipulation and behavioral modification. Although they could not see very far down that road, they nevertheless anticipated the utility of their findings for an eager audience of surveillance capitalists:

> With the ability to infer a user's personality, social media websites, e-commerce retailers, and even ad servers can be tailored to reflect the user's personality traits and present information such that users will be most receptive to it.... The presentation of Facebook ads could be adjusted based on the personality of the user.... Product reviews from authors with personality traits similar to the user could be highlighted to increase trust and perceived usefulness....[52]

The new capabilities also proved robust when applied to other sources of social media meta-data. Later that year, the Maryland team published findings that used publicly available Twitter data to predict scores on each of the five personality dimensions to within 11–18 percent of their actual value. Similar research findings would become central to the progress of rendering Facebook profiles as behavior for new caches of surplus from the depths.[53]

In the UK a team of researchers, including Cambridge University's Michal Kosinski and the deputy director of Cambridge's Psychometrics Centre, David Stillwell, built on this line of research.[54] Stillwell had already developed the myPersonality database, a "third-party" Facebook application that allows users to take psychometric tests, like those based on the five-factor model, and receive feedback on their results. Launched in 2007 and hosted at the Psychometrics Centre, by 2016 the database contained more than six million personality profiles complemented by four million individual Facebook profiles. Once regarded as a unique, if offbeat, source of psychological data, myPersonality had become the database of choice for the scoping, standardization, and validation of the new models capable of *predicting* personality values from ever-smaller samples of Facebook data and meta-data. Later, it would become the model for the work of a small consultancy called Cambridge Analytica, which used the new caches of behavioral surplus for an onslaught of politically inspired behavioral micro-targeting.

In a 2012 paper Kosinski and Stillwell concluded that "user personality can be easily and effectively predicted from public data" and warned that social media users are dangerously unaware of the vulnerabilities that follow their innocent but voluminous personal disclosures. Their discussion specifically cited Facebook CEO Mark Zuckerberg's unilateral upending of established privacy norms in 2010, when he famously announced that Facebook users no longer have an expectation of privacy. Zuckerberg had described the corporation's decision to unilaterally release users' personal information, declaring, "We decided that these would be the social norms now, and we just went for it."[55]

Despite their misgivings, the authors went on to suggest the relevance of their findings for "marketing," "user interface design," and recommender systems.[56] In 2013 another provocative study by Kosinski, Stillwell, and Microsoft's Thore Graepel revealed that Facebook "likes" could "automatically and accurately estimate a wide range of personal attributes that people would typically assume to be private," including sexual orientation, ethnicity, religious and political views, personality traits, intelligence, happiness, use of addictive substances, parental separation, age, and gender.[57]

The authors appeared increasingly ambivalent about the social implications of their work. On the one hand, they announced that these new predictive capabilities could be used to "improve numerous products and services." They concluded that online businesses can adjust their behavior to match each user's personality, with marketing and product recommendations psychologically tailored to each individual. But the authors also warned that automated prediction engines run by companies, governments, or Facebook itself can compute millions of profiles without individual consent or awareness, discovering facts "that an individual may not have intended to share." The researchers cautioned that "one can imagine situations in which such predictions, even if incorrect, could pose a threat to an individual's well-being, freedom, or even life."[58]

Despite these ethical quandaries, by 2015 Kosinski had moved to Stanford University (first to the Computer Science Department and then to the Graduate School of Business), where his research quickly attracted funding from the likes of Microsoft, Boeing, Google, the National Science Foundation, and the Defense Advanced Research Projects Agency (DARPA).[59] Kosinksi and a variety of collaborators, often including Stillwell, went on to publish a succession of articles that elaborated and extended the capabilities demonstrated in the early papers, refining procedures that "quickly and cheaply assess large groups of participants with minimal burden."[60]

A paper published in 2015 broke fresh ground again by announcing that the accuracy of the team's computer predictions had equaled or outpaced that of human judges, both in the use of Facebook "likes" to assess personality traits based on the five-factor model and to predict "life outcomes" such as "life satisfaction," "substance use," or "depression."[61] The study made clear that the real breakthrough of the Facebook prediction research was the achievement of economies in the exploitation of these most-intimate behavioral depths with "automated, accurate, and cheap personality assessment tools" that effectively target a new class of "objects" once known as your "personality."[62] That these economies can be achieved outside the awareness of unrestrained animals makes them even more appealing; as one research team emphasizes, "The traditional method for personality evaluation is extremely costly in terms of time and labor, and it cannot acquire customer personality information without their awareness…."[63]

Personality analysis for commercial advantage is built on behavioral surplus—the so-called meta-data or mid-level metrics—honed and tested by researchers and destined to foil anyone who thinks that she is in control of the "amount" of personal information that she reveals in social media. In the name of, for example, affordable car insurance, we must be coded as conscientious, agreeable, and open. This is not easily faked because the surplus retrieved for analysis is necessarily opaque to us. We are not scrutinized for substance but for form. The price you are offered does not derive from *what* you write about but *how* you write it. It is not what is in your sentences but in their length and complexity, not *what* you list but *that* you list, not the picture but the choice of filter and degree of saturation, not *what* you disclose but how you share or fail to, not *where* you make plans to see your friends but *how* you do so: a casual "later" or a precise time and place? Exclamation marks and adverb choices operate as revelatory and potentially damaging signals of your self.

That the "personality" insights themselves are banal should not distract us from the fact that the volume and depth of the new surplus supplies enabled by these extraction operations are unprecedented; nothing like this has ever been conceivable.[64] As Kosinski told an interviewer in 2015, few people understand that companies such as "Facebook, Snapchat, Microsoft, Google and others have access to data that scientists would not ever be able to collect."[65] Data scientists have successfully predicted traits on the five-factor personality model with surplus culled from Twitter profile pictures (color, composition, image type, demographic information, facial presentation, and expression...), selfies (color, photographic styles, visual texture...), and Instagram photos (hue, brightness, saturation...). Others have tested alternative algorithmic models and personality constructs. Another research team demonstrated the ability to predict "satisfaction with life" from Facebook messages.[66] This new world has no manager on his or her hands and knees scissoring pages of computer conferencing messages into thematically organized piles of paper scraps. It's not the office floor that is crawled. It is you.

In his 2015 interview, Kosinski observed that "all of our interactions are being mediated through digital products and services which basically means that everything is being recorded." He even characterized his own work as "pretty creepy": "I actually want to stress that I think that many of the things that...one *can* do should certainly *not be done* by corporations or governments without users' consent." Recognizing the woefully asymmetric division of learning, he lamented the refusals of Facebook and the other internet firms to share their data with the "general public," concluding that "it's not because they're evil, but because the general public is bloody stupid...as a society we lost the ability to convince large companies that have enormous budgets and enormous access to data to share this goodness with us....We should basically grow up finally and stop it."[67]

In capitalism, though, latent demand summons suppliers and supplies. Surveillance capitalism is no different. The prediction imperative unleashes the surveillance hounds to stalk behavior from the depths, and well-intentioned researchers unwittingly oblige, leaving a trail of cheap, push-button raw meat for surveillance capitalists to hunt and devour. It did not take long. By early 2015, IBM announced that its Watson Personality Service was open for business.[68] The corporation's machine intelligence tools are even more complex and invasive than those used in most academic studies. In addition to the five-factor personality model, IBM assesses each individual across twelve categories of "needs," including "Excitement, Harmony, Curiosity, Ideal, Closeness, Self-expression, Liberty, Love, Practicality, Stability, Challenge, and Structure."

It then identifies "values," defined as "motivating factors which influence a person's decision-making across five dimensions: Self-transcendence/Helping others, Conservation/Tradition, Hedonism/Taking pleasure in life, Self-enhancement/Achieving success, and Open to change/Excitement."[69]

IBM promises "limitless" applications of its new surplus supplies and "deeper portraits of individual customers." As we would expect, these operations are tested among captive employees who, once habituated, can become docile members of a behaviorally purified society. "Personality correlates" can now be identified that predict the precise ways in which each customer will react to marketing efforts. Who will redeem a coupon? Who will purchase which product? The corporation says that "social media content and behavior" can be used to "capitalize on targeted revenue-generating opportunities" with "mapping rules from personality to behavior." The messaging and approach of customer service agents, insurance agents, travel agents, real estate agents, investment brokers, and so on can be "matched" to the "personality" of the customer, with those psychological data displayed to the agent at the precise moment of contact.[70] IBM's research demonstrates that agents who express personality traits associated with "agreeableness" and "conscientiousness" produce significantly higher levels of customer satisfaction. It is common sense, except that now these interactions are measured and monitored in real time and at scale, with a view to rewarding or extinguishing behavior according to its market effect.[71]

Thanks to rendition, a handful of now-measurable personal characteristics, including the "need for love," predict the likelihood of "liking a brand."[72] In a Twitter targeted-ad experiment, IBM found that it could significantly increase click-through rates and "follow" rates by targeting individuals with high "openness" and low "neuroticism" scores on the five-factor personality analysis. In another study, IBM rendered behavioral data from 2,000 Twitter users to establish metrics such as response rates, activity levels, and elapsed time between tweets, in addition to psycholinguistic analyses of tweets and five-factor personality analysis. IBM "trained" its predictive model by asking the 2,000 users either location-related or product-related questions. The findings showed that personality information predicted the likelihood of responses. People whom the machines rated as moral, trusting, friendly, extroverted, and agreeable tended to respond, compared to low response rates from people rated as cautious and anxious. Many of the characteristics that we try to teach our children and model in our own behavior are simply repurposed as dispossession opportunities for hidden machine processes of rendition. In this new world, paranoia and anxiety function as sources of protection from machine invasion for profit. Must we teach our children to be anxious and suspicious?

IBM is not alone, of course. An innovative breed of personality mercenaries quickly set to work institutionalizing the new supply operations. Their efforts suggest how quickly we lose our bearings as institutionalization first establishes a sense of normalcy and social acceptance and then gradually produces the numbness that accompanies habituation. This process begins with business plans and marketing messages, new products and services, and journalistic representations that appear to accept the new facts as given.[73]

Among this new cohort of mercenaries was Cambridge Analytica, the UK consulting firm owned by the reclusive billionaire and Donald Trump backer Robert Mercer. The firm's CEO, Alexander Nix, boasted of its application of personality-based "micro-behavioral targeting" in support of the "Leave" and the Trump campaigns during the ramp-up to the 2016 Brexit vote and the US presidential election.[74] Nix claimed to have data resolved "to an individual level where we have somewhere close to four or five thousand data points on every adult in the United States."[75] While scholars and journalists tried to determine the truth of these assertions and the role that these techniques might have played in both 2016 election upsets, the firm's new chief revenue officer quietly announced the firm's less glamorous but more lucrative postelection strategy: "After this election, it'll be full-tilt into the commercial business." Writing in a magazine for car dealers just after the US election, he tells them that his new analytic methods reveal "how a customer wants to be sold to, what their personality type is, and which methods of persuasion are most effective.... What it does is change people's behavior through carefully crafted messaging that resonates with them.... It only takes small improvements in conversion rates for a dealership to see a dramatic shift in revenue."[76]

A leaked Facebook document acquired in 2018 by the *Intercept* illustrates the significance of data drawn from the depths in the fabrication of Facebook's prediction products, confirms the company's primary orientation to its behavioral futures markets, and reveals the degree to which Cambridge Analytica's controversial practices reflected standard operating procedures at Facebook.[77] The confidential document cites Facebook's unparalleled "machine learning expertise" aimed at meeting its customers' "core business challenges." To this end it describes Facebook's ability to use its unrivaled and highly intimate data stores "to predict future behavior," targeting individuals on the basis of how they will behave, purchase, and think: now, soon, and later. The document links prediction, intervention, and modification. For example, a Facebook service called "loyalty prediction" is touted for its ability to analyze behavioral surplus in order to predict individuals who are "at risk" of shifting their brand allegiance. The idea is that these predictions can trigger advertisers to intervene promptly, targeting aggressive messages to stabilize loyalty and thus achieve guaranteed outcomes by altering the course of the future.

Facebook's "prediction engine" is built on a new machine intelligence platform called "FBLearner Flow," which the company describes as its new "AI backbone" and the key to "personalized experiences" that deliver "the most relevant content." The machine learning system "ingests trillions of data points every day, trains thousands of models—either offline or in real time—and then deploys them to the server fleet for live predictions." The company explains that "since its inception, more than a million models have been trained, and our prediction service has grown to make more than 6 million predictions per second."[78]

As we have already seen, "personalization" derives from prediction, and prediction derives from ever richer sources of behavioral surplus and therefore ever more ruthless rendition operations. Indeed, the confidential document cites some of the key raw materials fed into this high-velocity, high-volume, and deeply scoped manufacturing operation, including not only location, Wi-Fi network details, and device information but also data from videos, analyses of affinities, details of friendships, and similarities with friends.

It was probably no coincidence that the leaked Facebook presentation appeared around the same time that a young Cambridge Analytica mastermind-turned-whistleblower, Chris Wylie, unleashed a torrent of information on that company's secret efforts to predict and influence individual voting behavior, quickly riveting the world on the small political analytics firm and the giant source of its data: Facebook. There are many unanswered questions about the legality of Cambridge Analytica's complex subterfuge, its actual political impact, and its relationship with Facebook. Our interest here is restricted to how its machinations shine a bright light on the power of surveillance capitalism's mechanisms, especially the determination to render data from the depth dimension.

Kosinski and Stillwell had called attention to the commercial value of their methods, understanding that surplus from the depths afforded new possibilities for behavioral manipulation and modification. Wylie recounts his fascination with this prospect, and, through a complicated chain of events, it was he who persuaded Cambridge Analytica to use Kosinski and Stillwell's data to advance its owner's political aims. The objective was "behavioral micro-targeting... influencing voters based not on their demographics but on their personalities...."[79] When negotiations with Kosinski and Stillwell broke down, a third Cambridge academic, Alexander Kogan, was hired to render a similar cache of Facebook personality data.

Kogan was well-known to Facebook. He had collaborated with its data scientists on a 2013 project in which the company provided data on 57 billion "friendships." This time, he paid approximately 270,000 people to take a personality quiz. Unknown to these participants, Kogan's app enabled him to access their Facebook profiles and, on average, the profiles of about 160 of each of the test takers' friends, "none of whom would have known or had reason to suspect" this invasion.[80] It was a massive rendition operation from which Kogan successfully produced psychological profiles of somewhere between 50 and 87 million Facebook users, data that he then sold to Cambridge Analytica.[81] When Facebook questioned him about his application, Kogan vowed that his research was solely for academic purposes. Indeed, mutual respect between the two parties was sufficiently robust that Facebook hired one of Kogan's assistants to join its in-house team of research psychologists.[82]

"We exploited Facebook to harvest millions of people's profiles," Wylie admitted, "and built models to exploit what we knew about them and target their inner demons." His summary of Cambridge Analytica's accomplishments is a précis of the surveillance capitalist project and a rationale for its determination to render from the depths. These are the very capabilities that have gathered force over the nearly two decades of surveillance capitalism's incubation in lawless space. These practices produced outrage around the world, when in fact they are routine elements in the daily elaboration of surveillance capitalism's methods and goals, both at Facebook and within other surveillance capitalist companies. Cambridge Analytica merely reoriented the surveillance capitalist machinery from commercial markets in behavioral futures toward guaranteed outcomes in the political sphere. It was Eric Schmidt, not Wylie, who first pried open this Pandora's box, paving the way for the transfer of surveillance capitalism's core mechanisms to the electoral process as he cemented the mutual affinities that produced surveillance exceptionalism. In fact, Wylie enjoyed his early training under Obama's "director of targeting."[83] Schmidt's now-weaponized innovations have become the envy of every political campaign and, more dangerously, of every enemy of democracy.[84]

In addition to employing surveillance capitalism's foundational mechanisms—rendition, behavioral surplus, machine intelligence, prediction products, economies of scale, scope, and action—Cambridge Analytica's dark adventure also exemplifies surveillance capitalism's tactical requirements. Its operations were designed to produce ignorance through secrecy and the careful evasion of individual awareness. Wylie calls this "information warfare," correctly acknowledging the asymmetries of knowledge and power that are essential to the means of behavioral modification:

> I think it's worse than bullying, because people don't necessarily know it's being done to them. At least bullying respects the agency of people because they know...if you do not respect the agency of people, anything that you're doing after that point is not conducive to a democracy. And fundamentally, information warfare is not conducive to democracy.[85]

This "warfare" and its structure of invasion and conquest represent surveillance capitalism's standard operating procedures to which billions of innocents are subjected each day, as rendition operations violate all boundaries and modification operations claim dominion over all people. Surveillance capitalism imposes this quid pro quo of "agency" as the price of information and connection, continuously pushing the envelope of rendition to new frontiers. Along the way, companies such as Facebook and Google employ every useful foot soldier, including social scientists such as Kogan who are willing to put their shoulders to the wheel as they help the company learn, perfect, and integrate the cutting-edge methods that can conquer the next frontier, a phenomenon that we will visit in more depth in Chapter 10.

Irrespective of Cambridge Analytica's actual competence and its ultimate political impact, the plotting and planning behind its ambitions are testament to the pivotal role of rendition from the depths in the prediction and modification of behavior, always in pursuit of certainty. Billionaires such as Zuckerberg and Mercer have discovered that they can muscle their way to dominance of the division of learning in society by setting their sights on these rendition operations and the fortunes they tell. They aim to be unchallenged in their power to know, to decide who knows, and to decide who decides. The rendition of "personality" was an important milestone in this quest: a frontier, yes, but not the final frontier.

III. Machine Emotion

In 2015 an eight-year-old startup named Realeyes won a 3.6 million euro grant from the European Commission for a project code-named "SEWA: Automatic Sentiment Analysis in the Wild." The aim was "to develop automated technology that will be able to read a person's emotion when they view content and then establish how this relates to how much they liked the content." The director of video at AOL International called the project "a huge leap forward in video ad tech" and "the Holy Grail of video marketing."[86] Just a year later, Realeyes won the commission's Horizon 2020 innovation prize thanks to its "machine learning-based tools that help market researchers analyze the impact of their advertising and make it more relevant."[87]

The SEWA project is a window on a burgeoning new domain of rendition and behavioral surplus supply operations known as "affective computing," "emotion analytics," and "sentiment analysis." The personalization project descends deeper toward the ocean floor with these new tools, where they lay claim to yet a new frontier of rendition trained not only on your personality but also on your emotional life. If this project of surplus from the depths is to succeed, then your unconscious—where feelings form before there are words to express them—must be recast as simply one more source of raw-material supply for machine rendition and analysis, all of it for the sake of more-perfect prediction. As a market research report on affective computing explains, "Knowing the real-time emotional state can help businesses to sell their product and thereby increase revenue."[88]

Emotion analytics products such as SEWA use specialized software to scour faces, voices, gestures, bodies, and brains, all of it captured by "biometric" and "depth" sensors, often in combination with imperceptibly small, "unobtrusive" cameras. This complex of machine intelligence is trained to isolate, capture, and render the most subtle and intimate behaviors, from an inadvertent blink to a jaw that slackens in surprise for a fraction of a second. Combinations of sensors and software can recognize and identify faces; estimate age, ethnicity, and gender; analyze gaze direction and blinks; and track distinct facial points to interpret "micro-expressions," eye movements, emotions, moods, stress, deceit, boredom, confusion, intentions, and more: all at the speed of life.[89] As the SEWA project description says,

> Technologies that can robustly and accurately analyse human facial, vocal and verbal behaviour and interactions in the wild, as observed by omnipresent webcams in digital devices, would have profound impact on both basic sciences and the industrial sector. They... measure behaviour indicators that heretofore resisted measurement because they were too subtle or fleeting to be measured by the human eye and ear....[90]

These behaviors also elude the conscious mind. The machines capture the nanosecond of disgust that precedes a rapid-fire sequence of anger, comprehension, and finally joy on the face of a young woman watching a few frames of film, when all she can think to say is "I liked it!" A Realeyes white paper explains that its webcams record people watching videos in their homes "so we can capture genuine reactions." Algorithms process facial expressions, and "emotions are detected, aggregated, and reported online in real time, second by second...enabling our clients to make better business decisions." Realeyes emphasizes its own "proprietary metrics" to help marketers "target audiences" and "predict performance."[91]

Once again, a key theme of machine intelligence is that quality is a function of quantity. Realeyes says that its data sets contain over 5.5 million individually annotated frames of more than 7,000 subjects from all over the world: "We are continuously working to build the world's largest expression and behaviour datasets by increasing the quality and volume of our already-existing categories, and by creating new sets—for other expressions, emotions, different behavioral clues or different intensities.... Having automated this process, it can then be scaled up to simultaneously track the emotions of entire audiences."[92] Clients are advised to "play your audience emotions to stay on top of the game."[93] The company's website offers a brief review of the history of research on human emotions, concluding that "the more people feel, the more they spend.... Intangible 'emotions' translate into concrete social activity, brand awareness, and profit."[94]

The chair of SEWA's Industry Advisory Board is frank about this undertaking, observing that unlocking the meaning of "the non-spoken language of the whole body and interpreting complex emotional response... will be wonderful for interpreting reactions to marketing materials," adding that "it is simply foolish not to take emotional response into account when evaluating all marketing materials." Indeed, these "nonconscious tools" extract rarified new qualities of behavioral surplus from your inner life in order to predict what you will buy and the precise moment at which you are most vulnerable to a push. SEWA's advisory chair says that emotional analytics are "like identifying individual musical notes." Each potential customer, then, is a brief and knowable composition: "We will be able to identify chords of human response such as 'liking,' boredom, etc.... We will ultimately become masters of reading each other's feelings and intent."[95]

This is not the first time that the unconscious mind has been targeted as an instrument of others' aims. Propaganda and advertising have always been designed to appeal to unacknowledged fears and yearnings. These have relied more on art than science, using gross data or professional intuition for the purpose of mass communication.[96] Those operations cannot be compared to the scientific application of today's historic computational power to the micro-measured, continuous rendition of your more-or-less actual feelings. The new toolmakers do not intend to rob you of your inner life, only to surveil and exploit it. All they ask is to know more about you than you know about yourself.

Although the treasures of the unconscious mind have been construed differently across the millennia—from spirit to soul to self—the ancient priest and the modern psychotherapist are united in an age-old reverence for its primal healing power through self-discovery, self-mastery, integration, restoration, and transcendence. In contrast, the conception of emotions as observable behavioral data first took root in the mid-1960s with the work of Paul Ekman, then a young professor at the University of California, San Francisco. From his earliest papers, Ekman argued that "actions speak louder than words."[97] Even when a person is determined to censor or control his or her emotional communications, Ekman postulated that some types of nonverbal behaviors "escape control and provide leakage."[98] Early on, he recognized the potential utility of a "categorical scheme" that reliably traced the effects of expression back to their causes in emotion,[99] and in 1978 Ekman, along with frequent collaborator Wallace Friesen, published the seminal Facial Action Coding System (FACS), which provided that scheme.

FACS distinguishes the elemental movements of facial muscles, breaking them down into twenty-seven facial "action units," along with more for the head, eyes, tongue, and so on. Later, Ekman concluded that six "basic emotions" (anger, fear, sadness, enjoyment, disgust, and surprise) anchored the wider array of human emotional expression.[100] FACS and the six-emotion model became the dominant paradigm for the study of facial expression and emotion, in much the same way that the five-factor model came to dominate studies of personality.

The program of emotional rendition began innocently enough with MIT Media Lab professor Rosalind Picard and the new field of computer science that she called "affective computing." She was among the first to recognize the opportunity for a computational system to automate the analysis of Ekman's facial configurations and correlate micro-expressions with their emotional causality.[101] She aimed to combine facial expression with the computation of vocal intonation and other physiological signals of emotion that could be measured as behavior. In 1997 she published *Affective Computing*, which posited a practical solution to the idea that some emotions are available to the conscious mind and can be expressed "cognitively" (*"I feel scared"*), whereas others may elude consciousness but nevertheless be expressed physically in beads of sweat, a widening of the eyes, or a nearly imperceptible tightening of the jaw.

The key to affective computing, Picard argued, was to render both conscious and unconscious emotion as observable behavior for coding and calculation. A computer, she reasoned, would be able to render your emotions as behavioral information. Affect recognition, as she put it, is "a pattern recognition problem," and "affect expression" is pattern synthesis. The proposition was that "computers can be given the ability to recognize emotions as well as a third-person human observer."

Picard imagined her affective insights being put to use in ways that were often good-hearted or, at the very least, benign. Most of the applications she describes conform to the logic of the Aware Home: any knowledge produced would belong to the subject to enhance his or her reflexive learning. For example, she envisioned a "computer-interviewing agent" that could function as an "affective mirror" coaching a student in preparation for a job interview or a date and an automatic agent that could alert you to hostile tones in your own prose before you press "send." Picard anticipated other tools combining software and sensors that she believed could enhance daily life in a range of situations, such as helping autistic children develop emotional skills, providing software designers with feedback on users' frustration levels, assigning points to video game players to reward courage or stress reduction, producing learning modules that stimulate curiosity and minimize anxiety, and analyzing emotional dynamics in a classroom. She imagined software agents that learn your preferences and find you the kinds of news stories, clothing, art, or music that make you smile.[102] Whatever one's reaction to these ideas might be, they share one key pattern: unlike SEWA's model, Picard's data were intended to be *for* you, not merely *about* you.

Back in 1997, Picard acknowledged the need for privacy "so that you remain in control over who gets access to this information." Importantly for our analysis, in the final pages of her book she expressed some concerns, writing that "there are good reasons not to broadcast your affective patterns to the world...you might flaunt your good mood in front of friends...you probably do not want it picked up by an army of sales people who are eager to exploit mood-based buying habits, or by advertisers eager to convince you that you'd feel better if you tried their new soft drink right now." She noted the possibility of intrusive workplace monitoring, and she voiced reservations about the possibility of a dystopian future in which "malevolent" governmental forces use affective computing to manipulate and control the emotions of populations.[103]

Despite these few paragraphs of apprehension, her conclusions were bland. Every technology arrives with its "pros and cons," she wrote. The concerns are not "insurmountable" because "safeguards can be developed." Picard was confident that technologies and techniques could solve any problem, and she imagined "wearable computers" that "gather information strictly for your own use...." She stressed the importance of ensuring "that the wearer retains ultimate control over the devices he chooses to wear, so that they are tools of helpful empowerment and not of harmful subjugation."[104]

In a pattern that is by now all too familiar, safeguards lagged as surveillance capitalism flourished. By early 2014, Facebook had already applied for an "emotion detection" patent designed to implement each of Picard's fears.[105] The idea was "one or more software modules capable of detecting emotions, expressions or other characteristic of a user from image information." As always, the company was ambitious. Its list of detectable emotions "by way of example and not limitation" included expressions such as "a smile, joy, humor, amazement, excitement, surprise, a frown, sadness, disappointment, confusion, jealousy, indifference, boredom, anger, depression, or pain." The hope was that "over time" their module would be able to assess "a user's interest in displayed content" for the purposes of "customization based on emotion type."[106]

By 2017, exactly twenty years after the publication of Picard's book, a leading market research firm forecast that the "affective computing market," including software that recognizes speech, gesture, and facial expressions along with sensors, cameras, storage devices, and processors, would grow from $9.35 billion in 2015 to $53.98 billion in 2021, predicting a compounded annual growth rate of nearly 35 percent. What happened to cause this explosion? The report concludes that heading up the list of "triggers" for this dramatic growth is "rising demand for mapping human emotions especially by the marketing and advertising sector...."[107] Picard's good intentions were like so many innocent iron filings in the presence of a magnet as the market demand exerted by the prediction imperative drew affective computing into surveillance capitalism's powerful force field.

Picard would eventually become part of this new dispossession industry with a company called Affectiva that was cofounded with Rana el Kaliouby, an MIT Media Lab postdoctoral research scientist and Picard protégé. The company's transformation from doing good to doing surveillance capitalism is a metaphor for the fate of the larger undertaking of emotion analysis as it is rapidly drawn into the competitive maelstrom for surveillance revenues.

Both Picard and Kaliouby had shared a vision of applying their research in medical and therapeutic settings. The challenges of autistic children seemed a perfect fit for their discoveries, so they trained a machine system called Mind-Reader to recognize emotions using paid actors to mimic specific emotional responses and facial gestures. Early on, MIT Media Lab corporate sponsors Pepsi, Microsoft, Bank of America, Nokia, Toyota, Procter and Gamble, Gillette, Unilever, and others had overwhelmed the pair with queries about using their system to measure customers' emotional responses. Kaliouby describes the women's hesitation and their determination to focus on "do-good" applications. According to her account, the Media Lab encouraged the two women to "spin off" their work into the startup they called Affectiva, imagined as a "baby IBM for emotionally intelligent machines."[108]

It wasn't long before the new company found itself fielding significant interest from ad agencies and marketing firms itching for automated rendition and analysis from the depths. Describing that time, as Picard told one journalist, "Our CEO was absolutely not comfortable with the medical space." As a result, Picard was "pushed out" of the firm three years after its founding. As an Affectiva researcher recounted, "We began with a powerful set of products that could assist people who have a very difficult time with perceiving affect.... Then they started to emphasize only the face, to focus on advertisements, and on predicting whether someone likes a product, and just went totally off the original mission."[109]

Companies such as market research firm Millward Brown and advertising powerhouse McCann Erickson, competing in a new world of targeted "personalized" ads, already craved access to the inarticulate depths of the consumer response. Millward Brown had even formed a neuroscience unit but found it impossible to scale. It was Affectiva's analysis of one particularly nuanced ad for Millward Brown that dazzled its executives and decisively turned the tide for the startup. "The software was telling us something we were potentially not seeing," one Millward Brown executive said. "People often can't articulate such detail in sixty seconds."[110]

By 2016, Kaliouby was the company's CEO, redefining its business as "Emotion AI" and calling it "the next frontier of artificial intelligence."[111] The company had raised $34 million in venture capital, included 32 Fortune 100 companies and 1,400 brands from all over the world among its clients, and claimed to have the largest repository of emotion data in the world, with 4.8 million face videos from 75 countries, even as it continued to expand its supply routes with data sourced from online viewing, video game participation, driving, and conversation.[112]

This is the commercial context in which Kaliouby came to feel that it is perfectly reasonable to assert that an "emotion chip" will become the base operational unit of a new "emotion economy." She speaks to her audiences of a chip embedded in all things everywhere, running constantly in the background, producing an "emotion pulse" each time you check your phone: "I think in the future we'll assume that every device just knows how to read your emotions."[113] At least one company, Emoshape, has taken her proposition seriously. The firm, whose tagline is "Life Is the Value," produces a microchip that it calls "the industry's first emotion synthesis engine," delivering "high performance machine emotion awareness." The company writes that its chip can classify twelve emotions with up to 98 percent accuracy, enabling its "artificial intelligence or robot to experience 64 trillion possible distinct emotional states."[114]

Kaliouby imagines that pervasive "emotion scanning" will come to be as taken for granted as a "cookie" planted in your computer to track your online browsing. After all, those cookies once stirred outrage, and now they inundate every online move. For example, she anticipates YouTube scanning its viewers' emotions as they watch videos. Her confidence is buoyed by demand that originates in the prediction imperative: "The way I see it, it doesn't matter that your Fitbit doesn't have a camera, because your phone does, and your laptop does, and your TV will. All that data gets fused with biometrics from your wearable devices and builds an emotional profile for you." As a start,

Affectiva pioneered the notion of "emotion as a service," offering its analytics on demand: "Just record people expressing emotion and then send those videos or images to us to get powerful emotion metrics back."[115]

The possibilities in the depth dimension seem endless, and perhaps they will be if Affectiva, its clients, and fellow travelers are free to plunder our selves at will. There are indications of more far-reaching ambitions in which "emotion as a service" expands from observation to modification. "Happiness as a service" seems to be within reach. "I do believe that if we have information about your emotional experiences we can help you be in a positive mood," Kaliouby says. She imagines emotion-recognition systems issuing reward points for happiness because, after all, happy customers are more "engaged."[116]

IV. When They Come for My Truth

Rendition is by now a global project of surveillance capital, and in the depth dimension we see it at its most pernicious. Intimate territories of the self, like personality and emotion, are claimed as observable behavior and coveted for their rich deposits of predictive surplus. Now the personal boundaries that shelter inner life are officially designated as bad for business by a new breed of mercenaries of the self determined to parse and package inner life for the sake of surveillance revenues. Their expertise disrupts the very notion of the autonomous individual by rewarding "boundarylessness" with whatever means are available—offers of elite status, bonuses, happiness points, discounts, "buy" buttons pushed to your device at the precise moment predicted for maximum success—so that we might strip and surrender to the pawing and prying of the machines that serve the new market cosmos.

I want to deliberately sidestep a more detailed discussion of what is "personality" or "emotion," "conscious" or "unconscious," in favor of what I hope is a less fractious truth thrown into relief by this latest phase of incursion. Experience is not what is given to me but rather what I make of it. The same experience that I deride may invite your enthusiasm. The self is the inward space of lived experience from which such meanings are created. In that creation I stand on the foundation of personal freedom: the "foundation" because I cannot live without making sense of my experience.

No matter how much is taken from me, this inward freedom to create meaning remains my ultimate sanctuary. Jean-Paul Sartre writes that "freedom is nothing but the *existence* of our will," and he elaborates: "Actually it is not enough to will; it is necessary to will to will."[117] This rising up of *the will to will* is the inner act that secures us as autonomous beings who project choice into the world and exercise the qualities of self-determining moral judgment that are civilization's necessary and final bulwark. This is the sense behind another of Sartre's insights: "Without bearings, stirred by a nameless anguish, the words labor.... The voice is born of a risk: either to lose oneself or win the right to speak in the first person."[118]

As the prediction imperative drives deeper into the self, the value of its surplus becomes irresistible, and cornering operations escalate. What happens to the right to speak in the first person from and as my self when the swelling frenzy of institutionalization set into motion by the prediction imperative is trained on cornering my sighs, blinks, and utterances on the way to my very thoughts as a means to others' ends? It is no longer a matter of surveillance capital wringing surplus from what I search, buy, and browse. Surveillance capital wants more than my body's coordinates in time and space. Now it violates the inner sanctum as machines and their algorithms decide the meaning of my breath and my eyes, my jaw muscles, the hitch in my voice, and the exclamation points that I offered in innocence and hope.

What happens to my *will to will* myself into the first person when the surrounding market cosmos disguises itself as my mirror, shape-shifting according to what it has decided I feel or felt or will feel: ignoring, goading, chiding, cheering, or punishing me? Surveillance capital cannot keep from wanting all of me as deep and far as it can go. One firm that specializes in "human analytics" and affective computing has this headline for its marketing customers: "Get Closer to the Truth. Understand the 'Why.'" What happens when they come for my "truth" uninvited and determined to march through my self, taking the bits and pieces that can nourish their machines to reach their objectives? Cornered in my self, there is no escape.[119]

It appears that questions like these may have come to trouble Picard. In a 2016 lecture she gave in Germany titled "Towards Machines That Deny Their Maker," the bland assertions of her 1997 book that "safeguards can be developed," that additional technologies and techniques could solve any problem, and that "wearable computers" would "gather information strictly for your own use" as "tools of helpful empowerment and not of harmful subjugation"[120] had given way to new reflections. "Some organizations want to sense human emotions without people knowing or consenting," she said, "A few scientists want to build computers that are vastly superior to humans, capable of powers beyond reproducing their own kind...how might we make sure that new affective technologies make human lives better?"[121]

Picard did not foresee the market forces that would transform the rendition of emotion into for-profit surplus: means to others' ends. That her vision is made manifest in thousands of activities should be a triumph, but it is diminished by the fact that so many of those activities are now bound to the commercial surveillance project. Each failure to establish bearings contributes to habituation, normalization, and ultimately legitimation. Subordinated to the larger aims of surveillance capitalism, the thrust of the affective project changed as if distorted in a fun-house mirror.

This cycle calls to mind the words of another MIT professor, the computer scientist and humanist Joseph Weizenbaum, who spoke eloquently and often on the inadvertent collusion of computer scientists in the construction of terrifying weapons systems. I believe he would have shaken his spear in the direction of today's sometimes-unwitting and sometimes-intentional mercenaries of the self, and it is fitting to conclude here with his voice:

> I don't quite know whether it is especially computer science or its sub-discipline Artificial Intelligence that has such an enormous affection for euphemism. We speak so spectacularly and so readily of computer systems that understand, that see, decide, make judgments...without ourselves recognizing our own superficiality and immeasurable naiveté with respect to these concepts. And, in the process of so speaking, we anesthetize our ability to...become conscious of its end use....One can't escape this state without asking, again and again: "What do I actually do? What is the final application and use of the products of my work?" and ultimately, "Am I content or ashamed to have contributed to this use?"[122]

CHAPTER TEN

MAKE THEM DANCE

But hear the morning's injured weeping and know why:
Ramparts and souls have fallen; the will of the unjust
Has never lacked an engine; still all princes must
Employ the fairly-noble unifying lie.

—W. H. AUDEN
SONNETS FROM CHINA, XI

I. Economies of Action

"The new power is *action*," a senior software engineer told me. "The intelligence of the internet of things means that sensors can also be *actuators*." The director of software engineering for a company that is an important player in the "internet of things" added, "It's no longer simply about ubiquitous computing. Now the real aim is ubiquitous intervention, action, and control. The real power is that now you can *modify* real-time actions in the real world. Connected smart sensors can register and analyze any kind of behavior and then actually figure out how to change it. Real-time analytics translate into real-time action." The scientists and engineers I interviewed call this new capability "actuation," and they describe it as the critical though largely undiscussed turning point in the evolution of the apparatus of ubiquity.

This actuation capability defines a new phase of the prediction imperative that emphasizes *economies of action*. This phase represents the completion of the new *means of behavior modification,* a decisive and necessary evolution of the surveillance capitalist "means of production" toward a more complex, iterative, and muscular operational system. It is a critical achievement in the race to guaranteed outcomes. Under surveillance capitalism the objectives and operations of automated behavioral modification are designed and controlled by companies to meet their own revenue and growth objectives. As one senior engineer told me,

> Sensors are used to modify people's behavior just as easily as they modify device behavior. There are many great things we can do with the internet of things, like lowering the heat in all the houses on your street so that the transformer is not overloaded, or optimizing an entire industrial operation. But at the individual level, it also means the power to take actions that can override what you are doing or even put you on a path you did not choose.

The scientists and engineers whom I interviewed identified three key approaches to economies of action, each one aimed at achieving behavior modification. The first two I call "tuning" and "herding." The third is already familiar as what behavioral psychologists refer to as "conditioning." Strategies that produce economies of action vary according to the methods with which these approaches are combined and the salience of each.

"Tuning" occurs in a variety of ways. It may involve subliminal cues designed to subtly shape the flow of behavior at the precise time and place for maximally efficient influence. Another kind of tuning involves what behavioral economists Richard Thaler and Cass Sunstein call the "nudge," which they define as "any aspect of a choice architecture that alters people's behavior in a predictable way."[1] The term *choice architecture* refers to the ways in which situations are already structured to channel attention and shape action. In some cases these architectures are intentionally designed to elicit specific behavior, such as a classroom in which all the seats face the teacher or an online business that requires you to click through many obscure pages in order to opt out of its tracking cookies. The use of this term is another way of saying in behaviorist language that social situations are always already thick with tuning interventions, most of which operate outside our awareness.

Behavioral economists argue a worldview based on the notion that human mentation is frail and flawed, leading to irrational choices that fail to adequately consider the wider structure of alternatives. Thaler and Sunstein have encouraged governments to actively design nudges that adequately shepherd individual choice making toward outcomes that align with their interests, as perceived by experts. One classic example favored by Thaler and Sunstein is the cafeteria manager who nudges students to healthier food choices by prominently displaying the fruit salad in front of the pudding; another is the automatic renewal of health insurance policies as a way of protecting individuals who overlook the need for new approvals at the end of each year.

Surveillance capitalists adapted many of the highly contestable assumptions of behavioral economists as one cover story with which to legitimate their practical commitment to a unilateral commercial program of behavior modification. The twist here is that nudges are intended to encourage choices that accrue to the architect, not to the individual. The result is data scientists trained on economies of action who regard it as perfectly normal to master the art and science of the "digital nudge" for the sake of their company's commercial interests. For example, the chief data scientist for a national drugstore chain described how his company designs automatic digital nudges that subtly push people toward the specific behaviors favored by the company: "You can make people do things with this technology. Even if it's just 5% of people, you've made 5% of people do an action they otherwise wouldn't have done, so to some extent there is an element of the user's loss of self-control."

"Herding" is a second approach that relies on controlling key elements in a person's immediate context. The uncontract is an example of a herding technique. Shutting down a car's engine irreversibly changes the driver's immediate context, herding her out the car door. Herding enables remote orchestration of the human situation, foreclosing action alternatives and thus moving behavior along a path of heightened probability that approximates certainty. "We are learning how to write the music, and then we let the music make them dance," an "internet of things" software developer explains, adding,

> We can engineer the context around a particular behavior and force change that way. Context-aware data allow us to tie together your emotions, your cognitive functions, your vital signs, etcetera. We can know if you shouldn't be driving, and we can just shut your car

down. We can tell the fridge, "Hey, lock up because he shouldn't be eating," or we tell the TV to shut off and make you get some sleep, or the chair to start shaking because you shouldn't be sitting so long, or the faucet to turn on because you need to drink more water.

"Conditioning" is a well-known approach to inducing behavior change, primarily associated with the famous Harvard behaviorist B. F. Skinner. He argued that behavior modification should mimic the evolutionary process, in which naturally occurring behaviors are "selected" for success by environmental conditions. Instead of the earlier, more simplistic model of stimulus/response, associated with behaviorists such as Watson and Pavlov, Skinner interpolated a third variable: "reinforcement." In his laboratory work with mice and pigeons, Skinner learned how to observe a range of naturally occurring behaviors in the experimental animal and then reinforce the specific action, or "operant," that he wanted the animal to reproduce. Ultimately, he mastered intricate designs or "schedules" of reinforcement that could reliably shape precise behavioral routines.

Skinner called the application of reinforcements to shape specific behaviors "operant conditioning." His larger project was known as "behavior modification" or "behavioral engineering," in which behavior is continuously shaped to amplify some actions at the expense of others. In the end the pigeon learns, for example, to peck a button twice in order to receive a pellet of grain. The mouse learns his way through a complicated maze and back again. Skinner imagined a pervasive "technology of behavior" that would enable the application of such methods across entire human populations.

As the chief data scientist for a much-admired Silicon Valley education company told me, "Conditioning at scale is essential to the new science of massively engineered human behavior." He believes that smartphones, wearable devices, and the larger assembly of always-on networked nodes allow his company to modify and manage a substantial swath of its users' behavior. As digital signals monitor and track a person's daily activities, the company gradually masters the schedule of reinforcements—rewards, recognition, or praise that can reliably produce the specific user behaviors that the company selects for dominance:

> The goal of everything we do is to change people's actual behavior at
> scale. We want to figure out the construction of changing a person's
> behavior, and then we want to change how lots of people are making
> their day-to-day decisions. When people use our app, we can capture
> their behaviors and identify good and bad [ones]. Then we develop
> "treatments" or "data pellets" that select good behaviors. We can test
> how actionable our cues are for them and how profitable certain
> behaviors are for us.

Although it is still possible to imagine automated behavioral modification without surveillance capitalism, it is not possible to imagine surveillance
capitalism without the marriage of behavior modification and the technological means to automate its application. This marriage is essential to economies
of action. For example, one can imagine a fitness tracker, a car, or a refrigerator whose data and operational controls are accessible exclusively to their
owners for the purposes of helping them to exercise more often, drive safely,
and eat healthily. But as we have already seen in so many domains, the rise of
surveillance capitalism has obliterated the idea of the simple feedback loop
characteristic of the behavioral value reinvestment cycle. In the end, it's not
the devices; it's Max Weber's "economic orientation," now determined by
surveillance capitalism.

The allure of surveillance revenues drives the continuous accumulation
of more and more predictive forms of behavioral surplus. The most predictive source of all is behavior that has already been modified to orient toward
guaranteed outcomes. The fusion of new digital means of modification and
new economic aims produces whole new ranges of techniques for creating and
cornering these new forms of surplus. A study called "Behavior Change Techniques Implemented in Electronic Lifestyle Activity Monitors" is illustrative.
Researchers from the University of Texas and the University of Central Florida studied thirteen such applications, concluding that the monitoring devices
"contain a wide range of behavior change techniques typically used in clinical behavior interventions." The researchers conclude that behavior-change
operations are proliferating as a result of their migration to digital devices
and internet connectivity. They note that the very possibility of a simple loop

designed by and for the consumer seems hopelessly elusive, observing that behavior-change apps "lend themselves ... to various types of surveillance" and that "official methods" of securely and simply transmitting data "do not appear to currently exist in these apps."[2]

Remember that Google economist Hal Varian extolled the "new uses" of big data that proceed from ubiquitous computer-mediated transactions. Among these he included the opportunity for "continuous experimentation." Varian noted that Google has its engineering and data science teams consistently running thousands of "A/B" experiments that rely on randomization and controls to test user reactions to hundreds of variations in page characteristics from layout to buttons to fonts. Varian endorsed and celebrated this self-authorizing experimental role, warning that all the data in the world "can only measure correlation, not causality."[3] Data tell what happened but not why it happened. In the absence of causal knowledge, even the best predictions are only extrapolations from the past.

The result of this conundrum is that the last crucial element in the construction of high-quality prediction products—i.e., those that approximate guaranteed outcomes—depends upon causal knowledge. As Varian says, "If you really want to understand causality, you have to run experiments. And if you run experiments continuously, you can continuously improve your system."[4]

Because the "system" is intended to produce predictions, "continuously improving the system" means closing the gap between prediction and observation in order to approximate certainty. In an analog world, such ambitions would be far too expensive to be practical, but Varian observes that in the realm of the internet, "experimentation can be entirely automated."

Varian awards surveillance capitalists the privilege of the experimenter's role, and this is presented as another casual fait accompli. In fact, it reflects the final critical step in surveillance capitalists' radical self-dealing of new rights. In this phase of the prediction imperative, surveillance capitalists declare their right to modify others' behavior for profit according to methods that bypass human awareness, individual decision rights, and the entire complex of self-regulatory processes that we summarize with terms such as *autonomy* and *self-determination*.

What follows now are two distinct narratives of surveillance capitalists as "experimenters" who leverage their asymmetries of knowledge to impose their will on the unsuspecting human subjects who are their users. The experimental insights accumulated through their one-way mirrors are critical to constructing, fine-tuning, and exploring the capabilities of each firm's for-profit means of behavioral modification. In Facebook's user experiments and in the augmented-reality game Pokémon Go (imagined and incubated at Google), we see the commercial means of behavioral modification evolving before our eyes. Both combine the components of economies of action and the techniques of tuning, herding, and conditioning in startling new ways that expose the Greeks secreted deep in the belly of the Trojan horse: the economic orientation obscured behind the veil of the digital.

II. Facebook Writes the Music

In 2012 Facebook researchers startled the public with an article provocatively titled "A 61-Million-Person Experiment in Social Influence and Political Mobilization," published in the scientific journal *Nature*.[5] In this controlled, randomized study conducted during the run-up to the 2010 US Congressional midterm elections, the researchers experimentally manipulated the social and informational content of voting-related messages in the news feeds of nearly 61 million Facebook users while also establishing a control group.

One group was shown a statement at the top of their news feed encouraging the user to vote. It included a link to polling place information, an actionable button reading "I Voted," a counter indicating how many other Facebook users reported voting, and up to six profile pictures of the user's Facebook friends who had already clicked the "I Voted" button. A second group received the same information but without the pictures of friends. A third control group did not receive any special message.

The results showed that users who received the social message were about 2 percent more likely to click on the "I Voted" button than did those who received the information alone and 0.26 percent more likely to click on polling place information. The Facebook experimenters determined that social messaging was an effective means of tuning behavior at scale because it "directly influenced political self-expression, information seeking and real-world voting behavior of millions of people," and they concluded that "showing familiar faces to users can dramatically improve the effectiveness of a mobilization message."

The team calculated that the manipulated social messages sent 60,000 additional voters to the polls in the 2010 midterm elections, as well as another 280,000 who cast votes as a result of a "social contagion" effect, for a total of 340,000 additional votes. In their concluding remarks, the researchers asserted that "we show the importance of social influence for effecting behavior change…the results suggest that online messages might influence a variety of offline behaviors, and this has implications for our understanding of the role of online social media in society…."[6]

The experiment succeeded by producing social cues that "suggested" or "primed" users in ways that tuned their real-world behavior toward a specific set of actions determined by the "experimenters." In this process of experimentation, economies of action are discovered, honed, and ultimately institutionalized in software programs and their algorithms that function automatically, continuously, ubiquitously, and pervasively to achieve economies of action. Facebook's surplus is aimed at solving one problem: how and when to intervene in the state of play that is your daily life in order to modify your behavior and thus sharply increase the predictability of your actions now, soon, and later. The challenge for surveillance capitalists is to learn how to do this effectively, automatically, and, therefore, economically, as a former Facebook product manager writes:

> Experiments are run on every user at some point in their tenure on the site. Whether that is seeing different size ad copy, or different marketing messages, or different call-to-action buttons, or having their feeds generated by different ranking algorithms.… The fundamental purpose of most people at Facebook working on data is to influence and alter people's moods and behavior. They are doing it all the time to make you like stories more, to click on more ads, to spend more time on the site. This is just how a website works, everyone does this and everyone knows that everyone does this.[7]

The Facebook study's publication evoked fierce debate as experts and the wider public finally began to reckon with Facebook's—and the other internet companies'—unprecedented power to persuade, influence, and ultimately manufacture behavior. Harvard's Jonathan Zittrain, a specialist in internet law, acknowledged that it was now possible to imagine Facebook quietly engineering an election, using means that its users could neither detect nor control. He described the Facebook experiment as a challenge to "collective rights" that could undermine "the right of people as a whole...to enjoy the benefits of a democratic process...."[8]

Public concern failed to destabilize Facebook's self-authorizing practice of behavior modification at scale. Even as the social influence experiment was being debated in 2012, a Facebook data scientist was already collaborating with academic researchers on a new study, "Experimental Evidence of Massive-Scale Emotional Contagion Through Social Networks," submitted to the prestigious *Proceedings of the National Academy of Sciences* in 2013, where it was edited by a well-known Princeton social psychologist, Susan Fiske, and published in June 2014.

This time the experimenters "manipulated the extent to which people ($N = 689,003$) were exposed to emotional expressions in their News Feed."[9] The experiment was structured like one of those allegedly benign A/B tests. In this case one group was exposed to mostly positive messages in their news feed and the other to predominantly negative messages. The idea was to test whether even subliminal exposure to specific emotional content would cause people to change their own posting behavior to reflect that content. It did. Whether or not users felt happier or sadder, the tone of their expression changed to reflect their news feed.

The experimental results left no doubt that once again Facebook's carefully designed, undetectable, and uncontestable subliminal cues reached beyond the screen into the daily lives of hundreds of thousands of naive users, predictably actuating specific qualities of emotional expression through processes that operate outside the awareness of their human targets, just as Stuart MacKay had originally prescribed for Galapagos turtles and Canadian elk (see Chapter 7). "Emotional states can be transferred to others via emotional contagion, leading people to experience the same emotions without their awareness," the researchers proclaimed. "Online messages influence our experience of emotions, which may affect a variety of offline behaviors." The team celebrated its work as "some of the first experimental evidence to support the controversial claims that emotions can spread throughout a network," and they reflected on the fact that even their relatively minimal manipulation had a measurable effect, albeit a small one.[10]

What Facebook researchers failed to acknowledge in either experiment is that a person's susceptibility to subliminal cues and his or her vulnerability to a "contagion" effect is largely dependent upon empathy: the ability to understand and share in the mental and emotional state of another person, including feeling another's feelings and being able to take another's point of view—sometimes characterized as "affective" or "cognitive" empathy. Psychologists have found that the more a person can project himself or herself into the feelings of another and take the other's perspective, the more likely he or she is to be influenced by subliminal cues, including hypnosis. Empathy orients people toward other people. It allows one to get absorbed in emotional experience and to resonate with others' experiences, including unconsciously mimicking another's facial expressions or body language. Contagious laughing and even contagious yawning are examples of such resonance.[11]

Empathy is considered essential to social bonding and emotional attachment, but it can also trigger "vicarious anxiety" for victims or others who are genuinely distressed. Some psychologists have called empathy a "risky strength" because it predisposes us to experience others' happiness but also their pain.[12] The successful tuning evident in both Facebook experiments is the result of the effective exploitation of the natural empathy present in its population of users.

The Facebook researchers claimed that the results suggested two inferences. First, in a massive and engaged population such as Facebook users, even small effects "can have large aggregated consequences." Second, the authors invited readers to imagine what might be accomplished with more-significant manipulations and larger experimental populations, noting the importance of their findings for "public health."

Once again, public outcry was substantial. "If Facebook can tweak emotions and make us vote, what else can it do?" the *Guardian* asked. The *Atlantic* quoted the study's editor, who had processed the article for publication despite her apparent misgivings.[13] She told the magazine that as a private company, Facebook did not have to adhere to the legal standards for experimentation required of academic and government researchers.

These legal standards are known as the "Common Rule." Designed to protect against the abuse of the experimenter's power, these standards must be adhered to by all federally funded research. The Common Rule enforces procedures for informed consent, avoidance of harm, debriefing, and transparency, and it is administered by panels of scientists, known as "internal review boards," appointed within every research institution. Fiske acknowledged that she had been persuaded by Facebook's argument that the experimental manipulation was an unremarkable extension of the corporation's standard practice of manipulating people's news feeds. As Fiske recounted, "They said... that Facebook apparently manipulates people's News Feeds all the time.... Who knows what other research they're doing."[14] In other words, Fiske recognized that the experiment was merely an extension of Facebook's standard practices of behavioral modification, which already flourish without sanction.

Facebook data scientist and principal researcher Adam Kramer was deluged with hundreds of media queries, leading him to write on his Facebook page that the corporation really does "care" about its emotional impact. One of his coauthors, Cornell's Jeffrey Hancock, told the New York Times that he didn't realize that manipulating the news feeds, even modestly, would make some people feel violated.[15] The Wall Street Journal reported that the Facebook data science group had run more than 1,000 experiments since its inception in 2007 and operated with "few limits" and no internal review board. Writing in the Guardian, psychology professor Chris Chambers summarized that "the Facebook study paints a dystopian future in which academic researchers escape ethical restriction by teaming up with private companies to test increasingly dangerous or harmful interventions."[16]

A month after the emotional contagion study's publication, the editor-in-chief of the Proceedings of the National Academy of Sciences, Inder M. Verma, published an "editorial expression of concern" regarding the Facebook research. After acknowledging the standard defense that Facebook is technically exempt from the "Common Rule," Verma added, "It is nevertheless a matter of concern that the collection of the data by Facebook may have involved practices that were not fully consistent with the principles of obtaining informed consent and allowing participants to opt out."[17]

Among US scholars, University of Maryland law professor James Grimmelmann published the most comprehensive argument in favor of holding Facebook and other social media companies accountable to the standards represented by the Common Rule. Corporate research is more likely than academic research to be compromised by serious conflicts of interests, he reasoned, making common experimental standards critical and not something to be left to individual ethical judgment. Grimmelmann imagined "Internal Review Board laundering," in which academics could "circumvent research ethics regulations whenever they work just closely enough with industry partners. The exception would swallow the Common Rule."[18]

Despite his conviction on this point, Grimmelmann acknowledged in the final pages of his analysis that even the most rigorous imposition of the Common Rule would do little to curb the immense power of a company such as Facebook that routinely manipulates user behavior at scale, using means that are indecipherable and therefore uncontestable. Like Fiske, Grimmelmann sensed the larger project of economies of action just beyond the reach of established law and social norms.

The journal *Nature* drew attention with a strongly worded letter defending the Facebook experiment, authored by bioethicist Michelle Meyer together with five coauthors and on behalf of twenty-seven other ethicists. The letter argued that the need to codify new knowledge about the online environment justifies experimentation even when it does not or cannot abide by accepted ethical guidelines for human subjects research. But Meyer's defense turned on a prescient note of warning that "the extreme response to this study...could result in such research being done in secret....If critics think that the manipulation of emotional content in this research is sufficiently concerning to merit regulation... then the same concern must apply to Facebook's standard practice...."[19]

The experiment's critics and supporters agreed on little but this: Facebook could easily turn rogue, threatening to retreat to secrecy if regulators attempted to intervene in its practices. The academic community shared a sense of threat in the face of known facts. Facebook owns an unprecedented means of behavior modification that operates covertly, at scale, and in the absence of social or legal mechanisms of agreement, contest, and control. Even the most stringent application of the "Common Rule" would be unlikely to change these facts.

As scholars promised to convene panels to consider the ethical issues raised by Facebook research, the corporation announced its own plans for better self-regulation. The corporation's chief technology officer, Mike Schroepfer, confessed to being "unprepared" for the public reaction to the emotional contagion study, and he admitted that "there are things we should have done differently." The company's "new framework" for research included clear guidelines, an internal review panel, a capsule on research practices incorporated into the company's famous "boot camp" orientation and training program for new hires, and a website to feature published academic research. These self-imposed "regulations" did not challenge the fundamental facts of Facebook's online community as the necessary developmental environment and target for the firm's economies of action.

A document acquired by the Australian press in May 2017 would eventually reveal this fact. Three years after the publication of the contagion study, the *Australian* broke the story on a confidential twenty-three-page Facebook document written by two Facebook executives in 2017 and aimed at the company's Australian and New Zealand advertisers. The report depicted the corporation's systems for gathering "psychological insights" on 6.4 million high school and tertiary students as well as young Australians and New Zealanders already in the workforce. The Facebook document detailed the many ways in which the corporation uses its stores of behavioral surplus to pinpoint the exact moment at which a young person needs a "confidence boost" and is therefore most vulnerable to a specific configuration of advertising cues and nudges: "By monitoring posts, pictures, interactions, and Internet activity, Facebook can work out when young people feel 'stressed,' 'defeated,' 'overwhelmed,' 'anxious,' 'nervous,' 'stupid,' 'silly,' 'useless,' and a 'failure.'"[20]

The report reveals Facebook's interest in leveraging this affective surplus for the purpose of economies of action. It boasts detailed information on "mood shifts" among young people based on "internal Facebook data," and it claims that Facebook's prediction products can not only "detect sentiment" but also predict how emotions are communicated at different points during the week, matching each emotional phase with appropriate ad messaging for the maximum probability of guaranteed outcomes. "Anticipatory emotions are more likely to be expressed early in the week," the analysis counsels, "while reflective emotions increase on the weekend. Monday–Thursday is about building confidence; the weekend is for broadcasting achievements."

Facebook publicly denied these practices, but Antonio Garcia-Martinez, a former Facebook product manager and author of a useful account of Silicon Valley titled *Chaos Monkeys*, described in the *Guardian* the routine application of such practices and accused the corporation of "lying through their teeth." He concluded, "The hard reality is that Facebook will never try to limit such use of their data unless the public uproar reaches such a crescendo as to be un-mutable."[21] Certainly the public challenge to Facebook's insertion of itself into the emotional lives of its users, as expressed in the contagion study, and its pledge to self-regulate did not quell its commercial interest in users' emotions or the corporation's compulsion to systematically exploit that knowledge on behalf of and in collaboration with its customers. It did not, because it cannot, not as long as the company's revenues are bound to economies of action under the authority of the prediction imperative.

Facebook's persistence warns us again of the dispossession cycle's stubborn march. Facebook had publicly acknowledged and apologized for its overt experimental *incursions* into behavior modification and emotional manipulation, and it promised *adaptations* to curb or mitigate these practices. Meanwhile, a new threshold of intimate life had been breached. Facebook's potential mastery of emotional manipulation became discussable and even taken for granted as *habituation* set in. From Princeton's Fiske to critic Grimmelmann and supporter Meyer, the experts believed that if Facebook's activities were to be forced into a new regulatory regime, the corporation would merely continue in secret. The Australian documents opened one door on these covert practices, suggesting the completion of the cycle with the *redirection* of action into clandestine zones protected by opacity and indecipherability, just as these scholars had anticipated.

Facebook's political mobilization experimenters discovered that they could manipulate users' vulnerabilities to social influence in order to create a motivational condition ("I want to be like my friends") that increases the probability that a relevant priming message—the "I Voted" button—will produce action. The emotional contagion study exploited the same underlying social influence orientation. In this case, Facebook planted subliminal cues in the form of positive or negative affective language, which combined with the motivational state triggered by social comparison—"I want to be like my friends"—to produce a measurable, if weak, contagion effect. Finally, the Australian ad-targeting document points to the seriousness and complexity of the backstage effort to strengthen this effect by specifying motivational conditions at a granular level. It reveals not only the scale and scope of Facebook's behavioral surplus but also the corporation's interest in leveraging its surplus to precisely determine the ebb and flow of a user's predisposition for real-time targeting by the branded cues that are most likely to achieve guaranteed outcomes.

Facebook's experimental success demonstrates that tuning through suggestion can be an effective form of telestimulation at scale. The evasion of individual and group awareness was critical to Facebook's behavior-modification success, just as MacKay had stipulated. The first paragraph of the research article on emotional contagion celebrates this evasion: "Emotional states can be transferred to others via emotional contagion, leading people to experience the same emotions without their *awareness*." Nor do the young adults of Australia's great cities suspect that the precise measure of their fears and fantasies is exploited for commercial result at the hour and moment of their greatest vulnerability.

This evasion is neither accidental nor incidental, but actually essential to the structure of the whole surveillance capitalist project. Individual awareness is the enemy of telestimulation because it is the necessary condition for the mobilization of cognitive and existential resources. There is no autonomous judgment without awareness. Agreement and disagreement, participation and withdrawal, resistance or collaboration: none of these self-regulating choices can exist without awareness.

A rich and flourishing research literature illuminates the antecedents, conditions, consequences, and challenges of human self-regulation as a universal need. The capacity for self-determination is understood as an essential foundation for many of the behaviors that we associate with critical capabilities such as empathy, volition, reflection, personal development, authenticity, integrity, learning, goal accomplishment, impulse control, creativity, and the sustenance of intimate enduring relationships. "Implicit in this process is a self that sets goals and standards, is *aware* of its own thoughts and behaviors, and has the capacity to change them," write Ohio State University professor Dylan Wagner and Dartmouth professor Todd Heatherton in an essay about the centrality of self-awareness to self-determination: "Indeed, some theorists have suggested that the primary purpose of self awareness is to enable self-regulation." Every threat to human autonomy begins with an assault on awareness, "tearing down our capacity to regulate our thoughts, emotions, and desires."[22]

The salience of self-awareness as a bulwark against self-regulatory failure is also underscored in the work of two Cambridge University researchers who developed a scale to measure a person's "susceptibility to persuasion." They found that the single most important determinant of one's ability to resist persuasion is what they call "the ability to premeditate."[23] This means that people who harness self-awareness to think through the consequences of their actions are more disposed to chart their own course and are significantly less vulnerable to persuasion techniques. Self-awareness also figures in the second-highest-ranking factor on their scale: commitment. People who are consciously committed to a course of action or set of principles are less likely to be persuaded to do something that violates that commitment.

We have seen already that democracy threatens surveillance revenues. Facebook's practices suggest an equally disturbing conclusion: human consciousness itself is a threat to surveillance revenues, as awareness endangers the larger project of behavior modification. Philosophers recognize "self-regulation," "self-determination," and "autonomy" as "freedom of will." The word *autonomy* derives from the Greek and literally means "regulation by the self." It stands in contrast to *heteronomy*, which means "regulation by others." The competitive necessity of economies of action means that surveillance capitalists must use all means available to supplant autonomous action with heteronomous action.

In one sense there is nothing remarkable in observing that capitalists would prefer individuals who agree to work and consume in ways that most advantage capital. We need only to consider the ravages of the subprime mortgage industry that helped trigger the great financial crisis of 2008 or the daily insults to human autonomy at the hands of countless industries from airlines to insurance for plentiful examples of this plain fact.

However, it would be dangerous to nurse the notion that today's surveillance capitalists simply represent more of the same. This structural requirement of economies of action turns the means of behavioral modification into an engine of growth. At no other time in history have private corporations of unprecedented wealth and power enjoyed the free exercise of economies of action supported by a pervasive global architecture of ubiquitous computational knowledge and control constructed and maintained by all the advanced scientific know-how that money can buy.

Most pointedly, Facebook's declaration of experimental authority claims surveillance capitalists' prerogatives over the future course of others' behavior. In declaring the right to modify human action secretly and for profit, surveillance capitalism effectively exiles us from our own behavior, shifting the locus of control over the future tense from "I will" to "You will." Each one of us may follow a distinct path, but economies of action ensure that the path is already shaped by surveillance capitalism's economic imperatives. The struggle for power and control in society is no longer associated with the hidden facts of class and its relationship to production but rather by the hidden facts of automated engineered behavior modification.

III. Pokémon Go! Do!

It had been a particularly grueling July afternoon in 2016. David had directed hours of contentious insurance testimony in a dusty New Jersey courtroom, where a power surge the night before had knocked out the building's fragile air-conditioning system. Then the fitful Friday commute home was cursed by a single car disabled by the heat that turned the once-hopeful flow of traffic into sludge. Finally home, he slid the car into his garage and made a beeline for the side door that opened to the laundry room and kitchen beyond. The cool air hit him like a dive into the ocean, and for the first time all day he took a deep breath. A note on the table said his wife would be back in a few minutes. He gulped down some water, made himself a drink, and climbed the stairs, heading for a long shower.

The doorbell rang just as the warm water hit his aching back muscles. Had she forgotten her key? Shower interrupted, he threw on a tee and shorts and ran downstairs, opening the front door to a couple of teenagers waving their cell phones in his face. "Hey, you've got a Pokémon in your backyard. It's ours! Okay if we go back there and catch it?"

"A what?" He had no idea what they were talking about, but he was about to get educated.

David's doorbell rang four more times that evening: perfect strangers eager for access to his yard and disgruntled when he asked them to leave. Throughout the days and evenings that followed, knots of Pokémon seekers formed on his front lawn, some of them young and others long past that excuse. They held up their phones, pointing and shouting as they scanned his house and garden for the "augmented-reality" creatures. Looking at this small slice of world through their phones, they could see their Pokémon prey but only at the expense of everything else. They could not see a family's home or the boundaries of civility that made it a sanctuary for the man and woman who lived there. Instead, the game seized the house and the world around it, reinterpreting all of it in a vast equivalency of GPS coordinates. Here was a new kind of commercial assertion: a for-profit declaration of eminent domain in which reality is recast as an unbounded expanse of blank spaces to be sweated for others' enrichment. David wondered, *When will this end? What gives them the right? Whom do I call to make this stop?*

Without knowing it, he had been yanked from his shower to join the villagers in Broughton, England, who had taken to their streets in 2009 protesting the invasion of Google's Street View camera cars. Like them, he had been abruptly thrust into contest with surveillance capitalism's economic imperatives, and like them he would soon understand that there was no number to call, no 9-1-1 to urgently inform the appropriate authorities that a dreadful mistake had blossomed on his lawn.

Back in 2009, as we saw in Chapter 5, Google Maps product vice president and Street View boss John Hanke ignored the Broughton protestors, insisting that only he and Google knew what was best, not just for Broughton but for all people. Now here was Hanke again at surveillance capitalism's next frontier, this time as the founder of the company behind Pokémon Go, Niantic Labs. Hanke, you may recall, nursed an abiding determination to own the world by mapping it. He had founded Keyhole, the satellite mapping startup funded by the CIA and later acquired by Google and rechristened as Google Earth. At Google, he was a vice president for Google Maps and a principal in its controversial push to commandeer public and private space through its Street View project.

Hanke recounts that Pokémon Go was born out of Google Maps, which also supplied most of the game's original development team.[24] Indeed, Street View's mystery engineer, Marius Milner, had joined Hanke in this new phase of incursion. By 2010, Hanke had set up his own launch pad, Niantic Labs, inside the Google mother ship. His aim was the development of "parallel reality" games that would track and herd people through the very territories that Street View had so audaciously claimed for its maps. In 2015, following the establishment of the Alphabet corporate structure and well after the development of Pokémon Go, Niantic Labs was formally established as an independent company with $30 million in funding from Google, Nintendo (the Japanese company that originally hosted Pokémon on its "Game Boy" devices in the late 1990s), and the Pokémon Company.[25]

Hanke had long recognized the power of the game format as a means to achieve economies of action. While still at Google he told an interviewer, "More than 80% of people who own a mobile device claim that they play games on their device…games are often the number 1 or number 2 activity… so for Android as an operative system, but also for Google, we think it's important for us to innovate and to be a leader in…the future of mobile gaming."[26]

It is worth noting that Hanke chose to name his group after a nineteenth-century merchant sailing vessel undone by greed. The *Niantic* had been sold and repurposed for the more lucrative whaling trade when it set sail for San Francisco and the northern Pacific whaling grounds in 1849. The ship's captain made an unplanned stop in Panama to board hundreds of pilgrims bound for the California Gold Rush, all of them eager to pay top dollar for cramped, smelly quarters on the whaler. The captain's avarice proved fatal to the ship's prospects when those passengers infected the ship's crew with gold fever. The sailors abandoned captain and vessel upon docking in San Francisco, heading instead for gold country. Unable to continue the journey, the captain was forced to sell the ship for a pittance, leaving it wedged deep in the sandy shallows at the foot of Clay and Montgomery streets. In 2016 Hanke took up the quest of that rebellious crew. His Niantic was bound for a new century's gold rush at the frontier of the prediction imperative's next wave of conquest: economies of action.

Hanke's Pokémon Go launched in July 2016 as a different answer to the question confronting the engineers and scientists shaping the surveillance capitalist project: how can human behavior be actuated quickly and at scale, while driving it toward guaranteed outcomes? At its zenith in the summer of 2016, Pokémon Go was a surveillance capitalist's dream come true, fusing scale, scope, and actuation; yielding continuous sources of behavioral surplus; and providing fresh data to elaborate the mapping of interior, exterior, public, and private spaces. Most important, it provided a living laboratory for telestimulation at scale as the game's owners learned how to automatically condition and herd collective behavior, directing it toward real-time constellations of behavioral futures markets, with all of this accomplished just beyond the rim of individual awareness. In Hanke's approach, economies of action would be achieved through the dynamics of a game.

Niantic designed the new game to be "played" in the real world, not on a screen. The idea is that players should be "going outside" for "adventures on foot" in the open spaces of cities, towns, and suburbs.[27] The game relies on "augmented reality" and is structured like a treasure hunt. Once you download the app from Niantic, you use GPS and your smartphone camera to hunt virtual creatures called Pokémon. The figures appear on your smartphone screen as if they are located beside you in your real-life surroundings: an unsuspecting man's backyard, a city street, a pizzeria, a park, a drugstore. Captured Pokémon are rewarded with game currencies, candies, and stardust, and are employed to battle other users. The ultimate goal is to capture a comprehensive array of the 151 Pokémon, but along the way players earn "experience points," rising to successive levels of expertise. At level five, players can join one of three teams to battle Pokémon at designated sites referred to as "gyms."

The ramp-up had begun years earlier with Ingress, Niantic's first mobile game designed for real-world play. Released in 2012, Ingress was a precursor and test bed for the capabilities and methods that would define Pokémon Go. The game drove its users through their cities and towns to find and control designated "portals" and capture "territory" as the game masters relied on

GPS to track users' movements and map the territories through which they roamed.

Hanke reflected on what he and his team had learned from Ingress. Most important was the Niantic team's "surprise" as they observed how much "the behavior of the players changes."[28] Hanke grasped that the seeds of behavior modification were planted in the game's rules and social dynamic: "If you want to turn the world into your game board, the places you want people to interact with have to have certain characteristics.... There should be a reason for the player to go there.... The game is enabling them and nudging you to have those interactions."[29] One user whose Ingress name was "Spottiswoode" provides an example: "As I cycle home, I stop near a location I'd scouted out previously, one with a weak enemy portal. I attack, using built-up XM ('exotic matter') to destroy the enemy infrastructure.... On Ingress's built-in chat client, a player called Igashu praises my handiwork. 'Good job, Spottiswoode,' he says. I feel proud and move on, plotting my next assault upon the enemy's portals."[30] According to Hanke, Pokémon Go would be designed to leverage what the team now understood as the key sources of motivation that induce players to change their behavior: a social gaming community based on real-world action.[31]

All games circumscribe behavior with rules, rewarding some forms of action and punishing others, and Niantic is not the first to employ the structure of a game as a means of effecting behavior change in its players. Indeed, "gamification" as an approach to behavioral engineering is a subject of intense interest that has produced a robust academic and popular literature.[32] According to Wharton professor Kevin Werbach, games include three tiers of action. At the highest level are the "dynamics" that drive the motivational energy of the game. These can be emotions aroused by competition or frustration, a compelling narrative, a structure of progression that creates the experience of development toward a higher goal, or relationships that produce feelings such as team spirit or aggression. Next are the "mechanics." These are the procedural building blocks that drive the action and also build engagement. For example, a game may be structured as a competition or a solo challenge, as turn taking and cooperation, as transactions and winner take all, as team sport or individual conquest. Finally, there are the game "components" that operationalize the mechanics. These are the most-visible aspects of a game: points to represent progress, quests laid out as predefined challenges, "badges" to represent achievements, "leader boards" to visually display all players' progress, "boss fights" to mark the culmination of a level, and so forth.[33]

Most research on games concludes that these structures can be effective at motivating action, and researchers generally predict that games will increasingly be used as the methodology of choice to change individual behavior.[34] In practice, this has meant that the power of games to change behavior is shamelessly instrumentalized as gamification spreads to thousands of situations in which a company merely wants to tune, herd, and condition the behavior of its customers or employees toward its own objectives. Typically, this involves importing a few components, such as reward points and levels of advancement, in order to engineer behaviors that serve the company's immediate interests, with programs such as customer loyalty schemes or internal sales competitions. One analyst compiled a survey of more than ninety such "gamification cases," complete with return-on-investment statistics.[35] Ian Bogost, a professor of interactive computing at Georgia Tech and a digital culture observer, insists that these systems should be called "exploitationware" rather than games because their sole aim is behavior manipulation and modification.[36]

Pokémon Go takes these capabilities in a wholly new direction, running game players through the real world, but not for the sake of the game they think they are playing. Hanke's unique genius is to point the game's behavior-modification efforts toward a target that occupies an unexplored zone beyond the boundaries of players' awareness. It aims to shape behavior in an even larger game of surveillance capitalism.

Pokémon Go was first unveiled to the *Wall Street Journal* in September 2015, shortly after Niantic's spin-off from Google. The game masters told the reporter that the game would not include ads. Instead, revenues would accrue from "microtransactions," presumably in-game purchases of virtual paraphernalia, although Niantic "declined to say" exactly what would be for sale. Niantic also promised a location-tracking bracelet that "vibrates and lights up" when a person approaches a Pokémon. It was clear that Pokémon Go would at least be a fresh source of surplus for refining and expanding the maps upon which the game depended.[37]

Released in the US, Australia, and New Zealand on July 6, 2016, Pokémon Go became the most downloaded and highest-grossing app in the US within only a week, quickly achieving as many active Android users as Twitter. More than 60 percent of the app's downloads were in daily use, and by July 8 that translated into a daily average of about 43.5 minutes per user.[38] With Niantic's servers groaning under the strain, the game's European rollout was delayed until July 13. By that time, however, Niantic had proved the value of its approach to economies of action, demonstrating unprecedented effectiveness in traversing that last tortured mile to guaranteed outcomes.

The unprecedented pattern was faintly discernible within days of the game's launch. A Virginia bar offered a discount to a Pokémon Go team; a tea shop in San Francisco offered a "buy one get one free" to the game's players.[39] The owner of a pizza bar in Queens, New York, paid about $10 for "Lure Modules," a bit of virtual game paraphernalia intended to attract Pokémon to a specific location, successfully producing virtual creatures on bar stools and in bathroom stalls. During the first weekend of game play, the bar's food and drink sales shot up by 30 percent and later were reported to be 70 percent above average. Bloomberg reporters gushed that the game had achieved the retailers' elusive dream of using location tracking to drive foot traffic: "It's easy to imagine a developer selling ads within the game world to local merchants, or even auctioning off the promise to turn specific shops and restaurants into destinations for players."[40] Hanke hinted to the *New York Times* that these real-world, real-time markets had been the plan all along. "Niantic has cut deals like that for Ingress," the paper reported, "and Mr. Hanke said the company would announce sponsored locations for Pokémon Go in the future."[41]

The future came quickly. Within a week the basic elements of surveillance capitalism's logic of accumulation were in place and were heralded as brilliant. As Hanke explained, "The game relies on a lot of modern cell phone and data technology to power the augmented reality, but that traffic generated by the game also changes what happens in the real world."[42] By July 12, the *Financial Times* exulted that "speculation has surged over the game's future power as a cash cow to retailers and other cravers of footfall." Nintendo shares were up 52 percent, adding $10.2 billion to its market capitalization.[43]

Earlier promises that the game would not serve ads turned out to be a technical claim that required careful parsing. In fact, the surveillance-based logic of online advertising had not disappeared. Rather, it had morphed into its physical-world mirror image, just as Sidewalk Labs' Dan Doctoroff had imagined for the "Google city," a precise extension of the methods and purposes honed in the online world but now amplified in "reality" under pressure from the prediction imperative (see Chapter 7).

By July 13, Hanke admitted to the *Financial Times* that in addition to "in-app payments" for game kit, "there is a second component to our business model at Niantic, which is the concept of *sponsored locations.*" He explained that this new revenue stream had always been in the plan, noting that companies will "pay us to be locations within the virtual game board—the premise being that it is an inducement that drives foot traffic." These sponsors, Hanke explained, would be charged on a *"cost per visit"* basis, similar to the *"cost per click"* used in Google's search advertising.[44]

The notion of "sponsored locations" is a euphemism for Niantic's behavioral futures markets, ground zero in Hanke's new gold rush. The elements and dynamics of the game, combined with its novel augmented-reality technology, operate to herd populations of game players through the real-world monetization checkpoints constituted by the game's actual customers: the entities who pay to play on the real-world game board, lured by the promise of guaranteed outcomes.

For a while it seemed that everyone was making money. Niantic inked a deal with McDonald's to drive game users to its 30,000 Japanese outlets. A British mall owner commissioned "recharging teams" to roam his malls with portable rechargers for game users. Starbucks announced that it would "join in with the fun," with 12,000 of its US stores becoming official "Pokéstops" or "gyms," along with a new "Pokémon Go Frappuccino…the perfect treat for any Pokémon trainer on the go." Another deal with Sprint would convert 10,500 Sprint retail and service outlets into Pokémon hubs. The music streaming company Spotify reported a tripling of Pokémon-related music sales. A UK insurance company offered special coverage for mobile phones, warning, "Don't let accidental damage get in the way of catching them all." Disney admitted that it was disappointed with its own strategies for "the blending of physical and digital to create new kinds of connected play experiences" and planned to transform its mammoth toy business "in a direction similar to Pokémon Go."[45]

The zeal for Pokémon Go gradually diminished, but the impact of Hanke's accomplishments is indelible. "We've only just scratched the surface," Hanke told a crowd of fans.[46] The game had demonstrated that it was possible to achieve economies of action on a global scale while simultaneously directing specific individual actions toward precise local market opportunities where high bidders enjoy an ever-closer approximation of guaranteed outcomes.

Niantic's distinctive accomplishment was to manage gamification as a way to guarantee outcomes for its actual customers: companies participating in the behavioral futures markets that it establishes and hosts. Hanke's game proved that surveillance capitalism could operate in the real world much as it does in the virtual one, using its unilateral knowledge (scale and scope) to shape your behavior now (action) in order to more accurately predict your behavior later. The logical inference is that real-world revenues will increase in proportion to the company's ability to match persons with locations, just as Google learned to wield surplus as a means of targeting online ads to specific individuals.

These requirements suggest that Niantic would conduct its operations in ways that establish substantial surplus supply chains aimed at scale and scope. Indeed, the company's "surveillance policy" signals its demand for behavioral data in excess of what is reasonable for effective game operations. Just six days after the game's release in July 2016, *BuzzFeed* reporter Joseph Bernstein advised Pokémon users to check how much data the app was collecting from their phones. According to his analysis, "Like most apps that work with the GPS in your smartphone, Pokémon Go can tell a lot of things about you based on your movement as you play: where you go, when you went there, how you got there, how long you stayed, and who else was there. And, like many developers who build those apps, Niantic keeps that information." Whereas other location-based apps might collect similar data, Bernstein concluded that "Pokémon Go's incredibly granular, block-by-block map data, combined with its surging popularity, may soon make it one of, if not *the most*, detailed location-based social graphs ever compiled."[47]

The industry news site *TechCrunch* raised similar concerns regarding the game's data-collection practices, questioning "the long list of permissions the app requires." Those permissions included the camera, yes, but also permission to "read your contacts" and "find accounts on device." Niantic's "surveillance policy" notes that it may share "aggregated information and non-identifying information with third parties for research and analysis, demographic profiling, and other similar purposes." *TechCrunch* noted the game's "precise location tracking" and "ability to perform audio fingerprinting" through its access to your camera and microphone, concluding, "So it's prudent to expect some of your location data to end up in Google's hands."[48] The Electronic Privacy Information Center noted in a letter of complaint to the Federal Trade Commission that Niantic had failed to provide compelling reasons for the "scope" of the information that it routinely gathers from users' phones and Google profiles. Nor had it set limits on how long it would retain, use, or share location data. As the letter concluded, "There is no evidence that Niantic's collection and retention of location data is necessary to the function of the game or otherwise provides a benefit to consumers that outweighs the privacy and safety harms it creates."[49]

By mid-July 2016, Niantic received a letter from US Senator Al Franken querying the company's privacy practices.[50] Niantic's late-August response is instructive, a marvel of misdirection and secrecy that focuses on the game's mechanics and discloses nothing about its business model or the more comprehensive logic of accumulation behind the model: "Pokémon Go has already been praised by public health officials, teachers, mental health workers, parents, park officials, and ordinary citizens around the world as an app that promotes healthy play and discovery." Though acknowledging the range of data it collects as a condition of play—location services, photos, media, files, camera, contacts, and network provider data—Niantic insists that data are used "to provide and improve" its services. However, it does not acknowledge that its services operate on two levels: game services for players and prediction services for Niantic's customers. The company concedes that it uses third-party services, including Google's, to "collect and interpret data," but it is careful to sidestep the aims of those analyses.[51]

The seven-page letter mentions "sponsored locations" only once, noting that sponsors receive reports about visits and game actions. There is no reference to "cost per visit" or the surplus that will be required to drive that metric, in the same way that Google's "cost per click" depended upon behavioral surplus drawn from online activity. Niantic's self-presentation carefully conceals its objectives in the design and development of economies of action that drive real-world, real-time behavior toward Niantic's behavioral futures markets.

The genius of Pokémon Go was to transform the game you see into a higher-order game of surveillance capitalism, a game about a game. The players who took the city as their board, roaming its parks and pizzerias, unwittingly constituted a wholly different kind of human game board for this second and more consequential game. The players in this other *real* game could not be found in the clot of enthusiasts waving their phones at the edge of David's lawn. In the real game, prediction products take the form of protocols that impose forms of telestimulation intended to prod and herd people across real-world terrains to spend their real-world money in the real-world commercial establishments of Niantic's flesh-and-blood behavioral futures markets.

Niantic itself is like a tiny probe rising from the immensity of Google's mapping capabilities, surplus flows, means of production, and vast server farms as it constructs and tests the prototype of a global means of behavior modification owned and operated by surveillance capitalism. Niantic discovered that in the rapture of engaging competitive social play, the dreaded friction of individual will voluntarily gives way to game protocols that set the conditions for "natural selection." In this way the game automatically elicits and breeds the specific behaviors sought by the high rollers in Niantic's behavioral futures markets. With this second game board in motion, the players in the real game vie for proximity to the wake of cash that follows each smiling member of the herd.

In the end we recognize that the probe was designed to explore the next frontier: the means of behavioral modification. The game about the game is, in fact, an experimental facsimile of surveillance capitalism's design for our future. It follows the prediction imperative to its logical conclusion, in which data about us in scale and scope combine with actuation mechanisms that align our behavior with a new market cosmos. All the flows of surplus from all the spaces, all the things, all the bodies, all the laughter, and all the tears are finally aimed at this triumph of certain outcomes and the revenue that it can unleash.

IV. What Were the Means of Behavioral Modification?

The new global means of behavioral modification that we see under construction at Facebook and Niantic represent a new regressive age of *autonomous capital* and *heteronomous individuals,* when the very possibilities of democratic flourishing and human fulfillment depend upon the reverse. This unprecedented state of affairs rises above debates about the Common Rule. It goes to the heart of our allegiance to the ideals of a democratic society, with full knowledge of the challenges that burden those ideals.

What has been forgotten here is that the Common Rule was the product of a similar challenge to principles of individual autonomy and democratic fidelity. It was one result of a deeply contested struggle in which democratically minded public officials joined with social activists, scholars, and litigators to resist the design, development, and deployment of behavioral modification as a mode of governmental power. It was not long ago that US society mobilized to resist, regulate, and control the means of behavioral modification, and it is this history that we can now draw upon to rediscover our bearings and rouse our awareness.

In 1971 the Senate Subcommittee on Constitutional Rights, led by North Carolina Senator Sam Ervin and including luminaries from across the political spectrum such as Edward Kennedy, Birch Bayh, Robert Byrd, and Strom Thurmond, undertook what would become a multiyear investigation into "a variety of programs designed to predict, control, and modify human behavior." Ervin was a conservative Democrat and constitutional expert who became an unlikely civil liberties hero for his defense of democracy during the Watergate crisis as chair of the Senate Watergate Committee. In this case the Subcommittee on Constitutional Rights would subject the principles and applications of behavior modification to intense constitutional scrutiny for the first time, questioning and ultimately rejecting the use of behavioral modification as an extension of state power.

The Senate investigation was triggered by a growing sense of public alarm at the spread of psychological techniques for behavior control. There were many points of origin, but most salient was the influence of the cold war and the range of psychological techniques and programs for behavior modification that it had bred. The Korean War had publicized communist "brainwashing" techniques that, according to then newly appointed CIA Director Allen Dulles, had reduced US prisoners of war to a state of robotic passivity, in which the victim's brain "becomes a phonograph playing a disc put on its spindle by an outside genius over which it has no control."[52] America's enemies appeared to be on the verge of mastering the art and science of "mind control" with psychological and pharmacological methods unknown to the US military. There were reports of Chinese and Soviet achievements in the remote alteration of a subject's mental capacities and the elimination of his "free will."[53] Dulles committed the agency to rapid research in and development of "mind control" capabilities, from "de-patterning" and "rewiring" an individual to shaping an entire country's attitudes and actions.[54]

Thus began a morbidly fascinating and often bizarre chapter in the history of American spy craft.[55] Much of the new work was conducted in the context of the CIA's highly classified MKUltra project, which was tasked with "research and development of chemical, biological, and radiological materials capable for employment in clandestine operations to control human behavior." According to testimony in the 1975 Senate investigation of covert CIA Foreign and Military Intelligence operations, a 1963 Inspector General's report on MKUltra noted several reasons for the program's secrecy, but chief among them was the fact that behavior modification was seen as illegitimate. "Research in the manipulation of human behavior is considered by many authorities in medicine and related fields to be professionally unethical, therefore the reputation of professional participants in the MKUltra program are on occasion in jeopardy," the report began. It also noted that many of the program's activities were illegal, violated the rights and interests of US citizens, and would alienate public opinion.[56]

Key to our interests is the growth and elaboration of behavioral modification as an extension of political power. To this end, CIA "demand" summoned an ever-more-audacious supply of behavioral-modification research and practical applications from academic psychologists. Scientists from the fields of medicine and psychology set out to demystify Chinese brainwashing techniques, reinterpreting them through the established frameworks of behavior modification.

Their research concluded that "mind control" was better understood as a complex system of conditioning based on unpredictable schedules of reinforcement, consistent with B. F. Skinner's important discoveries on operant conditioning. According to Harvard historian Rebecca Lemov, the "mind control" researchers had a powerful effect on the CIA and other branches of the military. The notion that "human material was changeable"—that one's personality, identity, awareness, and capacity for self-determining behavior could be crushed, eliminated, and replaced by external control—incited a new sense of panic and vulnerability: "If indeed the world was rife with threats to the inner as much as the outer man, then experts in these realms were needed more than ever. Many good and well-meaning professors—self-described or de facto human engineers—participated in the CIA's programs to bring about slow or rapid change in the minds and behavior of people."[57]

By the time the senators on the Constitutional Rights Subcommittee convened in 1971, the migration of behavior-modification practices from military to civilian applications was well underway. Behavior-modification techniques had dispersed from government-funded (typically CIA) psychology labs and military psyops to a range of institutional applications, each driven by a mission to re-engineer the defective personalities of captive individuals in settings that offered "total control" or close to it, including prisons, psychiatric wards, classrooms, institutions for the mentally challenged, schools for the autistic, and factories.

The subcommittee was emboldened to act when public concern bubbled over into outrage at the mainstreaming of these behavior-modification programs. Historian of psychology Alexandra Rutherford observes that Skinnerian behavior-modification practices expanded rapidly during the 1960s and 1970s, achieving some "remarkable successes" but also exposing practitioners to the scrutiny of an often-hostile public. A number of journalistic accounts raised alarms about the zealousness with which behavior-modification techniques were applied and the sense that they debased their subjects, violated ethical considerations, and infringed on fundamental civil liberties.[58]

Another factor was the 1971 publication of B. F. Skinner's incendiary social meditation *Beyond Freedom & Dignity*. Skinner prescribed a future based on behavioral control, rejecting the very idea of freedom (as well as every tenet of a liberal society) and cast the notion of human dignity as an accident of self-serving narcissism. Skinner imagined a pervasive "technology of behavior" that would one day enable the application of behavior-modification methods across entire human populations.

The ensuing storm of controversy made *Beyond Freedom & Dignity* an international best seller. "Skinner's science of human behavior, being quite vacuous, is as congenial to the libertarian as to the fascist," wrote Noam Chomsky in a widely read review of the book. "It would be not absurd but grotesque to argue that since circumstances can be arranged under which behavior is quite predictable—as in a prison, for example, or...concentration camp....therefore there need be no concern for the freedom and dignity of 'autonomous man.'"[59] (In the mid-1970s graduate department at Harvard where I studied and Skinner professed, many students referred to the book as *Toward Slavery and Humiliation*.)

From the first lines of the preface of the subcommittee's 1974 report, authored by Senator Ervin, it should be evident to any twenty-first-century captive of surveillance capitalism that US society has undergone a social discontinuity more profound than the mere passage of decades suggests. It is worth reading Ervin's own words, to grasp the passion with which he located the subcommittee's work at the heart of the Enlightenment project, pledging to defend the liberal ideals of freedom and dignity:

> When the founding fathers established our constitutional system of government, they based it on their fundamental belief in the sanctity of the individual....They understood that self-determination is the source of individuality, and individuality is the mainstay of freedom....Recently, however, technology has begun to develop new methods of behavior control capable of altering not just an individual's actions but his very personality and manner of thinking...the behavioral technology being developed in the United States today touches upon the most basic sources of individuality and the very core of personal freedom...the most serious threat...is the power this technology gives one man to impose his views and values on another....Concepts of freedom, privacy and self-determination inherently conflict with programs designed to control not just physical freedom, but the source of free thought as well....The question

becomes even more acute when these programs are conducted, as they are today, in the absence of strict controls. As disturbing as behavior modification may be on a theoretical level, the unchecked growth of the practical technology of behavior control is cause for even greater concern.[60]

The report's critique of behavior modification has unique relevance for our time. It begins by asking a question that we must also ask: "How did they get away with it?" Their answer invokes the "exceptionalism" of that era. Just as surveillance capitalism was initially able to root and flourish under the protection of a so-called "war against terror" and the compulsion for certainty that it stirred, in the middle of the twentieth century the means of behavioral modification migrated from the lab to the world at large primarily under the cover of cold-war anxieties. Later, the behavior-change professionals of the 1960s and 1970s were summoned into civilian practice by a society turned fearful after years of urban riots, political protests, and rising levels of crime and "delinquency." The senators reasoned that calls for "law and order" had motivated the search for "immediate and efficient means to control violence and other forms of anti-social behavior. The interest in controlling violence replaced more time-consuming attempts to understand its sources."

With so many behavior-modification programs aimed at involuntary populations in state prisons and mental institutions, the senators recognized that the means of behavioral modification had to be reckoned with as a form of state power and questioned the government's constitutional right to "control" citizens' behavior and mentation. In its survey of government agencies, the subcommittee found "a wide variety of behavior modification techniques... presently employed in the United States under the auspices of the federal government" and observed that "with the rapid proliferation of behavior modification techniques, it is all the more disturbing that few real efforts have been made to consider the basic issues of individual freedom involved and... fundamental conflicts between individual rights and behavior technology."[61]

The senators reserved their most vivid rebukes for what they considered as the two most extreme and pernicious behavior-modification techniques. The first was psychosurgery. The second was "electrophysiology," defined as "the use of mechanical devices to control various aspects of human behavior." The report notes with special horror the example of "devices" designed to be worn by a subject "constantly to monitor and control his behavior through a computer" and to "prevent a suspected behavior from occurring."

The First Amendment, the subcommittee argued, "must equally protect the individual's right to generate ideas," and the right to privacy should protect citizens from intrusions into their thoughts, behavior, personality, and identity lest these concepts "become meaningless." It was in this context that Skinnerian behavioral engineering was singled out for critical examination: "A major segment of the emerging behavior control technology is concerned with conditioning, through which various forms of persuasion are used to stimulate certain types of behaviors while suppressing others."[62]

In anticipation of future gamification techniques as means of behavioral modification, the subcommittee report also noted with apprehension more "benign" approaches that relied on "positive reinforcement," from "gold-star incentives" to elaborate reward systems, in order "to restructure personality through artificially applied techniques." The generalized obsession with controlling violence had also produced methods of "behavior prediction" that "raise profound questions with respect to due process, privacy, and individual liberties." A psychologist writing in the American Psychological Association's journal *Monitor* in 1974 sounded the alert, warning colleagues who touted their ability to "control behavior" that they were now "viewed with increasing suspicion, if not revulsion, and threatened with restriction... *The social control of behavior control is underway.*"[63]

The subcommittee's work had enduring consequences. Not only did prisoners' and patients' rights groups gain momentum in their efforts to end the behavioral oppression suffered in public institutions, but psychologists also began to discuss the need to professionalize their discipline with clear ethical standards, accreditation procedures, training programs, and career ladders.[64] The National Research Act, passed in 1974, stipulated the creation of institutional review boards and laid the foundation for the evolution and institutionalization of the Common Rule for the ethical treatment of human subjects, from which Facebook famously held itself exempt. That same year, Congress established the National Commission for the Protection of Human Subjects of Biomedical and Behavioral Research. When the commission published its findings five years later in the "Belmont Report," it became the professional standard for imposing ethical guidelines on all federally funded research with human subjects in the US.[65]

The clamorous rights consciousness of the 1970s drove behavior modification from civilian life, or at least dimmed its star. A Federal Bureau of Prisons official recommended that program leaders avoid using "the term 'behavioral modification' but to talk about positive reward and reinforcements for the type of behavior we are attempting to instill." Another said, "We're doing what we always did…but to call it 'behavior modification' just makes things more difficult."[66] Skinner's 1976 "primer," titled *About Behaviorism,* motivated by what he believed were public misconceptions stirred by the wave of harsh reaction to *Beyond Freedom & Dignity,* failed to capture much public attention. According to Skinner's biographer, "the battle had climaxed." The public had made *Beyond Freedom & Dignity* a bestseller "but had just as surely rejected Skinner's argument that there were cultural matters more important than preserving and extending individual freedom."[67]

Most fascinating is that throughout these years of anxiety and debate, it was impossible to imagine the means of behavioral modification as anything other than owned and operated by the government: a privileged modality of state power. A 1966 *Harvard Law Review* article addressed issues of electronic tracking, surveillance, and behavioral control, reasoning that it would "consider governmental attempts to change conduct, *since these seem more likely than private attempts.*"[68] The democratic impulse of US society, repelled by the excesses of its intelligence agencies, their support of criminal activities undertaken by the Nixon administration, and the migration of behavior modification as a means of disciplinary control in state institutions, led to the rejection of behavioral modification as an extension of governmental power.

Unknown to the senators, scholars, rights activists, litigators, and many other citizens who stood against the antidemocratic incursions of the behavioral engineering vision, these methods had not died. The project would resurface in a wholly unexpected incarnation as *a creature of the market,* its unprecedented digital capabilities, scale, and scope now flourishing under the flag of surveillance capitalism. During the same years that US democratic forces combined to resist behavior modification as a form of state power, the energies of capitalist counterinsurgency were already at work within society.

The corporation was to enjoy the rights of personhood but be free of democratic obligations, legal constraints, moral calculations, and social considerations. Certainly in the US case, a weakened state in which elected officials depend upon corporate wealth in every election cycle has shown little appetite to contest behavior modification as a market project, let alone to stand for the moral imperatives of the autonomous individual.

In its latest incarnation, behavioral modification comes to life as a global digital market architecture unfettered by geography, independent of constitutional constraints, and formally indifferent to the risks it poses to freedom, dignity, or the sustenance of the liberal order that Ervin's subcommittee was determined to defend. This contrast is even more distressing in light of the fact that in the mid-twentieth century, the means of behavior modification were aimed at individuals and groups who were construed as "them": military enemies, prisoners, and other captives of walled disciplinary regimes.

Today's means of behavioral modification are aimed unabashedly at "us." Everyone is swept up in this new market dragnet, including the psychodramas of ordinary, unsuspecting fourteen-year-olds approaching the weekend with anxiety. Every avenue of connectivity serves to bolster private power's need to seize behavior for profit. Where is the hammer of democracy now, when the threat comes from your phone, your digital assistant, your Facebook login? Who will stand for freedom now, when Facebook threatens to retreat into the shadows if we dare to be the friction that disrupts economies of action that have been carefully, elaborately, and expensively constructed to exploit our natural empathy, elude our awareness, and circumvent our prospects for self-determination? If we fail to take notice now, how long before we are numb to this incursion and to all the incursions? How long until we notice nothing at all? How long before we forget who we were before they owned us, bent over the old texts of self-determination in the dim light, the shawl around our shoulders, magnifying glass in hand, as if deciphering ancient hieroglyphs?

Throughout these chapters we have returned to the essential questions that define the division of learning: *Who knows? Who decides? Who decides who decides?* As to who knows, we have seen the titanic agglomerations of knowledge about our behavior in the shadow text, from vast patterns across populations to the intimate detail of individual lives. These new information territories are private and privileged, known only to the machines, their priests, and the market participants who pay to play in these new spaces. Although it is obviously the case that we are excluded because the knowledge is not for us, these chapters have revealed a deeper structural basis for our exclusion. Now we know that *surveillance capitalists' ability to evade our awareness is an essential condition for knowledge production.* We are excluded because we are friction that impedes the elaboration of the shadow text and with it surveillance capitalism's knowledge dominance.

As to who decides, this division of learning has been decided by the declarations and incursions of the owners of private surveillance capital as another essential condition of accumulation, enabled by the reluctance of the state to assert democratic oversight in this secret realm. Finally, there is the question of who decides who decides. So far, it is the asymmetrical power of surveillance capital unencumbered by law that decides who decides.

The commodification of behavior under the conditions of surveillance capitalism pivots us toward a societal future in which an exclusive division of learning is protected by secrecy, indecipherability, and expertise. Even when knowledge derived from your behavior is fed back to you in the first text as a quid pro quo for participation, the parallel secret operations of the shadow text capture surplus for crafting into prediction products destined for other marketplaces that are *about you* rather than *for you.* These markets do not depend upon you except first as a source of raw material from which surplus is derived, and then as a target for guaranteed outcomes. We have no formal control because we are not essential to the market action. In this future we are exiles from our own behavior, denied access to or control over knowledge derived from our experience. Knowledge, authority, and power rest with surveillance capital, for which we are merely "human natural resources."

CHAPTER ELEVEN

THE RIGHT TO THE
FUTURE TENSE

But He had planned such future for this youth:
Surely, His duty now was to compel,
To count on time to bring true love of truth
And, with it, gratitude. His eagle fell.

—W. H. AUDEN

SONNETS FROM CHINA, IX

I. *I Will to Will*

I wake early. The day begins before I open my eyes. My mind is in motion.
Words and sentences have streamed through my dreams, solving problems
on yesterday's pages. The first work of the day is to retrieve those words that
lay open a puzzle. Only then am I ready to awaken my senses. I try to discern
each birdcall in the symphony outside our windows: the phoebe, redwing,
blue jay, mockingbird, woodpecker, finch, starling, and chickadee. Soaring
above all their songs are the cries of geese over the lake. I splash warm water
on my face, drink cool water to coax my body into alertness, and commune
with our dog in the still-silent house. I make coffee and bring it into my
study, where I settle into my desk chair, call up my screen, and begin. I think.
I write these words, and I imagine you reading them. I do this every day of
every week—as I have for several years—and it is likely that I will continue to
do so for one or two years to come.

I watch the seasons from the windows above my desk: first green, then
red and gold, then white, and then back to green again. When friends come

to visit, they peek into my study. There are books and papers stacked on every surface and most of the floor. I know they feel overwhelmed at this sight, and sometimes I sense that they silently pity me for my obligation to this work and how it circumscribes my days. I do not think that they realize how free I am. In fact, I have never felt more free. How is this possible?

I made a promise to complete this work. This promise is my flag planted in the future tense. It represents my commitment to construct a future that cannot come into being should I abandon my promise. This future will not exist without my capacity first to imagine its facts and then to will them into being. I am an inchworm moving with determination and purpose across the distance between now and later. Each tiny increment of territory that I traverse is annexed to the known world, as my effort transforms uncertainty into fact. Should I renege on my promise, the world would not collapse. My publisher would survive the abrogation of our contract. You would find many other books to read. I would move on to other projects.

My promise, though, is an anchor that girds me against the vagaries of my moods and temptations. It is the product of my will to will and a compass that steers my course toward a desired future that is not yet real. Events may originate in energy sources outside my will and abruptly alter my course in ways that I can neither predict nor control. Indeed, they have already done so. Despite this certain knowledge of uncertainty, I have no doubt that I am free. I can promise to create a future, and I can keep my promise. If the book that I have imagined is to exist in the future, it must be because I will to will it so. I live in an expansive landscape that already includes a future that only I can imagine and intend. In my world, this book I write already exists. In fulfilling my promise, I make it manifest. *This act of will is my claim to the future tense.*

To make a promise is to predict the future; to fulfill a promise through the exercise of will turns that prediction into fact. Our hearts pump our blood, our kidneys filter that blood, and our wills create the future in the patient discovery of each new sentence or step. This is how we claim our right to speak in the first person as the author of our futures. The philosopher Hannah Arendt devoted an entire volume to an examination of will as the "organ for the future" in the same way that memory is our mental organ for the past. The power of will lies in its unique ability to deal with things, "visibles and invisibles, that have never existed at all. Just as the past always presents itself to the mind in the guise of certainty, the future's main characteristic is its basic uncertainty, no matter how high a degree of probability prediction may attain." When we refer to the past, we see only objects, but the view to the future brings "projects," things that are yet to be. With freedom of will we undertake action that is entirely contingent on our determination to see our project through. These are acts that we could have "left undone" but for our commitment. "A will that is not free," Arendt concludes, "is a contradiction in terms."[1]

Will is the organ with which we summon our futures into existence. Arendt's metaphor of will as the "mental organ of our future" suggests that it is something built into us: organic, intrinsic, inalienable. Moral philosophers have called this "free will" because it is the human counterpoint to the fear of uncertainty that suffocates original action. Arendt describes promises as "islands of predictability" and "guideposts of reliability" in an "ocean of uncertainty." They are, she argues, the only alternative to a different kind of "mastery" that relies on "domination of one's self and rule over others."[2]

Centuries of debate have been levied on the notion of free will, but too often their effect has been to silence our own declarations of will, as if we are embarrassed to assert this most fundamental human fact. I recognize my direct experience of freedom as an inviolate truth that cannot be reduced to the behaviorists' formulations of life as necessarily accidental and random, shaped by external stimuli beyond my knowledge or influence and haunted by irrational and untrustworthy mental processes that I can neither discern nor avoid.[3]

The American philosopher John Searle, whose work on the "declaration" we discussed in Chapter 6, comes to a similar conclusion in his examination of "free will." He points to the "causal gap" between the reasons for our actions and their enactment. We may have good reasons to do something, he observes, but that does not necessarily mean it will be done. "The traditional name of this gap in philosophy is 'the freedom of the will.'" In response to the "sordid history" of this concept, he reasons, "even if the gap is an illusion it is one we cannot shake off.... The notion of making and keeping promises presupposes the gap.... [It] requires consciousness and a sense of freedom on the part of the promise-making and promise-keeping agent."[4]

The freedom of will is the existential bone structure that carries the moral flesh of every promise, and my insistence on its integrity is not an indulgence in nostalgia or a random privileging of the pre-digital human story as somehow more truly human. This is the only kind of freedom that we can guarantee ourselves, no matter the weight of entropy or inertia, and irrespective of the forces and fears that attempt to collapse time into an eternity of shadowboxing now, and now, and now. These bones are the necessary condition for the possibility of civilization as a "moral milieu" that favors the dignity of the individual and respects the distinctly human capacities for dialogue and problem solving. Any person, idea, or practice that breaks these bones and tears this flesh robs us of a self-authored and we-authored future.

These principles are not quaint accessories, as Hal Varian and others suggest. Rather, they are hard-won achievements that have crystallized over millennia of human contest and sacrifice. Our freedom flourishes only as we steadily will ourselves to close the gap between making promises and keeping them. Implicit in this action is an assertion that through my will I can influence the future. It does not imply total authority over the future, of course, only over my piece of it. *In this way, the assertion of freedom of will also asserts the right to the future tense as a condition of a fully human life.*

Why should an experience as elemental as this claim on the future tense be cast as a human right? The short answer is that it is only necessary now because it is imperiled. Searle argues that such elemental "features of human life" rights are crystallized as formal human rights only at that moment in history when they come under systematic threat. So, for example, the ability to speak is elemental. The concept of "freedom of speech" as a formal right emerged only when society evolved to a degree of political complexity that the freedom to speak came under threat. The philosopher observes that speech is not more elemental to human life than breathing or being able to move one's body. No one has declared a "right to breathe" or a "right to bodily movement" because these elemental rights have not come under attack and therefore do not require formal protection. What counts as a basic right, Searle argues, is both "historically contingent" and "pragmatic."[5]

I suggest that we now face the moment in history when the elemental *right to the future tense* is endangered by a panvasive digital architecture of behavior modification owned and operated by surveillance capital, necessitated by its economic imperatives, and driven by its laws of motion, all for the sake of its guaranteed outcomes.

II. We Will to Will

Most simply put, there is no freedom without uncertainty; it is the medium in which human will is expressed in promises. Of course, we do not only make promises to ourselves; we also make promises to one another. When we join our wills and our promises, we create the possibility of collective action toward a shared future, linked in determination to make our vision real in the world. This is the origin of the institution we call "contract," beginning with the ancient Romans.[6]

Contracts originated as shared "islands of predictability" intended to mitigate uncertainty for the human community, and they still retain this meaning. "The simplest way to state the point of contract law is that it supports and shapes the social practice of making and keeping promises and agreements," concludes one eminent scholar. "Contract law focuses on problems of cooperation," summarizes another. "Contract law...reflects a moral ideal of equal respect for persons. This fact explains why contract law can produce genuine legal obligations and is not just a system of coercion," observes a third.[7]

It is in this context that the destructiveness of the uncontract is most clearly revealed. Recall Varian's assertion that if someone stops making monthly car payments, "Nowadays it's a lot easier just to instruct the vehicular monitoring system not to allow the car to be started and to signal the location where it can be picked up." Varian calls this new capability a "new contract form," when in reality the uncontract abandons the human world of legally binding promises and substitutes instead the positivist calculations of automated machine processes.[8] Without so much as a tip of the cap and a "fare-thee-well," Varian's uncontract disposes of several millennia of societal evolution during which Western civilization institutionalized the contract as a grand achievement of shared will.

It is no secret that the institution of the contract has been twisted and abused in every age, from the *Requirimiento* to the "slave contract," as incumbent power imposes painful inequalities that drain the meaning, and indeed the very possibility, of mutual promising.[9] For example, Max Weber warned that the great achievements of contractual freedom create opportunities to exploit property ownership as a means to "the achievement of power over others."[10]

However, today's uncontracts are unprecedented in their ability to impose unilateral power. They leverage the apparatus to combine pervasive monitoring and remote actuation for an internet-enabled "new economics" that bypasses human promises and social engagement.[11] The uncontract aims instead for a condition that the economist Oliver Williamson describes as "contract utopia": a state of perfect information known to perfectly rational people who always perform exactly as promised.[12] The problem is, as Williamson writes, "*All complex contracts are unavoidably incomplete*... parties will be confronted with the need to adapt to unanticipated disturbances that arise by reason of gaps, errors and omissions in the original contract."[13]

If you have ever seen a house built according to architectural plans, then you have a good idea of what Williamson means. There is no blueprint that sufficiently details everything needed to convert drawings and specifications into an actual house. No plan anticipates every problem that might arise, and most do not come close. The builders' skills are a function of how they collaborate to invent the actions that fulfill the intention of the drawings as they solve the unexpected but inevitable complications that arise along the way. They work together to construct a reality from the uncertainty of the plan.

Like builders, people in contractual agreements undertake this kind of collaboration. It's not simply finding the way through a maze to an already agreed-upon end point, but rather the continuous refinement and clarification of ends and means in the face of unanticipated obstacles. This sociality of contract may entail conflict, frustration, oppression, or anger, but it can also produce trust, cooperation, cohesion, and adaptation as the means through which human beings confront an unknowable future.

Were "contract utopia" to exist, Williamson says, it would best be described as a "plan" that, like other "utopian modes," requires "deep commitment to collective purposes" and "personal subordination." Subordination to what? To the plan. Contract in this context of perfect rationality is what Williamson describes as "a world of planning." Such planning was the basic institution of socialist economics, where the "new man" was idealized as possessing "a high level of cognitive competence" and therefore, it was espoused, could design highly effective plans.[14] Varian deftly swaps out socialism's "new man" and installs instead a market defined by surveillance capitalism's economic imperatives, expressed through a ubiquitous computational architecture, the machine intelligence capabilities to which data are continuously supplied, the analytics that discern patterns, and the algorithms that convert them into rules. This is the essence of the uncontract, which transforms the human, legal, and economic risks of contracts into plans constructed, monitored, and maintained by private firms for the sake of guaranteed outcomes: less contract utopia than *uncontract dystopia*.

In November 2016 the experience of three people in the small Illinois town of Belleville was testimony to what we lose when we subordinate ourselves to the dystopian rule of the uncontract. Pat and Stanford Kipping owed their credit union $350 on their 1998 Buick. Once again, they could not make their monthly $95 payment. The Kippings' credit union enlisted a local repo man, Jim Ford, to take away their car.

When Ford visited the Kippings' Belleville home, he was disturbed to find an elderly couple who were forced to choose between buying medicine and making their car payments. Ford's initial response was to waive his repossession fees. The Kippings generously thanked him, invited him in for tea, and shared their story. That's when Ford decided to bridge the gap between uncertain reality and the stipulations of their contract. He did the human thing, calling the credit union and offering to pay the couple's debt.

The credit union manager insisted that Ford had to follow the "process." Ford continued to invoke the ancient social principles of the contract, seeking a way through the maze to something that felt like justice. Ultimately the manager agreed to "work with" the couple to see what could be done. It didn't end there. Within twenty-four hours, an online fund-raising appeal produced enough to pay off the Kippings' car, detail it, purchase a Thanksgiving turkey, and give the couple an additional gift of $1,000.

Most interesting is that when a local paper picked up the story, it quickly went viral across the web and traditional media. Millions of people read and responded to this drama, presumably because it stirred memories of something precious and necessary but now threatened with extinction. Jim Ford reminded us of the most cherished requirements of a civilized life: our shared assertion of rights to the future tense and its expression in the joining of wills in mutual commitment to dialogue, problem solving, and empathy. He was eloquent on this point: "Just be nice to people. It's not that hard. The fact that this has gone so crazy is kind of sad. This should be a daily thing, a normal thing."[15]

In the dystopia of the uncontract, this daily human thing is not normal. What if the Kippings' credit union employed Spireon's telematics and merely had to instruct the vehicular monitoring system to disable the car? There would be no loan manager engaging in a give-and-take with customers. The algorithm tasked to eliminate the messy, unpredictable, untrustworthy eruptions of human will would have seized the old Buick. There would have been no shared tea time with the Kippings and no one to listen to their story. There would have been no opportunity to find an alternate route through the maze, no opportunity to build trust, no occasion for collective action, no heartwarming holiday story of kindness, no glimmer of hope for a human future in which the best of our institutions is preserved and fortified, no shared challenge of uncertainty, and no shared freedom.

In the dystopia of the uncontract, surveillance capitalism's drive toward certainty fills the space once occupied by all the human work of building and replenishing social trust, which is now reinterpreted as unnecessary friction in the march toward guaranteed outcomes. The deletion of uncertainty is celebrated as a victory over human nature: our cunning and our opportunism. All that's left to matter are the rules that translate reasons into action, the objective measures of behavior, and the degree of conformance between the two. Social trust eventually withers, a kind of vestigial oddity like a third nipple or wisdom teeth: traces of an evolutionary past that no longer appear in operational form because their context and therefore their purpose have vanished.[16]

The uncontract and the for-profit circuits of behavior modification in which it executes its objectives construe society as an acrid wasteland in which mistrust is taken for granted. By positing our lives together as already failed, it justifies coercive intervention for the sake of certainty. Against this background of the gradual normalization of the automated plan and its planners, the human response of one repo man bears simple witness to precisely what surveillance capitalism must extinguish.

Human replenishment from the failures and triumphs of choosing the future in the face of uncertainty gives way to the blankness of perpetual compliance. The word *trust* lingers, but its referent in human experience dissolves into reminiscence, an archaic footnote to a barely remembered dream of a dream that has long since faded for the sake of a new dictatorship of market reasons. As the dream dies, so too does our sense of astonishment and protest. We grow numb, and our numbness paves the way for more compliance. A pathological division of learning forged by unprecedented asymmetries of knowledge and power fixes us in a new inequality marked by the tuners and the tuned, the herders and herded, the raw material and its miners, the experimenters and their unwitting subjects, those who will the future and those who are shunted toward others' guaranteed outcomes.

So let us establish our bearings. Uncertainty is not chaos but rather the necessary habitat of the present tense. We choose the fallibility of shared promises and problem solving over the certain tyranny imposed by a dominant power or plan because this is the price we pay for the freedom to will, which founds our right to the future tense. In the absence of this freedom, the future collapses into an infinite present of mere behavior, in which there can be no subjects and no projects: only *objects*.

In the future that the surveillance capitalism prepares for us, my will and yours threaten the flow of surveillance revenues. Its aim is not to destroy us but simply to author us and to profit from that authorship. Such means have been imagined in the past, but only now are they feasible. Such means have been rejected in the past, but only now have they been allowed to root. We are ensnared without awareness, shorn of meaningful alternatives for withdrawal, resistance, or protection.

The promise of the promise and the will to will run deeper than these deformities. They remind us of that place again where we humans heal the breach between the known and the unknowable, navigating the seas of uncertainty in our vessels of shared promises. In the real world of human endeavor, there is no perfect information and no perfect rationality. Life inclines us to take action and to make commitments even when the future is unknown. Anyone who has brought a child into the world or has otherwise given his or her heart in love knows this to be true.

Gods know the future, but we move forward, take risks, and bind ourselves to others despite the fact that we can't know everything about the present, let alone the future. This is the essence of our freedom, expressed as the elemental right to the future tense. With the construction and ownership of the new means of behavioral modification, the fate of this right conforms to a pattern that we already have identified. It is not extinguished, but rather it is usurped: commandeered and accumulated by surveillance capital's exclusive claims on our futures.

III. How Did They Get Away with It?

In the course of the last ten chapters I have argued that surveillance capitalism represents an unprecedented logic of accumulation defined by new economic imperatives whose mechanisms and effects cannot be grasped with existing models and assumptions. This is not to say that the old imperatives—a compulsion toward profit maximization along with the intensification of the means of production, growth, and competition—have vanished. However, these must now operate through the novel aims and mechanisms of surveillance capitalism. I briefly review the new imperatives here, both as a summary of the ground that we have covered and as prelude to the question *How did they get away with it?*

Surveillance capitalism's new story begins with *behavioral surplus* discovered more or less ready-made in the online environment, when it was realized that the "data exhaust" clogging Google's servers could be combined with its powerful analytic capabilities to produce predictions of user behavior. Those *prediction products* became the basis for a preternaturally lucrative sales process that ignited new *markets in future behavior*.

Google's "machine intelligence" improved as the volume of data increased, producing better prediction products. This dynamic established *the extraction imperative*, which expresses the necessity of *economies of scale in surplus accumulation* and depends upon automated systems that relentlessly track, hunt, and induce more behavioral surplus. Google imposed the logic of conquest, defining human experience as free for the taking, available to be rendered as data and claimed as surveillance assets. The company learned to employ a range of rhetorical, political, and technological strategies to obfuscate these processes and their implications.

The need for scale drove a relentless search for new high-volume supplies of behavioral surplus, producing competitive dynamics aimed at cornering these supplies of raw material and seeking lawless undefended spaces in which to prosecute these unexpected and poorly understood acts of dispossession. All the while, surveillance capitalists stealthily but steadfastly habituated us to their claims. In the process, our access to necessary information and services became hostage to their operations, our means of social participation fused with their interests.

Lucrative *prediction products* depend upon behavioral surplus, and competition drove the supply challenges to a new level, expressed in the *prediction imperative.* More-powerful prediction products required *economies of scope* as well as scale, variation as well as volume. This variation occurs along two dimensions. The first is *extension* across a wide range of activities; the second is the *depth* of predictive detail within each activity.

In this new phase of competitive intensity, surveillance capitalists are forced from the virtual world into the real one. This migration necessitates new machine processes for the *rendition* of all aspects of human experience into behavioral data. Competition now occurs in the context of a rapidly evolving global architecture of ubiquitous computation and therefore ubiquitous supply opportunities, as prediction products are increasingly expected to approximate certainty and therefore to guarantee behavioral outcomes.

In a third phase of competitive intensity, surveillance capitalists discovered the necessity of *economies of action* based on new methods that go beyond tracking, capturing, analyzing, and predicting behavior in order to intervene in the state of play and actively shape behavior at the source. The result is that the means of production are subordinated to an elaborate new *means of behavioral modification,* which relies upon a variety of machine processes, techniques, and tactics (tuning, herding, conditioning) to shape individual, group, and population behavior in ways that continuously improve their approximation to guaranteed outcomes. Just as industrial capitalism was driven to the continuous intensification of the means of production, so surveillance capitalists are now locked in a cycle of continuous intensification of the means of behavioral modification.

Surveillance capitalists' interests have shifted from using automated machine processes to know about your behavior to using machine processes to shape your behavior according to their interests. In other words, this decade-and-a-half trajectory has taken us from *automating information flows about you* to *automating you.* Given the conditions of increasing ubiquity, it has become difficult if not impossible to escape this audacious, implacable web.

In order to reestablish our bearings, I have asked for a rebirth of astonishment and outrage. Most of all, I have asked that we reject the Faustian pact of participation for dispossession that requires our submission to the means of behavioral modification built on the foundation of the Google declarations. I am also mindful, though, that when we ask *How did they get away with it?* there are many compelling reasons to consider, no one of which stands alone. Instead of simple cause and effect, the answers to our question constitute a broad landscape of history, contingency, quicksand, and coercion.

Our question is even more vexing in light of the fact that in the great majority of surveys designed to probe public attitudes toward the loss of privacy and other elements of surveillance capitalist practices, few of us favor the status quo. In forty-six of the most prominent forty-eight surveys administered between 2008 and 2017, substantial majorities support measures for enhanced privacy and user control over personal data. (Only two early surveys were somewhat less conclusive, because so many participants indicated that they did not understand how or what personal information was being gathered.) Indeed, by 2008 it was well established that the more knowledge one has about "internet privacy practices," the more one is likely to be very concerned about privacy.[17]

Although the surveys vary in terms of their specific focus and questions, the general consistency of responses over the decade is noteworthy. For example, an important 2009 survey found that when Americans are informed of the ways that companies gather data for targeted online ads, 73–86 percent reject such advertising. Another substantial survey in 2015 found 91 percent of respondents disagreeing that the collection of personal information "without my knowing" is a fair trade-off for a price discount. Fifty-five percent disagreed that it was a fair exchange for improved services. In 2016 Pew Research reported only 9 percent of respondents as very confident in trusting social media sites with their data and 14 percent as very confident about trusting companies with personal data. More than 60 percent wanted to do more to protect their privacy and believed there should be more regulations to protect privacy.[18]

Surveillance capitalist firms have tended to dismiss these survey results, pointing instead to the spectacular growth of users and revenue. This discrepancy has confounded research and public policy. With so many people rejecting the practices of surveillance capitalism, even considering how little most of us actually know about these practices, how is it that this market form has been able to succeed? The reasons are plentiful:

1. *Unprecedented:* Most of us did not resist the early incursions of Google, Facebook, and other surveillance capitalist operations because it was impossible to recognize the ways in which they differed from anything that had gone before. The basic operational mechanisms and business practices were so new and strange, so utterly sui generis, that all we could see was a gaggle of "innovative" horseless carriages. Most significantly, anxiety and vigilance have been fixed on the known threats of surveillance and control associated with state power. Earlier incursions of behavior modification at scale were understood as an extension of the state, and we were not prepared for the onslaught from private firms.

2. *Declaration as invasion:* The lack of precedence left us disarmed and charmed. Meanwhile, Google learned the art of invasion by declaration, taking what it wanted and calling it theirs. The corporation asserted its rights to bypass our awareness, to take our experience and transform it into data, to claim ownership of and decisions over the uses of those data, to produce strategies and tactics that keep us ignorant of its practices, and to insist on the conditions of lawlessness required for these operations. These declarations institutionalized surveillance capitalism as a market form.

3. *Historical context:* Surveillance capitalism found shelter in the neoliberal zeitgeist that equated government regulation of business with tyranny. This "paranoid style" favored self-management regimes that imposed few limits on corporate practices. In a parallel development, the "war on terror" shifted the government's attention from privacy legislation to an urgent interest in the rapidly developing skills and technologies of Google and other rising surveillance capitalists. These "elective affinities" produced a trend toward surveillance exceptionalism, which further sheltered the new market form from scrutiny and nurtured its development.

4. *Fortifications:* Google aggressively protected its operations by establishing its utilities in the electoral process, strong relationships with elected

and appointed officials, a revolving door of staffers between Washington and Silicon Valley, lavish lobbying expenditures, and a steady "soft-power" campaign of cultural influence and capture.

5. *The dispossession cycle:* First at Google and later at Facebook and other firms, surveillance capitalist leaders mastered the rhythms and stages of dispossession. Audacious incursions are pursued until resistance is met, followed by a range of tactics from elaborate public relations gambits to legal combat, all designed to buy time for gradual habituation to once-outrageous facts. A third stage features public demonstrations of adaptability and even retreat, while in the final stage resources are redirected to achieve the same objectives camouflaged by new rhetoric and tactics.

6. *Dependency:* The free services of Google, Facebook, and others appealed to the latent needs of second-modernity individuals seeking resources for effective life in an increasingly hostile institutional environment. Once bitten, the apple was irresistible. As surveillance capitalism spread across the internet, the means of social participation become coextensive with the means of behavioral modification. The exploitation of second-modernity needs that enabled surveillance capitalism from the start eventually imbued nearly every channel of social participation. Most people find it difficult to withdraw from these utilities, and many ponder if it is even possible.

7. *Self-Interest:* New markets in future behavior give rise to networks of fellow travelers, partners, collaborators, and customers whose revenues depend on the prediction imperative. Institutional facts proliferate. The pizzeria owner on the Pokémon Go map, the merchant who saturates his shop with beacons, and the insurance companies vying for behavioral data unite in the race toward guaranteed outcomes and surveillance revenues.

8. *Inclusion:* Many people feel that if you are not on Facebook, you do not exist. People all over the world raced to participate in Pokémon Go. With so much energy, success, and capital flowing into the surveillance capitalist domain, standing outside of it, let alone against it, can feel like a lonely and risky prospect.

9. *Identification:* Surveillance capitalists aggressively present themselves as heroic entrepreneurs. Many people identify with and admire the financial success and popularity of the surveillance capitalists and regard them as role models.

10. *Authority:* Many also regard these corporations and their leaders as authorities on the future: geniuses who can see farther than the rest of us. It is easy to fall prey to the naturalistic fallacy, which suggests that because the companies are successful, they must also be right. As a result, many of us are respectful of these leaders' expert status and are eager to participate in innovations that anticipate the future.

11. *Social persuasion:* As we have seen repeatedly, there is an endless cascade of beguiling rhetoric aimed at persuading people of the wonders associated with surveillance capitalist innovations: targeted advertising, personalization, and digital assistants. Beyond that, economies of action are intentionally designed to persuade us to follow one another along prescribed courses of action.

12. *Foreclosed alternatives:* The "dictatorship of no alternatives" is in full force here. We have seen that the behavioral value reinvestment cycle is increasingly rare. The Aware Home gave way to the Google Home. Surveillance capitalism spread across the internet, and the drive toward economies of scope and action has forced it out into the real world. From apps to devices to the One Voice, it is ever more difficult to identify avenues of escape, let alone genuine alternatives.

13. *Inevitabilism:* The Trojan horse of computer mediation—devices, apps, connection—enters the scene in a relentless deluge of inevitabilist rhetoric, successfully distracting us from the highly intentional and historically contingent surveillance capitalism within. New institutional facts proliferate and stabilize the new practices. We fall into resignation and a sense of helplessness.

14. *The ideology of human frailty:* In addition to inevitabilism, surveillance capitalism has eagerly weaponized behavioral economics' ideology of human frailty, a worldview that frames human mentation as woefully irrational and incapable of noticing the regularity of its own failures. Surveillance capitalists employ this ideology to legitimate their means of behavior modification: tuning, herding, and conditioning individuals and populations in ways that are designed to elude awareness.

15. *Ignorance:* This remains a salient explanation. Surveillance capitalists dominate an abnormal division of learning in which they know things that we cannot know while compelled to conceal their intentions and practices

in secret backstage action. It is impossible to understand something that has been crafted in secrecy and designed as fundamentally illegible. These systems are intended to ensnare us, preying on our vulnerabilities bred by an asymmetrical division of learning and amplified by our scarcity of time, resources, and support.

16. ***Velocity:*** Surveillance capitalism rose from invention to domination in record time. This reflects its ability to attract capital and its laws of motion, but it also reflects a specific strategy in which velocity is consciously deployed to paralyze awareness and freeze resistance while distracting us with immediate gratifications. Surveillance capitalism's velocities outrun democracy even as they outrun our ability to understand what is happening and consider the consequences. This strategy is borrowed from a long legacy of political and military approaches to the production of speed as a form of violence, most recently known as "shock and awe."[19]

These sixteen answers suggest that in the nearly two decades since the invention of surveillance capitalism, existing laws, largely centered on privacy and antitrust, have not been sufficient to disrupt its growth. We need laws that reject the fundamental legitimacy of surveillance capitalism's declarations and interrupt its most basic operations, including the illegitimate rendition of human experience as behavioral data; the use of behavioral surplus as free raw material; extreme concentrations of the new means of production; the manufacture of prediction products; trading in behavioral futures; the use of prediction products for third-order operations of modification, influence, and control; the operations of the means of behavioral modification; the accumulation of private exclusive concentrations of knowledge (the shadow text); and the power that such concentrations confer.

The rejection of these new institutions of surveillance capital and the declarations upon which they are built would signify a withdrawal of social agreement to surveillance capitalism's aims and methods in the same way that we once withdrew agreement to the antisocial and antidemocratic practices of raw industrial capitalism, righting the balance of power between employers and workers by recognizing workers' rights to collective bargaining and outlawing child labor, hazardous working conditions, excessive work hours, and so on.

The withdrawal of agreement takes two broad forms, a distinction that will be useful as we move into Part III. The first is what I call the *counter-declaration.* These are defensive measures such as encryption and other privacy tools, or arguments for "data ownership." Such measures may be effective in discrete situations, but they leave the opposing facts intact, acknowledging their persistence and thus paradoxically contributing to their legitimacy. For example, if I "opt out" of tracking, I opt out for me, but my action does not challenge or alter the offending practice. The second form of disagreement is what I call the *synthetic declaration*. If the declaration is "check," the counter-declaration is "checkmate," and the synthetic declaration changes the game. It asserts an alternative framework that transforms the opposing facts. We bide our time with counter-declarations and make life more tolerable, but only a synthetic alternative vision will transform raw surveillance capitalism in favor of a digital future that we can call home.

I turn to the history of the Berlin Wall as an illustration of these two forms of disagreement. From 1961 through the early 1980s, courageous East Berliners carved seventy-one tunnels through the sandy soil beneath the city, affording several hundred people a means of escape to West Berlin.[20] The tunnels are testament to the necessity of counter-declarations, but they did not bring down the wall or the power that sustained it.

The synthetic declaration gathered force over decades, but its full expression would have to wait until near midnight on November 9, 1989, when Harald Jäger, the senior officer on duty that night at the Bornholmer Street passage, gave the order to open the gates, and twenty thousand people surged across the wall into West Berlin. As one historian describes that event, "By the night of November 9, when the people appeared at the Berlin Wall and demanded to know of the border officials, *Will you let us pass?*, those people had become so certain of themselves, and the officials so unsure of themselves, that the answer was *We will.*"[21]

IV. Prophecy

Nearly seventy years ago, the economic historian Karl Polanyi reflected on the ways in which industrial capitalism's market dynamics would, if left unchecked, destroy the very things that it aimed to buy and sell: "The commodity fiction disregarded the fact that leaving the fate of soil and people to the market would be tantamount to annihilating them."[22] In the absence of synthetic declaration, Polanyi's prophecy appears headed for fulfillment, and this fact alone must put us on alert. What does Polanyi's prophecy augur for our time?

Industrial capitalism followed its own logic of shock and awe, taking aim at nature to conquer "it" in the interests of capital; now surveillance capitalism has *human* nature in its sights. We have only gradually come to understand that the specific methods of domination employed by industrial capitalism for more than two centuries have fundamentally disoriented the conditions that support life on Earth, violating the most basic precepts of civilization. Despite the many benefits and immense accomplishments of industrial capitalism, it has left us perilously close to repeating the fate of the Easter Islanders, who wrecked the ground that gave them life, then fashioned statues to scan the horizon for the aid and succor that would never come. *If industrial capitalism dangerously disrupted nature, what havoc might surveillance capitalism wreak on human nature?*

The answer to this question requires a return to imperatives. Industrial capitalism brought us to the brink of epic peril, but not as a consequence of an evil lust for destruction or runaway technology. Rather, this result was ineluctably driven by its own inner logic of accumulation, with its imperatives of profit maximization, competition, the relentless drive for labor productivity through the technological elaboration of production, and growth funded by the continuous reinvestment of surplus.[23] It is Weber's "economic orientation" that matters, and how that orientation merges with the specific form of capitalism that rises to dominance in each age.

The logic of industrial capitalism exempted the enterprise from responsibility for its destructive consequences, unleashing the destabilization of the climate system and the chaos it spells for all creatures. Polanyi understood that raw capitalism could not be cooked from within. He argued that it was up to society to impose those obligations on capitalism by insisting on measures that tether the capitalist project to the social, preserving and sustaining life and nature.

Similarly, the meaning of Polanyi's prophecy for us now can be grasped only through the lens of surveillance capitalism's economic imperatives as they frame its claim to human experience. If we are to rediscover our sense of astonishment, then let it be here: *if industrial civilization flourished at the expense of nature and now threatens to cost us the Earth, an information civilization shaped by surveillance capitalism will thrive at the expense of human nature and threatens to cost us our humanity.* Polanyi's prophecy requires us to ask if we may yet avert this fate with our own synthetic declarations.

Parts I and II have been devoted to understanding the origins of surveillance capitalism and identifying, naming, and scrutinizing its foundational mechanisms and economic imperatives. The idea from the start was that naming and taming are inextricable, that fresh and careful naming can better equip us to intercept these mechanisms of dispossession, reverse their action, produce urgently needed friction, challenge the pathological division of learning, and ultimately synthesize new forms of information capitalism that genuinely meet our needs for effective life. Social participation and individual effectiveness should not require the sacrifice of our right to the future tense, which comprises our will to will, our autonomy, our decision rights, our privacy, and, indeed, our human natures.

However, it would be wrong to suppose that surveillance capitalism can be grasped solely through the prism of its economic action or that the challenges we face are restricted to discerning, containing, and transforming its foundational mechanisms. The consequences of this new logic of accumulation have already leaked and continue to leak beyond commercial practices into the fabric of our social relations, transforming our relationships to ourselves and to one another. These transformations provide the soil in which surveillance capitalism has flourished: an invasive species that creates its own food supply. In transforming us, it produces nourishment for its own march forward.

It is easier, perhaps, to see these dynamisms by looking to the past. The difference between industrial capitalism and industrial civilization is the difference between the economic operation and the societies it produced. The variant of industrial capitalism that rose to dominance in the late nineteenth and early twentieth centuries produced a specific kind of moral milieu that we sense intuitively even when we do not name it.

Industrial capitalism was marked by the specialized division of labor, with its historically specific characteristics: the conversion of craft work to mass production based on standardization, rationalization, and the interchangeability of parts; the moving assembly line; volume production; large populations of wage earners concentrated in factory settings; professionalized administrative hierarchies; managerial authority; functional specialization; and the distinction between white-collar work and blue-collar work.

The list is illustrative, not exhaustive, but enough to remind us that industrial civilization was drawn from these expressions of the economic imperatives that ruled industrial expansion. The division of labor shaped culture, psychology, and social experience. The shift from craft to hourly wages created new populations of employees and consumers, men and women wholly dependent on the means of production owned and operated by private firms.

This was the crucible of mass society, its hierarchical authority, and its centralized bureaucratic forms of public and private power, all of it haunted by the specters of conformity, obedience, and human standardization. Lives were defined by institutions that mirrored industrial organization: schools, hospitals, and even aspects of family and domestic life, in which ages and stages were understood as functions of the industrial system, from training to retirement.

At a time when surveillance capitalism has emerged as the dominant form of information capitalism, we must ask the question: what kind of civilization does it foretell? The chapters that follow in Part III are intended as an initial contribution to this urgent conversation. I have said that there can be no commitment to "guaranteed outcomes" without the power to make it so. What is the nature of this new power? How will it transform our societies? What solution for a third modernity does it proffer? What novel struggles will haunt these new days, and what do they portend for a digital future that we can call home? These are the questions that guide us into Part III.

PART III

INSTRUMENTARIAN POWER FOR A THIRD MODERNITY

CHAPTER TWELVE

TWO SPECIES OF POWER

So an age ended, and its last deliverer died
In bed, grown idle and unhappy; they were safe:
The sudden shadow of a giant's enormous calf
Would fall no more at dusk across their lawns outside.

—W. H. AUDEN

SONNETS FROM CHINA, X

I. A Return to the Unprecedented

Under surveillance capitalism, the "means of production" serves the "means of behavioral modification." Machine processes replace human relationships so that certainty can replace trust. This new assembly relies upon a vast digital apparatus, world-historic concentrations of advanced computational knowledge and skill, and immense wealth. The arc of behavioral modification at scale integrates the many operations that we have examined: ubiquitous extraction and rendition, actuation (tuning, herding, conditioning), behavioral surplus supply chains, machine-intelligence–based manufacturing processes, fabrication of prediction products, dynamic behavioral futures markets, and "targeting," which leads to fresh rounds of tuning, herding, conditioning, and the coercions of the uncontract, thus renewing the cycle.

This assembly is a market project: its purpose is to fabricate predictions, which become more valuable as they approach certainty. The best predictions feed on totalities of data, and on the strength of this movement toward totality, surveillance capitalists have hijacked the division of learning in society. They command knowledge from the decisive pinnacle of the social order, where they nourish and protect the shadow text: the urtext of certainty. This is the market net in which we are snared.

In Parts I and II we examined the conditions, mechanisms, and operations that construct this private knowledge kingdom and its lucrative predictions that evolve toward certainty in order to guarantee market players the outcomes that they seek. As I wrote in Chapter 7, there can be no *guarantee* of outcomes without the power to make it so. This is the dark heart of surveillance capitalism: a new type of commerce that reimagines us through the lens of its own distinctive power, mediated by its means of behavioral modification. What is this power, and how does it remake human nature for the sake of its lucrative certainties?

As to this species of power, I name it *instrumentarianism,* defined as *the instrumentation and instrumentalization of behavior for the purposes of modification, prediction, monetization, and control.* In this formulation, "instrumentation" refers to the puppet: the ubiquitous connected material architecture of sensate computation that renders, interprets, and actuates human experience. "Instrumentalization" denotes the social relations that orient the puppet masters to human experience as surveillance capital wields the machines to transform us into means to others' market ends. Surveillance capitalism forced us to reckon with an unprecedented form of capitalism. Now the instrumentarian power that sustains and enlarges the surveillance capitalist project compels a second confrontation with the unprecedented.

When scholars, civil society leaders, journalists, public figures, and, indeed, most of us speak out courageously against this new power, invariably we look to Orwell's Big Brother and more generally the specter of totalitarianism as the lens through which to interpret today's threats. Google, Facebook, and the larger field of commercial surveillance are frequently depicted as "digital totalitarianism."[1] I admire those who have stood against the incursions of commercial surveillance, but I also suggest that the equation of instrumentarian power with totalitarianism impedes our understanding as well as our ability to resist, neutralize, and ultimately vanquish its potency. There is no historical precedent for instrumentarianism, but there is vivid precedent for this kind of encounter with an unprecedented new species of power.

In the years before totalitarianism was named and formally analyzed, its critics appropriated the language of imperialism as the only framework at hand with which to articulate and resist the new power's murderous threats. Now surveillance capitalism has cast us adrift in another odd, dark sea of novel and thus indiscernible dangers. As scholars and citizens did before us, it is we who now reach for familiar vernaculars of twentieth-century power like lifesaving driftwood.

We are back to the syndrome of the horseless carriage, where we attach our new sense of peril to old, familiar facts, unaware that the conclusions to which they lead us are necessarily incorrect. Instead, we need to grasp the specific inner logic of a conspicuously twenty-first-century conjuring of power for which the past offers no adequate compass. Totalitarianism was bent on the reconstruction of the human species through the dual mechanisms of genocide and the "engineering of the soul." Instrumentarian power, as we shall see, takes us in a sharply different direction. Surveillance capitalists have no interest in murder or the reformation of our souls. Although their aims are in many ways just as ambitious as those of totalitarian leaders, they are also utterly distinct. The work of naming a strange form of power unprecedented in the human experience must begin anew for the sake of effective resistance and the creative power to insist on a future of our own making.

The remainder of this chapter prepares the way. The first task is to develop our understanding of what instrumentarian power is *not,* so in the section that follows we briefly consider key elements of twentieth-century totalitarianism. Most important is the fact that, like instrumentarianism, totalitarian power was also unprecedented. It literally defied human comprehension. There is much that we can learn from the struggles and missteps of scholars, journalists, and citizens as they found themselves overwhelmed by a force that they could neither fathom nor resist. Once we have tackled these questions, we will be equipped to bear down on an exploration of instrumentarianism's origins in a field of intellectual endeavor that came to be known as "radical behaviorism," most notably championed by B. F. Skinner and his dream of a "technology of behavior." In Chapter 13 we integrate our insights to consider the unique aims and strategies of instrumentarian power.

II. Totalitarianism as a New Species of Power

The word *totalitarianism* first appeared in the early-twentieth-century work of Italian philosopher Giovanni Gentile and came into wider use later with Mussolini's *The Doctrine of Fascism,* cowritten in 1932 with Gentile, who was by then Italy's premier philosopher of fascism.[2] Italy had entered the twentieth century as a second-rate country, ignored on the world stage, nursing a sense of failure and humiliation, and unable to support its own population as millions emigrated in search of a better life. In the first decade of the twentieth century a new generation of intellectuals and avant-garde futurists began to weave the dream of a "new Italy." Gentile dedicated his philosophical talents to this revival of nationalist zeal.

At the heart of Gentile's political philosophy is the concept of the "total."[3] The state was to be understood as an inclusive organic unity that transcends individual lives. All separateness and difference are surrendered to the state for the sake of this superordinate totality. In 1932 Mussolini charged Gentile with writing the philosophical introduction to his book, while Mussolini authored the social and political principles that would define the fascist worldview.[4] The *Doctrine* begins by declaring the fascist attitude as, above all, "a spiritual attitude" that penetrates the most intimate redoubt of each human participant:

> To know men one must know man.... Fascism is totalitarian and the Fascist State—a synthesis and a unit inclusive of all values—interprets, develops, and potentates the whole life of a people.... [It] is an inwardly accepted standard and rule of conduct, a discipline of the whole person; it permeates the will no less than the intellect... sinking deep down into his personality; it dwells in the heart of the man of action and of the thinker, of the artist, and of the man of science: soul of the soul.... It aims at refashioning not only the forms of life but their content—man, his character, and his faith... entering into the soul and ruling with undisputed sway.[5]

That year, the refashioning of the soul as the hallmark of the totalitarian impulse was immortalized by Stalin on a glittering, champagne-soaked

Moscow evening. The setting was an auspicious literary gathering hosted by a compliant Maxim Gorky in the sprawling mansion that Stalin had presented to the revered author upon his return to Russia from a self-imposed Italian exile. Stalin took the floor for a toast as the room fell silent. "Our tanks are worthless if the souls who must steer them are made of clay. This is why I say: The production of souls is more important than that of tanks.... Man is reshaped by life itself, and those of you here must assist in reshaping his soul. That is what is important, the production of human souls. And that is why I raise my glass to you, writers, to the *engineers of the soul*."[6] The authors assembled around Stalin that evening raised their glasses to his toast, persuaded perhaps by memories of less adaptive colleagues already exiled or executed, including the 1929 torture and murder of artists and writers in the Solovetsky Islands' aptly named Church of the Beheading.[7]

By 1933, the term *totalitarianism* had begun to circulate widely in Germany. Minister of Propaganda Joseph Goebbels adopted it, and German intellectuals declared the "turn to totalitarianism." Nazism also shifted the doctrine in an important way, asserting the "movement," not the "state," as the spiritual center of German totalitarianism, a relationship summarized during Hitler's first years as chancellor in National Socialism's popular slogan "The movement gives orders to the state."[8]

That totalitarianism was a new species of power had confounded its analysis from the start, as both its Russian and German variants swept through those societies, challenging the foundations of Western civilization. Although these totalitarian regimes began to take root years before World War II—first in Russia in 1929 with Stalin's ascension to power and then in Germany in 1933 with Hitler's installation as chancellor—they eluded systematic study until the end of the war. Analysis was impeded in part by the sheer mystery and perpetual movement of the whole enterprise: the secret plans executed by secret police, the silent complicities and hidden atrocities, the ceaseless transformation of who or what was up or down, the intentional torsion of facts into anti-facts accompanied by a perpetual deluge of propaganda, misinformation, euphemism, and mendacity. The authoritative leader, or "egocrat," to use the French philosopher Claude Lefort's term, displaces the rule of law and "common" sense to become the quixotic judge of what is just or unjust, truth or lie, at each moment.[9]

Western publics, especially in the US, were genuinely unable to grasp the enormity of what was underway. It literally boggled minds. This intellectual paralysis is immortalized in the pages of a cultural icon of that era, *Look* magazine. Its August 15, 1939, issue featured an article titled "What's Going On in Russia?" written by *New York Times* former Moscow bureau chief and Pulitzer Prize winner Walter Duranty.[10] The piece appeared just months after the conclusion of the Great Terror, when, between 1937 and 1938, Stalin ordered the murders of whole sectors of the Soviet population, from poets to diplomats, generals to political loyalists. According to Soviet historian Robert Conquest, that two-year period saw seven million arrests, one million executions, two million deaths in labor camps, one million people imprisoned, and another seven million people still in camps by the end of 1938.[11]

Despite the immediacy of catastrophic evil, Duranty's article describes the Constitution of the USSR as one of the "the most democratic in the world...a foundation on which a future democracy may be built." In addition to praiseworthy descriptions of the Red Army, free education and medical care, communal housing, and equality of the sexes, there is an upbeat commentary in which the "great purge" is breezily described as "one of the periodic cleansings of the Communist Party." Duranty reports that this "cleansing" is "now over," and people are "repairing the damage," as if the country were tidying up after a particularly nasty winter storm. In fact, the Stalinist pattern of violence, imprisonment, exile, and execution merely shifted its focus with a swift and terrifying ferocity to the Baltic and eastern Poland. Among the many atrocities between 1939 and 1941, hundreds of thousands of Poles were marched to northern labor camps,[12] and tens of thousands of members of the Polish Communist Party were murdered.[13] Just one week after Duranty's article, Stalin signed a nonaggression pact with Hitler, attacked Poland in September, and in November the Red Army invaded Finland.[14] In 1940 Stalin ordered the massacre of 15,000 Polish nationalists taken as prisoners of war in the 1939 attack.[15]

The most startling feature of Duranty's essay is its characterization of Stalin himself. There, sandwiched between a celebratory feature on a newly released film called *The Wizard of Oz* and a lengthy spread on embarrassing celebrity pictures, such as the famed ventriloquist's dummy Charlie McCarthy with a cigarette in his wooden mouth, is a photo of a handsome smiling

Joseph Stalin captioned "Stalin, chairman of the inner circle of the Communist Party…does not lay down the law as Lenin did. Stalin prefers to hear the views of his associates before he makes his own decision."[16] Stalin's 1939 lionization in *Look* as an exemplar of participatory management was followed a few months later with his ascension to *Time* magazine's cover as "Man of the Year." Indeed, between 1930 and his death in 1953, Stalin made a total of ten appearances on the cover of *Time*. All of this provides some sense of totalitarianism's elaboration and institutionalization long before it was identified and analyzed as a coherent new form of power that, as many scholars would conclude, posed history's greatest threat to civilization.[17]

With a few important exceptions, it was only after the Nazi defeat that the program of naming began in earnest. "Plenty of information was available contradicting the official picture," writes Conquest. He asks why "journalists, sociologists, and other visitors" were taken in by the Soviet regime's lies. One reason is that the Soviet government went to a great deal of trouble to present a false picture, including "model prisons" that betrayed no trace of the immense state machinery of torture and death. Another reason was the credulity of the observers themselves. In some cases, like Duranty's, they were blinded by an ideological allegiance to the idea of a socialist state.[18]

The most compelling reason of all is that in most of these cases, journalists, scholars, and Western governments had a difficult time reckoning the full weight of totalitarianism's monstrous achievements because the actual facts were so "improbable" that it was difficult even for specialists to grasp their truth. "The Stalin epoch," writes Conquest, "is replete with what appear as improbabilities to minds unfitted to deal with the phenomena."[19] This failure of comprehension has immediate significance for us as we learn to reckon with surveillance capitalism and its new instrumentarian power.

The confrontation with totalitarianism's impossibility is reflected in the moving accounts of the first scholars determined to lift the veil on that era's gruesome truths. Nearly every intellectual who turned to this project in the period immediately following the war cites the feeling of astonishment at the suddenness with which, as Harvard political scientist Carl Friedrich put it, totalitarianism had "burst upon mankind…unexpected and unannounced."[20] Its manifestations were so novel and unanticipated, so shocking, rapid, and unparalleled, that all of it eluded language, challenging every tradition, norm, value, and legitimate form of action. The systematic accretion of violence and complicity that engulfed whole populations at extreme velocity invoked a kind of bewilderment that ended in paralysis, even for many of the greatest minds of the twentieth century.

Friedrich was among the first scholars of totalitarianism to address this experience of improbability, writing in 1954 that "virtually no one before 1914 anticipated the course of development which has overtaken Western civilization since then...none of the outstanding scholars in history, law, and the social sciences discerned what was ahead...which culminated in totalitarianism. To this failure to foresee corresponds a difficulty in comprehending."[21] Not even the most farsighted of the early-century interpreters of industrial society, thinkers such as Durkheim and Weber, had anticipated this murderous turn. Hannah Arendt described the defeat of Nazi Germany as "the first chance to try to tell and to understand what had happened...still in grief and sorrow and...a tendency to lament, but no longer in speechless outrage and impotent horror."[22]

Ultimately, a courageous and brilliant body of scholarship would evolve to meet the challenge of comprehension. It yielded different models and schools of thought, each with distinct emphasis and insights, but these shared common purpose in finally naming the great evil. "Totalitarianism has discovered a means of dominating and terrorizing human beings from within," wrote Arendt, the German-born philosopher who would spend the six years after World War II writing her extraordinary study of totalitarian power, published in 1951 as *The Origins of Totalitarianism*.[23]

Arendt's was a detailed disclosure and a pioneering attempt to theorize what had just occurred. "Comprehension," she said, is the necessary response to the "truly radical nature of Evil" disclosed by totalitarianism. "It means... examining and bearing consciously the burden which our century has placed on us—neither denying its existence nor submitting meekly to its weight." Totalitarianism was bent on the "destruction of humanity" and "the essence of man," and, she insisted, "to turn our backs on the destructive forces of the century is of little avail."[24] Essential to totalitarianism was the deletion of all ties and sources of meaning other than "the movement": "Total loyalty—the psychological basis for domination—can be expected only from the completely isolated human being who, without any other social ties to family, friends, comrades, or even mere acquaintances, derives his sense of having a place in the world only from his belonging to a movement, his membership in the party."[25]

Mid-century scholars such as Friedrich, Adorno, Gurian, Brzezinski, and Aron added to these themes, recognizing totalitarianism's insistence on domination of the human soul.[26] To command populations right down to their souls requires unimaginable effort, which was one reason why totalitarianism was unimaginable. It requires henchmen, and their henchmen, and their henchmen, all willing to roll up both sleeves and thrust both hands into the blood and shit of actual living persons whose bodies stink and sweat and cry out in terror, grief, and pain.[27] It measures success at the cellular level, penetrating to the quick, where it subverts and commands each unspoken yearning in pursuit of the genocidal vision that historian Richard Shorten calls "the experiment in reshaping humanity."[28]

The destruction and reconstruction of society and the purification of the human species were prosecuted in the name of "class" in Stalin's Soviet Union and "race" in Hitler's Germany. Each regime invented "out-groups" slated for murder—the Jewish people, Romanies, homosexuals, and revolutionaries in Germany and Eastern Europe, whole segments of the population in Stalin's Russia—and "in-groups" required to submit body and soul to the regime.[29] In this way totalitarian regimes could achieve their fantastical aim of the "People-as-one," as Claude Lefort describes it: "Social unanimity corresponds to inner unanimity, held in place by hatred activated toward the 'enemies of the people.'"[30]

Totalitarian power cannot succeed by remote control. Mere conformity is insufficient. Each individual inner life must be claimed and transformed by the perpetual threat of punishment without crime. Mass murder warrants economies of scale—the camps, massacres, and gulags—but for the rest it would be a handmade terror that aims to remake every aspect of the individual from the inside out: heart, mind, sexuality, personality, and spirit. This craftwork requires the detailed orchestration of isolation, anxiety, fear, persuasion, fantasy, longing, inspiration, torture, dread, and surveillance. Arendt describes the relentless process of "atomization" and fusion in which terror destroys the ordinary human bonds of law, norms, trust, and affection, "which provide the living space for the freedom of the individual." The "iron band" of terror "mercilessly presses men...against each other so that the very space of free action...disappears." Terror "fabricates the oneness of all men."[31]

III. *An Opposite Horizon*

Instrumentarian power moves differently and toward an opposite horizon. Totalitarianism operated through the means of violence, but instrumentarian power operates through the means of behavioral modification, and this is where our focus must shift. Instrumentarian power has no interest in our souls or any principle to instruct. There is no training or transformation for spiritual salvation, no ideology against which to judge our actions. It does not demand possession of each person from the inside out. It has no interest in exterminating or disfiguring our bodies and minds in the name of pure devotion. It welcomes data on the behavior of our blood and shit, but it has no interest in soiling itself with our excretions. It has no appetite for our grief, pain, or terror, although it eagerly welcomes the behavioral surplus that leaches from our anguish. It is profoundly and infinitely indifferent to our meanings and motives. Trained on measurable action, it only cares that whatever we do is *accessible* to its ever-evolving operations of rendition, calculation, modification, monetization, and control.

Although it is not murderous, instrumentarianism is as startling, incomprehensible, and new to the human story as totalitarianism was to its witnesses and victims. Our encounter with unprecedented power helps to explain why it has been difficult to name and know this new species of coercion, shaped in secret, camouflaged by technology and technical complexity, and obfuscated by endearing rhetoric. Totalitarianism was a political project that converged with economics to overwhelm society. Instrumentarianism is a market project that converges with the digital to achieve its own unique brand of social domination.

It is not surprising, therefore, that instrumentarianism's specific "viewpoint of observation" was forged in the controversial intellectual domain known as "radical behaviorism" and its antecedents in turn-of-the-century theoretical physics. In the remainder of this chapter, our examination of power in the time of surveillance capitalism pivots to this point of origin far from totalitarianism's murder and mayhem. It takes us to laboratories and classrooms and the realms of thought spun by men who regarded freedom as a synonym for ignorance and human beings as distant organisms imprisoned in patterns of behavior beyond their own comprehension or control, such as ants, bees, or Stuart MacKay's herds of elk.

IV. The Other-One

In a 1971 cover story, *Time* magazine described Burrhus Frederic "B. F." Skinner as "an institution at Harvard...the most influential of living American psychologists, and the most controversial contemporary figure in the science of human behavior, adored as a messiah and abhorred as a menace."[32] Skinner spent most of his career in the Psychology Department at Harvard University, and some of my most vivid memories of graduate school are the times I spent with him in close debate. I admit that those conversations did little to alter our respective views, but they left me with an indelible sense of fascination with a way of construing human life that was—and is—fundamentally different from my own.

As an academic psychologist, Skinner was famous for the ingenious tools and techniques he invented to study animal behavior—first in the ancient murky basement of Memorial Hall and later on the seventh floor of the newly constructed William James Hall—as well as the insights that he and his students developed into the shaping of that behavior: pigeons and levers, pigeons and pellets, pigeons on hotplates, rats in mazes. His early research broke new ground as he engineered variable "schedules of reinforcement" to produce detailed patterns of activity that were foreign to an animal's original behavioral repertoire, work that he called "operant conditioning."

Inspired by his effort during World War II to transform a flock of pigeons into the conditioned intelligence for guided missiles (one that did not ultimately come to fruition), Skinner set out on a new path defined by the promise of behavioral engineering. In 1947 he told a symposium, "It is not a matter of bringing the world into the laboratory, but of extending the practices of an experimental science *to the world at large*. We can do this as soon as we wish to do it."[33] The missile project had cast the meaning of his experimental work "in a new light," he reflected years later in his autobiography. "It was no longer merely an experimental analysis. It had given rise to a technology."[34]

Skinner was eager to apply his laboratory insights to the world's ills despite precious few grounds for his inferential leaps. As a public intellectual, he spent nearly seven decades trying to persuade the public that his radical behaviorism offered the principles of social organization necessary to defend civilization from cataclysm. He brashly extrapolated from the conduct of beleaguered animals to grand theories of social behavior and human evolution in books such as his 1948 "utopian" novel *Walden Two* and his 1971 social philosophy *Beyond Freedom & Dignity*. In 1974 Skinner published *About Behaviorism*, another explanation of the radical behaviorist project, this time addressed to a general readership. It aimed to counter the opposition to his views that had grown even more virulent since the unusual—and, to many, repugnant—arguments advanced in *Beyond Freedom & Dignity*. He regarded such opposition as the result of an "extraordinary misunderstanding," and he was tireless in his efforts to reverse public opinion. He believed that once people correctly understood his meaning, they would certainly agree with his message.

In the very first pages of *About Behaviorism*, Skinner ignores the outrage generated by *Beyond Freedom & Dignity*, instead delving into behaviorism's roots and its earliest theorists and practitioners. He places much of the blame for the antipathy toward behaviorism on the man who is widely regarded as its founder, John B. Watson.[35] It was Watson who famously announced in 1913 the behaviorist point of view: "Psychology as the behaviorist views it is a purely objective experimental branch of natural science. Its theoretical goal is the prediction and control of behavior. Introspection forms no essential part of its methods.... The behaviorist...recognizes...no dividing line between man and brute."[36] But Watson turned out to be as much showman as scientist, and Skinner bitterly criticizes his extreme claims and "shortage of facts" that cast lasting doubt on radical behaviorism.

Having identified Watson as the main author of the problem, Skinner then credits the work of Max Meyer, an obscure early-twentieth-century German-trained experimental psychologist who spent most of his career at the University of Missouri, as key to the solution. Meyer had pursued his doctoral studies at the University of Berlin, where his dissertation advisor, Max Planck, was destined to become one of the most celebrated physicists of all time. Planck insisted on the unity of the physical world and the discoverability of the natural laws that would disclose their secrets only through mathematical analysis, including the laws of human behavior.[37] "The outside world is something independent from man, something absolute," Planck wrote. "The quest for the laws which apply to this absolute appeared...as the most sublime scientific pursuit in life."[38] Meyer carried Planck's teaching into his quest for the principles that would finally elevate the study of human behavior to true scientific status.

According to Skinner, Meyer succeeded in achieving the breakthrough that finally allowed psychology to assume its rightful place alongside the disciplines of physics, chemistry, and biology.[39] Why did Skinner extol this work that had been largely ignored, even in its own time? Skinner singles out for special praise a 1921 textbook that bore the ominous-sounding title *Psychology of the Other-One*. It received scant attention when it was published—Meyer had written it primarily for his own students—and had since fallen into total obscurity.[40] Nonetheless, Skinner praised the book for establishing the epistemological and methodological foundations of modern behaviorism: "consider only those facts which can be objectively observed in the behavior of one person in its relation to his prior environmental history."[41] In Skinner's formulation, Meyer's book was a turning point, bravely combining psychology and physics in the quest for absolutes. It asserted the essence of the behaviorist's point of view, where "the world within the skin of the Other-One loses its preferred status."[42]

The phrase that captured the new scientific perspective was the "Other-One." Human behavior would yield to scientific research only if psychologists learned to view humans as *others*. This "viewpoint of observation" was an absolute requirement for an "objective science of human behavior" that ceased to confuse inner experience with external action.[43] Central to this new viewpoint was his notion of the human being as *organism*. The human being is recast as an "it," an "other," a "they" of organisms: an "organism among organisms," distinguishable from a lettuce, a moose, or an inchworm only in degree of complexity.[44] A scientific psychology would restrict its interests to the social and therefore visible behaviors of this "organism as an organism." It would be "the study of the life of the Other-One—but of his life in so far as it is of social significance rather than as it is of significance for himself.... We are studying the Other-One in preference to Our-Selves."[45]

The logical consequences of the new viewpoint necessitated a reinterpretation of the higher-order human experiences that we call "freedom" and "will." Meyer echoed Planck in positing that "freedom of action in the animal world signifies the same that is meant by accidents in the world of physics."[46] Such accidents are simply phenomena for which there are insufficient information and understanding. And so it goes with freedom. The liberal idea of freedom persists in an inverse relationship to the growth of scientific knowledge, especially in the field of psychological science. Knowledge and freedom are necessarily adversaries. As Meyer wrote, "The Other-One's conduct is free, uncaused, only in the same sense in which the issue of a disease, the outcome of a war, the weather, the crops are free and uncaused; that is, in the sense of general human ignorance of the particular causes of the particular outcome."[47]

Decades later, this worldview would define the core of the controversial social philosophy espoused in *Beyond Freedom & Dignity,* in which Skinner argues that knowledge does not make us free but rather releases us from the illusion of freedom. In reality, he writes, freedom and ignorance are synonyms. The acquisition of knowledge is heroic in that it rescues us from ignorance, but it is also tragic because it necessarily reveals the impossibility of freedom.

For Meyer and for Skinner, our attachments to notions such as freedom, will, autonomy, purpose, and agency are defense mechanisms that protect us from the uncomfortable facts of human ignorance. I think of Dickens's Scrooge when he first encounters the doleful, chain-dragging ghost of his deceased partner Jacob Marley and denies the apparition, saying, "You may be an undigested bit of beef, a blot of mustard, a crumb of cheese, a fragment of an underdone potato." So it is with freedom: an undigested bit of fear, a crumb of denial that, once metabolized, will dispel the apparition and deliver us to reality. The environment determines behavior, and our ignorance of precisely how it does so is the void that we fill with the fantasy of freedom.

Meyer took great pains to insist that the significance of human inwardness—"soul," "self," "mind," "consciousness"—is restricted to the subjective life of the individual. It can have no scientific value because it cannot be observed and measured: "We do not deny the soul, but we do not devote our time to it. We find enough, and more than enough, to do studying the body." The soul is "your own business," an intimate experience and irreducible mystery that are formally exempt from scientific inquiry: "Human societies can thus be understood as originating from natural laws, not in the sense of groups of souls tho, but in the sense of groups of organisms."[48]

Meyer argued that the future of the social sciences and of civilization itself rested on this shift from soul to other, inside to outside, lived experience to observable behavior. The otherization of humanity was to be the road to a new kind of political liberation. History's grim cavalcade of repression, torture, massacre, enslavement, and genocide had been prosecuted in the name of the domination of the human soul for the sake of religious or political power. From Meyer's vantage point in the Missouri of 1921 during those years after the Great War, his solution for an efficient and rational modernity must have felt like a matter of life and death:

> He whose interest is centered in souls thinks, when he has forced others to speak his prayer, pronounce his creed, kneel before his altar, that he has saved their souls, and fails to admit that he has merely forced their bodies.... Political terrorism, too, has its main and inexhaustible source in the human tendency to think of other beings, not as bodies open to scientific investigation, but as souls, as mysterious beings, to be governed either by magic, or, if magic fails as it naturally must, by torture and death.... Recall the horrors of the torture... of such courts as the Spanish inquisition or the witch-craft courts of the 17th century.... These atrocities were due to the fact that the judge was serving as a mind reader, and that the accused was regarded primarily as a soul.[49]

In Meyer's thinking, the shift in perspective from "the human being as a soul" to "the human being as an organism" explained "why the trend of history is in the direction of democracy." As science overtakes civilization, Meyer assumed a new global recognition of equality and democratic fellowship rooted in the basic fact of our overwhelming similarity as fellow organisms. The divisions that haunt society, politics, and economics based on class, wealth, liberality, race, and so on would become ridiculous: "In real world-wide human life, the differences among individuals are entirely swamped by their likenesses. To him who accepts the scientific view that human society is a group of organisms, it is an absurd proposition to divide... into...classes...."[50]

Meyer believed that every social science that aspired to true scientific status would embrace the viewpoint of the Other-One—psychology, of course, but also sociology, economics, criminology, history, and the psychology of religion: "Christ going among his fellow men, an organism among organisms...."[51] Otherizing would pave the way to a rational future, with its bleak satisfactions that resign humanity to the forfeit of freedom as the price of knowledge.

V. Against Freedom

Skinner's commitment to the viewpoint of the Other-One was unshakable, and it is through his elaboration of this viewpoint that we can begin to grasp the essence of instrumentarian power. From the opening page of Skinner's first book, *The Behavior of Organisms*, published in 1938, he sounds Meyer's (and Planck's) caution: freedom is merely ignorance waiting to be conquered. "Primitive systems of behavior" assign causality to "entities beyond man." Equally inadequate are the "advanced systems of behavior" that ascribe control to vague fictions such as the "self" or "free will." "The inner organism," he wrote, "may in resignation be called free, as in the case of 'free will,' when no further investigation is held to be possible."[52]

Skinner called his work "radical behaviorism," insisting that the only meaningful object of behavioral study was observation of action devoid of subjective attributions. That's what made it radical. "Behavior is what an organism is doing—or more accurately what it is observed by another organism to be doing," he declared in his first book's opening pages. "Operant" behavior was his term for this active, observable "doing." The vocabulary for rendering descriptions of operant behavior was to be purged of inwardness: an organism cannot be said to "see" but rather to "look toward." Only such objective descriptions can render measurable behavioral facts that, in turn, lead to patterns and ultimately to the documentation of causal relationships between environment and behavior.[53]

Skinner published *Science and Human Behavior* in 1951, positing that all observation, even of one's own behavior, must be enacted from the viewpoint of the Other-One. This discipline makes it possible to take almost anything as an object of behavioral analysis, including inferred behaviors such as "making choices" or "problem solving," the very perspective that would later be amply exploited by the new discipline of behavioral economics:

> When a man controls himself, chooses a course of action, thinks out the solution to a problem, or strives toward an increase in self-knowledge, he is *behaving*. He controls himself precisely as he would control the behavior of anyone else—through the manipulation of variables of which behavior is a function. His behavior in so doing is a proper object of analysis, and eventually it must be accounted for with variables lying outside the individual himself.[54]

In nearly every book and article, Skinner declared the truth that Planck taught Meyer and Meyer imparted to his students, the single truth that can be grasped only through the viewpoint of the Other-One: *freedom is ignorance.* The felt experience of free will is but a bit of undigested denial, produced by a lack of information about the actual determinants of behavior. Like Meyer and Planck before him, Skinner regarded freedom as an "accident," arguing that the very notion of "accident" is illusory, a snapshot in time that reveals a lacuna waiting to be filled and eventually transformed by advancing knowledge into an expression of a lawful, predictable pattern. Under the behaviorist's gaze, these lacunae of ignorance that we mistake for free will are queuing up for explanation, rather like someone who consigns his or her body to cryonic preservation in the hope of some future awakening and cure.

In the most audacious of Skinner's literary efforts, the extended philosophical essay published in 1971 as *Beyond Freedom & Dignity*, he repeats: "There is no virtue in the accidental nature of an accident."[55] That entire work was trained on what Skinner continued to regard as the chief impediment to social progress: the conceptual confusion that cloaks our deepest ignorance in the sacred robes of freedom and dignity. Skinner argued that our allegiance to these lofty notions is simply the way we protect ourselves from the hard truths of "unsuspected controlling relations between behavior and environment."[56] They are a psychological "escape route" that slowly closes "as new evidences of the predictability of human behavior are discovered. Personal exemption from a complete determinism is revoked as a scientific analysis progresses...the achievements for which a person himself is to be given credit seem to approach zero...the behavior we admire is therefore the behavior we cannot yet explain."[57]

Richard Herrnstein, one of Skinner's most-accomplished students, later his colleague in Harvard's Psychology Department, and a luminary of radical behaviorism, once explained to me that any action regarded as an expression of free will is simply one for which "the vortex of stimuli" that produced it cannot yet be adequately specified. We merely lack the means of observation and calculation. I was a twenty-three-year-old student, and the term was new and startling to me. I never forgot that conversation, perhaps because it came as close as any to summarizing the behaviorist's conception of God. Indeed, there was a time when if you took the elevator to Skinner's lab on the seventh floor of the Psychology Department, the first thing you would see was a sign reading "God is a VI," a variable interval of behavioral reinforcement plucked from the vortex.

In this view, "freedom" or "accidents" shrink as our developing capabilities in measurement and computation provide more information about the vortex of stimuli. Ignorance about human behavior is like a melting iceberg on a warming planet, destined to yield to the mounting heat as we invent the means and methods smart enough to first decipher and then control the vortex of stimuli that shapes each fact of human behavior. Skinner pointed to weather forecasting as the iconic example of this transformation from ignorance to law, just as Meyer had done decades earlier:

> The problems imposed by the complexity of a subject matter must be dealt with as they arise. Apparently hopeless cases often become manageable in time. It is only recently that any sort of lawful account of the weather has been possible.... Self-determination does not follow from complexity.... Difficulty in calculating the orbit of the fly

does not prove capriciousness, though it may make it impossible to prove anything else.[58]

VI. A Technology of Human Behavior

Through six decades of academic and popular writing, Skinner would insist that "further investigation" is always possible. In the first pages of *Beyond Freedom & Dignity* he calls for a technological solution to ignorance: "We need to make vast changes in human behavior, and we cannot make them with the help of nothing more than physics or biology, no matter how hard we try...what we need is a technology of behavior...comparable in power and precision to physical and biological technology...."[59]

Skinner imagined technologies that would pervasively institutionalize the viewpoint of the Other-One as they observed, computed, analyzed, and automatically reinforced behavior to accomplish the "vast changes" that he believed were necessary. In this way the laws of human action would finally be illuminated so that behavior could be effectively predicted and shaped, just as other technologies had enabled physics and biology to change the world: "The difference is that the instruments and methods they use are of commensurate complexity. The fact that *equally powerful instruments and methods are not available in the field of human behavior* is not an explanation; it is only part of the puzzle."[60]

Skinner concluded that the literature of freedom and dignity "stands in the way of further human achievement."[61] He argued that the missing puzzle piece holding back the urgent development of the "instruments and methods" essential for a technology of behavior was the stubborn allegiance to these antique notions among people determined to preserve "due credit" for their actions. The belief in "autonomous man" is a regressive source of resistance to a rational future, an "alternative explanation of behavior" that obstructs the advancement of society.

The professor believed that humanity desperately needed a plan, and he imagined powerful new instruments that could engineer behavior in every domain. As early as 1953, he anticipated today's digitally engineered casino environments, whose sophistication in the precise shaping of gamblers' behavior has made them a testing ground for state security agencies and surveillance capitalists alike:[62] "But with proper instrumentation it should be possible to improve upon established practices in all these fields. Thus gambling devices could be 'improved'—*from the point of view of the proprietor*—by introducing devices which would pay off on a variable-interval basis, but only when the rate of play is exceptionally high."[63]

Technologies of behavioral engineering would not be restricted to "devices" but would also encompass organizational systems and procedures designed to shape behavior toward specific ends. In 1953 Skinner anticipated innovations such as Michael Jensen's incentive systems designed to maximize shareholder value and the "choice architectures" of behavioral economics designed to "nudge" behavior along a preferred path: "Schedules of pay in industry, salesmanship, and the professions, and the use of bonuses, incentive wages, and so on, could also be improved from the point of view of generating maximal productivity."[64]

Skinner understood that the engineering of behavior risked violating individual sensibilities and social norms, especially concerns about privacy. In order to allay these anxieties he advised that observation must be unobtrusive, ideally remaining outside the awareness of the organism: "Behavior may also be observed with a minimum of interaction between subject and scientist, and this is the case with which one naturally tries to begin."[65] But there would be challenges. New technologies of behavior would have to continually push the envelope of the public-private divide in order to access all the data relevant to behavioral prediction and control. In this he anticipated today's rendition frontier as new detection systems plumb the depths of personalities and emotions: "But we are still faced with events which occur at the private level and which are important to the organism without instrumental amplification. How the organism reacts to these events will remain an important question, even though the events may some day be made accessible to everyone."[66]

Such conflicts would be resolved, Skinner reckoned, by the gradual retreat of privacy norms as they fall to the advance of knowledge: "The line between public and private is not fixed." Like today's surveillance capitalists, he was confident that the slow drip of technological invention would eventually push privacy to the margins of human experience, where it would join "freedom" and other troublesome illusions. All these would be replaced by the viewpoint of the Other-One embodied in new instruments and methods: "The boundary shifts with every discovery of a technique for making private events public.... The problem of privacy may, therefore, eventually be solved by technical advances."[67]

Although privacy advocates and many other critics of surveillance capitalism are quick to appropriate Orwellian language in the search for meaning and metaphor that capture the sense of new menace, surveillance capital's instrumentarian power is best understood as the precise antithesis of Orwell's Big Brother. We now turn to this stark contrast, most vividly represented in the distinct conceptions of utopia that attach to each of these two species of power.

VII. *Two Utopias*

In the blood-soaked wake of World War II, both Skinner and the journalist and novelist George Orwell aimed curative, "utopian" novels at the mayhem of modernity's diminished prospects produced by an incomprehensible scale of violence. Viewed from a distance, Skinner's *Walden Two*, published in 1948, and Orwell's *1984*, released the following year, had much in common. Each elaborated a complete conception of a distinct logic of power, each imagined a society defined by the full flourishing of that power, and each was utopian from the point of view of the form of power that it described.[68] However, their public reception could not have been more different: *Walden Two* was dismissed as a dystopian nightmare and ignored by the general public for more than a decade.[69] Orwell's *1984* was immediately canonized as a dystopian masterpiece and the distillation of the twentieth century's worst nightmares.

The two utopias have often been confused with each other in their content and aims: *Time* magazine's 1971 cover story on Skinner described *Walden Two* as raising "the specter of a 1984 Orwellian society that might really come to pass." The great historian and literary critic Lewis Mumford once described *Walden Two* as a "totalitarian utopia" and a depiction of "hell," but in fact these characterizations are a persistent and, as we shall see, dangerous confusion.[70] Although both books have been described as depictions of totalitarianism, the forms of power that each describes are profoundly different. In most respects, they are precise opposites.

Meyer's prescription for modernity was founded on the scientific objectification of human experience and its reduction to observable measurable behavior. If we take that as our benchmark here, then Orwell's utopia is the "before" case; it precedes Meyer as a nightmare of the prescientific compulsion to dominate the soul. Skinner's utopia is "after" Meyer's reimagining of modernity, as channeled from the great physicist, Planck. *Walden Two* is built on Meyer's scientific viewpoint of observation, the Other-One, and represents the full flower of Meyer's hope for a scientifically induced global harmony of organisms in which freedom is necessarily forfeit to knowledge. Orwell laid bare the disease, and Skinner asserted the antidote.

The totalitarian power elaborated in *1984* is something wholly unprecedented, concocted in the twentieth century from the collision of industrialism and despair, a form utterly new to the human story. Orwell did more than simply fictionalize and extrapolate the totalitarian project. He sounded an alert by drawing a terrifying line of consequence from the recent German past and persistent Soviet present to an imagined but all-too-possible future. His genius was to craft a story that embodied totalitarianism's essence: the ruthless insistence upon the absolute possession of each individual human being, not as a distant other known only by its behavior, but rather from the inside out.

Big Brother's vigilance is not restricted to the grand continents of armies and statecraft or the observable flows of bodies and crowds. Big Brother is a panvasive consciousness that infects and possesses each individual soul, displacing all attachments once formed in romantic love and good fellowship. The essence of its operation is not simply that it knows every thought and feeling but rather the ruthless tenacity with which it aims to annihilate and replace unacceptable inward experience. "We are not content with negative obedience, nor even with the most abject submission," the cunning henchman O'Brien tells the rebellious Winston:

> We shall crush you down to the point from which there is no coming
> back.... When finally you surrender to us, it must be of your own
> free will. We do not destroy the heretic because he resists us: so long

as he resists us we never destroy him. We convert him, we capture his inner mind, we reshape him.... We bring him over to our side, not in appearance, but genuinely, heart and soul. We make him one of ourselves before we kill him.[71]

Ultimately, as every reader knows, Winston's stubborn soul is successfully "engineered." Orwell's chilling final passages fulfill the life of that dry seed planted at the turn of the century in Italy's impoverished soil and nourished by war, deprivation, and humiliation to flower in the nightmare of Nazi Germany and the apocalypse of Stalin's Russia, finally to bear fruit in Orwell's imagination: a testament for all time to what Mussolini had called the "ferocious totalitarian will" and the souls on which it feeds. Winston basks in serene elation, "his soul white as snow.... He had won the victory over himself. He loved Big Brother."[72]

In contrast, *Walden Two* was not intended as a warning but rather as the antidote to totalitarianism and, more broadly, a practical recipe for the challenges of rebuilding Western societies after the war. Skinner understood his utopia as a methodological cure for the nightmare of crushed souls, a cure that, he insisted, was superior to any of the conventional political, economic, or spiritual remedies on offer. He scoffed at the notion that "democracy" held the solution because it is a political system that merely perpetuates the illusion of freedom while impeding the dominion of science. The promise of the "free market" as the curative for postwar society was an equally empty dream, he believed, because it rewards destructive competitiveness between people and classes. Skinner also rejected existentialism's new humanism, seeing it as a breeding ground for passivity, and he regarded religion as the worst cure of all, enshrining ignorance and crippling the advance of science.

Skinner's cure was different and unique: a utopia of technique that promised a future of social equality and dispassionate harmony founded on the viewpoint of the Other-One, the "organism among organisms," as the object of "behavioral engineering." It is the future of Meyer's dreams, in which Frazier, the founder-leader of the fictional Walden Two and Skinner's outspoken avatar, describes that ideal community as a "superorganism" that can be shaped and controlled "as smoothly and efficiently as champion football teams."[73]

Skinner's utopia was meant to illustrate the possibility of a successful social order that transcends the use of force and rejects the need to dominate human souls. The Walden Two community equally disdains the practices of democratic politics and representative government. Its laws are derived from a science of human behavior, specifically Skinner's own radical behaviorism, founded on the physicist's ideal of the Other-One. His utopia was a vehicle for other ambitions as well, intended to illustrate the behavioral solutions that are essential for improvement in every domain of modern life: the nuclear threat, pollution, population control, urban growth, economic equality, crime, education, health care, the development of the individual, effective leisure. It aimed to cultivate "the good life," for which all the ideals of a liberal society—freedom, autonomy, privacy, a people's right to self-rule—must be forfeit.

Walden Two's fictional format provided the cover that Skinner needed to extrapolate from Meyer's methodological principles of otherness and his own research on animal behavior to a utopian community in which behavior has replaced the human spirit as the locus of control. Frazier laments that people "have been kept in their places," not just by external forces, "but much more subtly by a system of beliefs implanted within their skins. It's sometimes an almost hopeless task to take the shackles off their souls, but it can be done.... You can't in the long run enforce anything. We don't use force! All we need is adequate behavioral engineering."[74]

These two utopias reflect two distinct species of power, and each novel was intent on rescuing the future from the twentieth-century nightmare of the soul. Orwell was able to draw on the recent past, but Skinner imagined a future that he would not live long enough to enjoy. If surveillance capitalism and its instrumentarian power continue to flourish, then it is we who may see the *Walden Two* vision realized, as freedom falls to others' knowledge—but now in the service of others' wealth and power.

Skinner's vision is brought to life in the relentless pursuit of surveillance capitalism's economic imperatives and the ubiquitous digital apparatus that surveillance capitalism creates and harnesses to its novel aims. Instrumentarian power bends the new digital apparatus—continuous, autonomous, omnipresent, sensate, computational, actuating, networked, internet-enabled—to the interests of the surveillance capitalist project, finally fulfilling Skinner's call for the "instruments and methods" of "a behavioral technology comparable in power and precision to physical and biological technology." The result is a panvasive means of behavioral modification whose economies of action are designed to maximize surveillance revenues.

Until the rise of surveillance capitalism, the prospect of instrumentarian power was relegated to a gauzy world of dream and delusion. This new species of power follows the logic of Planck, Meyer, and Skinner in the forfeit of freedom for knowledge, but those scientists each failed to anticipate the actual terms of this surrender. The knowledge that now displaces our freedom is proprietary. The knowledge is *theirs,* but the lost freedom belongs solely to *us.*

With this origin story in hand, Chapter 13 turns to a close examination of instrumentarian power as it inscribes a sharp new asymmetry on the human community: the knowledge to which we sacrifice our freedom is constructed to advance surveillance capitalists' commercial interests, not our own. This is a stark departure from the technical origins of the apparatus in MacKay's principles of telemetry, which traded animals' freedom for the sake of scientific knowledge intended to benefit the animals themselves. Instead, surveillance capitalism's behavioral market regime finally has at its disposal the instruments and methods that can impose Skinner's technology of behavior across the varied domains of everyday life right down to our depths, now conceived as capital's global laboratory.

CHAPTER THIRTEEN

BIG OTHER AND THE RISE OF INSTRUMENTARIAN POWER

He was their servant (some say he was blind),
Who moved among their faces and their things:
Their feeling gathered in him like a wind
And sang. They cried "It is a God that sings."

—W. H. AUDEN
SONNETS FROM CHINA, VII

I. Instrumentarianism as a New Species of Power

Surveillance capitalism is the puppet master that imposes its will through the medium of the ubiquitous digital apparatus. I now name the apparatus *Big Other:* it is the sensate, computational, connected puppet that renders, monitors, computes, and modifies human behavior. Big Other combines these functions of knowing and doing to achieve a pervasive and unprecedented *means of behavioral modification.* Surveillance capitalism's economic logic is directed through Big Other's vast capabilities to produce instrumentarian power, replacing the engineering of souls with the engineering of behavior.

Instrumentarian power cultivates an unusual "way of knowing" that combines the "formal indifference" of the neoliberal worldview with the observational perspective of radical behaviorism (see Figure 4 on pages 396–397). Thanks to Big Other's capabilities, instrumentarian power reduces human experience to measurable observable behavior while remaining steadfastly indifferent to the meaning of that experience. I call this new way of knowing *radical indifference.* It is a form of *observation without witness* that yields the obverse of an intimate violent political religion and bears an utterly different signature of havoc: the remote and abstracted contempt of impenetrably complex systems and the interests that author them, carrying individuals on a fast-moving current to the fulfillment of others' ends. What passes for social relations and economic exchange now occurs across the medium of this robotized veil of abstraction.

Instrumentarianism's radical indifference is operationalized in Big Other's dehumanized methods of evaluation that produce *equivalence without equality.* These methods reduce individuals to the lowest common denominator of sameness—an organism among organisms—despite all the vital ways in which we are not the same. From Big Other's point of view we are strictly Other-Ones: *organisms that behave.* Big Other encodes the viewpoint of the Other-One as a global presence. There is no brother here of any kind, big or little, evil or good; there are no family ties, however grim. There is no relationship between Big Other and its otherized objects, just as there was no relationship between B. F. Skinner's "scientists and subjects." There is no domination of the soul that displaces all intimacy and attachment with terror—far better to let a multitude of relationships bloom. Big Other does not care what we think, feel, or do as long as its millions, billions, and trillions of sensate, actuating, computational eyes and ears can observe, render, datafy, and instrumentalize the vast reservoirs of behavioral surplus that are generated in the galactic uproar of connection and communication.

In this new regime, objectification is the moral milieu in which our lives unfold. Although Big Other can mimic intimacy through the tireless devotion of the One Voice—Amazon-Alexa's chirpy service, Google Assistant's reminders and endless information—do not mistake these soothing sounds for anything other than the exploitation of your needs. I think of elephants, that most majestic of all mammals: Big Other poaches our behavior for surplus and leaves behind all the meaning lodged in our bodies, our brains, and our beating hearts, not unlike the monstrous slaughter of elephants for ivory. Forget the cliché that if it's free, "You are the product." You are not the product; you are the abandoned carcass. The "product" derives from the surplus that is ripped from your life.

Big Other finally enables the universal technology of behavior that, as Skinner, Stuart MacKay, Mark Weiser, and Joe Paradiso each insisted, accomplishes its aims quietly and persistently, using methods that intentionally bypass our awareness, disappearing into the background of all things. Recall that Alphabet/Google's Eric Schmidt provoked uproar in 2015 when in response to a question on the future of the web, he said, "The internet will disappear." What he really meant was that "The internet will disappear into Big Other."

Thanks to Big Other's capabilities, instrumentarian power aims for a condition of *certainty without terror* in the form of "guaranteed outcomes." Because it does not claim our bodies for some grotesque regime of pain and murder, we are prone to undervalue its effects and lower our guard. Instead of death, torture, reeducation, or conversion, instrumentarianism effectively exiles us from our own behavior. It severs our insides from our outsides, our subjectivity and interiority from our observable actions. It lends credibility to the behavioral economists' hypothesis of the frailty of human reason by making it so, as otherized behavior takes on a life of its own that delivers our futures to surveillance capitalism's aims and interests.

In an instrumentarian utopia, Big Other simulates the vortex of stimuli, transforming "natural selection" into the "unnatural selection" of variation and reinforcement authored by market players and the competition for surveillance revenues. We may confuse Big Other with the behaviorist god of the vortex, but only because it effectively conceals the machinations of surveillance capital that are the wizard behind the digital curtain. The seductive voice crafted on the yonder side of this veil—*Google, is that you?*—gently nudges us along the path that coughs up the maximum of behavioral surplus and the closest approximation to certainty. Do not slumber in this opiated fog at the network's edge. That knowing voice is underwritten by the aims and rules of the very place we once hoped to flee, with its commercialized rituals of competition, contempt, and humiliation. Take one wrong step, one deviation from the path of seamless frictionless predictability, and that same voice turns acid in an instant as it instructs "the vehicular monitoring system not to allow the car to be started."

Under the regime of instrumentarian power, the mental agency and self-possession of the right to the future tense are gradually submerged beneath a new kind of automaticity: a lived experience of stimulus-response-reinforcement aggregated as the comings and goings of mere organisms. Our conformity is irrelevant to instrumentarianism's success. There is no need for mass submission to social norms, no loss of self to the collective induced by terror and compulsion, no offers of acceptance and belonging as a reward for bending to the group. All of that is superseded by a digital order that thrives within things and bodies, transforming volition into reinforcement and action into conditioned response.

In this way instrumentarian power produces endlessly accruing knowledge for surveillance capitalists and endlessly diminishing freedom for us as it continuously renews surveillance capitalism's domination of the division of learning in society. False consciousness is no longer produced by the hidden facts of class and their relation to production but rather by the hidden facts of instrumentarian power's command over the division of learning in society as it usurps the rights to answer the essential questions: *Who knows? Who decides? Who decides who decides?* Power was once identified with the ownership of the means of production, but it is now identified with ownership of the means of behavioral modification that is Big Other.

II. A Market Project of Total Certainty

Big Other and the instrumentarian power that it enables fulfill Skinner's vision for humankind. As early as 1948 in *Walden Two,* Skinner had pined for the new measurement and computational capabilities that would unlock the mysteries of the vortex of stimuli and illuminate those acts of ignorance that we foolishly value as free will. "I didn't say that behavior is always predictable, any more than the weather is always predictable," says Frazier, *Walden Two*'s protagonist. "There are often too many factors to be taken into account. We can't measure them all accurately, and we couldn't perform the mathematical operations needed to make a prediction if we had the measurements."[1]

It was Skinner's hard destiny to foresee the possibility of instrumentarian power and its operationalization in Big Other before the instruments existed to fulfill that vision. His lab had always been a fantasy world of engineering innovations for his behavioral experiments: mazes and boxes for animal conditioning, measurement tools, and recording devices. A fully operational technology of behavior was the prize that would elude Skinner in his lifetime, a source of frustration that was palpable in every article and lecture right to the end.

Though confident that science would eventually overcome the practical challenges of a behavioral technology, Skinner was more troubled by the cultural impediments to a human science of behavioral prediction and control founded on the viewpoint of the Other-One. He resented the friction produced by human beings in their stubborn attachment to the values and ideals of freedom inherited from the eighteenth-century philosophers, and he equally despised the existential project of the postwar philosophies that planted authenticity, free will, and autonomous action at the heart of second-modernity yearning.

In his penultimate essay, written in 1990, barely three years before his death, Skinner mourned the prospects of behavioral prediction as the foundation of a new society built on scientific certainty: "To say that a person is simply a place in which something happens seems even more threatening when it raises questions about what we are likely to do rather than what we have done."[2] In those last years he seemed more resigned to the tenacity of human friction and its stubborn allegiance to something like free will, his voice less outrageous and aggressive than the author of *Beyond Freedom & Dignity*, two decades earlier. Anger and contempt had shaded into weariness and despair with his final reflections:

> It may be said that this is a discouraging view of human behavior and that we shall be more likely to do something about the future if we continue to believe that our destiny is in our hands. That belief has survived for many centuries and has led to remarkable achievements, but the achievements were only the immediate consequences of what was done. We now know that other consequences have followed and that they are threatening our future. What we have done with our destiny may not be a testament we wish to leave to the world.[3]

In our time of surveillance exceptionalism, as astonishment succumbs to helplessness and resignation, the resistance that Skinner lamented appears to be waning. The belief that we can choose our destiny is under siege, and, in a dramatic reversal, the dream of a technology of behavioral prediction and control—for which Skinner had endured such public scorn—is now a flourishing fact. This prize now attracts immense capital, human genius, scientific elaboration, governmental protection, ecosystems of institutionalization, and the glamour that always has and always will attach to power.

The rise of instrumentarianism follows the path of "illuminating events" that, as Hannah Arendt writes, reveal "a beginning in the past which had hitherto been hidden."[4] It is in the nature of instrumentarian power to operate remotely and move in stealth. It does not grow through terror, murder, the suspension of democratic institutions, massacre, or expulsion. Instead, it grows through declaration, self-authorization, rhetorical misdirection, euphemism, and the quiet, audacious backstage moves specifically crafted to elude awareness as it replaces individual freedom with others' knowledge and replaces society with certainty. It does not confront democracy but rather erodes it from within, eating away at the human capabilities and self-understanding required to sustain a democratic life.

The narratives of Facebook's experimental maneuvers, Pokémon Go's prototype of a behavioral-futures-market–dominated society, and the endless examples of digital innovation crushed under the heel of the surveillance creed may be as close as we get to a public announcement of history-illuminating change that moves through us and among us, irreversibly altering life as we have known it. There is no violence here, only the steady displacement of the will to will that has been embodied in self-determination, expressed in the first-person voice, and nourished in the kind of sanctuary that depends upon the possibility of private life and the promise of public freedom.

Instrumentarian power, like Goethe's Faust, is morally agnostic. The only moral imperative here is distilled from the point of view of a thin utopian gruel. If there is sin, it is the sin of autonomy: the audacity to reject the flows that herd us all toward predictability. Friction is the only evil. Obstruction in law, action, or rhetoric is simply reactionary. The norm is submission to the supposed iron laws of technological inevitability that brook no impediment. It is deemed only rational to surrender and rejoice in new conveniences and harmonies, to wrap ourselves in the first text and embrace a violent ignorance of its shadow.

Totalitarianism was a transformation of the state into a project of total possession. Instrumentarianism and Big Other signal the transformation of the market into a project of total certainty, an undertaking that is unimaginable outside the digital milieu, but also unimaginable outside the logic of accumulation that is surveillance capitalism. This new power is the spawn of an unprecedented convergence: the surveillance and actuation capabilities of Big Other in combination with the discovery and monetization of behavioral surplus. It is only in the context of this convergence that we can imagine economic principles that instrumentalize and control human experience to systematically and predictably shape behavior toward others' profitable ends.

Instrumentarian power operates from the vantage point of the Other-One to reduce human persons to the mere animal condition of behavior shorn of reflective meaning. It sees only organisms bent to serve the new laws of capital now imposed on all behavior. Arendt anticipated the destructive potential of behaviorism decades ago when she lamented the devolution of our conception of "thought" to something that is accomplished by a "brain" and is therefore transferable to "electronic instruments":

> The last stage of the laboring society, the society of jobholders, demands of its members a sheer automatic functioning, as though individual life had actually been submerged in the over-all life process of the species and the only active decision still required of the individual were to let go, so to speak, to abandon his individuality, the still individually sensed pain and trouble of living, and acquiesce in a dazed, "tranquilized," functional type of behavior.
>
> The trouble with modern theories of behaviorism is not that they are wrong but that they could become true, that they actually are the best possible conceptualization of certain obvious trends in modern society. It is quite conceivable that the modern age— which began with such an unprecedented and promising outburst of human activity—may end in the deadliest, most sterile passivity history has ever known.[5]

Is this to be our home: the automation of the self as the necessary condition of the automation of society, and all for the sake of others' guaranteed outcomes?

III. This Century's Curse

One place to begin our reckoning with this question is in Arendt's "Concluding Remarks" in the first edition of *The Origins of Totalitarianism,* words that have haunted me since I first read them many years ago. They were written at a time when totalitarianism had been defeated in Europe but remained unchallenged in Stalin's USSR. It was a time when much of the world was united in the urgency of understanding and remembering, not only as testament but also as a vaccine against future terror.

Arendt's reflections summarize not only totalitarianism's "futility" and "ludicrousness" but also her sense of the "disturbing relevance of totalitarian regimes." She warned that totalitarianism could not be written off as an accidental turn toward tragedy but rather that it must be seen as "deeply connected with the crisis of this century." She concluded, "The fact is that the true problems of our time cannot be understood, let alone solved, without acknowledgement that totalitarianism became this century's curse only because it so terrifyingly took care of its problems."[6]

Now imagine, decades hence, another thinker meditating on the "disturbing relevance" of instrumentarian power, observing that "the true problems of our time cannot be understood, let alone solved, without acknowledgement that instrumentarianism became this century's curse only because it so terrifyingly took care of its problems."

What problems? I have argued that surveillance capitalism and its instrumentarian power feed on the volatile "conditions of existence" that I summarize as the "collision." Surveillance capitalism offers solutions to individuals in the form of social connection, access to information, time-saving convenience, and, too often, the illusion of support. These are the resources of the first text. More significantly, it offers solutions to institutions in the form of omniscience, control, and certainty. The idea here is not to heal instability—the corrosion of social trust and its broken bonds of reciprocity, dangerous extremes of inequality, regimes of exclusion—but rather to exploit the vulnerabilities produced by these conditions.

It is useful to note that despite the much-touted social advantages of always-on connection, social trust in the US declined precipitously during the same period that surveillance capitalism flourished. According to the US

General Social Survey's continuous measurement of "interpersonal trust attitudes," the percentage of Americans who "think that most people can be trusted" remained relatively steady between 1972 and 1985. Despite some fluctuations, 46 percent of Americans registered high levels of interpersonal trust in 1972 and nearly 50 percent in 1985. As the neoliberal disciplines began to bite, that percentage steadily declined to 34 percent in 1995, just as the public internet went live. The late 1990s through 2014 saw another period of steady and decisive decline to only 30 percent.[7]

Societies that display low levels of interpersonal trust also tend to display low levels of trust toward legitimate authority; indeed, levels of trust toward the government have also declined substantially in the US, especially during the decade and a half of growing connectivity and the spread of surveillance capitalism. More than 75 percent of Americans said that they trusted the government most or all of the time in 1958, about 45 percent in 1985, close to 20 percent in 2015, and down to 18 percent in 2017.[8] Social trust is highly correlated with peaceful collective decision making and civic engagement. In its absence, the authority of shared values and mutual obligations slips away. The void that remains is a loud signal of societal vulnerability. Confusion, uncertainty, and distrust enable power to fill the social void. Indeed, they welcome it.

In the age of surveillance capitalism it is instrumentarian power that fills the void, substituting machines for social relations, which amounts to *the substitution of certainty for society*. In this imagined collective life, freedom is forfeit to others' knowledge, an achievement that is only possible with the resources of the shadow text.

The private institutions of capital led the way in this ambitious reformation of collective life and individual experience, but they found necessary support from public institutions, especially as the declaration of a "war on terror" legitimated every inclination to enshrine machine-produced certainty as the ultimate solution to societal uncertainty. These mutual affinities assured that instrumentarian power would not be a stepchild but rather an equal partner or even, with increasing regularity, the lord and master upon whom the state depends in its quest for "total awareness."

That instrumentarian power is regarded as the certain solution to uncertain societal conditions is evident in the ways in which it is called into action by the state. The varied and complex institutional patterns produced by these interactions are a crucial frontier for study and democratic debate. My aim right now is simply to point out a few examples that illustrate the state's continuous demands for the intensification of surveillance capitalism's production of instrumentarian power—expressed in the growth and elaboration of Big Other—as the preferred solution to social breakdown, mistrust, and uncertainty. Although we have become desensitized to a seemingly endless train of such examples, it is important to recognize that in these entanglements, state and market institutions demonstrate a shared commitment to a relentless drive toward guaranteed outcomes. Their mutual affinities can help us define the problem that threatens to make instrumentarian power our century's curse.

Unsurprisingly, instrumentarian power is consistently called into action as a solution, if not *the* solution, to the threat of terrorism. Acts of terror reject the authority of civilizational norms and reveal the impossibility of society without mutual trust. Governments now turn to instrumentarian power as the solution to this new source of societal uncertainty, demanding the certainty machines that promise direct, reliable means of detection, prediction, and even the automatic actuation of countermeasures.

During the sixteen years of the Bush and Obama administrations, "progress in information technology" was understood as the "most effective response" to threat. Peter Swire observes that public officials "know that the private sector is developing many new techniques for collecting and processing data and making decisions based on that data." The consequence is "a large and ongoing shift toward information-intensive strategies" that avail themselves of these market capabilities.[9]

This status quo was interrupted in 2013, when Edward Snowden revealed the hidden complicity between state security agencies and the tech companies. In the backlash that followed, surveillance capitalists confronted new public relations challenges in what they portrayed as an uneasy or even unwilling alliance between private power and state security needs. Nevertheless, new terrorist threats invariably orient public officials toward the intensification and deployment of Big Other and the instrumentarian power it signifies. However, their ability to access this immense power is fraught with tension. It is not simply theirs to command. They must work, at least in part, through the surveillance capitalists.

For example, in the aftermath of the December 2015 terror attacks in Paris, President Obama, US legislators, and public officials around the world exhorted the tech companies, especially Google, Facebook, and Twitter, to identify and remove terrorist content. The companies were reportedly reluctant to be, or at least to be perceived as, "tools of government."[10] Journalists noted that public officials developed "workarounds" aimed at achieving access to instrumentarian power without imposing new burdens on the companies' public standing. For example, a government agency could assert that offending online content violates the internet company's terms of service, thus initiating the quick removal of offending material "without the paper trail that would go with a court order." Similarly, Google expanded its "trusted flagger" program, through which officials and others could identify problematic content for immediate action.[11]

The companies responded with their own initiatives. Eric Schmidt suggested new instruments, including a "spell check for hate," to target and eliminate terrorist accounts, remove content before it spreads, and accelerate the dissemination of counter-messages.[12] Top Obama administration officials endorsed that prospect on a pilgrimage to Silicon Valley in January 2016 for a "terror summit" with tech leaders. The agenda included discussions on how to disrupt terror activities on the internet, amplify alternative content, disrupt paths to radicalization, and enable security agencies to prevent attacks.[13] A White House briefing memo encouraged the companies to develop a "radicalism algorithm" that would digest social media and other sources of surplus to produce something comparable to a credit score, but aimed at evaluating the "radicalness" of online content.[14]

The turn to instrumentarian power as the solution to uncertainty is not restricted to the US government. Terrorism triggers similar responses in Germany, France, the UK, and around the world. After the 2016 attack on a Berlin Christmas market, German officials announced plans to require suspected extremists to wear electronic tags for perpetual tracking.[15] In 2017 surveillance capitalists, including Facebook, Microsoft, Google, and Twitter, established the Global Internet Forum to Counter Terrorism. The objective was to tighten the net of instrumentarian power through "collaboration on engineering solutions to the problem of sharing content classification techniques," "counterspeech initiatives," and a shared database of "unique digital fingerprints" for violent terrorist imagery to accelerate the identification of terrorist accounts.[16] The 2017 joint report of five countries—Australia, Canada, New Zealand, the United Kingdom, and the United States—included four key commitments, the very first of which was engagement with the internet companies to address online terrorism activities and to support the industry forum led by Google and Facebook.[17] That year, the European Council announced its expectation that "industry" would live up to its responsibility "to develop new technology and tools to improve the automatic detection and removal of content that incites to terrorist acts."[18] Meeting in Hamburg in 2017, the G20 countries vowed to work with the internet companies, insisting on the need for better instruments to filter, detect, and remove content, and "encouraging" the industry to invest in the technology and human capital able to detect and eliminate terrorist activity online.[19]

There are other emerging configurations of instrumentarian and state power. For example, former US Director of National Intelligence James Clapper told Congress in 2016 that the intelligence services might use the "internet of things" for "identification, surveillance, monitoring, location tracking, and targeting for recruitment, or to gain access to networks or user credentials."[20] Indeed, a research report from Harvard's Berkman Klein Center for Internet & Society concluded that surveillance capitalism's wave of "smart" appliances and products, networked sensors, and the "internet of things" would open up "numerous avenues for government actors to demand access to real-time and recorded communications."[21]

That "smart" and "connected" signal new channels for commercial *and* government surveillance is neither conjecture nor limited to federal intelligence agencies. In a 2015 murder case, police used data from a "smart" utility meter, an iPhone 6s Plus, and audio files captured by an Amazon Echo device to identify a suspect.[22] In 2014 data from a Fitbit wristband were used in a personal injury case, and in 2017 police used data from a pacemaker to charge a man with arson and insurance fraud.[23]

In the US, local law enforcement has joined the queue of institutions seeking access to instrumentarian power. Surveillance-as-a-service companies eagerly sell their wares to local police departments also determined to find a shortcut to certainty. One startup, Geofeedia, specializes in detailed location tracking of activists and protesters, such as Greenpeace members or union organizers, and the computation of individualized "threat scores" using data drawn from social media. Law-enforcement agencies have been among Geofeedia's most prominent clients.[24] When the Boston Police Department announced its interest in joining this roster in 2016, the city's police commissioner described to the *Boston Globe* his belief in machine certainty as the antidote to social breakdown: "The attack…on the Ohio State University campus is just the latest illustration of why local law-enforcement authorities need every tool they can muster to stop terrorism and other violence before it starts."[25] An ACLU attorney countered that the government is using tech companies "to build massive dossiers on people" based on nothing more than their constitutionally protected speech.[26] Another, more prominent surveillance-as-a-service company, Palantir, once touted by *Bloomberg Businessweek* as "the war on terror's secret weapon," was found to be in a secret collaboration with the New Orleans Police Department to test its "predictive policing" technology. Palantir's software not only identified gang members but also "traced people's ties to other gang members, outlined criminal histories, analyzed social media, and predicted the likelihood that individuals would commit violence or become a victim."[27]

IV. The China Syndrome

It is now possible to imagine one logical conclusion of this trend toward the substitution of certainty for society as the Chinese government develops a comprehensive "social credit" system described by one China scholar as the "core" of China's internet agenda. The aim is "to leverage the explosion of personal data…in order to improve citizens' behavior.…Individuals and enterprises are to be scored on various aspects of their conduct—where you go, what you buy and who you know—and these scores will be integrated within a comprehensive database that not only links into government information, but also to data collected by private businesses."[28]

The system tracks "good" and "bad" behavior across a variety of financial and social activities, automatically assigning punishments and rewards to decisively shape behavior toward "building sincerity" in economic, social, and political life: "The aim is for every Chinese citizen to be trailed by a file compiling data from public and private sources…searchable by fingerprints and other biometric characteristics."[29]

Although China's social credit vision is invariably described as "digital totalitarianism" and is often compared to the world of Orwell's *1984*, it is better understood as the apotheosis of instrumentarian power fed by public and private data sources and controlled by an authoritarian state. The accounts of its pilot programs describe powerful examples of surveillance capitalism's economies of action and the intricate construction of superscale means of behavior modification. The aim is the automation of society through tuning, herding, and conditioning people to produce preselected behaviors judged as desirable by the state and thus able to "preempt instability," as one strategic studies expert put it.[30] In other words, the aim is to achieve guaranteed *social* rather than *market* outcomes using instrumentarian means of behavioral modification. The result is an emergent system that allows us to peer into one version of a future defined by a comprehensive fusion of instrumentarian and state power.

China's vision is intended as the solution to its own unique version of the curse of social dissolution. Writing in *Foreign Policy,* journalist Amy Hawkins explains that China's pandemic of social distrust is the problem to which the social credit system is addressed as the cure: "To be Chinese today is to live in a society of distrust, where every opportunity is a potential con and every act of generosity a risk of exploitation."[31] A fascinating empirical study of social trust in contemporary China actually finds high levels of "in-group" trust but discovers that these are correlated with negative health outcomes. The conclusion is that many Chinese trust only the people who are well-known to them. All other relationships are regarded with suspicion and anxiety, with obvious consequences for social trust as well as well-being.[32] This rampant distrust, typically assigned to the traumas of rapid modernization and the shift to a quasi-capitalist economy, is also the legacy of Chinese totalitarianism. The Chinese Communist Party dismantled traditional domains of affiliation, identity, and social meaning—family, religion, civil society, intellectual discourse, political freedom—recalling Arendt's description of the "atomization" that destroys bonds of trust.[33] As Hawkins writes, "But rather than promoting the organic return of traditional morality to reduce the gulf of distrust, the Chinese government has preferred to invest its energy in technological fixes…and it's being welcomed by a public fed up of not knowing who to trust…in part because there's no alternative."[34] The Chinese government intends to commandeer instrumentarian power to replace a broken society with certain outcomes.

In 2015 the Chinese central bank announced a pilot project in which the top e-commerce companies would pioneer the data integration and software development for personal credit scoring. Among the largest of the pilots was Alibaba's Ant Financial and its "personal credit scoring" operation, "Sesame Credit." The Sesame Credit system produces a "holistic" rating of "character" with algorithmic learning that goes far beyond the timely payment of bills and loans. Algorithms evaluate and rank purchases (video games versus children's books), education degrees, and the quantity and "quality" of friends. One reporter's account of her Sesame Credit experience warns that the algorithm veers into "voodoo," considering the credit scores of her social contacts, the car she drives, her job, school, and a host of unspecified behavioral variables that supposedly "correlate with good credit." The shadow text remains out of reach, and users are left to guess how to improve their scores, including shedding friends with low scores and bulking up on high-scoring individuals who, some believe, can boost one's own rank.[35]

The company's CEO boasts that the scoring system "will ensure that the bad people in society don't have a place to go, while good people can move freely and without obstruction." Those with high scores receive honors and rewards from Sesame Credit's customers in its behavioral futures markets. They can rent a car without a deposit, receive favorable terms on loans and apartment rentals, receive fast-tracking for visa permits, enjoy being showcased on dating apps, and a host of other perks. However, one report warns that the privileges linked to a high personal credit score can suddenly tumble for reasons unrelated to consumption behavior, such as cheating on a university exam.[36]

In 2017 the central bank retracted its support for the private-sector personal credit programs, perhaps because they were too successful, their concentrations of knowledge and power too great. Sesame Credit had acquired more than 400 million users in just two years, staking a claim to just about every aspect of those users' lives.[37] A journalist who wrote a book on Ant Financial anticipates that the government is preparing to assert control over the whole system: "The government doesn't want this very important infrastructure of the people's credit in one big company's hands." The Chinese government appears to understand that power accrues to the owners of the means of behavioral modification. It is the owners who set the behavioral parameters that define guaranteed outcomes. Thus, fusion advances.

A sense of the kind of social world that might be produced by the fusion of instrumentarian and state power begins with the "judgment defaulter's list," described by the *Economist* as the heart of the social credit system and perhaps the best indicator of its larger ambitions. The list includes debtors and anyone who has ever defied a court order:

> People on the list can be prevented from buying aeroplane, bullet-train or first- or business-class rail tickets; selling, buying or building a house; or enrolling their children in expensive fee-paying schools. There are restrictions on offenders joining or being promoted in the party and army, and on receiving honours and titles. If the defaulter is a company, it may not issue shares or bonds, accept foreign investment or work on government projects.[38]

According to a report in *China Daily*, debtors on the list were automatically prevented from flying 6.15 million times since the blacklist was launched in 2013. Those in contempt of court were denied sales of high-speed train tickets 2.22 million times. Some 71,000 defaulters have missed out on executive positions at enterprises as a result of their debts. The Industrial and Commercial Bank of China said it had refused loans worth more than 6.97 billion yuan ($1.01 billion) to debtors on the list.[39] No one is sent to a reeducation camp, but they may not be allowed to purchase luxury goods. According to the director of the Institute of the Chinese Academy of International Trade and Economic Cooperation, "Given this inconvenience, 10 percent of people on the list started to pay back the money they owed spontaneously. This shows the system is starting to work."[40] Economies of action were performing to plan.

For the 400 million users of Sesame Credit, the fusion of instrumentarian and state power bites hard. Those who might find themselves on the blacklist discover that the credit system is designed to thrust their scores into an inexorable downward spiral: "First your score drops. Then your friends hear you are on the blacklist and, fearful that their scores might be affected, quietly drop you as a contact. The algorithm notices, and your score plummets further."[41]

The Chinese government's vision may be impossibly ambitious: the big dream of total awareness and perfect certainty mediated by algorithms that filter a perpetual flood of data flows from private and public supplies, including online and offline experience culled from every domain and able to ricochet back into the individual lives of 1.5 billion people, automating social behavior as the algorithms reward, punish, and shape action right down to the latest bus ticket. So far the project is fragmented across many pilots, not only in the tech companies but also in cities and regions, so there is no real test of the scale that the government envisions. There are plenty of experts who believe that a single system of that scale and complexity will be difficult if not impossible to achieve.

There are other good reasons to discount the social credit system's relevance for our story. To state the obvious, China is not a democracy and its culture differs profoundly from Western culture. Syracuse University researcher Yang Wang observes that Chinese culture places less value on privacy than does Western culture and that most Chinese have accommodated to the certain knowledge of online government surveillance and censorship. The most common word for privacy, *yinsi*, didn't even appear in popular Chinese dictionaries until the mid-1990s.[42] Chinese citizens have accepted national ID cards with biometric chips, "birth permits," and now social credit rankings because their society has been saturated with surveillance and profiling for decades. For example, the "dang'an" is a wide-ranging personal dossier compiled on hundreds of millions of urban residents from childhood and maintained throughout life. This "Mao-era system for recording the most intimate details of life" is updated by teachers, Communist Party officials, and employers. Citizens have no rights to see its contents, let alone contest them.

The dossier is only one feature of long-institutionalized and pervasive administrative systems of behavioral control and surveillance in daily life that bestow honors on some and punishments on others. Social control programs have expanded with the growth of the internet. For example, the "Golden Shield" is an extensive online surveillance system. China's cyber-censors can suspend internet or social media accounts if their users send messages containing sensitive terms such as "Tibetan independence" or "Tiananmen Square incident."[43]

As distinct as our politics and cultures may be or have been, the emerging evidence of the Chinese social credit initiatives broadcasts the logic of surveillance capitalism and the instrumentarian power that it produces. Sesame Credit doubles down on every aspect of surveillance capitalist operations, with hundreds of millions of people caught in the gears of an automated behavioral modification machine and its bubbling behavioral futures markets dispersing perks and honors like Pokémon fairy dust in return for guaranteed outcomes.

Chinese users are rendered, classified, and queued up for prediction with every digital touch, and so are we. We are ranked on Uber, on eBay, on Facebook, and on many other web businesses, and those are only the rankings that we see. Chinese users are assigned a "character" score, whereas the US government urges the tech companies to train their algorithms for a "radicalism" score. Indeed, the work of the shadow text is to evaluate, categorize, and predict our behavior in millions of ways that we can neither know nor combat—these are *our* digital dossiers. When it comes to credit scoring, US and UK banks and financial services firms have floated business models based on the mining and analysis of social media data for credit scores. Facebook itself has hinted of its interest, even filing a patent.[44] These efforts receded only because the Federal Trade Commission threatened regulatory intervention.[45]

Oxford University China scholar Rogier Creemers, who translated some of the first documents on the social credit system, observes that "the trend towards social engineering and 'nudging' individuals towards 'better' behavior is also part of the Silicon Valley approach that holds that human problems can be solved once and for all through the disruptive power of technology.... In that sense, perhaps the most shocking element of the story is not the Chinese government's agenda, but how similar it is to the path technology is taking elsewhere."[46]

In 2017 a surveillance technology trade show held in Shenzhen was packed with US companies selling their latest wares, especially cameras equipped with artificial intelligence and facial recognition. Among the crowd was the managing director of CCTV direct, a UK distributor of surveillance equipment. He lamented "how far behind the Western countries are," compared to the skills and thrills of China's surveillance infrastructure, but he also comforted himself with this thought: "What starts here ends up in homes, airports, and businesses back in America."[47]

The difference between surveillance capitalism in the West and China's emerging social credit system pivots on the patterns of entanglement and engagement between instrumentarian and state power. There are structural differences. In the West, as we have seen, the patterns have taken on many forms. The state began as bosom and shelter, then eager student and envious cousin. Surveillance capitalism and its instruments have come of age now, producing a fitful but necessary partnership. Key instrumentarian capabilities are docked in the big surveillance capitalist firms, and the state must move with and through these firms to access much of the power it seeks.

In the Chinese context, the state will run the show and own it, not as a market project but as a political one, a machine solution that shapes a new society of automated behavior for guaranteed political and social outcomes: certainty without terror. All the pipes from all the supply chains will carry behavioral surplus to this new, complex means of behavioral modification. The state will assume the role of the behaviorist god, owning the shadow text and determining the schedule of reinforcements and the behavioral routines that it will shape. Freedom will be forfeit to knowledge, but it will be the state's knowledge that it exercises, not for the sake of revenue but for the sake of its own perpetuation.

V. A Fork in the Road

Recall Carl Friedrich's observation on the challenge of grasping the naked facts of totalitarianism: "Virtually no one before 1914 anticipated the course of development which has overtaken Western civilization since then.... To this failure to foresee corresponds a difficulty in comprehending."[48] Recall too the grinning, robust "Joe" Stalin planted among Hollywood luminaries on the glossy pages of a 1939 *Look* magazine. Will we suffer the same lack of foresight as those who could not comprehend totalitarianism's rise, paralyzed by the sheer power of Big Other and its infinite echoes of consequence, distracted by our needs and confused by its speed, secrecy, and success?

Astonishment is a necessary alarm. We need it, but it should not leave us frozen in disbelief. The steady drumbeat of Big Other's manifest destiny, its breathtaking velocities, and the obscurity of its aims and purpose are intended to disarm, disorient, and bewilder. Inevitabilist ideology works to equate surveillance capitalism and its instrumentarian power with nature: not a human construction but something more like a river or a glacier, a thing that can only be joined or endured. All the more reason to ask: Might the banalities of today's declarations ("instruct the vehicular monitoring system not to allow the car to be started") also reveal themselves in the fullness of time as the seeds of our century's greatest nightmare? What of the authors of the instrumentarian project? How will we appraise the smiling, robust faces of the tech titans when we revisit those images in the glossy pixels of some twenty-first-century version of *Look*? The road from Shenzhen to an American or European airport also leads to the Roomba vacuum cleaner mapping your living room and your breakfast with Alexa. It is the road to machine certainty imposed by instrumentarian power and produced by surveillance capitalism. This journey is not as long as you might think.

There is a fork in the road.

In one direction lies the possibility of a synthetic declaration for a third modernity based on the strengthening of democratic institutions and the creative construction of a double movement for our time. On this road we harness the digital to forms of information capitalism that reunite supply and demand in ways that are both genuinely productive of effective life and compatible with a flourishing democratic social order. The first step down this road begins with naming, establishing our bearings, reawakening our astonishment, and sharing a sense of righteous indignity.

If we follow the other road, the one that links us to Shenzhen, we find our way to surveillance capitalism's antidemocratic vision for a third modernity fashioned by instrumentarian power. It is a future of certainty accomplished without violence. The price we pay is not with our bodies but with our freedom. This future does not yet exist, but like Scrooge's dream of Christmas future, the materials are all in place and ready for assembly. Chapter 14 examines this next way station on the road that began with an unprecedented capitalism, turned toward an unprecedented power, and now leads to an unprecedented society, theorized and legitimated by a burgeoning intellectual ecosystem of thinkers, researchers, and practitioners. What is this new place that they would have us call home?

TWO SPECIES OF POWER

Comparative Elements	Totalitarianism	Instrumentarianism
Central Metaphor	Big Brother	Big Other
Totalistic Vision	Total Possession	Total Certainty
Transcendent Purpose	Perfection of Society/Species Defined by Class or Race	Automation of Market/Society for the Certainty of Guaranteed Outcomes
Locus of Power	Control of the Means of Violence	Control of the Division of Learning in Society
Means of Power	Hierarchical Administration of Terror	Ownership of the Means of Behavioral Modification
Foundational Mechanisms	Arbitrary Terror; Murder	Dispossession of Behavioral Surplus for Computation, Control, Prediction
Theory and Practice	Theory Legitimates Practice	Practice Conceals Theory

Shoshana Zuboff, The Age of Surveillance Capitalism

Figure 4: Two Species of Power

Comparative Elements	Totalitarianism	Instrumentarianism
Ideological Style	Political Religion	Radical Indifference
Social Strategies	Atomization and Division; Total Believers or Total Enemies	Otherization of Predictable Organisms
Core Social Processes	In-Group, Out-Group for Conformity and Obedience	Hive Mind; Social Comparison for Confluence and Predictability
Unit of Social Production	Mass (Political)	Population (Statistical)
Vector of Social Influence	"Re-Education" Exerts Control from Inside-Out	Behavioral Modification Exerts Control from Outside-In
Social Patterning	Radical Isolation	Radical Connection
Demands on Individual	Absolute Loyalty through Subjugation to State/Species	Absolute Transparency through Subjugation to Guaranteed Outcomes
Primary Source of Individual Attachment to Power	Thwarted Identity	Thwarted Effectiveness
Primary Mode of Individual Attachment to Power	Identification	Dependency

Shoshana Zuboff, The Age of Surveillance Capitalism

CHAPTER FOURTEEN

A UTOPIA OF CERTAINTY

So from the years their gifts were showered: each
Grabbed at the one it needed to survive;
Bee took the politics that suit a hive,
Trout finned as trout, peach molded into peach,
And were successful at their first endeavor.

—W. H. AUDEN

SONNETS FROM CHINA, I

I. Society as the Other-One

Although he did not name it, the visionary of ubiquitous computing, Mark Weiser, foresaw the immensity of instrumentarian power as a totalizing societal project. He did so in a way that suggests both its utter lack of precedent and the danger of confounding it with what has gone before: "hundreds of computers in every room, all capable of sensing people near them and linked by high-speed networks have the potential to make totalitarianism up to now seem like sheerest anarchy."[1] In fact, all those computers are not the means to a digital hyper-totalitarianism. They are, as I think Weiser sensed, the foundation of an unprecedented power that can reshape society in unprecedented ways. If instrumentarian power can make totalitarianism look like anarchy, then what might it have in store for us?

Seven decades ago, Skinner's proto-instrumentarian behavioral utopia, *Walden Two,* was met with revulsion. Today the real thing is inspirational fodder for surveillance capitalist rhetoric as leaders promote the tools and visions that will bring the old professor's ideas to life...to *our lives*. The processes of normalization and habituation have begun. We have already seen that surveillance capitalism's pursuit of certainty—the mandate of the prediction imperative—requires a continuous approximation to total information as the ideal condition for machine intelligence. On the trail of totality, surveillance capitalists enlarged their scope from the virtual to the real world. The reality business renders all people, things, and processes as computational objects in an endless queue of equivalence without equality. Now, as the reality business intensifies, the pursuit of totality necessarily leads to the annexation of "society," "social relations," and key societal processes as a fresh terrain for rendition, calculation, modification, and prediction.

Big Other's ubiquity is revered as inevitable, but that is not the endgame. The aim in this new phase is the comprehensive visibility, coordination, confluence, control, and harmonization of social processes in the pursuit of scale, scope, and action. Although instrumentarianism and totalitarianism are distinct species, they each yearn toward totality, though in profoundly different ways. Totalitarianism seeks totality as a political condition and relies on violence to clear its path. Instrumentarianism seeks totality as a condition of market dominance, and it relies on its control over the division of learning in society, enabled and enforced by Big Other, to clear its path. The result is the application of instrumentarian power to societal optimization for the sake of market objectives: a utopia of certainty.

Although they resonate in many respects with the instrumentarian social vision of China's political elite, surveillance capitalists have distinct objectives. In their view, instrumentarian society is a market opportunity. Any norms and values they impose are designed to further the certain fulfillment of market goals. Like human experience, society is subordinated to the market dynamic and reborn as objectified computational behavioral metrics available to surveillance capitalism's economies of scale, scope, and action in the pursuit of the most-lucrative supplies of behavioral surplus. In order to achieve these aims, surveillance capitalists have conjured a chilling vision. They aim to fashion a new society that emulates machine learning in much the same way that industrial society was patterned on the disciplines and methods of factory production. In their vision, instrumentarian power replaces social trust, Big Other substitutes certainty for social relations, and society as we know it shades into obsolescence.

II. Totality Includes Society

Like generals delivering a chest-thumping tally of their armies, surveillance capitalist leaders take care to assure allies of their great power. This is typically expressed in an inventory of the instrumentarian troops massed at the border, poised for the rendition of everything in pursuit of totality. This pursuit, it becomes clear, does not merely *have consequences* for society; it *includes* society.

In the spring of 2017, Microsoft CEO Satya Nadella bounded onstage to open the company's annual developers' conference, his slender profile accentuated by the requisite black polo shirt, black jeans, and trendy black high-tops. He quickly dazzled the audience with a roll call of his troops. Nadella recounted the 500 million Windows 10 devices; 100 million monthly users of its Office software; 140 million monthly users of the corporation's digital "assistant," Cortana; and more than 12 million organizations signed on to its cloud services, including 90 percent of the Fortune 500.

Nadella did not fail to remind his audience of the crushing velocity that drives the instrumentarian project in an explosion of shock and awe, especially in the years since surveillance capitalism came to dominate digital services: internet traffic increased by a factor of 17.5 million over 1992's 100 gigabytes per day; 90 percent of the data in 2017 was generated in the prior two years; a single autonomous car will generate 100 gigabytes per second; there will be an estimated 25 billion intelligent devices by 2020. "It's stunning to see the progress across the depth and breadth of our society and economy and how digital technology is so *pervasive....* It's about what you can do with that technology to have broad impact." His final exhortation to the assembled developers—"Change the world!"—earned a thunderous round of applause.[2]

In celebrating Google's ambitions with the company's developers in 2017, CEO Sundar Pichai ran parallel to Nadella, showcasing his troop strength as Google's battalions fan out to embrace every corner of social life, demonstrating the breadth and depth of the corporation's instrumentarian power with a zeal that would have made Professor Skinner glow. Pichai reports that seven of the company's most-salient "products and platforms" engage one billion monthly active users, including Gmail, Android, Chrome, Maps, Search,

YouTube, and the Google Play Store; two billion active Android devices; 800 million monthly active users of Google Drive with three billion objects uploaded each week; 500 million Photos users uploading 1.2 billion photos each day; 100 million devices using Google Assistant. Every device is recast as a vehicle for Assistant, which will be available "throughout the day, at home and on the go" for every kind of task or social function. Pichai wants even more, telling his team, "We must go deeper." Assistant should be wherever "people might want to ask for help." Google executives share the enthusiasm. "Technology is now on the cusp of taking us into a magical age," writes Eric Schmidt, "solving problems today that we simply couldn't solve on our own."[3] Machine learning, he says, will do everything from curing blindness to saving animals from extinction. Above all, however, it is founder Larry Page who has long had his sights set on the transformation of society.

"The societal goal is our primary goal," Page told the *Financial Times* in 2016.[4] "We need revolutionary change, not incremental change," he told another interviewer that year. "We could probably solve a lot of the issues we have as humans."[5] Much of Page's future vision turns out to be stock utopian fare, themes that have been repeated for millennia. Page anticipates machine intelligence that restores humankind to the Garden of Eden, lifting us from toil and struggle into a new realm of leisure and fulfillment. He foresees, for example, a future society graced by "abundance" in all things, where employment is but a "crazy" distant memory.[6]

Most unusual, however, is that Page portrays Google's totalistic ambitions as a logical consequence of its commitment to the perfection of society. From his point of view, we should welcome the opportunity to lean on Big Other and willingly subordinate all knowledge and decision rights to Google's plan. For the sake of the plan, the totality of society—every person, object, and process—must be corralled into the supply chains that feed the machines, which, in turn, spin the algorithms that animate Big Other to manage and mitigate our frailty:

> What you should want us to do is to really build amazing products
> and to really do that...we have to understand apps and we have to
> understand things you could buy, and we have to understand airline
> tickets. We have to understand anything you might search for. And

> people are a big thing you might search for.... We're going to have
> people as a first class object in search... if we're going to do a good
> job meeting your information needs, we actually need to understand
> things and we need to understand things pretty deeply.[7]

Total knowledge is sold as a requirement of the "preemptive" services
that lead to the solution of solutions in the AI-powered, omniscient "Google
Assistant":

> It's really trying to understand everything in the world and make
> sense of it.... A lot of queries are actually about places, so we need to
> understand places.... A lot of the queries are about content we can't
> find. We did books, and so on.... So, we've been gradually expanding
> that... maybe you don't want to ask a question. Maybe you want to
> just have it answered for you before you ask it. That would be better.[8]

Google originated in the prospect of optimally organizing the world's information, but Page wants the corporation to optimize the organization of
society itself: "In my very long-term worldview," he said in 2013, "our software understands deeply what you're knowledgeable about, what you're
not, and how to organize the world so that the world can solve important
problems."[9]

Facebook CEO Mark Zuckerberg shares these totalistic ambitions, and
he is increasingly frank about "society," not just the individuals within it, as
subordinate to Facebook's embrace. His "three big company goals" include
"connecting everyone; understanding the world; and building the knowledge economy, so that every user will have 'more tools' to share 'different
kinds of content.'"[10] Zuckerberg's keen appreciation of second-modernity
instabilities—and the yearning for support and connection that is among its
most-vivid features—drives his confidence, just as it did for Google economist
Hal Varian. The corporation would know every book, film, and song a person
had ever consumed. Predictive models would enable the corporation to "tell
you what bar to go to" when you arrive in a strange city. The vision is detailed:
when you arrive at the bar, the bartender has your favorite drink waiting, and
you're able to look around the room and identify people just like you.

Zuckerberg described the flow of behavioral surplus as "growing at an exponential rate... that lets us project into the future... two years from now people are going to be sharing twice as much... four years, eight times as much...." And in a nod to the already pressing competition for totality, Zuckerberg anticipated that Facebook's social graph will "start to be a better map of how you navigate the web than the traditional link structure."[11]

To that end, the CEO told investors that Facebook would bring affordable internet access "to every person in the world" so that every user will have "more tools" to share "different kinds of content."[12] Nothing was likely to impede the corporation's progress on the societal front, he asserted, because "humans have such a deep desire to express themselves."[13]

In 2017 Zuckerberg went even further in articulating his societal ambitions, this time aiming straight at the heart of second-modernity anxieties: "People feel unsettled. A lot of what was settling in the past doesn't exist anymore." Zuckerberg believes that he and his company can provide a future "that works for everyone" and fulfills "personal, emotional, and spiritual needs" for "purpose and hope," "moral validation," and "comfort that we are not alone." "Progress now requires humanity coming together not just as cities or nations," Zuckerberg urged, "but also as a global community... the most important thing we at Facebook can do is develop the social infrastructure... to build a global community...." Citing Abraham Lincoln, Facebook's founder located his company's mission in the evolutionary time line of civilization, during which humanity organized itself first in tribes, then cities, then nations. The next phase of social evolution would be "global community," and Facebook was to lead the way, constructing the means and overseeing the ends.[14]

Speaking at Facebook's 2017 developers' conference, Zuckerberg linked his assertion of the company's historic role in establishing a "global community" to the standard myth of the modern utopia, assuring his followers, "In the future, technology is going to... free us up to spend more time on the things we all care about, like enjoying and interacting with each other and expressing ourselves in new ways.... A lot more of us are gonna do what today is considered the arts, and that's gonna form the basis of a lot of our communities."[15]

As Nadella and other surveillance capitalists spin their utopian dreams, the surveillance capitalists fail to mention that the magical age they envision comes at a price: Big Other must expand toward totality as it deletes all boundaries and overwhelms every source of friction in the service of its economic imperatives. All power yearns toward totality, and only authority stands in the way: democratic institutions; laws; regulations; rights and obligations; private governance rules and contracts; the normal market constraints exercised by consumers, competitors, and employees; civil society; the political authority of the people; and the moral authority of individual human beings who have their bearings.

This point was made in Goethe's fable of the sorcerer's apprentice, when, in the absence of the sorcerer's authority to guide and check the action, the apprentice transforms the broom into a demonic force of pure unrelenting power:

> *Ah, the word with which the master*
> *Makes the broom a broom once more!*
> *Ah, he runs and fetches faster!*
> *Be a broomstick as before!*
> *Ever new the torrents*
> *That by him are fed,*
> *Ah, a hundred currents*
> *Pour upon my head!*[16]

III. Applied Utopistics

Instrumentarian power, like the apprentice's broom, has flourished in the sorcerer's absence with little authority to check its action, and the surveillance capitalists' appetite for totality has grown with this success. The utopian rhetoric of a magical age has been critical to this progress. The notion that Big Other will solve all of humanity's problems while empowering each individual is usually dismissed as mere "techno utopianism," but it would be a mistake for us to ignore this rhetoric without examining its purpose. Such discourse is no mere hogwash. It is the minesweeper that precedes the foot soldiers and the canny diplomat sent ahead to disarm the enemy and smooth the way for a quiet surrender. The promise of a magical age plays a critical strategic role, simultaneously distracting us from and legitimating surveillance capitalism's totalistic ambitions that necessarily include "people" as a "first class object."

The "societal goal" articulated by the leading surveillance capitalists fits snugly into the notion of limitless technological progress that dominated utopian thought from the late eighteenth century through the late nineteenth century, culminating with Marx. Indeed, surveillance capitalists such as Nadella, Page, and Zuckerberg conform to five of the six elements with which the great scholars of utopian thought, Frank and Fritzie Manuel, define the classic profile of the most ambitious modern utopianists: (1) a tendency toward highly focused tunnel vision that simplifies the utopian challenge, (2) an earlier and more trenchant grasp of a "new state of being" than other contemporaries, (3) the obsessive pursuit and defense of an *idée fixe*, (4) an unshakable belief in the inevitability of one's ideas coming to fruition, and (5) the drive for total reformation at the level of the species and the entire world system.[17]

The Manuels observe a sixth characteristic of the future-oriented modern visionary, and this is where the men and the corporations examined here represent powerful exceptions to the rule: "Often a utopian foresees the later evolution and consequences of technological development already present in an embryonic state; he may have antennae sensitive to the future. *His gadgets, however, rarely go beyond the mechanical potentialities of his age. Try as he may to invent something wholly new, he cannot make a world out of nothing.*"[18] In our time, however, surveillance capitalists can and do make such a world—a genuinely historic deviation from the norm.

Individually and collectively, the knowledge, power, and wealth that surveillance capitalists command would be the envy of any ancient potentate, just as they are now coveted by the modern state. With 2017 balance sheets reporting $126 billion in cash and securities for Microsoft, $92 billion for Google, and about $30 billion for Facebook, and the financial markets endorsing their ever-expanding instrumentarian regimes with more than $1.6 trillion in market capitalization in mid-2017, these are the rare utopianists who can oversee the translation of their imaginations into fact without soldiers to pave the way in blood.[19]

In this respect, the surveillance capitalist leaders are sui generis utopianists. Marx grasped the world with his thickly articulated theory, but with only the power of his ideas, he could not implement his vision. Long after the publication of Marx's theories, men such as Lenin, Stalin, and Mao applied them to real life. Indeed, the Manuels describe Lenin as a specialist in "applied utopistics."[20] In contrast, the surveillance capitalists seize the world in practice. Their theories are thin—at least this is true of the thinking that they share with the public. The opposite is true of their power, which is monumental and largely unimpeded.

When it comes to theory and practice, the usual sequence is that theory is available to inspect, interrogate, and debate before action is initiated. This allows observers an opportunity to judge a theory's worthiness for application, to consider unanticipated consequences of application, and to evaluate an application's fidelity to the theory in which it originates. The unavoidable gap between theory and practice creates a space for critical inquiry. For example, we can question whether a law or governmental practice is consistent with a nation's constitution, charter of rights, and governing principles because we can inspect, interpret, and debate those documents. If the gap is too great, citizens act to close the gap by challenging the law or practice.

The surveillance capitalists reverse the normal sequence of theory and practice. Their practices move ahead at high velocity in the absence of an explicit and contestable theory. They specialize in displays of instrumentarianism's unique brand of shock and awe, leaving onlookers dazed, uncertain, and helpless. The absence of a clear articulation of their theory leaves the rest of us to ponder its practical effects: the vehicular monitoring system that shuts down your engine; the destination that appears with the route; the suggested purchase that flashes on your phone the moment your endorphins peak; Big Other's continuous tracking of your location, behavior, and mood; and its cheerful herding of city dwellers toward surveillance capitalism's customers.

However meager and secretive the surveillance capitalists' theories might be, the instrumentarian power they wield can make their dreams come true, or at least inflict a whirlwind of consequences as they try. The only way to grasp the theory advanced in their applied utopistics is to reverse engineer their operations and scrutinize their meaning, as we have done throughout these chapters.

Applied utopistics are on the move at Facebook, Google, and Microsoft as the frontier of behavioral surplus extraction moves into realms of life traditionally understood as societal and elaborated under some combination of civil institutions and public leadership. Zuckerberg's 2017 mission statement for Facebook, introduced as "building global community," announced a new phase of applied utopistics: "Overall, it is important that the governance of our community scales with the complexity and demands of its people. We are committed to always do better, even if that involves building a worldwide voting system to give you more voice and control. Our hope is that this model provides examples of how collective decision-making may work in other aspects of the global community."[21] Later that year, Zuckerberg told an audience of developers that "we have a full roadmap of products to help build groups and community, help build a more informed society, help keep our communities safe, and we have a lot more to do here."[22]

Back on that stage in the spring of 2017, Microsoft's Nadella encouraged his developers: "Whether it's precision medicine or precision agriculture, whether it's digital media or the industrial internet, the opportunity for us as developers to have broad, deep impact on all parts of society and all parts of the economy has never been greater."[23] The vision that Nadella unveiled that day is emblematic of the wider surveillance capitalist template for our futures. Where do they think they are taking us?

IV. Confluence as Machine Relations

In order to decipher the true measure of an instrumentarian society, let's set aside the razzle-dazzle of a "magic age" and focus instead on the practices of applied utopistics and the social vision they imply. Nadella provided a valuable opportunity when he unveiled a series of practical applications that imply a sweeping new vision of machine relations as the template for a new era's social relations.

The reveal begins with Nadella's account of a Microsoft collaboration with a 150-year-old Swedish manufacturer of high-precision metal-cutting equipment that has reinvented itself for the twenty-first century. The project is a state-of-the-art illustration of what Nadella describes as the "fundamental change in the paradigm of the apps that we are building, a change in the worldview that we have…from…a mobile-first, cloud-first world to a new world that is going to be made up of an intelligent cloud and an intelligent edge." Artificial intelligence, he says, "learns from information and interacts with the physical world," thus citing the capabilities required for economies of action.[24]

Nadella first describes the machines linked by telemetry in the new factory setting as they continuously stream data to the "IoT hub" in the "cloud," where Microsoft's analyses search out anomalies that could put the machines at risk. Each anomaly is traced back through the data stream to its cause, and machine intelligence in the hub learns to identify the causal patterns so that it can preemptively shut down a threatened piece of equipment in about two seconds, before a potentially damaging event can occur.

Then Nadella describes the new "breakthrough capability" in which a computational actuating sensor is embedded directly in the machine, dramatically reducing the time to a preemptive shutdown: "That logic is now running locally, so there's no cloud loop." The "edge" knows immediately when the machine experiences an event that predicts a future anomaly, and it shuts down the equipment within 100 milliseconds of this computation, a "20X improvement." This is celebrated as "the power of the cloud working in harmony with an intelligent edge" to *anticipate and preempt variations from the norm* "before they happen."[25]

The power of machine learning develops exponentially as the devices learn from one another's experiences, feeding into and drawing upon the intelligence of the hub. In this scenario it's not that the whole is greater than the sum of the parts; it's more like there are no parts. The whole is everywhere, fully manifest in each device embedded in each machine. Nadella translates these facts into their practical application, observing that once you have lots of devices around, an "ad hoc data center" is created "on a factory floor, at home, or anywhere else.…You can turn any place into a safe, AI-driven place."[26]

With this statement it finally becomes clear that "safe" means "automatically anomaly-free." In Nadella's factory, machine knowledge instantaneously replaces ignorance, herding all machine behavior to preestablished norms. Rather than concern for the multiplication of risk and the contagion of failure should machine learning go awry, Nadella celebrates the synchrony and universality of certain outcomes, as every machine is the same machine marching to the same song.

Just as a century ago the logic of mass production and its top-down administration provided the template for the principles of industrial society and its wider civilizational milieu, so too is Nadella's new-age factory revealed as the proving ground for his social vision—surveillance capitalism's vision—of an instrumentarian society enabled by a new form of collective action. Machine learning is rendered here as a collective mind—a hive mind—in which each element learns and operates in concert with every other element, a model of collective action in which all the machines in a networked system move seamlessly toward confluence, all sharing the same understanding and operating in unison with maximum efficiency to achieve the same outcomes. Confluent action means that the "freedom" of each individual machine is surrendered to the knowledge that they share. Just as the behavioral theorists Planck, Meyer, and Skinner predicted, this sacrifice amounts to an all-out war on accidents, mistakes, and randomness in general.

Nadella takes this template of new machine relations and applies it to a more complex illustration of a human and machine system, though still in the "economic domain." This time it's a construction site, where human and machine behaviors are tuned to preestablished parameters determined by superiors and referred to as "policies." Algorithmic uncontracts apply rules and substitute for social functions such as supervision, negotiation, communication, and problem solving. Each person and piece of equipment takes a place among an equivalence of objects, each one "recognizable" to "the system" through the AI devices distributed across the site.

For example, each individual's training, credentials, employment history, and other background information are instantly on display to the system. A "policy" might declare that "only credentialed employees can use jackhammers." If an employee who is not accredited for jackhammer use approaches that tool, the possibility of an impending violation is triggered, and the jackhammer screams an alert, instantly disabling itself.

Significantly, it is not only the unified action of things on the site that are mobilized in alignment with policies. Confluent human action is also mobilized, as social influence processes are triggered in the preemptive work of anomaly avoidance. In the case of the at-risk jackhammer, the humans at the site are mobilized to swarm toward the location of the AI-anticipated anomalous offense in order to "quickly resolve it." "The intelligent edge," Microsoft developers are told, "is the interface between the computer and the real world...you can search the real world for people, objects and activities, and *apply policies to them....*"[27]

Once people and their relationships are rendered as otherized, equivalent "things in the cloud," 25 billion computational actuating devices can be mobilized to shape behavior around safe and harmonious "policy" parameters. The most "profound shift," Nadella explained, is that "people and their relationship with other people is now *a first-class thing* in the cloud. It's not just people but it's their relationships, it's their relationships to all of the work artifacts, their schedules, their project plans, their documents; all of that now is manifest in this Microsoft Graph." These streams of total information are key to optimizing "the future of productivity," Nadella exulted.[28]

In Microsoft's instrumentarian society, the factories and workplaces are like Skinner's labs, and the machines replace his pigeons and rats. These are the settings where the architecture and velocities of instrumentarian power are readied for translation to society in a digital-age iteration of *Walden Two* in which machine relations are the model for social relations. Nadella's construction site exemplifies the grand confluence in which machines and humans are united as objects in the cloud, all instrumented and orchestrated in accordance with the "policies." The magnificence of "policies" lies precisely in the fact that they appear on the scene as guaranteed outcomes to be automatically imposed, monitored, and maintained by the "system." They are baked into Big Other's operations, an infinity of uncontracts detached from any of the social processes associated with private or public governance: conflict and negotiation, promise and compromise, agreement and shared values, democratic contest, legitimation, and authority.

The result is that "policies" are functionally equivalent to plans, as Big Other directs human and machine action. It ensures that doors will be locked or unlocked, car engines will shut down or come to life, the jackhammer will scream "no" in suicidal self-sacrifice, the worker will adhere to norms, the group will swarm to defeat anomalies. We will all be safe as each organism hums in harmony with every other organism, less a society than a population that ebbs and flows in perfect frictionless confluence, shaped by the means of behavioral modification that elude our awareness and thus can neither be mourned nor resisted.

Just as the division of labor migrated from the economic domain to society in the twentieth century, Nadella's construction site is the economic petri dish in which a new division of learning mutates into life, ready for translation to society. In the twentieth century the critical success factors of industrial capitalism—efficiency, productivity, standardization, interchangeability, the minute division of labor, discipline, attention, scheduling, conformity, hierarchical administration, the separation of knowing and doing, and so forth—were discovered and crafted in the workplace and then transposed to society, where they were institutionalized in schools, hospitals, family life, and personality. As generations of scholars have documented, society became more factory-like so that we might train and socialize the youngest among us to fit the new requirements of a mass production order.

We have entered this cycle anew, but now the aim is to remake twenty-first-century society as a "first-class thing" organized in the image of the machine hive for the sake of others' certainty. The connectedness that we once sought for personal sustenance and effectiveness is recast as the medium for a new species of power and the social confluence that translates into guaranteed outcomes.

V. Confluence as Society

Microsoft scientists have been working for years on how to take the same logic of automated preemptive control at the network's edge and transpose it to social relations. As Nadella observed in 2017, if "we" can do this in a "physical place," it can be done "everywhere" and "anywhere." He advised his audience of applied utopianists, "You could start reasoning about people, their relationship with other people, the things in the place...."[29]

The imaginative range of this new thinking is demonstrated in a 2013 Microsoft patent application updated and republished in 2016 and titled "User Behavior Monitoring on a Computerized Device."[30] With conspicuously thin theory complemented by thick practice, the patented device is designed to monitor user behavior in order to preemptively detect "any deviation from normal or acceptable behavior that is likely to affect the user's mental state. A prediction model corresponding to features of one or more mental states may be compared with features based upon current user behavior."

The scientists propose an application that can sit in an operating system, server, browser, phone, or wearable device continuously monitoring a person's behavioral data: interactions with other people or computers, social media posts, search queries, and online activities. The app may activate sensors to record voice and speech, videos and images, and movement, such as detecting "when the user engages in excessive shouting by examining the user's phone calls and comparing related features with the predication model."

All these behavioral data are stored for future historical analyses in order to improve the prediction model. If the user normally restrains the volume of his or her voice, then sudden excessive shouting may indicate a "psychosocial event." Alternatively, the behavior could be assessed in relation to a "feature distribution representing normal and/or acceptable behavior for an average member of a population…a statistically significant deviation from that behavior baseline indicates a number of possible psychological events." The initial proposition is that in the event of an anomaly, the device would alert "trusted individuals" such as family members, doctors, and caregivers. But the circle widens as the patent specifications unfold. The scientists note the utility of alerts for health care providers, insurance companies, and law-enforcement personnel. Here is a new surveillance-as-a-service opportunity geared to preempt whatever behavior clients choose.

Microsoft's patent returns us to Planck, Meyer, and Skinner and the viewpoint of the Other-One. In their physics-based representation of human behavior, anomalies are the "accidents" that are called freedom but actually denote ignorance; they simply cannot yet be explained by the facts. Planck/Meyer/Skinner believed that the forfeit of this freedom was the necessary price to be paid for the "safety" and "harmony" of an anomaly-free society in which all processes are optimized for the greater good. Skinner imagined that with the correct technology of behavior, knowledge could preemptively eliminate anomalies, driving all behavior toward preestablished parameters that align with social norms and objectives. "If we could show that our members preferred life in Walden Two," says Frazier-Skinner, "it would be the best possible evidence that we had reached a safe and productive social structure."[31]

In this template of social relations, behavioral modification operates just beyond the threshold of human awareness to induce, reward, goad, punish, and reinforce behavior consistent with "correct policies." Thus, Facebook learns that it can predictably move the societal dial on voting patterns, emotional states, or anything else that it chooses. Niantic Labs and Google learn that they can predictably enrich McDonald's bottom line or that of any other customer. In each case, corporate objectives define the "policies" toward which confluent behavior harmoniously streams.

The machine hive—the confluent mind created by machine learning—is the material means to the final elimination of the chaotic elements that interfere with guaranteed outcomes. Eric Schmidt and Sebastian Thrun, the machine intelligence guru who once directed Google's X Lab and helped lead the development of Street View and Google's self-driving car, make this point in championing Alphabet's autonomous vehicles. "Let's stop freaking out about artificial intelligence," they write.

Schmidt and Thrun emphasize the "crucial insight that differentiates AI from the way people learn."[32] Instead of the typical assurances that machines can be designed to be more like human beings and therefore less threatening, Schmidt and Thrun argue just the opposite: it is necessary for people to become more machine-like. Machine intelligence is enthroned as the apotheosis of collective action in which all the machines in a networked system move seamlessly toward confluence, all sharing the same understanding and thus operating in unison with maximum efficiency to achieve the same outcomes. The jackhammers do not independently appraise their situation; they each learn what they all learn. They each respond the same way to uncredentialed hands, their brains operating as one in service to the "policy." The machines stand or fall together, right or wrong together. As Schmidt and Thrun lament,

> When driving, people mostly learn from their own mistakes, but they rarely learn from the mistakes of others. People collectively make the same mistakes over and over again. As a result, hundreds of thousands of people die worldwide every year in traffic collisions. AI evolves differently. When one of the self-driving cars makes an error,

all of the self-driving cars learn from it. In fact, new self-driving cars are "born" with the complete skill set of their ancestors and peers. So collectively, these cars can learn faster than people. With this insight, in a short time self-driving cars safely blended onto our roads alongside human drivers, as they kept learning from each other's mistakes.... Sophisticated AI-powered tools will empower us to better learn from the experiences of others.... The lesson with self-driving cars is that we can learn and do more collectively.[33]

This is a succinct but extraordinary statement of the machine template for the social relations of an instrumentarian society. The essence of these facts is that first, *machines are not individuals,* and second, *we should be more like machines.* The machines mimic each other, and so must we. The machines move in confluence, not many rivers but one, and so must we. The machines are each structured by the same reasoning and flowing toward the same objective, and so must we be structured.

The instrumentarian future integrates this symbiotic vision in which the machine world and social world operate in harmony within in and across "species" as humans emulate the superior learning processes of the smart machines. This emulation is not intended as a throwback to mass production's Taylorism or Chaplin's hapless worker swallowed by the mechanical order. Instead, this prescription for symbiosis takes a different road on which human interaction mirrors the relations of the smart machines as individuals learn to think and act by emulating one another, just like the self-driving cars and the policy-worshipping jackhammers.

In this way, the machine hive becomes the role model for a new human hive in which we march in peaceful unison toward the same direction based on the same "correct" understanding in order to construct a world free of mistakes, accidents, and random messes. In this world the "correct" outcomes are known in advance and guaranteed in action. The same ubiquitous instrumentation and transparency that define the machine system must also define the social system, which in the end is simply another way of describing the ground truth of instrumentarian society.

In this human hive, individual freedom is forfeit to collective knowledge and action. Nonharmonious elements are preemptively targeted with high doses of tuning, herding, and conditioning, including the full seductive force of social persuasion and influence. We march in certainty, like the smart machines. We learn to sacrifice our freedom to collective knowledge imposed by others and for the sake of their guaranteed outcomes. This is the signature of the third modernity offered up by surveillance capital as its answer to our quest for effective life together.

CHAPTER FIFTEEN

THE INSTRUMENTARIAN COLLECTIVE

So an age ended, and its last deliverer died
In bed, grown idle and unhappy; they were safe:
The sudden shadow of a giant's enormous calf
Would fall no more at dusk across their lawns outside.

—W. H. AUDEN
SONNETS FROM CHINA, X

I. The Priests of Instrumentarian Power

Applied utopianist executives such as Page, Nadella, and Zuckerberg do not say much about their theories. At best the information we have is episodic and shallow. But a cadre of data scientists and "computational social scientists" has leapt into this void with detailed experimental and theoretical accounts of the gathering momentum of instrumentarian power, providing invaluable insight into the social principles of an instrumentarian society.

One outstanding example is the work of Alex Pentland, the director of the Human Dynamics Lab within MIT's Media Lab. Pentland is the rare applied utopianist who, in collaboration with his students and collaborators, has vigorously articulated, researched, and disseminated a theory of instrumentarian society in parallel to his prolific technical innovations and practical applications. The studies that this group has produced are a contemporary signal of an increasingly taken-for-granted worldview among data scientists whose computational theories and innovations exist in dynamic interaction with the progress of surveillance capitalism, as in the case of Picard's affective

416

computing and Paradiso's digital omniscience. However, few consider the social ramifications of their work with Pentland's insight and conviction, providing us with an invaluable opportunity to critically explore the governance assumptions, societal principles, and social processes that define an instrumentarian society. My aim is to infer the theory behind the practice, as surveillance capitalists integrate "society" as a "first class object" for rendition, computation, modification, monetization, and control.

Pentland is a prolific author or coauthor of hundreds of articles and research studies in the field of data science and is a prominent institutional actor who advises a roster of organizations, including the World Economic Forum, the Data-Pop Alliance, Google, Nissan, Telefonica, and the Office of the United Nations Secretary General. Pentland's research lab is funded by a who's who of global corporations, consultancies, and governments: Google, Cisco, IBM, Deloitte, Twitter, Verizon, the EU Commission, the US government, the Chinese government, "and various entities who are all concerned with why we don't know what's going on in the world...."[1]

Although Pentland is not alone in this field, he is something of a high priest among an exclusive group of priests. Unlike Hal Varian, Pentland does not speak of Google in the first-person plural, but his work is showcased in surveillance capitalist enclaves, where it provides the kind of material and intellectual support that helps to legitimate instrumentarian practices. Appearing for a presentation at Google, where Pentland is on the Advisory Board for the Advanced Technology and Projects Group, former Pentland doctoral student and top Google executive Brad Horowitz introduced his mentor as an "inspirational educator" with credentials across many disciplines and whose former students lead the computational sciences in theory and practice.[2]

Pentland is often referred to as the "godfather of wearables," especially Google Glass. In 1998 he predicted that wearables "can extend one's senses, improve memory, aid the wearer's social life and even help him or her stay calm and collected."[3] Thad Starner, one of Pentland's doctoral students, developed a primitive "wearable" device while at MIT and was hired by Sergey Brin in 2010 to continue that work at Google: a project that produced Google Glass. More than fifty of Pentland's doctoral students have gone on to spread the instrumentarian vision in top universities, in industry research groups, and in thirty companies in which Pentland participates as cofounder, sponsor, or advisor. Each one applies some facet of Pentland's theory, analytics, and inventions to real people in organizations and cities.[4]

Pentland's academic credentials and voluble intelligence help legitimate a social vision that repelled and alarmed intellectuals, public officials, and the general public just decades ago. Most noteworthy is that Pentland "completes" Skinner, fulfilling his social vision with big data, ubiquitous digital instrumentation, advanced mathematics, sweeping theory, numerous esteemed coauthors, institutional legitimacy, lavish funding, and corporate friends in high places without having attracted the worldwide backlash, moral revulsion, and naked vitriol once heaped on Harvard's outspoken behaviorist. This fact alone suggests the depth of psychic numbing to which we have succumbed and the loss of our collective bearings.

Like Skinner, Pentland is a designer of utopias and a lofty thinker quick to generalize from animals to the entire arc of humanity. He is also a hands-on architect of instrumentarianism's practical architecture and computational challenges. Pentland refers to his theory of society as "social physics," a conception that confirms him as this century's B. F. Skinner, by way of Planck, Meyer, and MacKay.[5] And although Pentland never mentions the old behaviorist, Pentland's book, *Social Physics,* summons Skinner's social vision into the twenty-first century, now fulfilled by the instruments that eluded Skinner in his lifetime. Pentland validates the instrumentarian impulse with research and theory that are boldly grounded in Skinner's moral reasoning and epistemology as captured by the viewpoint of the Other-One.

Professor Pentland began his intellectual journey as Skinner did, in the study of animal behavior. Where Skinner trained his reasoning on the detailed behaviors of blameless individual creatures, Pentland concerned himself with the gross behavior of animal populations. As a part-time researcher at NASA's Environmental Research Institute while still an undergraduate, he developed a method for assessing the Canadian beaver population from space by counting the number of beaver ponds: "You're watching the lifestyle, and you get an indirect measure."[6]

The experience appears to have hooked Pentland on the distant detached gaze, which he would later embrace as the "God's eye view." You may have experienced the sensation of the God view from the window seat of an airplane as it lifts you above the city, transforming all the joys and woes below into the mute bustle of an anthill. Up there, any sense of "we" quickly dissolves into the viewpoint of the Other-One, and it is this angle of observation that founded Pentland's science as he learned to apply MacKay's principles of remote observation and telestimulation to humans: "If you think about people across the room talking, you can tell a lot....It's like watching beavers from outer space, like Jane Goodall watching gorillas. You observe from a distance."[7] (This is a slur on Goodall, of course, whose seminal genius was her ability to understand the gorillas she studied not as "other ones" but rather as "one of us.")

The God view would come to be essential to the conception of instrumentarian society, but a comprehensive picture emerged gradually over years of piecemeal experimentation. In the following section we track that journey as Pentland and his students learned to render, measure, and compute social behavior. With that foundation, we turn to Pentland's *Social Physics*, which aims to recast society as an instrumentarian hive mind—like Nadella's machines—but now extensively theorized and deeply evocative of Skinner's formulations, values, worldview, and vision of the human future.

II. When Big Other Eats Society: The Rendition of Social Relations

Skinner bitterly lamented the absence of "instruments and methods" for the study of human behavior comparable to those available to physicists. As if in response, Pentland and his students have spent the last two decades determined to invent the instruments and methods that can transform all of human behavior, especially social behavior, into highly predictive math. An early milestone was a 2002 collaboration with then-doctoral student Tanzeem Choudhury, in which the coauthors wrote, "As far as we know, there are currently no available methods to automatically model face-to-face interactions. This absence is probably due to the difficulty of obtaining reliable measurements from real-world interactions within a community....We believe sensing and modeling physical interactions among people is an untapped resource."[8] In other words, the "social" remained an elusive domain even as data and computers had become more commonplace.

The researchers' response was to introduce the "sociometer," a wearable sensor that combines a microphone, accelerometer, Bluetooth connection, analytic software, and machine learning techniques designed to infer "the structure and dynamic relationships" in human groups.[9] (Choudhury would eventually run the People Aware Computing group at Cornell University.) From that point onward, Pentland and his teams have worked to crack the code on the instrumentation and instrumentalization of social processes in the name of a totalistic social vision founded on a comprehensive means of behavior modification.

A 2005 collaboration with doctoral student Nathan Eagle reiterated the problem of insufficient data on human society, noting the "bias, sparsity of data, and lack of continuity" in social science's understanding of human behavior and the resulting "absence of dense continuous data that also hinders the machine learning and agent-based modeling communities from constructing more comprehensive predictive models of human dynamics."[10] Pentland had insisted that even the relatively new field of "data mining" could not capture the "real action" of conversations and face-to-face interactions necessary for a trenchant and comprehensive grasp of social behavior.[11] But he also recognized that a rapidly growing swath of human activity—from transactions to communication—was falling to computer mediation, largely as a result of the cell phone.

The team saw that it would be possible to exploit the increasingly "ubiquitous infrastructure" of mobile phones and combine those data with new streams of information from their wearable behavioral monitors. The result was a radical new solution that Pentland and Eagle called "reality mining." Mentor and student demonstrated how the data from cell phones "can be used to uncover regular rules and structure in the behavior of both individuals and organizations," thus furthering the progress of behavioral surplus capture and analysis and pointing the way toward the larger shift in the nature of behavioral dispossession from virtual, to actual, to social experience.[12] As a technological and cultural landmark, the researchers' announcement that "reality" was now fair and feasible game for surplus capture, search, extraction, rendition, datafication, analysis, prediction, and intervention helped to forge a path toward the new practices that would eventually become the "reality business."

Pentland and Eagle began with 100 MIT students and faculty centered at the Media Lab, equipping them with 100 Nokia phones preloaded with special software in a project that would become the basis of Eagle's doctoral dissertation. The two researchers exposed the revelatory power of continuously harvested behavioral data, which they validated with survey information collected directly from each participant. Their analyses produced detailed portraits of individual and group life: the "social system," as the authors called it. They were able to specify regular temporal and spatial patterns of location, activity, and communication-use patterns, which together enabled predictions of up to 90 percent accuracy regarding where someone was likely to be and what that person was likely to be doing within the hour, as well as highly accurate predictions about an individual's colleagues, casual friends, and close relationships. The team identified patterns of communication and interaction within work groups, as well as the broad "organizational rhythms and network dynamics" of the Media Lab. (Eagle became CEO of Jana, a mobile advertising company that offers free internet to emerging markets in exchange for behavioral surplus.)

As the theory and practice of reality mining continued to evolve in Pentland's lab, work projects, and theories, the *MIT Technology Review* singled out "reality mining" as one of its "10 Breakthrough Technologies" in 2008. "My students and I have created two behavior-measurement platforms to speed the development of this new science," Pentland said. "These platforms today produce vast amounts of quantitative data for hundreds of research groups around the world."[13]

This allegiance to speed is, as we know, not a casual description but rather a key element in the art and science of applied utopistics. Pentland understands the rapid encroachments of Big Other and instrumentarian power as a "light-speed, hyperconnected world" where virtual crowds of millions from anywhere in the world "can form in minutes." He views the MIT community as the avant-garde: brilliant pioneers of light speed, already in sync with its extreme velocities and thus a model for the rest of society. Reflecting on his students and colleagues, Pentland writes that "I have also gotten to see how creative cultures must change in order to thrive in the hyperconnected, warp-speed world that is MIT, an environment that the rest of the world is now entering."[14] Pentland reasons that his group's adaptation to MIT's norms of rapid deployment simply foreshadows what is in store for the rest of us.

In the *MIT Technology Review*'s enthusiastic 2008 tribute to "reality mining," it noted the then-still-new and disturbing facts of behavioral surplus: "Some people are nervous about trailing digital bread crumbs behind them. Sandy Pentland however revels in it." Pentland would like to see phones collect "even more information" about their users: "It's an interesting God's-eye view."[15] Indeed, Pentland regularly celebrates "the predictive power of digital breadcrumbs" in his articles, indulging in the euphemisms and thin rationalizations that are also standard fare for surveillance capitalists and that contribute to the normalization of the dispossession of human experience. He says, for example,

> As we go about our daily lives, we leave behind virtual breadcrumbs—digital records of the people we call, the places we go, the things we eat and the products we buy. These breadcrumbs tell a more accurate story of our lives than anything we choose to reveal about ourselves....Digital breadcrumbs...record our behavior as it actually happened.[16]

Pentland was among the first to recognize the commercial relevance of behavioral surplus. Although he does not discuss it explicitly, he appears to embrace the realpolitik of surveillance capitalism as the necessary condition for an instrumentarian society. Pentland's own companies are extensions of his applied utopistics: proving grounds for instrumentarian techniques and the habituation of populations to pervasive rendition, monitoring, and modification in pursuit of surveillance revenues.

From the start, Pentland understood reality mining as the gateway to a new universe of commercial opportunities. In 2004 he asserted that cell phones and other wearable devices with "computational horsepower" would provide the "foundation" for reality mining as an "exciting new suite of business applications." The idea was always that businesses could use their privileged grasp of "reality" to shape behavior toward maximizing business objectives. He describes new experimental work in which speech-recognition technology generated "profiles of individuals based on the words they use," thus enabling a manager to "form a team of employees with harmonious social behavior and skills."[17]

In their 2006 article, Pentland and Eagle explained that their data would be "of significant value in the workplace," and the two jointly submitted a patent for a "combined short range radio network and cellular telephone network for interpersonal communications" that would add to the stock of instruments available for businesses to mine reality.[18] Eagle told *Wired* that year that the reality mining study represented an "unprecedented data set about continuous human behavior" that would revolutionize the study of groups and offer new commercial applications. He was reported to be "in talks" with a large company that already wanted to apply his instruments and methods.[19] Pentland argued that information gathered by his sociometers—"unobtrusive wearable sensors" measuring communication, voice tones, and body language—"could help managers understand who is working with whom and infer the relationships between colleagues" and "would be an efficient way to find people who might work well together."[20]

In a 2009 collaboration with several graduate students, Pentland presented results on the design and deployment of a "wearable computing platform" based on the sociometric badge and its machine analytics. The goal, the authors said, was to make machines that can "monitor social communication and provide real-time intervention." To that end, twenty-two office employees were "instrumented" with the badge for one month in order to "automatically measure individual and collective patterns of behavior, predict human behavior from unconscious social signals, identify social affinity among individuals working in the same team, and enhance social interactions by providing feedback to the users of our system." The research provided credible results, revealing patterns of communication and behavior that the authors concluded "would not be available without the use of a device such as the *sociometric* badge. Our results... argue strongly for the use of automatic sensing data collection tools to understand social systems." They warned that organizations will become "truly *sensible*" only when they employ "hundreds or thousands of wireless environmental and wearable sensors capable of monitoring human behavior, extracting meaningful information, and providing managers with group performance metrics and employees with self-performance evaluations and recommendation."[21]

The 2002 invention was continuously elaborated and eventually shepherded from lab to market. In 2010 Pentland and his 2009 coauthors founded a company, Sociometric Solutions, to bring Skinner's longed-for "instruments and methods" to the marketplace. It was one of many companies that Pentland would create to apply the rigors of his social physics to captive populations of office workers.[22] Sociometric Solutions' CEO, Ben Waber, one of Pentland's doctoral students, calls his operation "people analytics," and in his book of the same name, he anticipates a future of "connection, collaboration, and data" with the badge or something like it "deployed across millions of individuals at different companies in countries all over the world for not minutes but years or decades.... Imagine what we could learn about to help people collaborate more effectively...."[23]

Pentland and his crew continued to develop the sociometer and its applications, and by 2013 the device had been used by dozens of research groups and companies, including members of the Fortune 1000. A 2014 study, authored with Waber and colleagues from Harvard and Northeastern University, quantified gender differences in interaction patterns. The success of the analysis occasioned this announcement: "It is now possible to actively instrument human behavior to collect detailed data on various dimensions of social interaction." The authors signaled their aim to employ MacKay's cardinal rule of unobtrusive surveillance for effective monitoring of herds, flocks, and packs, acknowledging that the continuous pervasive collection of human behavioral data could succeed only when conducted outside the boundaries of human awareness, thus eliminating possible resistance, just as we saw at Facebook. As the researchers enthused, "Electronic sensors can be used to complement or replace human observers altogether, and while they may convey a slight sense of surveillance this perception is likely reduced as sensors get smaller and smaller, and consequently less obtrusive." They concluded that "minimally invasive ways to instrument human behavior" would enable comprehensive data collection in "naturalistic settings."

By 2015, the company opted for euphemism in a rebranding effort, changing its name to Humanyze. Its technology is described as a platform that uses a "smart employee badge to collect employee behavioral data, which it links to specific metrics with the goal of improving business performance."[24] Waber portrays the work as "moneyball" for business, enabling any organization to manage its workers like a sports team based on measures that reveal how people move through the day, with whom they interact, their tone of voice, if they "lean in" to listen, their position in the social network across a variety of office situations, and much more, all of it to produce forty separate measures that are then integrated with a "business metric dashboard." The company does not identify its client organizations, although one account describes its work with 10,000 employees in Bank of America's customer service centers and a partnership with the consulting firm Deloitte.[25] Writing in *Scientific American* on the power of sociometric data, Pentland says, "I persuaded the manager of a Bank of America call center to schedule coffee breaks simultaneously. The goal was to promote more engagement between employees. This single change resulted in a productivity increase of $15 million a year."[26]

Of the nineteen commercial ventures currently listed in Pentland's MIT biography, many are surveillance-as-a-service companies. For example, Pentland cofounded Endor, which markets itself to business customers as a solution to the prediction imperative. Endor's website explains its origins in "the revolutionary new science" of social physics combined with a "proprietary technology" to produce a "powerful engine that is able to explain and predict any sort of human behavior...." The site explains that every human activity (e.g., phone call records, credit card purchases, taxi rides, web activity) contains a set of hidden mathematical patterns. On the strength of its analysis, "emerging behavioral patterns" can be detected before they can be observed "by any other technique.... We've been working with some of the world's top consumer brands to unravel the most demanding of data problems."[27]

In 2014 another Pentland company called Sense Networks was acquired by YP, two letters that once stood for "yellow pages" and now describe "North America's largest local search, media and advertising company connecting consumers with local businesses." The 2014 YP statement on its acquisition of Sense Networks portrays a familiar picture of the behavioral surplus land grab, describing the firm as a "sophisticated location data processing platform to deliver mobile audiences at scale. Sense's retargeting solution for retailers can identify and reach shoppers and prospects of the top retailers with relevant mobile ads when they are near the retailer...at home or work."[28]

Pentland understands his experiments and paid interventions in workplace settings as emblematic of the larger challenges of social relations in an instrumentarian society. Once again we see the intended path from the economic to the social domain. Those instrumented office workers function as living labs for the translation of instrumentarian relations to the wider society. Pentland appeared in 2016 at a conference organized by Singularity University, a Silicon Valley hub of instrumentarian ideology funded in part by Larry Page. An interviewer tasked to write about Pentland explains, "Though people are one of the most valuable assets in an organization, many companies are still approaching management with a 20th century mentality.... Pentland saw the factor that was always messing things up was—the people."[29] Like Nadella, Pentland described his aims as developing the social systems that would work along the same lines as the machine systems, using behavioral data flows to judge the "correctness" of action patterns and to intervene when it is necessary to change "bad" action to "correct" action. "If people aren't interacting correctly and information isn't spreading correctly," Pentland warns, "people make bad decisions.... What you're trying to do is make a human-machine symbiote, where the humans understand more about the network of interactions because of the computers, and the computers are able to understand more about how humans work." As the interviewer notes, "Pentland has found this data [from sociometric badges] goes a long way in helping organizations mend their 'broken behaviors.'"[30]

Pentland's vision of an instrumentarian society grew in proportion to his instruments, his ideas waxing increasingly ambitious as the new tools and methods from his lab merged with the contemporary swell of computer mediation, all of it on the path to Big Other's global ubiquity. Pentland articulated his ambitions for the capabilities and objectives of this new milieu in a series of papers, published primarily between 2011 and 2014, but one remarkable 2011 essay of which he is the sole author stands out: "Society's Nervous System: Building Effective Government, Energy, and Public Health Systems."[31]

Pentland begins the report by announcing the institutional bona fides of this work: "Drawing on a unique, multi-year collaboration with the heads of major IT, wireless, hardware, health, and financial firms, as well as the heads of American, EU, and other regulatory organizations, and a variety of NGOs [a footnote here indicates the World Economic Forum], I describe the potential for pervasive and mobile sensing and computing over the next decade...." From there, his reasoning leaps across a range of inferences to stitch together a crucial rationale for a totalistic society constructed, sustained, and directed by instrumentarian power. The initial premise is reasonable enough: industrial-age technology once revolutionized the world with reliable systems for water, food, waste, energy, transportation, police, health care, education, and so forth, but these systems are now hopelessly "old," "centralized," "obsolete," and "unsustainable."

New digital systems are required that must be "integrated," "holistic," "responsive," "dynamic," and "self-regulating": "We need a radical rethinking of societies' systems. We must create a nervous system for humanity that maintains the stability of our societies' systems throughout the globe." Referring to the progress of ubiquitous computational sensing devices able to govern complex machine processes and information flows, Pentland observes that the "sensing" technologies necessary for this nervous system are "already in place." Even in 2011, Pentland understood that the basic contours of Big Other were up and running, describing it as a "world-spanning living organism" in which "wireless traffic systems, security sensors, and especially mobile telephone networks are combining to become intelligent reactive systems with sensors serving as their eyes and ears...the evolution...will continue at a quickening speed...devices will have more sensors...."[32]

But Pentland saw a problem. Although ubiquitous technologies are well on the way to solving the technical challenges of a global nervous system, Big Other will not be complete until it also understands *human behavior* on a global scale: "What is missing...are the dynamic models of demand and reaction," along with an architecture that guarantees "safety, stability, and efficiency.... The models required must describe *human* demand and reactions, since humans are at the core of all of these systems...the necessary observations are observations of individual behavior...."[33]

Pentland had identified a dangerous void, which foreshadows the "profound shift" that Nadella extolled to Microsoft developers in 2017 when he said: "People and their relationship with other people is now a first-class thing in the cloud!" "People" would have to become part of Big Other's purview, lest they fall prey to "incorrect" behavior. Society's safety, stability, and efficiency hang in the balance. Fortunately, Pentland informs us, the instruments and methods to capture behavioral surplus for reality mining are uniquely suited to answer this call:

> For the first time in history, the majority of humanity is linked…. As a consequence, our mobile wireless infrastructure can be "reality mined" in order to…monitor our environments, and plan the development of our society…. Reality mining of the "digital breadcrumbs" left behind as we go about our daily lives offers potential for creating remarkable, second-by-second models of group dynamics and reactions over extended periods of time…. In short, we now have the capacity to collect and analyze data about people with a breadth and depth that was previously inconceivable.[34]

In a style reminiscent of Larry Page's rejection of "old laws," Pentland is equally critical of a range of concepts and frameworks inherited from the Enlightenment and political economics. Pentland insists that the "old" social categories of status, class, education, race, gender, and generation are obsolete, as irrelevant as the energy, food, and water systems that he wants to replace. Those categories describe societies through the lens of history, power, and politics, but Pentland prefers "populations" to societies, "statistics" to meaning, and "computation" to law. He sees the "stratification of the population" coded not by race, income, occupation, or gender but rather by "behavior patterns" that produce "behavior subgroups" and a new "behavior demographics" that can predict disease, financial risk, consumer preferences, and political views with "between 5 and 10 *times* the accuracy" of the standard measures.[35]

A final question is urgently posed: "how to get the humans in these systems to participate in the plan?" His answers do not lie in persuasion or education but in behavioral modification. He says we need "new predictive theories of human decision making" as well as "incentive mechanism design," an idea that is comparable to Skinner's "schedules of reinforcement." Regarding how to get humans to follow the plan, Pentland offers the principle of "social influence" to explain the design mechanisms through which millions of human beings can be herded toward the guaranteed outcomes of safety, stability, and efficiency. He refers to his own studies, in which "the problems of industry and government" can largely be explained by the pattern of information transfer, especially how people influence and mimic one another.

This notion of social influence is a significant piece in Pentland's puzzle that anticipates a great deal of what is to come. Pentland understands that Big Other is not only an architecture that monitors and controls things. Big Other's instrumentation and data flows also make people mutually visible to one another, from the updates on your breakfast to the population flows in cities. Back in 2011, Pentland enthused, "Revolutionary new…infrastructures are providing us with a God's eye view of ourselves."[36] The aim is a computer-mediated society where our mutual visibility becomes the habitat in which we attune to one another, producing social patterns based on imitation that can be manipulated for confluence, just as the logic of the machine hive suggests.

Regarding incentives, Pentland outlines a principle of "social efficiency," which means that participation must provide value to the individual but also *to the system as a whole*.[37] For the sake of this wholeness, it is believed, each of us will surrender to a totally measured life of instrumentarian order. Sounding ever so much like Eric Schmidt and Larry Page with their silky promises of Google's all-knowing preemptive magic, Pentland believes that what we stand to lose is more than compensated for by the social rewards of efficient corporations and governments and the individual rewards that are simply *magic,* as he baldly appeals to second-modernity stress:

> For society, the hope is that we can use this new in-depth under-standing of individual behavior to increase the efficiency and responsiveness of industries and governments. For individuals, the attraction is the possibility of a world where everything is arranged for your convenience—your health checkup is magically scheduled just as you begin to get sick, the bus comes just as you get to the bus stop, and there is never a line of waiting people at city hall. As these new abilities become refined by the use of more sophisticated statistical models and sensor capabilities, we could well see the creation of a quantitative, predictive science of human organizations and human society.[38]

III. The Principles of an Instrumentarian Society

Pentland's theory of instrumentarian society came to full flower in his 2014 book *Social Physics,* in which his tools and methods are integrated into an expansive vision of our futures in a data-driven instrumentarian society governed by computation. Pentland transforms Skinner's fusty, odd utopia into something that sounds sophisticated, magical, and plausible, largely because it resonates with the waves of applied utopistics that wash over our lives each day. In completing Skinner, Pentland fashions more than an updated portrait of a behaviorist utopia. He outlines the principles of a full-blown instrumentarian society based on the pervasive outfitting and measurement of human behavior for the purposes of modification, control, and—in light of surveillance capitalism's commercial dominance of the networked sphere—profit.

Pentland insists that "social phenomena are really just aggregations of billions of small transactions between individuals...." This is a key point because it turns out that in order for social physics to replace the old ways of thinking, total knowledge of these billions of small things is required: "Big Data give us a chance to view society in all its complexity, through the millions of networks of person-to-person exchanges. If we had a 'god's eye,' an all seeing view, then we could potentially arrive at a true understanding of how society works and take steps to fix our problems."[39]

Pentland is sanguine on this point: total knowledge is within reach. As he states, "In just a few short years *we are likely to have incredible rich data available about the behavior of virtually all of humanity—on a continuous basis. The data mostly already exists.*"[40] The right to the future tense—and with it social trust, authority, and politics—is surrendered to Big Other and the transcendent computational systems that rule society under the watchful eye of a group that Pentland calls "we." He never defines this "we," which imposes an us-them relationship, introducing the exclusivity of the shadow text and its one-way mirror. It is an omission that haunts his text. Does it refer to the priesthood of data scientists like Pentland? The priesthood in collaboration with the owners of the means of behavior modification?

The theory aims to establish laws of social behavior comparable to laws of physics, and Pentland introduces two such laws that, as he says, determine the success of every "social organism." The first is the quality of the "idea flow," characterized by "exploration" to find new ideas and "engagement" to coordinate behavior around the best ideas. The second is "social learning," in which people imitate one another until new ideas become population-wide habits. (Social learning is defined as a mathematical relationship derived from "how an entity's state impacts other entities' states and vice versa.") Pentland notes that social learning is "rooted in statistical physics and machine learning."[41] The social hive is meant to reproduce the machine hive, and to this end Pentland advocates methods by which social learning "can be accelerated and shaped by *social pressure*."[42]

The scientific aims of Pentland's social physics depend upon a tightly integrated set of new social norms and individual adaptations, which I summarize here as five overarching principles that describe the social relations of an instrumentarian society. These principles echo Skinner's social theory of a behaviorally controlled society, in which knowledge replaces freedom. In exploring each of the five, I compare Pentland's statements to Skinner's own formulations on these topics. As we shall see, Skinner's once reviled thinking now defines this frontier of instrumentarian power.

1. Behavior for the Greater Good

Skinner had emphasized the need for an urgent shift to a collective perspective and values. "The intentional design of a culture and the control of human behavior it implies are essential if the human species is to continue to develop," he wrote in *Beyond Freedom & Dignity*.[43] The imperative to shift human behavior toward the greater good was already clear in *Walden Two*, where Frazier, its protagonist, asserts, "The fact is, we not only *can* control human behavior, we *must*."[44] Ultimately, this challenge was understood as an engineering problem. "And what are the techniques, the engineering practices, that will shape the behavior of the members of a group so that they will function smoothly for the benefit of all?" Frazier asks.[45] Skinner advocated, via Frazier, that the virtue of a "planned society" is "to keep intelligence on the right track, for the good of society rather than of the intelligent individual.... It does this by making sure that the individual will not forget his personal stake in the welfare of society."[46]

Pentland understands instrumentarian society as an historical turning point comparable to the invention of the printing press or the internet. It means that for the first time in human history, "We will have the data required to really know ourselves and understand how society evolves."[47] Pentland says that "continuous streams of data about human behavior" mean that everything from traffic, to energy use, to disease, to street crime will be accurately forecast, enabling a "world without war or financial crashes, in which infectious disease is quickly detected and stopped, in which energy, water, and other resources are no longer wasted, and in which governments are part of the solution rather than part of the problem."[48] This new "collective intelligence" operates to serve the greater good as we learn to act "in a coordinated manner" based on "social universals."

"Great leaps in health care, transportation, energy, and safety are all possible," Pentland writes, but he laments the obstacles to these achievements: "The main barriers are privacy concerns and the fact that we don't yet have any consensus around the trade-offs between personal and social values." Like Skinner, he is emphatic that these attachments to a bygone era of imperfect knowledge threaten to undermine the prospect of a perfectly engineered future society: "We cannot ignore the public goods that such a nervous system could provide...."[49] Pentland avoids the question "Whose greater good?" How is the greater good determined when surveillance capitalism owns the machines and the means of behavioral modification? "Goodness" arrives already oriented toward the interests of the owners of the means of behavioral modification and the clients whose guaranteed outcomes they seek to achieve. The greater good is someone's, but it may not be ours.

2. Plans Replace Politics

Skinner yearned for the computational capabilities that would perfect behavioral prediction and control, enabling perfect knowledge to supplant politics as the means of collective decision making. In spite of his pre-digital limitations, Skinner had no difficulty in conceptualizing the necessary requirements for species salvation as a new "communal science." As Frazier explains, "We know almost nothing about the special capacities of the *group* ... the individual, no matter how extraordinary ... can't think thoughts big enough."[50]

Smooth operations leave no room for unreasonable or unintentional outcomes, and Skinner viewed the creative and often messy conflicts of politics, especially democratic politics, as a source of friction that threatens the rational efficiency of the community as a single, high-functioning "superorganism." He laments our inclination to try to change things with "political action," and he endorses what he perceives as a widespread loss of faith in democracy. In *Walden Two* Frazier insists that "I don't like the despotism of ignorance. I don't like the despotism of neglect, of irresponsibility, the despotism of accident, even. And I don't like the despotism of democracy!"[51]

Capitalism and socialism are equally tainted by their shared emphasis on economic growth, which breeds overconsumption and pollution. Skinner is intrigued by the Chinese system but rejects it on the grounds of the bloody revolution that any effort to convert Westerners would entail. "Fortunately," Skinner concludes in the preface to *Walden Two*, "there is another possibility." This option is Skinner's version of a behaviorist society that provides a way in which "political action is to be avoided." In *Walden Two* a "plan" replaces politics, overseen by a "noncompetitive" group of "Planners" who eschew power in favor of the dispassionate administration of the schedules of reinforcement aimed at the greater good.[52] Planners exercise unique control over society but "only because that control is necessary for the proper functioning of the community."[53]

Like Skinner, Pentland argues that computational truth must necessarily replace politics as the basis for instrumentarian governance. We recall Nadella's enthusiasm over persons and relationships as "objects in the cloud," when considering Pentland's allegiance to the notion that certainty machines will displace earlier forms of governance. "Having a mathematical, predictive science of society that includes both individual differences and the relationships between individuals," Pentland writes, "has the potential to dramatically change the way government officials, industry managers, and citizens think and act...."[54]

Pentland worries that our political-economic constructs such as "market" and "class" hail from an old, slow world of the eighteenth and nineteenth centuries. The new, "light-speed hyperconnected world" leaves no time for the kind of rational deliberation and face-to-face negotiation and compromise that characterized the social milieu in which such political concepts originated: "We can no longer think of ourselves as only individuals reaching carefully considered decisions; we must include the dynamic social effects that influence our individual decisions and drive economic bubbles, political revolutions, and the internet economy."[55]

The velocity of instrumentarian society leaves us no time to get our bearings, and that speed is repurposed here as a moral imperative demanding that we relinquish individual agency to the automated systems that can keep up the pace in order to quickly perceive and impose correct answers for the greater good. There is no room for politics in this instrumentarian society because politics means establishing and asserting our bearings. Individual moral and political bearings are a source of friction that wastes precious time and diverts behavior from confluence.

Instead of politics, markets, classes, and so on, Pentland reduces society to his laws of social physics: a reincarnation of Skinner's "communal science." Indeed, Pentland regards his work as the practical foundation of a new "computational theory of behavior" capable of producing a "causal theory of social structure…a mathematical explanation of why society reacts as it does and how these reactions may (or may not) solve human problems…." These new mathematical analyses not only reveal the deep "mechanisms of social interactions" (Skinner's "special capacities of the group") but also combine with "our newly acquired massive amounts of behavior data" in order to reveal the patterns of causality that make it possible to "engineer better social systems," all of it based on "unprecedented instrumentation."[56]

Computation thus replaces the political life of the community as the basis for governance. The depth and breadth of instrumentation make it possible, Pentland says, to calculate idea flow, social network structure, the degree of social influence between people, and even "individual susceptibilities to new ideas." Most important, instrumentation makes it possible for those with the God view to modify others' behavior. The data provide a "reliable prediction of how changing any of these variables will change the performance of all the people in the network" and thus achieve the optimum performance of Skinner's superorganism. This mathematics of idea flow is the basis for Pentland's version of a "plan" that dictates the targets and objectives of behavior change. Human behavior must be herded and penned within the parameters of the plan, just as behavior at Nadella's construction site was continuously and automatically molded to policy parameters. Pentland calls this "*tuning the network.*"

"Tuners" fill the role of Pentland's "we." He says, for example, that cities can be understood as "idea engines" and that "we can use the equations of social physics to begin to tune them to perform better."[57] Like Skinner's planners, Pentland's tuners oversee pesky anomalies that represent leakage from an old world of ignorance mistaken as freedom. Tuners tweak Big Other's operations to preemptively steer such misguided behavior back into the fold of harmonious confluence and optimum performance for the greater good of whomever or whatever owns the machines that perform the math and pays the tuners to decipher and impose its parameters. Pentland provides an example from one of his own "living labs":

> This mathematically derived concept of idea flow allows us to "tune" social networks in order to make better decisions and achieve better results.... Within the eToro digital finance world, we have found that we can shape the flows of ideas between people by providing small incentives or nudges to individuals, thus causing isolated traders to engage more with others and those who were too interconnected to engage less....[58]

3. Social Pressure for Harmony

In the community of *Walden Two,* reinforcement is precisely orchestrated to eliminate emotions that threaten cooperation. Only "productive and strengthening emotions—joy and love" are allowed. Feelings of sorrow and hate "and the high-voltage excitements of anger, fear, and rage" are considered "wasteful and dangerous" threats to "the needs of modern life." Any form of distinction between persons undermines the harmony of the whole and its capacity to bend to collective purpose. Frazier acknowledges that you cannot coerce people into doing the right thing. The solution is far more subtle and sophisticated, based upon scientifically calibrated schedules of reinforcement: "Instead you have to set up certain behavioral processes which will lead the individual to design his own 'good' conduct.... We call that sort of thing 'self-control.' But don't be misled, *the control always rests in the last analysis in the hands of society.*"[59]

Pentland's idea is comparable: "The social physics approach to getting everyone to cooperate" is "social network incentives," his version of "reinforcement." With such incentives, he explains, "we focus on changing the connections between people rather than focusing on getting people individually to change their behavior.... We can leverage those exchanges to generate social pressure for change."[60] Social media is critical to establishing these tuning capabilities, Pentland believes, because this is the environment in which social pressure can best be controlled, directed, manipulated, and scaled.[61]

In Pentland's view Facebook already exemplifies these dynamics. Its contagion experiments reveal active mastery of the ability to manipulate human empathy and attachment with tuning techniques such as priming and suggestion. Indeed, Pentland finds Facebook's "contagion" experiments particularly enlightening, seeing all sorts of practical insights in their complexities. For example, in the corporation's 61-million-person voting experiment, Pentland sees confirmation that social pressure can be effectively instrumentalized in social networks, especially among people with "strong ties": "The knowledge that our face-to-face friends had already voted generated enough social pressure that it convinced people to vote."[62] With this knowledge and more like it, Pentland's "we," the tuners, will be able to activate the "right incentives."

That Pentland's "we" is able "to generate social pressure for change" reflects his understanding of the superorganism. The God view persuades him that assessing human action really is comparable to counting beavers: "We can observe humans in just the same way we observe apes or bees and derive rules of behavior, reaction, and learning."[63] In all of these populations, the collective exerts pressure on each organism to go with the flow, stay with the herd, return to the hive, and take flight with the flock. Idea flows mimic the pattern of the machine hive, the edge conflates with the hub, identity yields to synchrony, the parts dissolve in the whole. He writes:

> I believe that we can we think of each stream of ideas as a swarm
> or collective intelligence, flowing through time, with all the humans
> in it learning from each other's experiences in order to jointly dis-
> cover the patterns of preferences and habits of action that best suit
> the surrounding physical and social environment. This is counter to
> the way most modern Westerners understand themselves, which is

as rational individuals, people who know what they want and who decide for themselves what actions to take in order to accomplish their goals.[64]

This shift from society to swarm and from individuals to organisms is the cornerstone upon which the structure of an instrumentarian society rests.

Pentland ignores the role of empathy in emulation because empathy is a felt experience that is not subject to the observable metrics required for computational governance. Instead, Pentland subscribes to the label *Homo imitans* to convey that it is mimicry, not empathy, and certainly not politics, which defines human existence. The term itself derives from studies of infant learning, but for Pentland it is a fitting explanation of all human behavior all the time: an assertion, like Skinner's, that control always rests with society. "The largest single factor driving adoption of new behaviors," he writes, is "the behavior of peers."[65]

Because we are born to imitate one another, Pentland argues, the whole species is attuned to social pressure as an efficient means of behavioral modification. This model of human learning is a throwback to bees and apes but also a forward pass to the machine hive. Machines do not learn by empathy; learning is automatically updated in a lockstep progression of collective intelligence.

4. Applied Utopistics

Both Skinner and Pentland believe in the authority of the utopianists to impose their plan. Instrumentarian society is a planned society, produced through total control of the means of behavioral modification. Neither Skinner's planners nor Pentland's tuners shrink from their responsibility to wield the power that shapes the superorganism.

Skinner never lost faith in the social vision of *Walden Two*. He understood utopia as a "total social environment" in which all parts work in harmony toward collective aims:

> The home does not conflict with the school or the street, religion does not conflict with government.... And if planned economies,

benevolent dictatorships, perfectionistic societies, and other utopian ventures have failed, we must remember that unplanned, undictated, and unperfected cultures have failed too. A failure is not always a mistake; it may simply be the best one can do under the circumstances. The real mistake is to stop trying.[66]

Pentland similarly conceives his social physics as both comprehensive and necessary. Nothing short of its totalistic rendering and control of all human behavior will serve civilization in a hyperconnected future, and there is no sign of hesitation to assert computational governance over the whole domain of human endeavor for the sake of a collective destiny. The politics and economics of that destiny, which is to say the authority and power that found and sustain it, do not require specification because the machines and their math transcend these once fundamental coordinates of human society. Instead, computation reveals the truth hidden in the data and therefore determines what is "correct." A new social class of tuners exercises perpetual vigilance to cure human nature of its weaknesses by ensuring that populations are tuned, herded, and conditioned to produce the most-efficient behaviors. The "tools of social network incentives" are all that is required "to establish new norms of behavior, rather than relying on regulatory penalties and market competition.... Given the well-known shortcomings of human nature, social efficiency is a desirable goal.... Our focus should be on providing the idea flow required for individuals to make correct decisions and develop useful behavioral norms...."[67] Finally, like Skinner, Pentland rejects the notion that his imagined "data-driven society" is merely a utopian fantasy, insisting instead that it is not only practical and feasible but also a moral imperative in which the benefits to the collective outweigh all other considerations.

5. The Death of Individuality

Individuality is a threat to instrumentarian society, troublesome friction that sucks energy from "collaboration," "harmony," and "integration." In an article titled "The Death of Individuality," Pentland insists that "instead of individual rationality, our society appears to be governed by a collective

intelligence that comes from the surrounding flow of ideas and examples....
It is time that we dropped the fiction of individuals as the unit of rationality
and recognised that our rationality is largely determined by the surrounding
social fabric...."[68]

Here again the behaviorist from Harvard had already sounded the first
and most eloquent iteration of this message, elevating the Other-One and de-
nouncing the autonomous self. In *Beyond Freedom & Dignity*, Skinner freely
displayed his contempt for this most transcendent ideal of the Sartrean age:
the will to will oneself into first-person voice and action. Skinner argued that
the differences between humans and other species are greatly exaggerated,
and he would have found Pentland entirely justified in his rejection of the individual in favor of the distant, computer-mediated gaze. Beavers or people,
the variance hardly matters once we shed the destructive fiction of individual
autonomy. The surrender of the individual to manipulation by the planners
clears the way for a safe and prosperous future built on the forfeit of freedom
for knowledge. Skinner was unrelenting on this point:

> What is being abolished is autonomous man—the inner man, the
> homunculus, the possessing demon, the man defended by the litera-
> tures of freedom and dignity. His abolition has long been overdue....
> He has been constructed from our ignorance, and as our understand-
> ing increases, the very stuff of which he is composed vanishes...and
> it must do so if it is to prevent the abolition of the human species. To
> man qua man we readily say good riddance. Only by dispossessing
> him can we turn...from the inferred to the observed, from the mirac-
> ulous to the natural, from the inaccessible to the manipulable.[69]

The long-overdue death of individuality finally dispels the distracting fictions that fetishize the notions of freedom and dignity. The twentieth-century
behaviorist from Harvard and the twenty-first-century data scientist from
MIT agree that the notion of free will is but another unfortunate hangover
from a dark age when science had not yet demonstrated that, as Skinner says,
we live "under the control of a social environment" that "millions of others...
have constructed." The blunt behaviorist delivers his final truth: "A person
does not act upon the world, the world acts upon him."[70]

In a lecture at Google that garnered enthusiastic applause, Pentland flattered the audience by signaling that the digital cognoscenti will easily accept the obsolescence of the individual as a necessary fate. "What about free will?" he asked the audience in Mountain View. "That may not have occurred to *you,* but that's a traditional thing to ask." He went on to explain that most human behavior—from political views to spending choices to the music that people listen to—is predicted by "what's cool to do...exposure to what other people do." Many people reject this idea, he noted, because "it's not the rhetoric in our society." Then he assured the Googlers, "You guys are the last people to be saying this to, because you guys are like the best and smartest in the world." For such people as these, Pentland appeared to say, the death of the individual is yesterday's news:

> So you've heard about rational individuals. And everybody rags on the rational part. I'm not going to do that. I'm going to rag on the individual part, OK? Because I don't think we are individuals. What we desire, the ways we learn to go about doing it, what's valuable, are consensual things...individual incentives...that's part of this mindset that comes from the 1700s...the action is not between our ears. The action is in our social networks, OK? We are a social species.[71]

Pentland's vision is Skinner's vision, now standing on the shoulders of Big Other with its Big Data and its Big Math. These are the resources of the smart machines required to divine the "correct" answers. Such is Pentland's resonance with Skinner's social theory that without ever mentioning the behaviorist's name, a later section of Pentland's book is titled "Social Physics Versus Free Will and Dignity."

If we are to annihilate and bury the individual as an existential reality, philosophical idea, and political ideal, then this death should at least merit the gravitas of an ancient Greek funerary ritual. The existence of the individual is, after all, an achievement carved from millennia of human suffering and sacrifice. Instead, Pentland brushes it aside as just another debugging of humanity's computer code, a much-needed upgrade to the outdated software that is the long human story.

Unlike Skinner, though, Pentland is careful to round the square, perhaps in the hope of evading a withering review by the likes of Noam Chomsky. (In "The Case Against B. F. Skinner," as you may recall from Chapter 10, Chomsky famously denounced Skinner as "vacuous" and "devoid of scientific content," and he assessed the work as burdened with misconceptions that "virtually guarantee failure."[72]) Pentland sidesteps the risks of Skinner's jeremiad by taking a softer tone: "Some people react negatively to the phrase social physics, because they feel that it implies that people are machines without free will and without the ability to move independently of our role in society."[73] Like Meyer, Pentland acknowledges that humans have a "capacity for independent thought" but insists that social physics "does not need to try to account for it." As Pentland sees it, the problem is not that "independent thought" is omitted from the picture but rather that "internal, unobservable" thought processes are just friction that "will occasionally emerge to defeat our best social physics models." Fortunately, the models are not really in danger because "the data tell us that deviations from our regular social patterns occur only a few percent of the time."[74] The autonomous individual is but a statistical blip, a slip of the pen that is easily overridden in the march toward confluent action and someone's greater good.

In this vein, Professor Pentland does not ignore issues like privacy and social trust. He actively advocates solutions to these problems, but the solutions he pursues are already tailored to the intensification of a "data-driven" instrumentarian society. Pentland's approach recalls the early conviction of his former doctoral student Rosalind Picard that societal challenges are not insurmountable, that new technical solutions will resolve any problems, and that "safeguards can be developed." Two decades later Picard's view had a darker cast, but Pentland expresses little trace of doubt. For example, Pentland works with influential institutions like the World Economic Forum to craft "a new deal on data" that favors individual "ownership" of personal information but does not question the ubiquitous rendition of such personal information in the first instance.[75] Data ownership, he believes, will create financial incentives for participation in a market-oriented instrumentarian society. Like Skinner, Pentland assumes that the sheer weight of incentives and ubiquitous connection, monitoring, and tuning will eventually wear down older sensibilities such as the interest in privacy. "The New Deal gives customers a stake in the new data economy; that will bring first greater stability and then eventually greater profitability as people become more comfortable sharing data."[76]

In Pentland's view of data ownership, certainty machines like blockchain, which relies on complex encryption and algorithms to create a decentralized tamper-proof database, are commandeered to bypass social trust. He advocates systems "that live everywhere and nowhere, protecting and processing the data of millions of people, and executing on millions of internet computers."[77] One important study of Bitcoin, the cryptocurrency that relies on blockchain, suggests that such machine solutions both express and contribute to the general erosion of the social fabric in ways that are both consistent with instrumentarianism and further pave the way for its success. Information scholars Primavera De Filippi and Benjamin Loveluck conclude that contrary to popular belief, "Bitcoin is neither anonymous nor privacy-friendly...anyone with a copy of the blockchain can see the history of all Bitcoin transactions...every transaction ever done on the Bitcoin network can be traced back to its origin." Such systems rely on "perfect information," but the kinds of coordination processes that build open democratic societies, such as "social trust" or "loyalty," are "expunged" in favor of "a profoundly market-driven approach."[78] Like Varian, Pentland does not acknowledge the social and political implications of such systems, which are in any case irrelevant to an instrumentarian future in which democracy and social trust are superseded by the certainty machines, their priests, and their owners.

Surveillance capitalism grew to dominance during the years that Pentland has argued for his "New Deal," even as it benefitted from his theoretical and commercial innovations. During those same years, as we have seen, Picard's "affective computing" fell to the surveillance paradigm. Nevertheless, Professor Pentland is sanguine that surveillance capitalism can easily be pushed aside by market forces, despite its concentrations of knowledge, rights, and power; its unilateral control of the shadow text; and its dominant position in the division of learning in society. "It simply requires that creative businesspeople harness the will of consumers in order to construct a value proposition better than the current steal-all-your-data paradigm. We've just got to push on through."[79] Power, politics, and law do not enter into the equation, presumably because they are already obsolete in the social vision under construction here.

IV. The Third Modernity of the Hive

It is no surprise that capitalism shapes social relations. A century ago it was the new means of mass production that fashioned mass society in its image. Today, surveillance capitalism offers a new template for our future: the machine hive in which our freedom is forfeit to perfect knowledge administered for others' profit. This is an unheralded social revolution that is difficult to discern in the fog of utopian rhetoric and high-speed applied utopistics conjured by leading surveillance capitalists and the many communities of practice—from developers to data scientists—that enable and sustain the dominance of the commercial surveillance project.

Surveillance capitalists work hard to camouflage their purpose as they master the uses of instrumentarian power to shape our behavior while evading our awareness. That is why Google conceals the operations that turn us into the objects of its search and Facebook distracts us from the fact that our beloved connections are essential to the profit and power that flow from its network ubiquity and totalistic knowledge.

Pentland's experimental work and theoretical analyses perform an important political and social function in piercing this fog. They map the tactical and conceptual pathways of instrumentarian society that place the means of behavior modification at the heart of this social system, founded on the scientific and technological control of collective behavior and administered by a specialist class. In China the state appears determined to "own" this complex, but in the West it is largely owned and operated by surveillance capital.

Instrumentarian society defines the ultimate institutionalization of a pathological division of learning. *Who knows? Who decides? Who decides who decides?* Here too the comparison with China is useful. An abnormal division of learning marks both China and the West. In China the state vies with its surveillance capitalists for control. In the US and Europe the state works with and through the surveillance capitalists to accomplish its aims. It is the private companies who have scaled the rock face to command the heights. They sit at the pinnacle of the division of learning, having amassed unprecedented and exclusive wealth, information, and expertise on the strength of their dispossession of our behavior. They are making their dreams come true. Not even Skinner could have aspired to this condition.

The social principles of instrumentarianism's third modernity represent a stark break with the legacies and ideals of the liberal order. Instrumentarian society is a topsy-turvy fun-house-mirror world in which everything that we have cherished is turned upside down and inside out. Pentland doubles down on the illiberality of behavioral economics. In his hands the ideology of human frailty is not merely cause for contempt but a justification for the death of individuality. Self-determination and autonomous moral judgment, generally regarded as the bulwark of civilization, are recast as a threat to collective well-being. Social pressure, well-known to psychologists for its dangerous production of obedience and conformity, is elevated to the highest good as the means to extinguish the unpredictable influences of autonomous thought and moral judgment.

These new architectures feed on our fellow feeling to exploit and ultimately to suffocate the individually sensed inwardness that is the wellspring of personal autonomy and moral judgment, the first-person voice, the will to will, and the sense of an inalienable right to the future tense. That we vibrate to one another should be a life-enhancing fact, but this third modernity amplifies our mutual vibration to an excruciating pitch. In the milieu of total instrumentation, it is less that we resonate to one another's presence and more that we drown in its inescapability.

Instrumentarianism reimagines society as a hive to be monitored and tuned for guaranteed outcomes, but this tells us nothing of the lived experience of its members. What are the consequences of life lived in the hive, where one is perceived as an "other" to the surveillance capitalists, designers, and tuners who impose their instruments and methods? How and when do we each become an organism among organisms to ourselves and to one another, and with what result? The answers to these questions are not all guesswork. We can begin by asking our children. Without knowing it, we sent the least formed and most vulnerable among us to scout the hive and settle its wilderness. Now their messages are filtering in from the frontier.

CHAPTER SIXTEEN

OF LIFE IN THE HIVE

All grew so fast his life was overgrown,
Till he forgot what all had once been made for:
He gathered into crowds but was alone...

—W. H. AUDEN

SONNETS FROM CHINA, VIII

I. Our Canaries in the Coal Mine

"I felt so lonely...I could not sleep well without sharing or connecting to others," a Chinese girl recalled. "Emptiness," an Argentine boy moaned. "Emptiness overwhelms me." A Ugandan teenager muttered, "I felt like there was a problem with me," and an American college student whimpered, "I went into absolute panic mode." These are but a few of the lamentations plucked from one thousand student participants in an international study of media use that spanned ten countries and five continents. They had been asked to abstain from all digital media for a mere twenty-four hours, and the experience released a planet-wide gnashing of teeth and tearing of flesh that even the study's directors found disquieting.[1] Capping the collective *cri de coeur,* a Slovakian university student reflected, "Maybe it is unhealthy that I can't be without knowing what people are saying and feeling, where they are, and what's happening."

The students' accounts are a message in a bottle for the rest of us, narrating the mental and emotional milieu of life in an instrumentarian society with its architectures of behavioral control, social pressure, and asymmetrical power. Most significantly, our children are harbingers of the emotional toll of the viewpoint of the Other-One as young people find themselves immersed in a hive life, where the other is an "it" to me, and I experience myself as the "it" that others see. These messages offer a glimpse of the instrumentarian future, like the scenes revealed by Dickens's Ghost of Christmas Yet to Come. So shaken was Scrooge by his glimpse of bitter destiny that he devoted the remainder of his life to altering its course. What will *we* do?

The question hangs over this chapter. Pentland celebrates Facebook as the perfect milieu for effective social pressure and tuning. In the sections that follow, we explore the mechanisms that Pentland admires. Why is it so difficult for young people to unplug? What are the consequences of that attachment for them and for all of us? Facebook has learned to bite hard on the psychological needs of young people, creating new challenges for the developmental processes that build individual identity and personal autonomy. The effects of these challenges are already evident in a parade of studies that document the emotional toll of social media on young people. As we shall see, the hive and its larger architecture of Big Other plunge us into an intolerable world of "no exit."

The international "unplug" study helps to set the stage, for it reveals a range of emotional anguish summarized in six categories: addiction, failure to unplug, boredom, confusion, distress, and isolation. The students' sudden disconnection from the network produced the kinds of cravings, depression, and anxiety that are characteristic of clinically diagnosed addictions. The result was that a majority in every country admitted that they could not last out the day unplugged. Their angst was compounded by the same Faustian pact with which we are all too familiar, as they discovered that nearly all daily logistical, communicative, and informational requirements were dependent upon their connected devices: "Meeting with friends became difficult or impossible, finding the way to a destination without an online map or access to the internet became a problem, and simply organizing an evening at home became a challenge." Worse yet, the students found it impossible to imagine even casual social participation without social media, especially Facebook: "Increasingly no young person who wants a social life can afford *not* to be active on the site, and being active on the site means living one's life on the site."

Business and tech analysts cite "network effects" as a structural source of Facebook's dominance in social media, but those effects initially derived from the demand characteristics of adolescents and emerging adults, reflecting the peer orientation of their age and stage. Indeed, Facebook's early advantage in this work arose in no small measure from the simple fact that its founders and original designers were themselves adolescents and emerging adults. They designed practices for an imagined universe of adolescent users and college students, and those practices were later institutionalized for the rest of us, reducing the social world to a tally of "friends" who are not friends and "likes" that provide a continuous ticker tape of one's value on the social market, stoking the anxieties of pre-adulthood and anticipating the mesmerizing social disciplines of the hive.[2]

The researchers concluded that their global study of students had "ripped back the curtain" on the loneliness and acute disorientation that overwhelm young people when faced with disconnection from social media. It wasn't simply that they didn't know what to do with themselves but rather that "they had problems articulating what they were feeling or even who they were if they couldn't connect." The students felt as though "they had lost part of themselves."[3]

These feelings of disorientation and isolation suggest a psychological dependency on the "others," and additional studies only deepen our understanding of how "Generation Z," the demographic cohort born in and after 1996—the first group of digital natives, with no memory of life before the rise of surveillance capitalism—relies on a range of social media for psychological sustenance as they bounce between four or five platforms more or less simultaneously. Consider first the older cohorts. A 2012 survey concluded that emerging adults devote more time to using media than any other daily activity, spending nearly twelve hours each day with media of some form.[4] By 2018 Pew Research reported that nearly 40 percent of young people ages 18–29 report being online "almost constantly," as do 36 percent of those ages 30–49. Generation Z intensifies the trend: 95 percent use smartphones, and 45 percent of teens say they are online "on a near-constant basis."[5] If that is how you spend your days and nights, then the findings of a 2016 study are all too logical, as 42 percent of teenage respondents said that social media affects how people see them, having adopted what the researchers call an *outside-looking-in* approach to how they express themselves. Their dependency penetrates deeply into their sense of well-being, affecting how they feel about themselves (42 percent) and their happiness (37 percent).[6]

In a subsequent elaboration on the psychological consequences of experiencing oneself from the "outside looking in," a 2017 survey of young British women ages 11–21 suggests that the social principles of instrumentarian society, so enthusiastically elaborated by Pentland and endorsed by surveillance capitalist leaders, appear to be working effectively.[7] Thirty-five percent of the women said that their biggest worry online was comparing themselves and their lives with others as they are drawn into "constant comparisons with often idealized versions of the lives, and bodies, of others."[8]

A director of the project observed that even the youngest girls in this cohort feel pressured to create a "personal brand," the ultimate in self-objectification, as they seek reassurance "in the form of likes and shares." When the *Guardian* tried to explore girls' reflections on these survey findings, the responses eloquently betray the plight of the organism among organisms. "I do feel I need to be perfect and compare myself to others all the time," says one. "You see other people's lives and what they are doing... you... see their 'perfect' lives and it makes you think yours isn't," says another.[9]

In light of these findings, one UK medical specialist comments on the young people in her practice: "People are growing up to want to be influencers and that is now a job role.... I am not sure if parents are fully aware of the pressure people face...."[10] Indeed, only 12 percent of respondents in that 2017 survey reckoned that their parents understood these pressures. The reports confirm that social pressure is well institutionalized as the means of online social influence, but contrary to Pentland's belief that "class" divisions would disappear, life in the hive produces new cleavages and forms of stratification: not only tune or be tuned but also pressure or be pressured.

Nothing summarizes young life in the hive better than the insights of Facebook's own North American marketing director, Michelle Klein, who told an audience in 2016 that while the average adult checks his or her phone 30 times a day, the average millennial, she enthusiastically reported, checks more than 157 times daily. Generation Z, we now know, exceeds this pace. Klein described Facebook's engineering feat: "a sensory experience of communication that helps us connect to others, *without having to look away*," noting with satisfaction that this condition is a boon to marketers. She underscored the design characteristics that produce this mesmerizing effect: design is narrative, engrossing, immediate, expressive, immersive, adaptive, and dynamic.[11]

If you are over the age of thirty, you know that Klein is not describing your adolescence, or that of your parents, and certainly not that of your grandparents. Adolescence and emerging adulthood in the hive are a human first, meticulously crafted by the science of behavioral engineering;

institutionalized in the vast and complex architectures of computer-mediated means of behavior modification; overseen by Big Other; directed toward economies of scale, scope, and action in the capture of behavioral surplus; and funded by the surveillance capital that accrues from unprecedented concentrations of knowledge and power. Our children endeavor to come of age in a hive that is owned and operated by the applied utopianists of surveillance capitalism and is continuously monitored and shaped by the gathering force of instrumentarian power. Is this the life that we want for the most open, pliable, eager, self-conscious, and promising members of our society?

II. The Hand and the Glove

The magnetic pull that social media exerts on young people drives them toward more automatic and less voluntary behavior. For too many, that behavior shades into the territory of genuine compulsion. What is it that mesmerizes the youngest among us, lashing them to this mediated world despite the stress and disquiet that they encounter there?

The answer lies in a combination of behavioral science and high-stakes design that is precision-tooled to bite hard on the felt needs of this age and stage: a perfectly fitted hand and glove. Social media is designed to engage and hold people of all ages, but it is principally molded to the psychological structure of adolescence and emerging adulthood, when one is naturally oriented toward the "others," especially toward the rewards of group recognition, acceptance, belonging, and inclusion. For many, this close tailoring, combined with the practical dependencies of social participation, turns social media into a toxic milieu. Not only does this milieu extract a heavy psychological toll, but it also threatens the course of human development for today's young and the generations that follow, all spirits of a Christmas Yet to Come.

The hand-and-glove relationship of technology addiction was not invented at Facebook, but rather it was pioneered, tested, and perfected with outstanding success in the gaming industry, another setting where addiction is formally recognized as a boundless source of profit. Skinner had anticipated the relevance of his methods to the casino environment, which executives and engineers have transformed into as vivid an illustration as one can muster of the startling power of behavioral engineering and its ability to exploit individual inclinations and transform them into closed loops of obsession and compulsion.

No one has mapped the casino terrain more insightfully than MIT social anthropologist Natasha Dow Schüll in her fascinating examination of machine gambling in Las Vegas, *Addiction by Design*. Most interesting for us is her account of the symbiotic design principles of a new generation of slot machines calculated to manipulate the psychological orientation of players so that first they never have to look away, and eventually they become incapable of doing so. Schüll learned that addictive players seek neither entertainment nor the mythical jackpot of cash. Instead, they chase what Harvard Medical School addiction researcher Howard Shaffer calls "the capacity of the drug or gamble to shift subjective experience," pursuing an experiential state that Schüll calls the "machine zone," a state of self-forgetting in which one is carried along by an irresistible momentum that feels like one is "played by the machine."[12] The machine zone achieves a sense of complete immersion that recalls Klein's description of Facebook's design principles—engrossing, immersive, immediate—and is associated with a loss of self-awareness, automatic behavior, and a total rhythmic absorption carried along on a wave of compulsion. Eventually, every aspect of casino machine design was geared to echo, enhance, and intensify the hunger for that subjective shift, but always in ways that elude the player's awareness.

Schüll describes the multi-decade learning curve as gaming executives gradually came to appreciate that a new generation of computer-based slot machines could trigger and amplify the compulsion to chase the zone, as well as extend the time that each player spends in the zone. These innovations drive up revenues with the sheer volume of extended play as each machine is transformed into a "personalized reward device."[13] The idea, as the casinos came to understand it, is to avoid anything that distracts, diverts, or interrupts the player's fusion with the machine; consoles "mold to the player's natural posture," eliminating the distance between the player's body and frictionless touch screens: *"Every feature of a slot machine—its mathematical structure, visual graphics, sound dynamics, seating and screen ergonomics—is calibrated to increase a gambler's 'time on device' and to encourage 'play to extinction.'"*[14] The aim is a kind of crazed machine sex, an intimate closed-loop architecture of obsession, loss of self, and auto-gratification. The key, one casino executive says in words that are all too familiar, "is figuring out how to leverage technology to act on customers' preferences [while making] it as invisible—or what I call auto-magic—as possible."[15]

The psychological hazards of the hand-glove fit have spread far beyond the casino pits where players seek the machine zone: they define the raw heart of Facebook's success. The corporation brings more capital, information, and science to this parasitic symbiosis than the gaming industry could ever muster. Its achievements, pursued in the name of surveillance revenues, have produced a prototype of instrumentarian society and its social principles, especially for the youngest among us. There is much that we can grasp about the lived experience of the hive in the challenges faced by the young people whose fate it is to come of age in this novel social milieu in which the forces of capital are dedicated to the production of compulsion. Facebook's marketing director openly boasts that its precision tools craft a medium in which users "never have to look away," but the corporation has been far more circumspect about the design practices that eventually make users, especially young users, *incapable* of looking away.

There are some chinks in the armor. For example, in 2017 Napster cofounder and one-time Facebook president Sean Parker frankly admitted that Facebook was designed to consume the maximum possible amount of users' time and consciousness. The idea was to send you "a little dopamine hit every once in a while"—a.k.a. "variable reinforcement—in the form of 'likes' and comments. The goal was to keep users glued to the hive, chasing those hits while leaving a stream of raw materials in their wake."[16]

Shaffer, the addiction researcher, has identified five elements that characterize this state of compulsion: frequency of use, duration of action, potency, route of administration, and player attributes. We already know quite a bit about the high frequency and long duration of young people's engagement in social media. What we need to understand is something of (1) the psychological attributes that draw them to social media in the first place (the hand), (2) the design practices that ratchet up potency in order to transform inclinations into unquenchable need (the glove), and (3) the mental and emotional consequences of Facebook's ever-more-exquisite ability to enmesh young people in chasing their own kind of zone.

Consider the final moments of a 2017 *Washington Post* profile on a thirteen-year-old girl, part of a series chronicling "what it's like to grow up in the age of likes, lols, and longing." It is the girl's birthday, and only one question will decide her happiness: do her friends like her enough to post pictures of her on their pages in appreciation of the occasion? "She scrolls, she waits. For that little notification box to appear."[17] Regardless of your age, who among us does not feel a painful blast of recognition? Adolescence has always been a time when acceptance, inclusion, and recognition from the "others" can feel like matters of life and death, and social media has not been required to make it so. Is adolescence really any different today than in any other era? The answer is yes...and no.

Adolescence was officially "discovered" in the United States in 1904 by G. Stanley Hall, and even then, Hall, the first doctor of psychology in the country, located the challenges of youth in the rapidly changing context of "our urbanized hothouse life that tends to ripen everything before its time."[18] While writing about teenagers in 1904, he observed that adolescence is a period of extreme orientation toward the peer group: "Some seem for a time to have no resource in themselves, but to be abjectly dependent for their happiness upon their mates."[19] He also pointed to the potential for cruelty within the peer group, a phenomenon that contemporary psychologists refer to as "relational aggression." Decades later, the central challenge of adolescence was famously characterized as "identity formation" by the developmental psychologist Erik Erikson, who did much to explain twentieth-century adolescence. Erikson emphasized the adolescent struggle to construct a coherent identity from the mutual "joinedness" of the adolescent clique. He described the "normative crisis" when fundamental questions of "right" and "wrong" require inner resources associated with "introspection" and "personal experimentation." The healthy resolution of that conflict between self and other leads to a durable sense of identity.[20]

Today, most psychologists agree that our longer lives combined with the challenges of an information-intensive society have further lengthened the time between childhood and adulthood. Many have settled on the notion of "emerging adulthood" to denote the years between eighteen and the late twenties as a new life stage: emerging adulthood is to the twenty-first century what adolescence was to the twentieth.[21] And although contemporary researchers embrace a diverse range of methods and paradigms, most concur that the essential challenge of emerging adulthood is the differentiation of a "self" from the "others."[22]

There is a broad consensus that our extended life spans often require us to revisit the core questions of identity more than once during our lives, but researchers agree that psychological success during emerging adulthood depends on at least some resolution of identity issues as the basis for the shift toward full adulthood. As one research scholar writes, "A prime challenge of emerging adulthood is to become the author of your own life."[23] Who among us does not recognize that call? This existential challenge is enduring, a source of continuity that links generations. What *has* changed are the circumstances in which young people today must meet this challenge.

III. Proof of Life

Emerging adulthood is "ground zero" in the struggle for the "relational autonomy" that prepares young people for the transition into adulthood, as Notre Dame psychologists Daniel Lapsley and Ryan Woodbury characterize it.[24] By "relational autonomy," they mean to underscore the idea that autonomy is not a simplistic cliché of "individualism," unencumbered by attachment or empathy, but instead it strikes a vital balance between the cultivation of inner resources and the capacity for intimacy and relationship. Emerging adulthood requires "hard bargaining" to establish a self that is separate from but still connected to others, and the quality of this inner bargain "gives emerging adulthood a sense of anticipation and urgency," aiding a successful transition to adulthood.[25]

Even with these insights, it remains difficult to fully grasp the felt experiences of young people who, as Hall aptly described more than a century ago, "seem...to have no resources in themselves." Perhaps the most difficult quality to capture is that in this period that precedes the hard bargaining, an "inner" sense of "self" simply does not yet exist. It is a time when "I" *am* whatever the "others" think of me, and how "I" *feel* is a function of how the "others" treat me. Instead of a stable sense of identity, there is only a chameleon that reinvents itself depending upon the social mirror into which it is drawn. In this condition, the "others" are not individuals but the audience for whom I perform. Who "I" am depends upon the audience. This state of existence in the mirror is pure "fusion," and it captures the meaning of a thirteen-year-old girl anxiously awaiting the appearance of the little notification box as a sign of her existence and her worth. The young person who has not yet carved out an inward space exists for herself only in the viewpoint of the Other-One. Without the "others," the lights go out. Anger is out of the question: one dare not alienate the others who are one's mirror and thus one's proof of life.

In this most elemental sense, the young person who feels compelled to use social media is more truly and accurately described as *hanging on for dear life,* alive in the gaze of others because it's the only life one has, even when it hurts. As developmental psychologist Robert Kegan described the adolescent experience long before the advent of Facebook, "There *is* no self independent of the context of 'other people liking.'"[26] This is not a moral or emotional shortcoming but a fact of life in this developmental moment, and it entails certain predictable consequences. For example, one tends to operate through social comparison. One can be easy prey to manipulation, with few defenses against social pressure and other forms of social influence. The fixed belief system of an established group can all too easily fill the inner void, substituting an externally sourced identity for the work of self-construction.[27]

Moving on from "fusion" means a transition from being someone who *is* their relationships to someone who *has* their relationships. It entails a deep reconfiguration of how we make sense of our experience. In Kegan's language, this means a shift away from a "culture of mutuality" to a more complex "culture of identity, self-authorship, and personal autonomy." This shift depends upon encountering people and life experiences that demand something more than our reflection in the mirror. It requires individuals and situations that insist on our first-person voice, provoking us to carve out our own unique response to the world.

This is an inner act that eludes rendition or datafication as we begin to compose an inward sense of valid truth and moral authority. This is the reference point from which we can say, "I think," "I feel," "I believe." Gradually, this "I" learns to feel authorship and ownership of its experiences. It can reflect on itself, know itself, and regulate itself with intentional choices and purposive action. Research shows that these big leaps in self-construction are stimulated by experiences such as structured reflection, conflict, dissonance, crisis, and failure. The people who help trigger this new inward connection refuse to act as our mirrors. They reject fusion in favor of genuine reciprocity.

"Who comes into a person's life," Kegan observes, "may be the single greatest factor to influence what that life becomes."[28]

What are the consequences of the failure to win a healthy balance between inner and outer, self and relationship? Clinical studies identify specific patterns associated with this developmental stagnation. Not surprisingly, these include an inability to tolerate solitude, the feeling of being merged with others, an unstable sense of self, and even an excessive need to control others as a way of keeping the mirror close. Loss of the mirror is the felt equivalent of extinction.[29]

The cultivation of inner resources is thus critical to the capacity for intimacy and relationship, challenges that have become more time-consuming with each new phase of the modern era. And while young people are bound as ever to the enduring existential task of self-making, our story suggests three critical ways in which this task now converges with history and the unique conditions of existence in our time.

First, the waning of traditional society and the evolution of social complexity have accelerated the processes of individualization. We must rely upon our self-making and inner resources more than at any time in the human story, and when these are thwarted, the sense of dislocation and isolation is bitter.

Second, digital connection has become a necessary means of social participation, in part because of a widespread institutional failure to adapt to the needs of a new society of individuals. The computer mediation of the social infrastructure simultaneously alters human communication, illuminating individual and collective behavior, as reflected in the undulating waves of tweets, likes, clicks, patterns of mobility, search queries, posts, and thousands of other daily actions.

Third, surveillance capitalism dominates and instrumentalizes digital connection. "What is different as a result of social media," writes researcher danah boyd in her examination of the social lives of networked teens, "is that teens' perennial desire for social connection and autonomy is now being expressed in *networked publics*."[30] It's true that for the sake of connection, the travails of identity are visible to a wider group. But the notion of "networked publics" is a paradox. In fact, our visibility is magnified and compelled not only by the public-ness of networked spaces but by the fact that they are privatized. Young life now unfolds in the spaces of private capital, owned and operated by surveillance capitalists, mediated by their "economic orientation," and operationalized in practices designed to maximize surveillance revenues. These private spaces are the media through which every form of social influence—social pressure, social comparison, modeling, subliminal priming—is summoned to tune, herd, and manipulate behavior in the name of surveillance revenues. This is where adulthood is now expected to emerge.

Whereas casino executives and slot machine developers can be garrulous and boastful, eager to share their "addiction by design" achievements, the surveillance capitalist project relies on secrecy. An entire discourse has thus sprung to life, trained on decoding the stealth design that first deters users from ever looking away and then makes them incapable of doing so. There are chat groups and endless query threads as people try to divine what Facebook actually does. Relevant design practices are discussed in journalistic accounts as well as in books with such titles as *Evil by Design, Hooked,* and *Irresistible,* all of which help to normalize the very methods they discuss. For example, *Evil by Design* author Chris Nodder, a user-experience consultant, explains that evil design aims to exploit human weakness by creating interfaces that "make users emotionally involved in doing something that benefits the designer more than them." He coaches his readers in psychic numbing, urging them to accept the fact that such practices have become the standard suggesting that consumers and designers find ways to "turn them to your advantage."[31]

If we are to judge coming of age in our time, then we have to understand something of the specific practices that turn social participation into a glove that doesn't simply embrace the hand but rather magnetizes and paralyzes the hand for the sake of economic imperatives. Facebook relies on specific practices that feed the inclinations of people, especially young people, to know themselves from "the outside looking in." Most critical is that the more the need for the "others" is fed, the less able one is to engage the work of self-construction. So devastating is the failure to attain that positive equilibrium between inner and outer life that Lapsley and Woodbury say it is "at the heart" of most adult personality disorders.[32]

For example, Nodder highlights Facebook's precocious mastery of "social proof": "Much of our behavior is determined by our impressions of what is the correct thing to do...based on what we observe others doing....This influence is known as *social proof*."[33] The company instrumentalizes this aspect of adolescent nature by using messages from "friends" to make a product, service, or activity feel "more personal and emotional." This ubiquitous tactic, much admired by Pentland, was used in the Facebook voting experiment. It fuels young people's needs to garner approval and avoid disapproval by doing what the others are doing.

Facebook's single most momentous innovation in behavioral engineering is the now equally ubiquitous "Like" button, adopted in 2009. According to contemporaneous blog posts by longtime Facebook executive Andrew Bosworth, the "Like" button had been debated internally for more than a year and a half before Zuckerberg's final decision to incorporate it. He had rejected the idea more than once, fearing that it would detract from other features intended to lift monetization, such as the controversial Beacon program. Significantly, the founder embraced the button only when new data revealed it as a powerful source of behavioral surplus that helped to ratchet up the magnetism of the Facebook News Feed, as measured by the volume of comments.[34]

Facebook's leadership appears to have realized only gradually that the button could transform the platform from a book into a blizzard of mirrors, a passive read into an active sea of mutual reflections that would glue users to their news feeds. On the supply side, the "Like" button was a planet-size one-way mirror capable of exponentially increasing raw-material supplies. The more that a user "liked," the more that she informed Facebook about the precise shape and composition of her "hand," thus allowing the company to continuously tighten the glove and increase the predictive value of her signals.

The protocols at Instagram, a Facebook property, provide another good example of these processes. Here one sees these tight linkages as compulsion draws more surplus to feed more compulsion. Instagram rivets its users with photos that appeal to their interests, so how does it select those photos from the millions that are available? The obvious, but incorrect, answer would be that it analyzes the contents of photos that you like and shows you more. Instead, Instagram's analytics are drawn from behavioral surplus: the shadow text. As one manager describes it, "You base predictions off an action, and then you do stuff around that action." Actions are signals like "following," "liking," and "sharing," now and in the past. The circle widens from there.

With whom did you share? Who do they follow, like, and share with? "Instagram is mining the multilayered social web between users," but that mining is based on observable, measurable behaviors moving through time: the dynamic surplus of the shadow text drawn from its own caches as well as Facebook's, not the content displayed in the public text.[35] In the end, the photos you see resonate with strange relevance for your life. More begets more.

On the demand side, Facebook's "likes" were quickly coveted and craved, morphing into a universal reward system or what one young app designer called "our generation's crack cocaine." "Likes" became those variably timed dopamine shots, driving users to double down on their bets "every time they shared a photo, web link, or status update. A post with zero likes wasn't just privately painful, but also a kind of public condemnation."[36] In fact, most users craved the reward more than they feared humiliation, and the "Like" button became Facebook's signature, spreading across the digital universe and actively fusing users in a new kind of mutual dependency expressed in a pastel orgy of giving and receiving reinforcement.

The "Like" button was only the start of what was to be an historic construction of a new social world that for many users is defined by fusion with the social mirror, especially among the young. Just as gamblers chase the zone of fusion with the machine, a young person embedded in the culture of mutuality chases the zone of fusion with the social mirror. For anyone already struggling with the challenge of the self-other balance, the "Like" button and its brethren continuously tip the scales toward regression.

The short history of Facebook's News Feed is further evidence of the efficacy of the ever-tightening feedback loops that aim to shape and sustain this fusion. When News Feed was first launched in 2006, it transformed Facebook from a site where users had to visit friends' pages to see their updates to having those messages automatically shared in a stream on each person's home page. Hundreds of thousands of users joined opposition groups, repelled by the company's unilateral invasion of privacy. "No one was prepared for their online activity to suddenly be fodder for mass consumption," recalled the tech news site *TechCrunch* on News Feed's tenth anniversary in 2016, as it offered readers "The Ultimate Guide to the News Feed," with instructions on "how you can get your content seen by more people," how to appear "prominently," and how to resonate with your "audience."[37] Ten years earlier a *TechCrunch* reporter had presciently noted, "Users who don't participate will quickly find that they are falling out of the attention stream, and I suspect will quickly add themselves back in."[38]

Playing to the fear of invisibility and abandonment worked in 2006, when Facebook had just 9.5 million users (and required a college e-mail address to sign up), and it has driven the acceptance of every subsequent tweak to News Feed as Facebook has amassed more than 2 billion users. News Feed grew to become the "epicenter" of the corporation's revenue success and "the most valuable billboard on Earth," as *Time* magazine stated in 2015, just three years after Facebook's IPO.[39]

News Feed is also the fulcrum of the social mirror. In the years between revulsion and reverence, News Feed became Facebook's most intensely scrutinized object of data science and the subject of extensive organizational innovation, all of it undertaken at a level of sophistication and capital intensity that one might more naturally associate with the drive to solve world hunger, cure cancer, or avert climate destruction.

In addition to Facebook's already complex computational machinery for targeting ads, by 2016 the News Feed function depended upon one of the world's most secretive predictive algorithms, derived from a God view of more than 100,000 elements of behavioral surplus that are continuously computed to determine the "personal relevancy" score of thousands of possible posts as it "scans and collects everything posted in the past week by each of your friends, everyone you follow, each group you belong to, and every Facebook page you've liked," writes Will Oremus in *Slate*. "The post you see at the top of your feed, then, has been chosen over thousands of others as the one most likely to make you laugh, cry, smile, click, like, share, or comment."[40] The glove tightens around the hand with closed feedback loops enabled by the God view, which favors posts from people with whom you have already interacted, posts that have drawn high levels of engagement from others, and posts that are like the ones with which you have already engaged.[41]

In 2015 the See First "curation tool" was introduced to channel direct data on the shape of a user's social mirror by soliciting his or her personal priorities for the News Feed. Facebook's chief product officer describes the corporation's interest in supplying what is "most meaningful" for you to know today from "everything that happened on Earth…published anywhere by any of your friends, any of your family, any news source."[42] Each post sequenced in the News Feed also now hosts a range of explicit feedback options: *I want more of this. I want none of that.* These direct surplus supply lines are important sources of innovation aimed at broadening the target of the fusion zone, increasing the tenacity of an ever-tightening glove. In 2016 Facebook's product director confirmed that this direct sourcing of surplus "led to an increase in overall engagement and time spent on the site."[43]

Facebook's science and design expertise aim for a closed loop that feeds on, reinforces, and amplifies the individual user's inclination toward fusion with the group and the tendency to over-share personal information. Although these vulnerabilities run deepest among the young, the tendency to over-share is not restricted to them. The difficulty of self-imposed discipline in the sharing of private thoughts, feelings, and other personal information has been amply demonstrated in social research and summarized in an important 2015 review by Carnegie Mellon professors Alessandro Acquisti, Laura Brandimarte, and George Loewenstein. They concluded that because of a range of psychological and contextual factors, "People are often unaware of the information they are sharing, unaware of how it can be used, and even in the rare situations when they have full knowledge of the consequences of sharing, uncertain about their own preferences…." The researchers cautioned that people are "easily influenced in what and how much they disclose. Moreover, what they share can be used to influence their emotions, thoughts, and behaviors…." The result is alteration in "the balance of power between those holding the data and those who are the subjects of that data."[44]

Facebook has Pentland's prized God view on its side, an unparalleled resource that is drawn upon to remake this naturally longed-for fusion into a space of no escape. Science and capital are united in this long-game project. Yesterday it was the "Like" button, today it is augmented reality, and tomorrow there will be new innovations added to this repertoire. The company's growth in user engagement, surplus capture, and revenue are evidence that these innovations have hit their marks.

Young people crave the hive, and Facebook gives it to them, but this time it's owned and operated by surveillance capital and scientifically engineered into a continuous spiral of escalating fusion, amply fulfilling Shaffer's five criteria for achieving an addictive state of compulsion. Potency is engineered according to a recipe dictated by the hidden attributes of those who crave valorization from the group to fill the void where a self must eventually stand.

These cravings may not be the sole motivations of Facebook's currently two billion users, but they aptly describe the attributes upon which Facebook's incentives are designed to bite the hardest. Climbing the mountain of the self-other balance is an adventure that we each must undertake: a journey of risk, conflict, uncertainty, and electrifying discovery. But what happens when the forces of surveillance capital turn the mountain into a mountain range? *Look at us! Yes, you are alive! Do not look away! Why would you? How could you? Today, we might "like" you!*

IV. The Next Human Nature

A growing body of evidence testifies to the psychic toll of life in the hive, where surveillance capital's behavioral engineering expertise collides with the centuries-in-the-making human impulse toward self-construction. Researchers are already providing answers to two key questions: What are the psychological processes that dominate the hive? What are the individual and societal consequences of these processes? According to the 302 most significant quantitative research studies on the relationships between social media use and mental health (most of them produced since 2013), the psychological process that most defines the Facebook experience is what psychologists call "social comparison."[45] It is usually considered a natural and virtually automatic process that operates outside of awareness, "effectively forced upon the individual by his social environment" as we apply evaluative criteria tacitly internalized from our society, community, group, family, and friends.[46] As one research review summarizes, "Almost at the moment of exposure, an initial holistic assessment of the similarity between the target and the self is made."[47] As we go through life being exposed to other people, we naturally compare ourselves along the lines of similarity and contrast—*I am like you. I am different from you*—subliminal perceptions that translate into judgments—*I am better than you. You are better than I.*

Researchers have come to appreciate the way in which these automatic human processes converge with the changing conditions of each historical era. For most of human history, people lived in small enclaves and were typically surrounded by others very much like themselves. Social comparisons with little variation are unlikely to entail great psychological risk. Research suggests that the diffusion of television in the second half of the twentieth century dramatically increased the intensity and negativity of social comparison, as it brought vivid evidence of varied and more-affluent lives dramatically different from one's own. One study found an increase in criminal larceny as television diffused across society, awakening an awareness of and desire for consumer goods. A related issue was that increased exposure to television programs depicting affluence led to "the overestimation of others' wealth and more dissatisfaction with one's own life."[48]

Social media marks a new era in the intensity, density, and pervasiveness of social comparison processes, especially for the youngest among us, who are "almost constantly online" at a time of life when one's own identity, voice, and moral agency are a work in progress. In fact, the psychological tsunami of social comparison triggered by the social media experience is considered unprecedented. If television created more life dissatisfaction, what happens in the infinite spaces of social media?

Both television and social media deprive us of real-life encounters, in which we sense the other's inwardness and share something of our own, thus establishing some threads of communality. Unlike television, however, social media entails active self-presentation characterized by "profile inflation," in which biographical information, photos, and updates are crafted to appear ever more marvelous in anticipation of the stakes for popularity, self-worth, and happiness.[49] Profile inflation triggers more negative self-evaluation among individuals as people compare themselves to others, which then leads to more profile inflation, especially among larger networks that include more "distant friends." As one study concluded, "Expanding one's social network by adding a number of distant friends through Facebook may be detrimental by stimulating negative emotions for users."[50]

One consequence of the new density of social comparison triggers and their negative feedback loops is a psychological condition known as FOMO ("fear of missing out"). It is a form of social anxiety defined as "the uneasy and sometimes all-consuming feeling that...your peers are doing, in the know about, or in possession of more or something better than you."[51] It's a young person's affliction that is associated with negative mood and low levels of life satisfaction. Research has identified FOMO with compulsive Facebook use: FOMO sufferers obsessively checked their Facebook feeds—during meals, while driving, immediately upon waking or before sleeping, and so on. This compulsive behavior is intended to produce relief in the form of social reassurance, but it predictably breeds more anxiety and more searching.[52]

Social comparison can make people do things that they might not otherwise do. Facebook's experiments and Pokémon Go's augmented reality each exploit mutual visibility and its inevitable release of social comparison processes for successful tuning and herding. Both of these illustrate the ways in which once-natural psychological processes are repurposed to heighten the effectiveness of Pentland's vaunted "social pressure," thus enabling behavior modification at scale. Social pressure is activated by "I want to be like you" as the risks of difference and exclusion threaten negative social comparison.

What do we know about the mental health consequences of social comparison as it ensnares Facebook users, especially the young? Most of the research aimed at a deeper grasp of cause and effect in the user experience has been conducted with college-age participants, and even a brief review of a few key studies tells a grim tale, as adolescents and emerging adults run naked through these digitally mediated social territories in search of proof of life. A 2011 study found that social media users exposed to pictures of "beautiful users" developed a more negative self-image than those who were shown less attractive profile pictures. Men who were shown profiles of high-career-status men judged their own pursuits as inadequate, compared to others who saw profiles of less successful men.[53] By 2013, researchers found that Facebook use could predict negative shifts in both how their young subjects felt moment to moment as well as their overall life satisfaction.[54] That year, German researchers found that the "astounding...wealth of social information" presented on Facebook produces "a basis for social comparison and envy on an unprecedented scale." Their work demonstrated that "passive following" on Facebook exacerbates feelings of envy and decreases life satisfaction. More than 20 percent of all recent experiences of envy reported by the students in the research study had been triggered by Facebook exposure.[55]

A three-phase investigation in 2014 found that spending a lot of time browsing profiles on Facebook produced a negative mood immediately afterward. Then, upon reflection, those users felt worse, reckoning that they had wasted their time. Instead of walking away, they typically chose to spend even more time browsing the network in the hope of feeling better, chasing the dream of a sudden and magical reversal of fortune that would justify past suffering. This cycle not only leads to more social comparison and more envy, but it can also predict depressive symptoms.[56]

The self-objectification associated with social comparison is also associated with other psychological dangers. First we present ourselves as data objects for inspection, and then we experience ourselves as the "it" that others see. One 2014 study demonstrated the deleterious effect of these loops on body consciousness. An analysis of young men and women who had used Facebook for at least six years concluded that, regardless of gender, more Facebook participation leads to more body surveillance. A sense of self-worth comes to depend on physical appearance and being perceived as a sex object. Body shame leads to constant rounds of manicuring self-portrayals for a largely unknown audience of "followers."[57]

Life in the hive favors those who most naturally orient toward external cues rather than toward one's own thoughts, feelings, values, and sense of personal identity.[58] When considered from the vantage point of the self-other balance, positive social comparisons are just as pernicious as negative comparisons. Both are substitutes for the "hard bargain" of carving out a self that is capable of reciprocity rather than fusion. Whether the needle moves up or down, social comparison is the flywheel that powers the closed loop between the inclination toward the social mirror and its reinforcement. Both ego gratification and ego injury drive the chase for more external cues.

Over time, studies increase in complexity as they try to identify the underlying mechanisms through which social comparison in social media is associated with symptoms of depression and feelings of social isolation.[59] One notable three-year study published in 2017 considered both the direct Facebook data of more than five thousand participants as well as self-reported data on their "real-world social networks." This approach enabled ongoing direct comparisons between real-world relationships and Facebook associations across four domains of self-reported well-being: physical health, mental health, life satisfaction, and body mass index. "Liking others' content and clicking links to posts by friends," the researchers summarized, "were consistently related to compromised well-being, whereas the number of status updates was related to reports of diminished mental health." So strong was this relationship that "a 1-standard-deviation increase in 'likes clicked'... 'links clicked'... or 'status updates' was associated with a decrease of 5%–8% of a standard deviation in self-reported mental health," even controlling for a person's initial state of well-being. The researchers' definitive conclusion? "Facebook use does not promote well-being.... Individual social media users might do well to curtail their use of social media and focus instead on real-world relationships."[60]

V. Homing to the Herd

This is not a rehearsal. This is the show. Facebook is a prototype of instrumentarian society, not a prophecy. It is the first frontier of a new societal territory, and the youngest among us are its vanguard. The frontier experience is an epidemic of the viewpoint of the Other-One, a hyper-objectification of one's own personhood shaped by the relentless amplification of life lived from the "outside looking in." The consequence is a pattern of overwhelming anxiety and disorientation in the simple act of digital disconnection, while connection itself is haunted by fresh anxieties that paradoxically leave too many feeling isolated, diminished, and depressed. One wants to say that the struggles of youth can be painful in any era and that it is simply the destiny of today's young people to encounter the work of self-construction in this milieu of digital connection and illumination, with its truly marvelous opportunities for voice, community, information, and exploration. One wants to say they will get through it, just as other generations survived the adolescent trials of their time and place.

But this time it is not a question of simply packing their lunch and crossing our fingers as they head into the school-day maze of adolescent cliques, or sending them off to college knowing that they may stumble or fall but eventually find their passions and their people as they find themselves. This time, we have sent them into the raw heart of a rogue capitalism that amassed its fortune and power through behavioral dispossession parlayed into behavior modification in the service of others' guaranteed outcomes.

They crave the hive, just as Hall's teenagers did in 1904, but the hive they encounter is not the unadulterated product of their natures and their culture of mutuality. It is a zone of asymmetrical power, constructed by surveillance capital as it operates in secrecy beyond confrontation or accountability. It is an artificial creation designed in the service of surveillance capital's greater good. When young people enter this hive, they keep company with a surveillance priesthood: the world's most-sophisticated data scientists, programmers, machine learning experts, and technology designers, whose single-minded mission to tighten the glove is mandated by the economic imperatives of surveillance capital and its "laws of motion."

Innocent hangouts and conversations are embedded in a behavioral engineering project of planetary scope and ambition that is institutionalized in Big Other's architectures of ubiquitous monitoring, analysis, and control. In their encounter with the self-other balance, teenagers step onto a playing field already tilted by surveillance capital to tip them into the social mirror and keep them fixed on its reflections. Everything depends upon feeding the algorithms that can effectively and precisely bite on him and bite on her and not let go. All those outlays of genius and money are devoted to this one goal of keeping users, especially young users, plastered to the social mirror like bugs on the windshield.

The research studies and first-person accounts that we have reviewed reveal the coercive underbelly of the instrumentarian's much revered "confluence," in which harmonies are achieved at the expense of the psychological integrity of participants. This is the world of Pentland's "social learning," his theory of "tuning" little more than the systematic manipulation of the rewards and punishments of inclusion and exclusion. It succeeds through the natural human inclination to avoid psychological pain. Just as ordinary consumers can become compulsive gamblers at the hands of the gaming industry's behavioral technologies, psychologically ordinary young people are drawn into an unprecedented vortex of social information that automatically triggers social comparison on an equally unprecedented scale. This mental and emotional milieu appears to produce a virus of insecurity and anxiety that drives a young person deeper into this closed loop of escalating compulsion as he or she chases relief in longed-for signals of valorization.

This cycle unnaturally exacerbates and intensifies the natural orientation toward the group. And although we all share in this disposition to varying degrees, it is most pronounced in the stages of life that we call adolescence and emerging adulthood. Ethologists call this orientation "homing to the herd," an adaptation of certain species, such as passenger pigeons and herring, that home to the crowd rather than to a particular territory. In the confrontation with human predators, however, this instinct has proven fatal.

For example, biologist Bernd Heinrich describes the fate of the passenger pigeons, whose "social sense was so strong that it drew the new predator, technologically equipped humans, from afar. It made them not only easy targets, but easily duped." Commercial harvesters followed the pigeons' flight and nesting patterns, and then used huge nets to catch thousands of pigeons at a time, shipping millions by rail each year to the markets from St. Louis to Boston. The harvesters used a specific technique, designed to exploit the extraordinary bonds of empathy among the birds and immortalized in the term "stool pigeon." A few birds would be captured first and attached to a perch with their eyes sewn shut. As these birds fluttered in panic, the flock would descend to "attend to them." This made it easy for the harvesters to "catch and slaughter" thousands at once. The last passenger pigeon died in the Cincinnati Zoo in 1914: "The pigeon had no home boundaries over which to spread itself and continued to orient only to itself, so it could be everywhere, even to the end.... To the pigeons, the only 'home' they knew was in the crowd, and now they had become victims of it... the lack of territorial boundaries of human predators had tipped the scales to make their adaptation their doom."[61]

Facebook, social media in general—these are environments engineered to induce and exaggerate this homing to the human herd, particularly among the young. We are lured to the social mirror, our attention riveted by its dark charms of social comparison, social pressure, social influence. "Online all day," "online almost all day." As we fixate on the crowd, the technologically equipped commercial harvesters circle quietly and cast their nets. This artificial intensification of homing to the herd can only complicate, delay, or impede the hard psychological bargain of the self-other balance. When we multiply this effect by hundreds of millions and distribute it across the globe, what might it portend for the prospects of human and societal development?

Facebook is the crucible of this new dark science. It aims to perfect the relentless stimulation of social comparison in which natural empathy is manipulated and instrumentalized to modify behavior toward others' ends. This synthetic hive is a devilish pact for a young person. In terms of sheer everyday effectiveness—contact, logistics, transactions, communications—turn away, and you are lost. And if you simply crave the fusion juice that is proof of life at a certain age and stage—turn away, and you are extinguished.

It is a new phenomenon to live continuously in the milieu of the gaze of others, to be followed by hundreds or thousands of eyes, augmented by Big Other's devices, sensors, beams, and waves rendering, recording, analyzing, and actuating. The unceasing pace, density, and volume of the gaze deliver a perpetual stream of evaluative metrics that raise or lower one's social currency with each click. In China, these rankings are public territory, shiny badges of honor and scarlet letters that open or shut every door. In the West, we have "likes," "friends," "followers," and hundreds of other secret rankings that invisibly pattern our lives.

The extension and depth of exposure include every data point but necessarily omit the latency within each person, precisely because it cannot be observed and measured. This is the latency of a possible self that awaits ignition from that one spark caused by the caring attention of another embodied human being. It is in that clash of oxygen and ember that the latent is perceived, comprehended, and yanked forward into existence. This is real life: fleshy, soft, uncertain, and replete with silence, risk, and, when fortune smiles, genuine intimacy.

Facebook entered the world bypassing old institutional boundaries, offering us freedom to connect and express ourselves at will. It is impossible to say what the Facebook experience might have been had the company chosen a path that did not depend upon surveillance revenues. Instead, we confront the sudden accretion of an instrumentarian power that spins our society in an unanticipated direction. Facebook's applied utopistics are a prototype of an instrumentarian future, showcasing feats of behavioral engineering that groom populations for the rigors of instrumentarianism's coercive harmonies. Its operations are designed to exploit the human inclination toward empathy, belonging, and acceptance. The system tunes the pitch of our behavior with the rewards and punishments of social pressure, herding the human heart toward confluence as a means to others' commercial ends.

From this vantage point, we see that the full scope of the Facebook operation constitutes a vast experiment in behavior modification designed not only to test the specific capabilities of its tuning mechanisms, as in its official "large-scale experiments," but also to do so on the broadest possible social and psychological canvas. Most significantly, the applied utopistics of social pressure, its flywheel of social comparison, and the closed loops that bind each user to the group system vividly confirm Pentland's theoretical rendering of the case. Instrumentarian social principles are evident here, not as hypotheses but as facts, the facts that currently constitute the spaces where our children are meant to "grow up."

What we witness here is a bet-the-farm commitment to the socialization and normalization of instrumentarian power for the sake of surveillance revenues. Just as Pentland stipulated, these closed loops are imposed outside the realm of politics and individual volition. They move in stealth, crafting their effects at the level of automatic psychological responses and tipping the self-other balance toward the pseudo-harmonies of the hive mind. In this process, the inwardness that is the necessary source of autonomous action and moral judgment suffers and suffocates. These are the preparatory steps toward the death of individuality that Pentland advocates.

In fact, this death devours centuries of individualities: (1) the eighteenth century's political ideal of the individual as the repository of inalienable dignity, rights, and obligations; (2) the early twentieth century's individualized human being called into existence by history, embarking on Machado's road because she must, destined to create "a life of one's own" in a world of ever-intensifying social complexity and receding traditions; and (3) the late twentieth century's psychologically autonomous individual whose inner resources and capacity for moral judgment rise to the challenges of self-authorship that history demands and act as a bulwark against the predations of power. The self-authorship toward which young people strive carries forward these histories, strengthening, protecting, and rejuvenating each era's claims to the sanctity and sovereignty of the individual person.

What we have seen in Facebook is a living example of the third modernity that instrumentarianism proffers, defined by a new collectivism owned and operated by surveillance capital. The God view drives the computations. The computations enable tuning. Tuning replaces private governance and public politics, without which individuality is merely vestigial. And just as the uncontract bypasses social mistrust rather than healing it, the post-political societal processes that bind the hive rely on social comparison and social pressure for their durability and predictive certainty, eliminating the need for trust. Rights to the future tense, their expression in the will to will, and their sanctification in promises are drawn into the fortress of surveillance capital. On the strength of that expropriation, the tuners tighten their grasp, and the system flourishes.

Industrial capitalism depended upon the exploitation and control of nature, with catastrophic consequences that we only now recognize. Surveillance capitalism, I have suggested, depends instead upon the exploitation and control of human nature. The market reduces us to our behavior, transformed into another fictional commodity and packaged for others' consumption. In the social principles of instrumentarian society, already brought to life in the experiences of our young, we can see more clearly how this novel capitalism aims to reshape our natures for the sake of its success. We are to be monitored and telestimulated like MacKay's herds and flocks, Pentland's beavers and bees, and Nadella's machines. We are to live in the hive: a life that is naturally challenging and often painful, as any adolescent can attest, but the hive life in store for us is not a natural one. "Men made it." Surveillance capitalists made it.

The young people we have considered in this chapter are the spirits of a Christmas Yet to Come. They live on the frontier of a new form of power that declares the end of a human future, with its antique allegiances to individuals, democracy, and the human agency necessary for moral judgment. Should we awaken from distraction, resignation, and psychic numbing with Scrooge's determination, it is a future that we may still avert.

VI. No Exit

When Samuel Bentham, brother of philosopher Jeremy, first designed the panopticon as a means of overseeing unruly serfs on the estate of Prince

Potemkin in the late eighteenth century, he drew inspiration from the architecture of the Russian Orthodox churches that dotted the countryside. Typically, these structures were built around a central dome from which a portrait of an all-powerful "Christ Pantokrator" stared down at the congregation and, by implication, all humanity. There was to be no exit from this line of sight. This is the meaning of the hand and glove. The closed loop and the tight fit are meant to create the conditions of *no exit*. Once, it was no exit from God's total knowledge and power. Today, it is no exit from the others, from Big Other, and from the surveillance capitalists who decide. This condition of no exit creeps on slippered feet. First we do not even have to look away, and later we cannot.

In the closing lines of Jean-Paul Sartre's existential drama *No Exit,* the character Garcin arrives at his famous realization, "Hell is other people." This was not intended as a statement of misanthropy but rather a recognition that the self-other balance can never be adequately struck as long as the "others" are constantly "watching." Another mid-century social psychologist, Erving Goffman, took up these themes in his immortal *The Presentation of Self in Everyday Life.* Goffman developed the idea of the "backstage" as the region in which the self retreats from the performative demands of social life.

The language of backstage and onstage, inspired by observations of the theater, became a metaphor for the universal need for a place of retreat in which we can "be ourselves." Backstage is where the "impression fostered by performance is knowingly contradicted" along with its "illusions and impressions." Devices such as the telephone are "sequestered" for "private" use. Conversation is "relaxed," "truthful." It is the place where "vital secrets" can be visible. Goffman observed that in work as in life, "control of the backstage" allows individuals "to buffer themselves from the deterministic demands that surround them." Backstage, the language is one of reciprocity, familiarity, intimacy, humor. It offers the seclusion in which one can surrender to the "uncomposed" face in sleep, defecation, sex, "whistling, chewing, nibbling, belching, and flatulence." Perhaps most of all, it is an opportunity for "regression," in which we don't have to be "nice": "The surest sign of backstage solidarity is to feel that it is safe to lapse into an asociable mood of sullen, silent irritability." In the absence of such respite where a "real" self can incubate and grow, Sartre's idea of hell begins to make sense.[62]

In a classroom of undergraduates, students discuss their strategies of self-presentation on Facebook. Scholars refer to these as "chilling effects": the continuous "curation" of one's photos, comments, and profile with deletions, additions, and modifications, all of it geared to the maximization of "likes" as the signal of one's value in this existential marketplace.[63] I ask if this twenty-first-century work of self-presentation is really that much different from what Goffman had described: have we just traded the real world for the virtual in constructing and performing our personas? There is a lull as the students reflect, and then a young woman speaks:

> The difference is that Goffman assumed a backstage where you could be your true self. For us, the backstage is shrinking. There is almost no place left where I can be my true self. Even when I am walking by myself, and I think I am backstage, something happens—an ad appears on my phone or someone takes a photo, and, I discover that I am onstage, and everything changes.[64]

The "everything" that changes is the sudden cognizance, part realization and part reminder, that Big Other knows no boundaries. Experience is seamlessly rendered across the once-reliable borders of the virtual and real worlds. This accrues to the immediate benefit of surveillance capital—"Welcome to McDonald's!" "Buy this jacket!"—but any worldly experience can just as immediately be delivered to the hive: a post here, a photo there. Ubiquitous connection means that the audience is never far, and this fact brings all the pressures of the hive into the world and the body.

Recent research has begun to turn to this dour fact that a team of British researchers describes as the "extended chilling effect."[65] The idea here is that people—especially, though not exclusively, young people—now censor and curate their real-world behavior in consideration of their own online networks as well as the larger prospect of the internet masses. The researchers conclude that participation in social media "is profoundly intertwined with the knowledge that information about our offline activities may be communicated online, and that the thought of displeasing 'imagined audiences' alters our 'real-life' behavior."

When I catch myself wanting to cheer the students who are anguished by connection and terrified of its loss, I consider the meaning of "no exit" as recounted in a personal recollection of the social psychologist Stanley Milgram regarding an experiment that demonstrated "the power of immediate circumstances on feelings and behavior."[66]

Milgram's class was studying the force with which social norms control behavior. He had the idea of examining the real-life phenomenon by having his students approach a person on the subway and, without providing any justification, simply look the person in the eye and ask for his or her seat. One afternoon, Milgram himself boarded the subway ready to make his contribution. Despite his years of observing and theorizing disturbing patterns of human behavior, it turned out that he was unprepared for his own moment of social confrontation. Assuming that it would be an "easy" caper, Milgram approached a passenger and was about to utter the "magical phrase" when "the words seemed lodged in my trachea and would simply not emerge. I stood there frozen, then retreated...I was overwhelmed by paralyzing inhibition." The psychologist eventually hectored himself into trying again. He recounts what occurred when he finally approached a passenger and "choked out" his request:

> "Excuse me, sir, may I have your seat?" A moment of stark anomic
> panic overcame me. But the man got right up and gave me the seat....
> Taking the man's seat, I was overwhelmed by the need to behave in a
> way that would justify my request. My head sank between my knees,
> and I could feel my face blanching. I was not role-playing. I actually
> felt as if I were going to perish.

Moments later the train pulled into the next station, and Milgram exited. He was surprised to discover that as soon as he left the train, "all the tension disappeared." Milgram left the subway, where he vibrated in tune with the "others," and that exit enabled a return to his "self."

Milgram identified three key themes in the subway experiment as he and his students debriefed their experiences. The first was a new sense of gravitas toward "the enormous inhibitory anxiety that ordinarily prevents us from breaching social norms." Second was that the reactions of the "breacher" are not an expression of individual personality but rather are "a compelled playing out of the logic of social relations." The intense "anxiety" that Milgram and others experienced in confronting a social norm "forms a powerful barrier that must be surmounted, whether one's action is consequential—disobeying an authority—or trivial, asking for a seat on the subway.... Embarrassment and the fear of violating apparently trivial norms often lock us into intolerable predicaments.... These are not minor regulatory forces in social life, but basic ones."

Finally, Milgram understood that any confrontation of social norms crucially depends upon the ability to escape. It was not an adolescent who boarded the subway that day. Milgram was an erudite adult and an expert on human behavior, especially the mechanisms entailed in obedience to authority, social influence, and conformity. The subway was just an ordinary slice of life, not a capital-intensive architecture of surveillance and behavior modification, not a "personalized reward device." Still, Milgram could not fight off the anxiety of the situation. The only thing that made it tolerable was the possibility of an exit.

Unlike Milgram, we face an intolerable situation. Like the gamblers in their machine wombs, we are meant to fuse with the system and play to extinction: not the extinction of our funds but rather the extinction of our selves. Extinction is a design feature formalized in the conditions of no exit. The aim of the tuners is to contain us within "the power of immediate circumstances" as we are compelled by the "logic of social relations" in the hive to bow to social pressure exerted in calculated patterns that exploit our natural empathy. Continuously tightening feedback loops cut off the means of exit, creating impossible levels of anxiety that further drive the loops toward confluence. What is to be killed here is the inner impulse toward autonomy and the arduous, exciting elaboration of the autonomous self as a source of moral judgment and authority capable of asking for a subway seat or standing against rogue power.

Inside the hive, it is easy to forget that every exit is an entrance. To exit the hive means to enter that territory beyond, where one finds refuge from the artificially tuned-up social pressure of the others. Exit leaves behind the point of view of the Other-One in favor of entering a space in which one's gaze can finally settle inward. To exit means to enter the place where a self can be birthed and nurtured. History has a name for that kind of place: *sanctuary*.

CHAPTER SEVENTEEN

THE RIGHT TO SANCTUARY

*Refuge and prospect are opposites: refuge is small and dark;
prospect is expansive and bright.... We need them both
and we need them together.*

—GRANT HILDEBRAND
"FINDING A GOOD HOME"
ORIGINS OF ARCHITECTURAL PLEASURE

I. Big Other Outruns Society

That summer night when our home was destroyed by a lightning strike, we watched in the driving rain as the gables and rambling porches exploded in fire. Within hours a smoldering field of black ash covered the ground where home had been. In the months and years that followed, my recollections of the house took an unexpected shape, less rooms and objects than shadow, light, and fragrance. I conjured in perfect clarity the rush of my mother's scent when I opened the drawer filled with her once-cherished scarves. I closed my eyes and saw the late-afternoon sun slicing through the velvety air by the bedroom fireplace with its ancient sloping mantle where our treasures were on display: a photo of my father and me, heads tilted toward each other, blending our two shocks of curly black hair; the miniature painted enamel boxes, discovered in a Parisian flea market years before the thought of motherhood, which later became the shelter for our children's milk teeth huddled like secret caches of seed pearls. It was impossible to explain the quality of this sadness and longing: how our selves and the life of our family had evolved symbiotically with those spaces that we called home. How our attachments transformed a house into a hallowed place of love, meaning, and commemoration.

My difficulty began to ease only when I discovered the work of Gaston Bachelard, an extraordinary man who had been a postal worker, physicist, philosopher, and ultimately a professor of philosophy at the Sorbonne:

> The old house, for those who know how to listen, is a sort of geometry of echoes. The voices of the past do not sound the same in the big room as in the little bed chamber....Among the most difficult memories, well beyond any geometry that can be drawn, we must recapture the quality of the light; then come the sweet smells that linger in the empty rooms....[1]

One of Bachelard's works in particular, *The Poetics of Space,* is instructive as we reckon with the prospects of life in the no-exit shadow of Big Other and its power brokers behind the curtain. In this book Bachelard elaborates his notion of "topoanalysis," the study of how our deepest relationships to inner self and outer world are formed in our experience of space, specifically the space we call "home":

> The house shelters daydreaming, the house protects the dreamer, the house allows one to dream in peace....The house is one of the greatest powers of integration for the thoughts, memories, and dreams of mankind....It is body and soul. It is the human being's first world. Before he is "cast into the world,"...man is laid in the cradle of the house....Life begins well, it begins enclosed, protected, all warm in the bosom of the house....[2]

Home is our school of intimacy, where we first learn to be human. Its corners and nooks conceal the sweetness of solitude; its rooms frame our experience of relationship. Its shelter, stability, and security work to concentrate our unique inner sense of self, an identity that imbues our day dreams and night dreams forever. Its hiding places—closets, chests, drawers, locks, and keys—satisfy our need for mystery and independence. Doors—locked, closed, half shut, wide open—trigger our sense of wonder, safety, possibility, and adventure.

Bachelard plumbs not only the imagery of the human house but also of nests and shells, the "primal images" of home that convey the absolute "primitiveness" of the need for a safe refuge: "Well-being takes us back to the primitiveness of the refuge. Physically, the creature endowed with a sense of refuge huddles up to itself, takes to cover, hides away, lies snug, concealed…a human being likes to 'withdraw into his corner'…it gives him physical pleasure to do so."[3]

The shelter of home is our original way of living in space, Bachelard discovers, shaping not only the existential counterpoint of "home" and "away" but also many of our most fundamental ways of making sense of experience: house and universe, refuge and world, inside and outside, concrete and abstract, being and nonbeing, this and that, here and elsewhere, narrow and vast, depth and immensity, private and public, intimate and distant, self and other.

Our family instinctively pursued these themes in imagining a new home. When we were finally able to undertake that project, we foraged for durable natural materials: old stone and scarred wooden beams that had weathered the storms of time. We were drawn to old furniture that had already lived many lives composing others' homes. This is how the walls of the new house came to be massive, nearly a foot deep and packed with insulation. The result is just as we had hoped: a lush and peaceful stillness. We know that nothing guarantees safety and certainty in this world, but we are comforted by the serenity of this home and its layered silences.

The days unfurl now within the embrace of these generous walls, where once again our spirits spread and root. This is how a house becomes a home and a home becomes a sanctuary. I feel this most acutely when I crawl into bed at night. I wait to hear my husband's breathing in syncopation with the muffled sighs of our beloved dog on the floor beside us as she sprints through her ecstatic dreams. I sense beyond to the dense envelope of our bedroom walls and listen to their lullaby of seclusion.

According to Big Other's architects, these walls must come down. There can be no refuge. The primal yen for nests and shells is kicked aside like so much detritus from a fusty human time. With Big Other, the universe takes up residence in our walls, no longer the sentinels of sanctuary. Now they are simply the coordinates for "smart" thermostats, security cameras, speakers, and light switches that extract and render our experience in order to actuate our behavior.

That our walls are dense and deep is of no importance now because the boundaries that define the very experience of home are to be erased. There can be no corners in which to curl up and taste the pleasures of solitary inwardness. There can be no secret hiding places because there can be no secrets. Big Other swallows refuge whole, along with the categories of understanding that originate in its elemental oppositions: house and universe, depth and immensity. Those ageless polarities in which we discover and elaborate our sense of self are casually eviscerated as immensity installs itself in my refrigerator, the world chatters in my toothbrush, elsewhere stands watch over my bloodstream, and the garden breeze stirs the chimes draped from the willow tree only to be broadcast across the planet. The locks? They have vanished. The doors? They are open.

In the march of institutional interests intent on implementing Big Other, the very first citadel to fall is the most ancient: the principle of sanctuary. The sanctuary privilege has stood as an antidote to power since the beginning of the human story. Even in ancient societies where tyranny prevailed, the right of sanctuary stood as a fail-safe. There was an exit from totalizing power, and that exit was the entrance to a sanctuary in the form of a city, a community, or a temple.[4] By the time of the Greeks, sanctuaries were sacred sites built across the ancient Greek world and consecrated to the purposes of asylum and religious sacrifice. The Greek word *asylon* means "unplunderable" and founds the notion of a sanctuary as an inviolable space.[5] The right of asylum survived into the eighteenth century in many parts of Europe, attached to holy sites, churches, and monasteries. The demise of the sanctuary privilege was not a repudiation but rather a reflection of social evolution and the firm establishment of the rule of law. One historian summarized this transformation: "justice as sanctuary."[6]

In the modern era the sacredness, inviolability, and reverence that once attached to the law of asylum reemerged in constitutional protections and declarations of inalienable rights. English common law retained the idea of the castle as an inviolable fortress and translated that to the broader notion of "home," a sanctuary free from arbitrary intrusion: unplunderable. The long thread of the sanctuary privilege reappeared in US jurisprudence. Writing in 1995, legal scholar Linda McClain argued that the equation of home with sanctuary has depended less on the sanctity of property rights than on a commitment to the "privacies of life." As she observed, "There is a strong theme of a proper realm of inaccessibility or secrecy with respect to the world at large as well as a recognition of the important social dimension of such protected inner space...."[7]

The same themes appear from the perspective of psychology. Those who would eviscerate sanctuary are keen to take the offensive, putting us off guard with the guilt-inducing question "What have you got to hide?" But as we have seen, the crucial developmental challenges of the self-other balance cannot be negotiated adequately without the sanctity of "disconnected" time and space for the ripening of inward awareness and the possibility of reflexivity: reflection on and by oneself. The real psychological truth is this: *If you've got nothing to hide, you are nothing.*

One empirical study makes the point. In "Psychological Functions of Privacy," Darhl Pedersen defines privacy as a "boundary control process" that invokes the decision rights associated with "restricting and seeking interaction."[8] Pedersen's research identifies six categories of privacy behaviors: solitude, isolation, anonymity, reserve, intimacy with friends, and intimacy with family. His study shows that these varied behaviors accomplish a rich array of complex psychological "privacy functions" considered salient for psychological health and developmental success: contemplation, autonomy, rejuvenation, confiding, freedom, creativity, recovery, catharsis, and concealment. These are experiences without which we can neither flourish nor usefully contribute to our families, communities, and society.

As the digital era intensifies and surveillance capitalism spreads, the centuries-old solution of "justice as sanctuary" no longer holds. Big Other outruns society and law in a self-authorized destruction of the right to sanctuary as it overwhelms considerations of justice with its tactical mastery of shock and awe. The facts of surveillance capitalism's dominance of the division of learning, the unrepentant momentum of its dispossession cycle, the institutionalization of its means of behavior modification, the convergence of these with the requirements of social participation, and the manufacture of prediction products for trade in behavioral futures markets are de facto evidence of a new condition that has not been tamed by law. The remainder of this chapter explores the implications of this failure. What will taming require? What kind of life is left to us if taming fails?

II. Justice at the New Frontier of Power

If sanctuary is to be preserved, synthetic declarations are required: alternative pathways that lead to a human future. It's the wall, not the tunnels, that requires our attention. So far US privacy laws have failed to keep pace with the march of instrumentarianism. Analyses of the "invasion of privacy," according to legal scholar Anita Allen, fall into "a handful of easily illustrated categories." Allen contrasts "physical privacy" (sometimes called "spatial privacy") with "informational privacy." She observes that physical privacy is violated "when a person's efforts to seclude or conceal himself or herself are frustrated." Information privacy is disturbed "when data, facts, or conversations that a person wishes to secret or anonymize are nonetheless acquired or disclosed."[9]

In the era of Big Other, though, these categories bend and break. Physical places, including our homes, are increasingly saturated with informational violations as our lives are rendered as behavior and expropriated as surplus. In some cases we inflict this on ourselves, typically because we do not grasp the backstage operations and their full implications. Other violations are simply imposed upon us, as in the case of the talking doll, the listening TV, the hundreds of apps programmed for secret rendition, and so on. We have surveyed many of the objects and processes already earmarked to be smart, sensate, actuating, connected, and internet-enabled by surveillance capital. By the time you read these pages, there will be more, and more after that. It's the sorcerer's apprentice cursed with the perpetual filling and refilling driven by an unbounded claim that asserts its right to everything.

When US scholars and jurists assess the ways in which digital capabilities challenge existing law, the focus is on Fourth Amendment doctrine as it circumscribes the relationship *between individuals and the state*. It is of course vital that Fourth Amendment protections catch up to the twenty-first century by protecting us from search and seizure of our information in ways that reflect contemporary realities of data production.[10] The problem is that even expanded protections from the state do not shield us from the assault on sanctuary wrought by instrumentarian power and animated by surveillance capitalism's economic imperatives.[11] The Fourth Amendment as currently construed does not help us here. There is no sorcerer in sight ready to command the surveillance capitalists, in Goethe's words, "Corner broom! Hear your doom."

Legal scholarship is just beginning to reckon with these facts. As a 2016 article on the "internet of things" by Fourth Amendment scholar Andrew Guthrie Ferguson concludes, "If billions of sensors filled with personal data fall outside of Fourth Amendment protections, a large-scale surveillance network will exist without constitutional limits."[12] As we have seen, it already does. Dutch scholars make a similar case for the inadequacy of Dutch law as it falls behind Big Other, no longer able to effectively assert the sanctity of the home from the invasive action of either industry or the state: "The walls no longer shield the individual effectively from the outside in the pursuing of... personal life without intrusion...."[13]

Many hopes today are pinned on the new body of EU regulation known as the General Data Protection Regulation (GDPR), which became enforceable in May 2018. The EU approach fundamentally differs from that of the US in that companies must justify their data activities within the GDPR's regulatory framework. The regulations introduce several key new substantive and procedural features, including a requirement to notify people when personal data is breached, a high threshold for the definition of "consent" that puts limits on a company's reliance on this tactic to approve personal data use, a prohibition on making personal information public by default, a requirement to use privacy by design when building systems, a right to erasure of data, and expanded protections against decision making authored by automated systems that imposes "consequential" effects on a person's life.[14] The new regulatory framework also imposes substantial fines for violations, which will rise to a possible 4 percent of a company's global revenue, and it allows for class-action lawsuits in which users can combine to assert their rights to privacy and data protection.[15]

These are vital and necessary accomplishments, and the question that is most important to our story is whether this new regulatory regime can be a springboard to challenging the legitimacy of surveillance capitalism and ultimately vanquishing its instrumentarian power. In time, the world will learn if the GDPR can move out in front of Big Other, reasserting a division of learning aligned with the values and aspirations of a democratic society. Such a victory would depend upon society's rejection of markets based on the dispossession of human experience as a means to the prediction and control of human behavior for others' profit.

Scholars and specialists debate the implications of the sweeping new regulations, some arguing the inevitability of decisive change and others arguing the likelihood of continuity over dramatic reversals of practice.[16] There are some things that we do know, however. Individuals each wrestling with the myriad complexities of their own data protection will be no match for surveillance capitalism's staggering asymmetries of knowledge and power. If the past two decades have taught us anything, it is that the individual alone cannot bear the burden of this fight at the new frontier of power.

This theme is illustrated in the odyssey of Belgian mathematician and data protection activist Paul-Olivier Dehaye, who in December 2016 initiated a request for his personal data collected through Facebook's Custom Audiences and tracking Pixel tools, which would reveal the web pages where Facebook had tracked him. Dehaye probably knew more about the rogue data operations of Cambridge Analytica than anyone in the world, outside of its own staff and masterminds. His aim was a bottom-up investigative approach to uncover the secrets of its illegitimate means of political behavior modification.

A first step was to determine what Facebook knew about him, especially the kind of data that would become relevant in an electoral context and thus make him, and others, vulnerable to the kinds of hidden maneuvers that Cambridge Analytica had employed. Dehaye wanted to understand how a citizen might come to ascertain the data that enabled, judging from the worldwide outrage over the revelations of secret online political manipulation, what many people consider a highly "consequential" threat. He carefully documented the twists and turns of his journey, hoping that his experience would be useful for journalists, citizens, and communities determined to understand the scope and political vulnerabilities of Facebook's practices. Dehaye writes:

> It is of course extremely difficult to talk to a company like Facebook
> as an individual, so by April 2017 I had to escalate the matter to the

Irish Data Protection Commissioner. By October 2017, after a lot of prodding, the Irish Data Protection Commissioner finally agreed to take the first step with my complaint, and ask Facebook for a comment. By December 2017, they had apparently received a response, but as of March 2018 they are still "assessing" it, despite frequent reminders. It is very hard not to see a problem here with respect to enforcement.[17]

In March 2018, fifteen months after his initial request, Dehaye finally received an e-mail from Facebook's Privacy Operations Team. He was told that the information he sought "is not available through our self-service tools" but is stored in "Hive," Facebook's "log storage area," where it is retained for "data analytics" and maintained as separate from "the data bases that power the Facebook site." The company insisted that accessing the data required it to surmount "huge technical challenges." "This data," the company writes, "*is also not used to directly serve the live Facebook website which users experience.*"[18]

In our language, the information that Dehaye sought required access to the "shadow text," specifically asking for, among other things, details on the targeting analyses that determined the ads he was shown on Facebook. The corporation's response indicates that Hive's data are part of this exclusive "second text," in which behavioral surplus is queued up for manufacture into predictions products.[19] This process is completely separate from the "first text," "which users experience."

Facebook makes clear that the shadow text is not available to users, despite the promotion of its self-service download tools that promise to give users access to their personal data retained by the company. Indeed, the competitive dynamics of surveillance capitalism make the shadow text a crucial proprietary source of advantage. Any attempt to breach its content will be experienced as an existential threat; no surveillance capitalist will *voluntarily* provide data from the shadow text. Only law can compel this challenge to the pathological division of learning.

In the wake of the Cambridge Analytica scandal in March 2018, Facebook announced it would expand the range of personal data that it allows users to download, but even these data remain wholly contained within the first text, composed largely of the information that users themselves have provided, including information that they have deleted: friends, photos, video, ads that have been clicked, pokes, posts, location, and so on. These data do not include behavioral surplus, prediction products, and the fate of those predictions as they are used for behavioral modification, bought, and sold. When you download your "personal information," you access the stage, not the backstage: the curtain, not the wizard.[20]

Facebook's response to Dehaye illustrates another consequence of the extreme asymmetries of knowledge at play. The company insisted that access to the requested data required it to surmount "huge technical challenges." As behavioral surplus flows converge in machine-learning–based manufacturing operations, the sheer volume of data inputs and methods of analysis moves beyond human comprehension. Consider something as trivial as the case of Instagram's machines selecting what images to show you. Its computations are based on varied streams of behavioral surplus from a subject user, then more streams from the friends in that user's network, then more from the activities of people who follow the same accounts as the subject user, then the data and social links from the user's Facebook activity. When it finally applies a ranking logic to predict what images the user will want to see next, that analysis also includes data on the past behavior of the subject user. Instagram has machines doing this "learning" because humans cannot.[21] In the case of more "consequential" analyses, the operations are likely to be equally or even more complex.

This recalls our discussion of Facebook's "prediction engine" FBLearner Flow, where the machines are fed tens of thousands of data points derived from behavioral surplus, diminishing the very notion of the right to contest "automatic decision making." If the algorithms are to be contestable in any meaningful way, it will require new countervailing authority and power, including machine resources and expertise to reach into the core disciplines of machine intelligence and construct new approaches that are available for inspection, debate, and combat. Indeed, one expert has already proposed the creation of a government agency—an "FDA for algorithms"—to oversee the development, distribution, sale, and use of complex algorithms, arguing that existing laws "will prove no match for the difficult regulatory puzzles algorithms pose."[22]

Dehaye's experience is but one illustration of the self-sustaining nature of a pathological division of learning and the insuperable burden placed on individuals moved to challenge its injustice. Dehaye is an activist, and his aim is not only to access data but also to document the arduousness and even absurdity of the undertaking. Given these realities, he suggests that the data-protection regulations are comparable to freedom-of-information laws. The procedures for requesting and receiving information under these laws are imperfect and onerous, typically undertaken by legal specialists, but nevertheless essential to democratic freedom.[23] Although effective contest will require determined individuals, the individual alone cannot shoulder the burden of justice, any more than an individual worker in the first years of the twentieth century could bear the burden of fighting for fair wages and working conditions. Those twentieth-century challenges required collective action, and so do our own.[24]

In her discussion of "the life of the law," anthropologist Laura Nader reminds us that law projects "possibilities of democratic empowerment" but that these are pulled forward into real life only when citizens actively contest injustice, using the law as a means to higher purpose. "The life of the law is the plaintiff," Nader writes, a truth we saw brought to life in the action of the Spanish citizens who claimed the right to be forgotten. "By contesting their injustices by means of law, plaintiffs and their lawyers can still decide the place of law in making history."[25] These plaintiffs do not stand alone; they stand for citizens bonded together as the necessary means of confronting collective injustice.

This brings us back to the GDPR and the question of its impact. The only possible answer is that everything will depend upon how European societies interpret the new regulatory regime in legislation and in the courts. It will not be the wording of the regulations but rather the popular movements on the ground that shape these interpretations. A century ago, workers organized for collective action and ultimately tipped the scales of power, and today's "users" will have to mobilize in new ways that reflect our own unique twenty-first-century "conditions of existence." We need synthetic declarations that are institutionalized in new centers of democratic power, expertise, and contest that challenge today's asymmetries of knowledge and power. This quality of collective action will be required if we are finally to replace lawlessness with laws that assert the right to sanctuary and the right to the future tense as essential for effective human life.

It is already possible to see a new awakening to empowering collective action, at least in the privacy domain. One example is None of Your Business (NOYB), a nonprofit organization led by privacy activist Max Schrems. After many years of legal contest, Schrems made history in 2015 when his challenge to Facebook's data-collection and data-retention practices—which he asserted were in violation of EU privacy law—led the Court of Justice of the European Union to invalidate the Safe Harbor agreement that governed data transfers between the US and the EU. In 2018 Schrems launched NOYB as a vehicle for "professional privacy enforcement." The idea is to push regulators to close the gap between written regulations and corporate privacy practices, leveraging the threat of significant fines to change a company's actual procedures. NOYB wants to become "a stable European enforcement platform" that unites groups of users and assists them through the litigation process while building coalitions and advancing "targeted and strategic litigation to maximize the impact 'on the right to privacy.'"[26] However this undertaking progresses, the key point for us is the way in which it points to a social void that must be filled with creative new forms of collective action, if the life of the law is to move against surveillance capitalism.

Only time will tell if the GDPR will be a catalyst for a new phase of combat that wrangles and tames an illegitimate marketplace in behavioral futures, the data operations that feed it, and the instrumentarian society toward which they aim. In the absence of new synthetic declarations, we may be disappointed by the intransigence of the status quo. If the past is a prologue, then privacy, data protection, and antitrust laws will not be enough to interrupt surveillance capitalism. The reasons that we examined in answering the question "How did they get away with it?" suggest that the immense and intricate structures of surveillance capitalism and its imperatives will require a more direct challenge.

This is at least one conclusion from the past decade: despite far more stringent privacy and data-protection laws in the EU as compared to the US, as well as a forceful commitment to antitrust, Facebook and Google have continued to flourish in Europe. For example, between 2010 and 2017 the compounded annual growth rate for Facebook daily active users was 15 percent in Europe compared to 9 percent in the US and Canada.[27] During that period the company's revenue grew by a compounded annual growth rate of

50 percent in both regions.[28] Between 2009 and the first quarter of 2018, Google's share of the search market in Europe declined by about 2 percent while increasing in the US by about 9 percent. (Google's European market share remained high, at 91.5 percent in 2018, compared to 88 percent in the US.) In the case of its Android mobile phones, however, Google's market share increased by 69 percent in Europe compared to 44 percent in the US. Google's Chrome browser increased its market share by 55 percent in Europe and 51 percent in the US.[29]

Those growth rates are not mere good fortune, as our list of "how they got away with it" suggests. In recognition of this fact, Europe's Data Protection Supervisor Giovanni Buttarelli told the *New York Times* that the GDPR's impact will be determined by regulators who "will be up against well-funded teams of lobbyists and lawyers."[30] Indeed, corporate lawyers were already honing their strategies for the preservation of business as usual and setting the stage for the contests ahead. For example, a white paper published by one prominent international law firm rallies corporations to the barricades of data processing, arguing that the legal concept of "legitimate interest" offers a promising opportunity to bypass new regulatory obstacles:

> Legitimate interest may be the most accountable ground for processing in many contexts, as it requires an assessment and balancing of the risks and benefits of processing for organisations, individuals and society. The legitimate interests of the controller or a third party may also include other rights and freedoms. The balancing test will sometimes also include…freedom of expression, right to engage in economic activity, right to ensure protection of IP rights, etc. These rights must also be taken into account when balancing them against the individuals' right to privacy.[31]

Surveillance capitalism's economic imperatives were already on the move in late April 2018, in anticipation of the GDPR taking effect that May. Earlier in April, Facebook's CEO had announced that the corporation would apply the GDPR "in spirit" across the globe. In practice, however, the company was making changes to ensure that the GDPR would not circumscribe the majority of its operations. Until then, 1.5 billion of its users, including those in

Africa, Asia, Australia, and Latin America, were governed by terms of service issued by the company's international headquarters in Ireland, meaning that these terms fell under the EU framework. It was in late April that Facebook quietly issued new terms of service, placing those 1.5 billion users under US privacy laws and thus eliminating their ability to file claims in Irish courts.[32]

III. Every Unicorn Has a Hunter

What life is left to us if taming fails? Without protection from surveillance capitalism and its instrumentarian power—their behavioral aims and societal goals—we are trapped in a condition of "no exit," where the only walls are made of glass. The natural human yearning for refuge must be extinguished and the ancient institution of sanctuary deleted.

"No exit" is the necessary condition for Big Other to flourish, and its flourishing is the necessary condition for all that is meant to follow: the tides of behavioral surplus and their transformation into revenue, the certainty that will meet every market player with guaranteed outcomes, the bypass of trust in favor of the uncontract's radical indifference, the paradise of effortless connection that exploits the needs of harried second-modernity individuals and transforms their lives into the means to others' ends, the plundering of the self, the extinction of autonomous moral judgment for the sake of frictionless control, the actuation and modification that quietly drains the will to will, the forfeit of your voice in the first person in favor of others' plans, the destruction of the social relations and politics of the old and slow and still-unfulfilled ideals of self-determining citizens bound to the legitimate authority of democratic governance.

Each of these exquisite unicorns has inspired the best that humanity has achieved, however imperfectly they have been fulfilled. But every unicorn has a hunter, and the ideals that have nurtured the liberal order are no exception. For the sake of this hunter, there can be no doors, no locks, no friction, no opposition between intimacy and distance, house and universe. There is no need for "topoanalysis" now because all spaces have collapsed into the one space that is Big Other. Seek not the petal-soft iridescent apex of the shell.

There is no purpose to curling up in its dark spire. The shell is just another connected node, and your daydream is already finding an audience in the pulsating net of this clamorous glass life.

In the absence of synthetic declarations that secure the road to a human future, the intolerability of glass life turns us toward a societal arms race of counter-declarations in which we search for and embrace increasingly complex ways to *hide in our own lives,* seeking respite from lawless machines and their masters. We do this to satisfy our enduring need for sanctuary and as an act of resistance with which to reject the instrumentarian disciplines of the hive, its "extended chilling effects," and Big Other's relentless greed. In the context of government surveillance, the practices of "hiding" have been called "privacy protests" and are well-known for drawing the suspicion of law-enforcement agencies.[33] Now, hiding is also invoked by Big Other and its market masters, whose reach is far and deep as they install themselves in our walls, our bodies, and on our streets, claiming our faces, our feelings, and our fears of exclusion.

I have suggested that too many of the best and brightest of a new generation devote their genius to the intensification of the click-stream. Equally poignant is the way in which a new generation of activists, artists, and inventors feels itself called to create the art and science of hiding.[34] The intolerable conditions of glass life compel these young artists to dedicate their genius to the prospects of human invisibility, even as their creations demand that we aggressively seek and find our bearings. Their provocations already take many forms: signal-blocking phone cases, false fingerprint prosthetics that prevent your fingertips from being "used as a key to your life," LED privacy visors to impede facial-recognition cameras, a quilted coat that blocks radio waves and tracking devices, a scent diffuser that releases a metallic fragrance when an unprotected website or network is detected on any of your devices, a "serendipitor app" to disrupt any surveillance "that relies on subjects maintaining predictable routines," a clothing line called "Glamouflage" featuring shirts covered with representations of celebrity faces to confuse facial-recognition software, anti-neuroimaging surveillance headgear to obstruct digital invasion of brain waves, and an anti-surveillance coat that creates a shield to block invasive signals. Chicago artist Leo Selvaggio produces

3-D–printed resin prosthetic masks to confound facial recognition. He calls his effort "an organized artistic intervention."[35]

Perhaps most poignant is the Backslash Tool Kit: "a series of functional devices designed for protests and riots of the future," including a smart bandana for embedding hidden messages and public keys, independently networked wearable devices, personal black-box devices to register abuse of law enforcement, and fast deployment routers for off-grid communication.[36] Backslash was created as part of a master's thesis project at New York University, and it perfectly reflects the contest for the third modernity that this generation faces. The designer writes that for young, digitally native protesters, "*connectivity is a basic human right.*" Yet, he laments, "the future of technology in protests looks dark" because of overwhelming surveillance. His tool kit is intended to create "a space to explore and research the tense relationship between protests and technology and a space to cultivate dialogue about freedom of expression, riots and disruptive technology." In a related development, students at the University of Washington have developed a prototype for "on-body transmissions with commodity devices." The idea here is that readily available devices "can be used to transmit information to only wireless receivers that are in contact with the body," thus creating the basis for secure and private communications independent of normal Wi-Fi transmissions, which can easily be detected.[37]

Take a casual stroll through the shop at the New Museum for Contemporary Art in Manhattan, and you pass a display of its bestseller: table-top mirrors whose reflecting surface is covered with the bright-orange message "Today's Selfie Is Tomorrow's Biometric Profile." This "Think Privacy Selfie Mirror" is a project of the young Berlin-based artist Adam Harvey, whose work is aimed at the problem of surveillance and foiling the power of those who surveil. Harvey's art begins with "reverse engineering…computer vision algorithms" in order to detect and exploit their vulnerabilities through camouflage and other forms of hiding. He is perhaps best known for his "Stealth Wear," a series of wearable fashion pieces intended to overwhelm, confuse, and evade drone surveillance and, more broadly, facial-recognition software. Silver-plated fabrics reflect thermal radiation, "enabling the wearer to avert overhead thermal surveillance." Harvey's fashions are inspired by traditional Islamic dress, which expresses the idea that "garments can provide a separation between man and God." Now he redirects that meaning to create garments that separate human experience from the powers that surveil.[38] Another Harvey project created an aesthetic of makeup and hairstyling—blue feathers suspended from thick black bangs, dreadlocks that dangle below the nose, cheekbones covered in thick wedges of black and white paint, tresses that snake around the face and neck like octopus tentacles—all designed to thwart facial-recognition software and other forms of computer vision.

Harvey is one among a growing number of artists, often *young* artists, who direct their work to the themes of surveillance and resistance. Artist Benjamin Grosser's Facebook and Twitter "demetricators" are software interfaces that present each site's pages with their metrics deleted: "The numbers of 'likes,' 'friends,' followers, retweets...all disappear." How is an interface that foregrounds our friend count changing our conceptions of friendship? he asks. "Remove the numbers and find out." Grosser's "Go Rando" project is a web browser extension that "obfuscates your feelings on Facebook" by randomly choosing an emoji each time you click "Like," thus undermining the corporation's surplus analyses as they compute personality and emotional profiles.[39] Trevor Paglen's richly orchestrated performance art combines music, photography, satellite imagery, and artificial intelligence to reveal Big Other's omnipresent knowing and doing. "It's trying to look inside the software that is running an AI...to look into the architectures of different computer vision systems and trying to learn what it is that they are seeing," Paglen says. Chinese artist Ai Weiwei's 2017 installation "Hansel & Gretel" created a powerful experience in which participants viscerally confront the surveillance implications of their own innocent picture taking, Instagramming, tweeting, texting, tagging, and posting.[40]

Our artists, like our young people, are canaries in the coal mine. That the need to make ourselves invisible is the theme of a brilliant artistic vanguard is another kind of message in a bottle cast from the front lines of mourning and revulsion. Glass life is intolerable, but so is fitting our faces with masks and draping our bodies in digitally resistant fabrics to thwart the ubiquitous lawless machines. Like every counter-declaration, hiding risks becomes an adaptation when it should be a rallying point for outrage. *These conditions are unacceptable. Tunnels under this wall are not enough. This wall must come down.*

The greatest danger is that we come to feel at home in glass life or in the prospect of hiding from it. Both alternatives rob us of the life-sustaining inwardness, born in sanctuary, that finally distinguishes us from the machines. This is the well from which we draw the capacities to promise and to love, without which both the private bonds of intimacy and the public bonds of society wither and die. If we do not alter this course now, we leave a monumental work for the generations that follow us. Industrial capitalism commandeered nature only to saddle the coming generations with the burden of a burning planet. Will we add to this burden with surveillance capitalism's invasion and conquest of human nature? Will we stand by as it subtly imposes the life of the hive while demanding the forfeit of sanctuary and the right to the future tense for the sake of its wealth and power?

Paradiso calls it a revolution, and Pentland says it is the death of individuality. Nadella and Schmidt advocate the machine hive as our role model, with its coercive confluence and preemptive harmonies. Page and Zuckerberg understand the transformation of society as a means to their commercial ends. There are dissenters among us, to be sure, but the declaration of life without walls has thus far failed to trigger a mass withdrawal of agreement. This is in part the result of our dependency and in part because we do not yet appreciate the breadth and depth of what the architects have in store, let alone the consequences that this "revolution" might entail.

Our sensibilities grow numb to the monstrosity of Big Other as its features are developed, tested, elaborated, and normalized. We become deaf to the lullaby of walls. Hiding from the machines and their masters drifts from the obsessions of the vanguard to a normal theme of social discourse and eventually our conversations around the dinner table. Each step down this path occurs as if in the fog of war: scattered fragments and incidents that appear abruptly and often in obscurity. There is little room to perceive the pattern, let alone its origins and meaning. Nonetheless, each deletion of the possibility of sanctuary leaves a void that is seamlessly and soundlessly filled by the new conditions of instrumentarian power.

CONCLUSION

CHAPTER EIGHTEEN

A COUP FROM ABOVE

He shook with hate for things he'd never seen,
Pined for a love abstracted from its object,
And was oppressed as he had never been.

—W. H. AUDEN
SONNETS FROM CHINA, III

Surveillance capitalism departs from the history of market capitalism in three startling ways. First, it insists on the privilege of unfettered freedom *and* knowledge. Second, it abandons long-standing organic reciprocities with people. Third, the specter of life in the hive betrays a collectivist societal vision sustained by radical indifference and its material expression in Big Other. In this chapter we explore each of these departures from historical norms and then face the question that they raise: is surveillance capitalism merely "capitalism"?

I. *Freedom* and *Knowledge*

Surveillance capitalists are no different from other capitalists in demanding freedom from any sort of constraint. They insist upon the "freedom to" launch every novel practice while aggressively asserting the necessity of their "freedom from" law and regulation. This classic pattern reflects two bedrock assumptions about capitalism made by its own theorists: The first is that markets are intrinsically *unknowable*. The second is that the ignorance produced by this lack of knowledge requires wide-ranging *freedom* of action for market actors.

The notion that ignorance and freedom are essential characteristics of capitalism is rooted in the conditions of life before the advent of modern systems of communication and transportation, let alone global digital networks, the internet, or the ubiquitous computational, sensate, actuating architectures of Big Other. Until the last few moments of the human story, life was necessarily local, and the "whole" was necessarily invisible to the "part."

Adam Smith's famous metaphor of the "invisible hand" drew on these enduring realities of human life. Each individual, Smith reasoned, employs his capital locally in pursuit of immediate comforts and necessities. Each one attends to "his own security...his own gain...led by an invisible hand to promote an end which was no part of his intention." That end is the efficient employ of capital in the broader market: the wealth of nations. The individual actions that produce efficient markets add up to a staggeringly complex pattern, a mystery that no one person or entity could hope to know or understand, let alone to direct: "The statesman, who should attempt to direct private people in what manner they ought to employ their capitals, would... assume an authority which could safely be trusted, not only to no single person, but to no council or senate whatever...."[1]

The neoliberal economist Friedrich Hayek, whose work we discussed briefly in Chapter 2 as the foundation for the market-privileging economic policies of the past half century, drew the most basic tenets of his arguments from Smith's assumptions about the whole and the part. "Adam Smith," Hayek wrote, "was the first to perceive that we have stumbled upon methods of ordering human economic cooperation that exceed the limits of our knowledge and perception. His 'invisible hand' had perhaps better have been described as an invisible or unsurveyable pattern."[2]

As with Planck, Meyer, and Skinner, both Hayek and Smith unequivocally link freedom and ignorance. In Hayek's framing, the mystery of the market is that a great many people can behave effectively while remaining ignorant of the whole. Individuals not only *can* choose freely, but they *must* freely choose their own pursuits because there is no alternative, no source of total knowledge or conscious control to guide them. "Human design" is impossible, Hayek says, because the relevant information flows are "beyond the span of the control of any one mind." The market dynamic makes it possible for people to operate in ignorance without "anyone having to tell them what to do."[3]

Hayek chose the market over democracy, arguing that the market system enabled not only the division of labor but also "the coordinated utilization of resources based on *equally divided knowledge*." This system, he argued, is the only one compatible with freedom. Perhaps some other kind of civilization might have been devised, he reckoned, "like the 'state' of the termite ants," but it would not be compatible with human freedom.[4]

Something is awry. It is true that many capitalists, including surveillance capitalists, vigorously employ these centuries-old justifications for their freedom when they reject regulatory, legislative, judicial, societal, or any other form of public interference in their methods of operation. However, Big Other and the steady application of instrumentarian power challenge the classic quid pro quo of freedom for ignorance.

When it comes to surveillance capitalist operations, the "market" is no longer invisible, certainly not in the way that Smith or Hayek imagined. The competitive struggle among surveillance capitalists produces the compulsion toward totality. Total information tends toward certainty and the promise of guaranteed outcomes. These operations mean that the supply and demand of behavioral futures markets are rendered in infinite detail. Surveillance capitalism thus replaces mystery with certainty as it substitutes rendition, behavioral modification, and prediction for the old "unsurveyable pattern." This is a fundamental reversal of the classic ideal of the "market" as intrinsically unknowable.

Recall Mark Zuckerberg's boast that Facebook would know every book, film, and song a person had ever consumed and that its predictive models would tell you what bar to go to when you arrive in a strange city, where the bartender would have your favorite drink waiting.[5] As the head of Facebook's data science team once reflected, "This is the first time the world has seen this scale and quality of data about human communication.... For the first time, we have a microscope that...lets us examine social behavior at a very fine level that we've never been able to see before...."[6] A top Facebook engineer put it succinctly: "We are trying to map out the graph of everything in the world and how it relates to each other."[7]

The same objectives are echoed in the other leading surveillance capitalist firms. As Google's Eric Schmidt observed in 2010, "You give us more information about you, about your friends, and we can improve the quality of our searches. We don't need you to type at all. We know where you are. We know where you've been. We can more or less know what you're thinking about."[8] Satya Nadella of Microsoft understands all physical and institutional spaces, people, and social relationships as indexable and searchable: all of it subject to machine reasoning, pattern recognition, prediction, preemption, interruption, and modification.[9]

Surveillance capitalism is not the old capitalism, and its leaders are not Smith's or even Hayek's capitalists. Under this regime, freedom and ignorance are no longer twin born, no longer two sides of the same coin called mystery. Surveillance capitalism is instead defined by an unprecedented convergence of freedom *and* knowledge. The degree of that convergence corresponds exactly to the scope of instrumentarian power. This unimpeded accumulation of power effectively hijacks the division of learning in society, instituting the dynamics of inclusion and exclusion upon which surveillance revenues depend. Surveillance capitalists claim the freedom to order knowledge, and then they leverage that knowledge advantage in order to protect and expand their freedom.

Although there is nothing unusual about the prospect of capitalist enterprises seeking every kind of knowledge advantage in a competitive marketplace, the surveillance capitalist capabilities that translate ignorance into knowledge are unprecedented because they rely on the one resource that distinguishes the surveillance capitalists from traditional utopianists: the financial and intellectual capital that permits the actual transformation of the world, materialized in the continuously expanding architectures of Big Other. More astonishing still is that surveillance capital derives from the dispossession of human experience, operationalized in its unilateral and pervasive programs of rendition: *our lives are scraped and sold to fund their freedom and our subjugation, their knowledge and our ignorance about what they know.*

This new condition unravels the neoliberal justification for the evisceration of the double movement and the triumph of raw capitalism: its free markets, free-market actors, and self-regulating enterprises. It suggests that surveillance capitalists mastered the rhetoric and political genius of the neoliberal ideological defense while pursuing a novel logic of accumulation that belies the most fundamental postulates of the capitalist worldview. It's not just that the cards have been reshuffled; the rules of the game have been transformed into something that is both unprecedented and unimaginable outside the digital milieu and the vast resources of wealth and scientific prowess that the new applied utopianists bring to the table.

We have carefully examined surveillance capitalism's novel foundational mechanisms, economic imperatives, gathering power, and societal objectives. One conclusion of our investigations is that surveillance capitalism's command and control of the division of learning in society are the signature feature that breaks with the old justifications of the invisible hand and its entitlements. The combination of knowledge and freedom works to accelerate the asymmetry of power between surveillance capitalists and the societies in which they operate. This cycle will be broken only when we acknowledge as citizens, as societies, and indeed as a civilization that *surveillance capitalists know too much to qualify for freedom.*

II. After Reciprocity

In another decisive break with capitalism's past, surveillance capitalists abandon the organic reciprocities with people that have long been a mark of capitalism's endurance and adaptability. Symbolized in the twentieth century by Ford's five-dollar day, these reciprocities hearken back to Adam Smith's original insights into the productive social relations of capitalism, in which firms rely on people as employees and customers. Smith argued that price increases had to be balanced with wage increases "so that the labourer may still be able to purchase that quantity of those necessary articles which the state of the demand for labour... requires that he should have."[10] The shareholder-value movement and globalization went a long way toward destroying this centuries-old social contract between capitalism and its communities, substituting formal indifference for reciprocity. Surveillance capitalism goes further. It not only jettisons Smith, but it also formally rescinds any remaining reciprocities with its societies.

First, surveillance capitalists no longer rely on people as consumers. Instead, the axis of supply and demand orients the surveillance capitalist firm to businesses intent on anticipating the behavior of populations, groups, and individuals. The result, as we have seen, is that "users" are sources of raw material for a digital-age production process aimed at a new business customer. Where individual consumers continue to exist in surveillance capitalist operations—purchasing Roomba vacuum cleaners, dolls that spy, smart vodka bottles, or behavior-based insurance policies, just to name a few examples—social relations are no longer founded on mutual exchange. In these and many other instances, products and services are merely hosts for surveillance capitalism's parasitic operations.

Second, by historical standards the large surveillance capitalists employ relatively few people compared to their unprecedented computational resources. This pattern, in which a small, highly educated workforce leverages the power of a massive capital-intensive infrastructure, is called "hyperscale." The historical discontinuity of the hyperscale business operation becomes apparent by comparing seven decades of GM employment levels and market capitalization to recent post-IPO data from Google and Facebook. (I have confined the comparison here to Google and Facebook because both were pure surveillance capitalist firms even before their public offerings.)

From the time they went public to 2016, Google and Facebook steadily climbed to the heights of market capitalization, with Google reaching $532 billion by the end of 2016 and Facebook reaching $332 billion, without Google ever employing more than 75,000 people or Facebook more than 18,000. General Motors took four decades to reach its highest market capitalization of $225.15 billion in 1965, when it employed 735,000 women and men.[11] Most startling is that GM employed more people during the height of the Great Depression than either Google or Facebook employs at their heights of market capitalization.

The GM pattern is the iconic story of the United States in the twentieth century, before globalization, neoliberalism, the shareholder-value movement, and plutocracy unraveled the public corporation and the institutions of the double movement. Those institutions rationalized GM's employment policies with fair labor practices, unionization, and collective bargaining, emblematic of stable reciprocities during the pre-globalization decades of the twentieth century. In the 1950s, for example, 80 percent of adults said that "big business" was a good thing for the country, 66 percent believed that business required little or no change, and 60 percent agreed that "the profits of large companies help make things better for everyone who buys their products or services."[12]

Although some critics blamed these reciprocities for GM's failure to adapt to global competition in the late 1980s, leading eventually to its bankruptcy in 2009, analyses have shown that chronic managerial complacency and doomed financial strategies bore the greatest share of responsibility for the firm's legendary decline, a conclusion that is fortified by the successes of the German automobile industry in the twenty-first century, where strong labor institutions formally share decision making authority.[13]

Hyperscale firms have become emblematic of modern digital capitalism, and as capitalist inventions they present significant social and economic challenges, including their impact on employment and wages, industry concentration, and monopoly.[14] In 2017, 24 hyperscale firms operated 320 data centers with anywhere between thousands and millions of servers (Google and Facebook were among the largest).[15]

Not all hyperscale firms are surveillance capitalists, however, and our focus here is restricted to the convergence of these two domains. The surveillance capitalists that operate at hyperscale or outsource to hyperscale operations dramatically diminish any reliance on their societies as sources of employees, and the few for whom they do compete, as we have seen, are drawn from the most-rarefied strata of data science.

The absence of organic reciprocities with people as either sources of consumers or employees is a matter of exceptional importance in light of the historical relationship between market capitalism and democracy. In fact, the origins of democracy in both America and Britain have been traced to these very reciprocities. In America the violation of consumer reciprocities awakened an unstoppable march toward liberty as economic power translated into political power. A half century later in Britain, a grudging, practical, self-interested respect for the necessary interdependence of capital and labor translated into new patterns of political power, expressed in the gradual expansion of the franchise and the nonviolent shift to more-inclusive democratic institutions. Even a brief glance at these world-altering histories can help us grasp the degree to which surveillance capitalism diverges from capitalism's past.

The American Revolution is the outstanding example of how the reciprocities of consumption contributed to the rise of democracy. Historian T. H. Breen argues in his pathbreaking study *The Marketplace of Revolution* that it was the violation of these reciprocities that set the Revolution into motion, uniting disparate provincial strangers into a radical new patriotic force. Breen explains that American colonists had come to depend upon the "empire of goods" imported from England and that this dependency instilled the sense of a reciprocal social contract: "For ordinary people, the palpable experience of participating in an expanding Anglo-American consumer market" intensified their sense of a "genuine partnership" with England.[16] Eventually, the British Parliament famously misjudged the rights and obligations of this partnership, imposing a series of taxes that turned imported goods such as cloth and tea into "symbols of imperial oppression." Breen describes the originality of a political movement born in the shared experience of consumption, the outrage at the violation of essential producer-consumer interdependencies, and the determination to make "goods speak to power."

The translation of consumer expectations into democratic revolution occurred in three waves, beginning in 1765, when the Stamp Act triggered popular protests, riots, and organized resistance finally expressed in the "nonimportation movement" (today we would call it a consumer boycott). As Breen tells it, the details of the Stamp Act were less important than the colonists' realization that England did not perceive them as political or economic equals bound in mutually beneficial reciprocities: "By compromising the Americans' ability to purchase the goods they desired, Parliament had revealed an intention to treat the colonists like second-class subjects," levying a heavy price "on the pursuit of material happiness."[17] The Stamp Act was experienced as a violation of the colonists' rights not only as subjects of the empire but also as consumers of the empire: it was the first translation of consumers' economic power into political power, a "radically new form of politics" in which the most ordinary members of colonial society experienced "an exhilarating surge of empowerment."[18] Parliament withdrew the Stamp Act before the nonimportation movement could effectively spread across the colonies, and it appeared that the principle of "no taxation without represen-tation" had prevailed.

When the Townshend Acts were passed just two years later, in 1767, this time imposing taxes on a range of imported goods, a new wave of out-rage mobilized people in every colony. Detailed nonimportation agreements turned consumer sacrifice into the front line of political resistance. The shared experience of violated expectations cut across regional, religious, and cultural differences, providing a new basis for social solidarity.[19] By 1770, the Townshend Acts were also repealed, and once again it seemed that a full-blown rebellion would be avoided.

The 1773 Tea Act plunged the colonies into a new phase of resistance that shifted the political focus from nonimportation, which depended upon merchants holding the line, to *nonconsumption,* which demanded the par-ticipation of all individuals in the unique solidarity of their shared status as "customers." It was in this context that Samuel Adams proclaimed that the cause of liberty "depended on the ability of the American people to free themselves from 'the Baubles of Britain.'"[20]

British goods had so thoroughly come to symbolize dependency and op-pression that when the tiny impoverished community of Harvard, Massa-chusetts, gathered to discuss the merchant vessels arriving in Boston Harbor loaded with chests of tea, they deemed it "a matter of as interesting and im-portant a nature when viewed in all its Consequences not only to this Town and Province, but to America in general, and that for ages and generations to come, as ever came under the deliberation of this Town."[21]

A year later, in 1774, the First Continental Congress convened in Phil-adelphia and produced a "grand scheme" to abolish trade with England. "It brought to fruition a brilliantly original strategy of consumer resistance to political oppression," Breen writes, "one that had invited Americans to think of themselves as Americans even before they entertained a thought of independence."[22]

In early-nineteenth-century Britain, as Daron Acemoglu and James A. Robinson have shown, the rise of democracy was inextricably bound to in-dustrial capitalism's dependency on the "masses" and their contribution to the prosperity necessitated by the new organization of production.[23] The rise of volume production and its wage-earning labor force established British workers' economic power and led to a growing appreciation of their political legitimacy and power. This produced a new sense of interdependence between ordinary people and elites.

Acemoglu and Robinson conclude that the "dynamic positive feedback" between "inclusive economic institutions" (i.e., industrial firms defined by employment reciprocities) and political institutions was critical to Britain's substantial and nonviolent democratic reforms. Inclusive economic institutions, they argue, "level the playing field," especially when it comes to the fight for power, making it more difficult for elites to "crush the masses" rather than accede to their demands. Reciprocities in employment produced and sustained reciprocities in politics: "Clamping down on popular demands and undertaking a coup against inclusive political institutions would...destroy...[economic] gains, and the elites opposing greater democratization and greater inclusiveness might find themselves among those losing their fortunes from this destruction."[24] In sharp contrast to the pragmatic concessions of Britain's early industrial capitalists, surveillance capitalists' extreme structural independence from people breeds exclusion rather than inclusion and lays the foundation for the unique approach that we have called "radical indifference."

III. The New Collectivism and Its Masters of Radical Indifference

The accumulation of freedom *and* knowledge combines with the lack of organic reciprocities with people to shape a third unusual feature of surveillance capitalism: a collectivist orientation that diverges from the long-standing values of market capitalism and market democracy, while also sharply departing from surveillance capitalism's origins in the neoliberal worldview. For the sake of its own commercial success, surveillance capitalism aims us toward the hive collective. This privatized instrumentarian social order is a new form of collectivism in which it is the market, not the state, which concentrates both knowledge and freedom within its domain.

This collectivist orientation is an unexpected development in light of surveillance capitalism's origins in a neoliberal creed conceived sixty years ago as a reaction to the collectivist totalitarian nightmares of the mid-twentieth century. Later, with the demise of the fascist and socialist threats, neoliberal ideology cunningly succeeded in redefining the modern democratic state as a fresh source of collectivism to be resisted by any and all means. Indeed, the evisceration of the double movement has been prosecuted in the name of defeating the supposed collectivist hazards of "too much democracy."[25] Now the hive emulates the "termite state," which even the democracy-despising Hayek derided as incompatible with human freedom.

The convergence of freedom and knowledge transforms surveillance capitalists into society's self-appointed masters. From their high perch in the division of learning, a privileged priesthood of "tuners" rules the connected hive, cultivating it as a source of continuous raw-material supply. Just as early-twentieth-century managers were once taught the "administrative point of view" as the mode of knowledge required for the hierarchical complexities of the new large-scale corporation, today's high priests practice the applied arts of radical indifference, a fundamentally asocial mode of knowledge. With the application of radical indifference, content is judged by its volume, range, and depth of surplus as measured by the "anonymous" equivalence of clicks, likes, and dwell times, despite the obvious fact that its profoundly dissimilar meanings originate in distinct human situations.

Radical indifference is a response to economic imperatives, and only occasionally do we catch an unobstructed view of its strict application as a managerial discipline. One such occasion was a 2016 internal Facebook memo acquired by *BuzzFeed* in 2018. Written by one of the company's long-standing and most influential executives, Andrew Bosworth, it provided a window into radical indifference as an applied discipline. "We talk about the good and the bad of our work often. I want to talk about the ugly," Bosworth began. He went on to explain how equivalence wins out over equality in the worldview of "an organism among organisms" that is essential to the march toward totality and thus the growth of surveillance revenues:

> We connect people. That can be good if they make it positive. Maybe someone finds love. Maybe it even saves the life of someone on the brink of suicide. So we connect more people. That can be bad if they make it negative. Maybe it costs a life by exposing someone to bullies.

> Maybe someone dies in a terrorist attack coordinated on our tools.
> And still we connect people. The ugly truth is that…anything that
> allows us to connect more people more often is de facto good. It is
> perhaps the only area where the metrics do tell the true story as far as
> we are concerned.…That's why all the work we do in growth is justi-
> fied. All the questionable contact importing practices. All the subtle
> language that helps people stay searchable by friends. All of the work
> we do to bring more communication in.…The best products don't
> win. The ones everyone uses win…make no mistake, growth tactics
> are how we got here.[26]

As Bosworth makes clear, from the viewpoint of radical indifference the positives and negatives must be viewed as equivalent, despite their distinct moral meanings and human consequences. From this perspective the only rational objective is the pursuit of products that snare "everyone," not "the best products."

A significant result of the systematic application of radical indifference is that the public-facing "first text" is vulnerable to corruption with content that would normally be perceived as repugnant: lies, systematic disinformation, fraud, violence, hate speech, and so on. As long as content contributes to "growth tactics," Facebook "wins." This vulnerability can be an explosive problem on the demand side, the user side, but it breaks through the fortifications of radical indifference only when it threatens to interrupt the flow of surplus into the second "shadow" text: the one that is for them but not for us. The norm is that information corruption is not catalogued as problematic unless it poses an existential threat to supply operations—Bosworth's imperative of connection—either because it might trigger user disengagement or because it might attract regulatory scrutiny. This means that any efforts toward "content moderation" are best understood as defensive measures, not as acts of public responsibility.

So far, the greatest challenge to radical indifference has come from Facebook and Google's overreaching ambitions to supplant professional journalism on the internet. Both corporations inserted themselves between publishers and their populations, subjecting journalistic "content" to the same categories of equivalence that dominate surveillance capitalism's other landscapes. In a formal sense, professional journalism is the precise opposite of radical indifference. The journalist's job is to produce news and analysis that separate truth from falsehood. This rejection of equivalence defines journalism's raison d'être as well as its organic reciprocities with its readers. Under surveillance capitalism, though, these reciprocities are erased. A consequential example was Facebook's decision to standardize the presentation of its News Feed content so that "all news stories looked roughly the same as each other... whether they were investigations in *The Washington Post*, gossip in the *New York Post*, or flat-out lies in the *Denver Guardian*, an entirely bogus newspaper."[27] This expression of equivalence without equality made Facebook's first text exceptionally vulnerable to corruption from what would come to be called "fake news."

This is the context in which Facebook and Google became the focus of international attention following the discovery of organized political disinformation campaigns and profit-driven "fake news" stories during the 2016 US presidential election and the UK Brexit vote earlier that year. Economists Hunt Allcott and Matthew Gentzkow, who have studied these phenomena in detail, define "fake news" as "distorted signals uncorrelated with the truth" that impose "private and social costs by making it more difficult... to infer the true state of the world...." They found that in the lead-up to the 2016 US election there were 760 million instances of a user reading these intentionally orchestrated lies online, or about three such stories for each adult American.[28]

As radical indifference would predict, however, "fake news" and other forms of information corruption have been perennial features of Google and Facebook's online environments. There are countless examples of disinformation that survived and even thrived because it fulfilled economic imperatives, and I point out just a few. In 2007 a prominent financial analyst worried that the subprime mortgage bust would harm Google's lucrative ad business. It seems a strange observation until you learn that in the years prior to the Great Recession, Google eagerly welcomed shady subprime lenders into its behavioral futures markets, anxious to net the lion's share of the $200 million in monthly revenue that mortgage lenders were spending on online advertising.[29] A 2011 Consumer Watchdog report on Google's advertising practices leading up to and during the Great Recession concluded that "Google has been a prominent beneficiary of the national home loan and foreclosure crisis...by accepting deceptive advertising from fraudulent operators who falsely promise unwary consumers that they can solve their mortgage and credit problems." Despite these increasingly public facts, Google continued to serve its fraudulent business customers until 2011, when the US Treasury Department finally required the company to suspend advertising relationships with "more than 500 internet advertisers associated with the 85 alleged online mortgage fraud schemes and related deceptive advertising."[30]

Only a few months earlier, the Department of Justice had fined Google $500 million, "one of the largest financial forfeiture penalties in history," for accepting ads from online Canadian pharmacies that encouraged Google's US users to illegally import controlled drugs, despite repeated warnings. As the US Deputy Attorney General told the press, "The Department of Justice will continue to hold accountable companies who in their bid for profits violate federal law and put at risk the health and safety of American consumers."[31]

Information corruption has also been a continuous feature of the Facebook environment. The turmoil associated with the 2016 US and UK political disinformation campaigns on Facebook was a well-known problem that had disfigured elections and social discourse in Indonesia, the Philippines, Columbia, Germany, Spain, Italy, Chad, Uganda, Finland, Sweden, Holland, Estonia, and the Ukraine. Scholars and political analysts had called attention to the harmful consequences of online disinformation for years.[32] One political analyst in the Philippines worried in 2017 that it might be too late to fix the problem: "We already saw the warning signs of this years ago....Voices that were lurking in the shadows are now at the center of the public discourse."[33]

The guiding principles of radical indifference are reflected in the operations of Facebook's hidden low-wage labor force charged with limiting the perversion of the first text. Nowhere is surveillance capitalism's outsized influence over the division of learning in society more concretely displayed than in this outcast function of "content moderation," and nowhere is the nexus of economic imperatives and the division of learning more vividly exposed than in the daily banalities of these rationalized work flows where the world's horrors and hate are assigned to life or death at a pace and volume that leave just moments to point thumbs up or down. It is only thanks to the determined reporting of a handful of investigative journalists and research scholars that we even have a glimpse of these highly secretive procedures, which now spread across a range of call centers, boutique firms, and "micro-labor" sites around the world. As one account notes, "Facebook and Pinterest, along with Twitter, Reddit, and Google, all declined to provide copies of their past or current internal moderation policy guidelines."[34]

Among the few reports that have managed to assess Facebook's operations, the theme is consistent. This secret workforce—some estimates reckon at least 100,000 "content moderators," and others calculate the number to be much higher—operates at a distance from the corporation's core functions, applying a combination of human judgment and machine learning tools.[35] Sometimes referred to as "janitors," they review queues of content that users have flagged as problematic. Although some general rules apply across the board, such as eliminating pornography and images of child abuse, a detailed rulebook aims to reject as little content as possible in the context of a local assessment of the minimum threshold of user tolerance. The larger point of the exercise is to find the point of equilibrium between the ability to pull users and their surplus into the site and the risk of repelling them. This is a calculation of radical indifference that has nothing to do with assessing the truthfulness of content or respecting reciprocities with users.[36] This tension helps to explain why disinformation is not a priority. One investigative report quotes a Facebook insider: "They absolutely have the tools to shut down fake news...."[37]

That radical indifference produces equivalence without equality also affects the high science of targeted advertising. For example, journalist Julia Angwin and her colleagues at *ProPublica* discovered that Facebook "enabled advertisers to direct their pitches to the news feeds of almost 2,300 people who expressed interest in the topics of 'Jew hater,' 'How to burn jews,' or 'History of why jews ruin the world.'"[38] As the journalists explained, "Facebook has long taken a hands-off approach to its advertising business.... Facebook generates its ad categories automatically based both on what users explicitly share with Facebook and what they implicitly convey through their online activity." Similarly, reporters at *BuzzFeed* discovered that Google enables advertisers to target ads to people who type racist terms into the search bar and even suggests ad placements next to searches for "evil jew" and "Jewish control of banks."[39]

In the 2017 postelection environments in the United States and the United Kingdom, as "fake news" dominated the spotlight, journalists discovered hundreds of examples in which prediction products had placed ads from legitimate brands, such as Verizon, AT&T, and Walmart, alongside heinous material, including disinformation sites, hate speech, extreme political content, and terrorist, racist, and anti-Semitic publications and videos.[40]

Most interesting was the assumed outrage and disbelief among surveillance capitalism's customers: the advertising agencies and their clients who long ago chose to sell their souls to radical indifference, turning Google and Facebook into the duopoly of the online ad market and driving the massive expansion of surveillance capitalism.[41] It had been nearly two decades since Google invented the formula that ceded ad placement to the equivalency metrics of click-through rates, supplanting earlier approaches that sought to align advertising placements with content that reflected the advertiser's brand values. Customers forfeited those established reciprocities in favor of the "auto magic" of Google's secret algorithms trained on proprietary behavioral surplus culled from unwitting users. Indeed, it was the radical indifference of click metrics that bred online displays of extremism and sensationalism in the first place, as prediction products favor content designed to magnetize engagement.

The election scandals shined a harsh spotlight on these settled practices to which the world had already become accustomed. In the heat of controversy, many top brands made a show of suspending their ads on Google and Facebook until the companies eliminated corrupt content or guaranteed acceptable ad placements. Politicians in Europe and the United States accused Google and Facebook of profiting from hatred and of weakening democracy with corrupt information. Initially, both companies seemed to assume that the noise would quickly fade. Mark Zuckerberg said it was "crazy"[42] to think that fake news had influenced the elections. Google responded to its advertising customers with vague platitudes, offering little in the way of change.

This was not the first time that the leading surveillance capitalists had been called to account by public and press.[43] In addition to the many cycles of outrage generated by Street View, Beacon, Gmail, Google Glass, News Feed, and other incursions, Edward Snowden's 2013 revelations of the tech companies' collusion with state intelligence agencies triggered an international eruption of loathing toward the surveillance capitalists. Google and Facebook learned to weather these storms with what I have called the "dispossession cycle," and close observation of this new crisis suggested that a fresh cycle was in full throttle. As the threat of regulatory oversight grew, the adaptation phase of the cycle set in with a vengeance. There were public apologies, acts of contrition, attempts at mollification, and appearances before the US Congress and the EU Parliament.[44] Zuckerberg "regretted" his "dismissive" attitude and prayed for forgiveness on the Jewish Day of Atonement, Yom Kippur.[45] Sheryl Sandberg told *ProPublica* that "we never intended or anticipated this functionality being used this way...."[46] Facebook conceded that it could do more to combat online extremism.[47] Google's European chief told customers, "We apologise. Whenever anything like that happens, we don't want it to happen and we take responsibility for it."[48]

Consistent with the aims of the adaptation phase of the cycle, *Bloomberg Businessweek* observed of Google, "The company is trying to fight fake news without making sweeping changes."[49] Although both Google and Facebook made modest operational adjustments to try to diminish economic incentives for disinformation and instituted warning systems to alert users to probable corruption, Zuckerberg also used his super-voting power to reject a shareholder proposal that would have required the company to report on its management of disinformation and the societal consequences of its practices, and Google executives successfully fought back a similar shareholder proposal that year.[50] Time would tell whether the companies' users and customers would inflict financial punishment, and if so, how sustained that punishment might be.

By early 2018, a quiet shift from adaptation to redirection at Facebook was already poised to transform crisis into opportunity. "Despite facing important challenges...we also need to keep building new tools to help people connect, strengthen our communities, and bring the world closer together," Zuckerberg told investors.[51] A Zuckerberg post followed up by a statement from the head of the company's News Feed declared that henceforth the

News Feed would favor posts from friends and family, especially posts that "spark conversations and meaningful interactions between people...we will predict which posts you might want to interact with your friends about.... These are posts that inspire back-and-forth discussion...whether that's a post from a friend seeking advice...a news article or video prompting lots of discussion.... Live videos often lead to discussion among viewers...six times as many interactions as regular videos."[52]

Radical indifference means that it doesn't matter what is in the pipelines as long as they are full and flowing. Camouflaged as a retreat from corruption, the new strategy doubled down on activities rich in behavioral surplus, especially the live videos that Zuckerberg had long coveted. In a *New York Times* report, advertisers were quick to observe that the new rules would fuel Facebook's "'long-held' video ambitions" and that the company had made clear its belief that its future lay in videos and video ads. One advertising executive commented that video content is "among the most shared and commented-upon content on the web."[53]

Beyond all the explanations for the scourge of disinformation in the surveillance capitalist online environment is a deeper and more intransigent fact: radical indifference is a permanent invitation to the corruption of the first text. It sustains the pathological division of learning in society by forfeiting the integrity of public knowledge for the sake of the volume and scope of the shadow text. Radical indifference leaves a void where reciprocities once thrived. For all their freedom and knowledge, this is one void that surveillance capitalists will not fill because doing so would violate their own logic of accumulation. It is obvious that the rogue forces of disinformation grasp this fact more crisply than do Facebook's or Google's genuine users and customers as those forces learn to exploit the blind eye of radical indifference and escalate the perversion of learning in an open society.

IV. What Is Surveillance Capitalism?

Surveillance capitalism's successful claims to freedom *and* knowledge, its structural independence from people, its collectivist ambitions, and the radical indifference that is necessitated, enabled, and sustained by all three now propel us toward a society in which capitalism does not function as a means to inclusive economic or political institutions. Instead, surveillance capitalism must be reckoned as a profoundly antidemocratic social force. The reasoning I employ is not mine alone. It echoes Thomas Paine's unyielding defense of the democratic prospect in *The Rights of Man*, the polemical masterpiece in which he contested the defense of monarchy in Edmund Burke's *Reflections on the Revolution in France*. Paine argued for the capabilities of the common person and against aristocratic privilege. Among his reasons to reject aristocratic rule was its lack of accountability to the needs of people, "because a body of men holding themselves accountable to nobody, ought not to be trusted by any body."[54]

Surveillance capitalism's antidemocratic and antiegalitarian juggernaut is best described as a market-driven coup from above. It is not a coup d'état in the classic sense but rather a *coup de gens*: an overthrow of the people concealed as the technological Trojan horse that is Big Other. On the strength of its annexation of human experience, this coup achieves exclusive concentrations of knowledge and power that sustain privileged influence over the division of learning in society: the privatization of the central principle of social ordering in the twenty-first century. Like the *adelantados* and their silent incantations of the *Requirimiento*, surveillance capitalism operates in the declarative form and imposes the social relations of a premodern absolutist authority. It is a form of tyranny that feeds on people but is not of the people. In a surreal paradox, this coup is celebrated as "personalization," although it defiles, ignores, overrides, and displaces everything about you and me that is personal.

"Tyranny" is not a word that I choose lightly. Like the instrumentarian hive, tyranny is the obliteration of politics. It is founded on its own strain of radical indifference in which every person, except the tyrant, is understood as an organism among organisms in an equivalency of Other-Ones. Hannah Arendt observed that tyranny is a perversion of egalitarianism because it treats all others as equally insignificant: "The tyrant rules in accordance with his own will and interest... the ruler who rules one against all, and the 'all' he oppresses are all equal, namely equally powerless." Arendt notes that classical political theory regarded the tyrant as "out of mankind altogether... a wolf in human shape...."[55]

Surveillance capitalism rules by instrumentarian power through its materialization in Big Other, which, like the ancient tyrant, exists out of mankind while paradoxically assuming human shape. Surveillance capitalism's tyranny does not require the despot's whip any more than it requires totalitarianism's camps and gulags. All that is needed can be found in Big Other's reassuring messages and emoticons, the press of the others not in terror but in their irresistible inducements to confluence, the weave of your shirt saturated with sensors, the gentle voice that answers your queries, the TV that hears you, the house that knows you, the bed that welcomes your whispers, the book that reads you.... Big Other acts on behalf of an unprecedented assembly of commercial operations that must modify human behavior as a condition of commercial success. It replaces legitimate contract, the rule of law, politics, and social trust with a new form of sovereignty and its privately administered regime of reinforcements.

Surveillance capitalism is a boundary-less form that ignores older distinctions between market and society, market and world, or market and person. It is a profit-seeking form in which production is subordinated to extraction as surveillance capitalists unilaterally claim control over human, societal, and political territories extending far beyond the conventional institutional terrain of the private firm or the market. Using Karl Polanyi's lens, we see that surveillance capitalism annexes human experience to the market dynamic so that it is reborn as behavior: the fourth "fictional commodity." Polanyi's first three fictional commodities—land, labor, and money—were subjected to law. Although these laws have been imperfect, the institutions of labor law, environmental law, and banking law are regulatory frameworks intended to defend society (and nature, life, and exchange) from the worst excesses of raw capitalism's destructive power. Surveillance capitalism's expropriation of human experience has faced no such impediments.

The success of this *coup de gens* stands as sour testimony to the thwarted needs of the second modernity, which enabled surveillance capitalism to flourish and still remains its richest vein for extraction and exploitation. In this context it is not difficult to understand why Facebook's Mark Zuckerberg offers his social network as *the* solution to the third modernity. He envisions a totalizing instrumentarian order—he calls it the new global "church"—that will connect the world's people to "something greater than ourselves." It will be Facebook, he says, that will address problems that are civilizational in scale and scope, building "the long-term infrastructure to bring humanity together" and keeping people safe with "artificial intelligence" that quickly understands "what is happening across our community."[56] Like Pentland, Zuckerberg imagines machine intelligence that can "identify risks that nobody would have flagged at all, including terrorists planning attacks using private channels, people bullying someone too afraid to report it themselves, and other issues both local and global."[57] When asked about his responsibility to shareholders, Zuckerberg told CNN, "That's why it helps to have control of the company."[58]

For more than three centuries, industrial civilization aimed to exert control over nature for the sake of human betterment. Machines were our means of extending and overcoming the limits of the animal body so that we could accomplish this aim of domination. Only later did we begin to fathom the consequences: the Earth overwhelmed in peril as the delicate physical systems that once defined sea and sky gyrated out of control.

Right now we are at the beginning of a new arc that I have called information civilization, and it repeats the same dangerous arrogance. The aim now is not to dominate *nature* but rather *human nature*. The focus has shifted from machines that overcome the limits of bodies to machines that modify the behavior of individuals, groups, and populations in the service of market objectives. This global installation of instrumentarian power overcomes and replaces the human inwardness that feeds the will to will and gives sustenance to our voices in the first person, incapacitating democracy at its roots.

The rise of instrumentarian power is intended as a bloodless coup, of course. Instead of violence directed at our bodies, the instrumentarian third modernity operates more like a taming. Its solution to the increasingly clamorous demands for effective life pivots on the gradual elimination of chaos, uncertainty, conflict, abnormality, and discord in favor of predictability, automatic regularity, transparency, confluence, persuasion, and pacification. We are expected to cede our authority, relax our concerns, quiet our voices, go with the flow, and submit to the technological visionaries whose wealth and power stand as assurance of their superior judgment. It is assumed that we will accede to a future of less personal control and more powerlessness, where new sources of inequality divide and subdue, where some of us are subjects and many are objects, some are stimulus and many are response.

The compulsions of this new vision threaten other delicate systems also many millennia in the making, but this time they are social and psychological. I am thinking here of the hard-won fruits of human suffering and conflict that we call the democratic prospect and the achievements of the individual as a source of autonomous moral judgment. Technological "inevitability" is the mantra on which we are trained, but it is an existential narcotic prescribed to induce resignation: a snuff dream of the spirit.

We've been alerted to the "sixth extinction" as vertebrate species disappear faster than at any time since the end of the dinosaurs. This cataclysm is the unintended consequence of the reckless and opportunistic methods, also exalted as inevitable, with which industrialization imposed itself on the natural world because its own market forms did not hold it to account. Now the rise of instrumentarian power as the signature expression of surveillance capitalism augurs a different kind of extinction. This "seventh extinction" will not be of nature but of what has been held most precious in human nature: the will to will, the sanctity of the individual, the ties of intimacy, the sociality that binds us together in promises, and the trust they breed. The dying off of this human future will be just as unintended as any other.

V. Surveillance Capitalism and Democracy

Instrumentarian power has gathered strength outside of mankind but also outside of democracy. There can be no law to protect us from the unprecedented, and democratic societies, like the innocent world of the Taínos, are vulnerable to unprecedented power. In this way, surveillance capitalism may be viewed as part of an alarming global drift toward what many political scientists now view as a softening of public attitudes toward the necessity and inviolability of democracy itself.

Many scholars point to a global "democratic recession" or a "deconsolidation" of Western democracies that were long considered impervious to antidemocratic threats.[59] The extent and precise nature of this threat are being debated, but observers describe the bitter *saudade* associated with rapid social change and fear of the future conveyed in the lament "My children will not see the life that I lived."[60] Such feelings of alienation and unease were expressed by many people around the world in a thirty-eight-nation survey published by Pew Research in late 2017. The findings suggest that the democratic ideal is no longer a sacred imperative, even for citizens of mature democratic societies. Although 78 percent of respondents say that representative democracy is "good," 49 percent also say that "rule by experts" is good, 26 percent endorse "rule by a strong leader," and 24 percent prefer "rule by the military."[61]

The weakening attachment to democracy in the United States and many European countries is of serious concern.[62] According to the Pew survey, only 40 percent of US respondents support democracy and *simultaneously* reject the alternatives. A full 46 percent find both democratic and nondemocratic alternatives to be acceptable, and 7 percent favor only the nondemocratic choice. The US sample trails Sweden, Germany, the Netherlands, Greece, and Canada in its depth of commitment to democracy, but other key Western democracies, including Italy, the UK, France, and Spain, along with Poland and Hungary, fall at or below the thirty-eight-country median of 37 percent that are exclusively committed to democracy.

Many have concluded from this turmoil that market democracy is no longer viable, despite the fact that the combination of markets and democracy has served humanity well, helping to lift much of humankind from millennia of ignorance, poverty, and pain. For some of these thinkers it is the markets that must go, and for others it is democracy that's slated for obsolescence. Repulsed by the social degradation and climate chaos produced by nearly four decades of neoliberal policy and practice, an important and varied group of scholars and activists argues that the era of capitalism is at end. Some propose more-humane economic alternatives,[63] some anticipate protracted decline,[64] and others, repelled by social complexity, favor a blend of elite power and authoritarian politics in closer emulation of China's authoritarian system.[65]

These developments alert us to a deeper truth: just as capitalism cannot be eaten raw, people cannot live without the felt possibility of homecoming.

Hannah Arendt explored this territory more than sixty years ago in *The Origins of Totalitarianism,* where she traced the path from a thwarted individuality to a totalizing ideology. It was the individual's experience of insignificance, expendability, political isolation, and loneliness that stoked the fires of totalitarian terror. Such ideologies, Arendt observed, appear as "a last support in a world where nobody is reliable and nothing can be relied upon."[66] Years later, in his moving 1966 essay "Education after Auschwitz," social theorist Theodor Adorno attributed the success of German fascism to the ways in which the quest for effective life had become an overwhelming burden for too many people: "One must accept that fascism and the terror it caused are connected with the fact that the old established authorities... decayed and were toppled, while the people psychologically were not yet ready for self-determination. They proved to be unequal to the freedom that fell into their laps."[67]

Should we grow weary of our own struggle for self-determination and surrender instead to the seductions of Big Other, we will inadvertently trade a future of homecoming for an arid prospect of muted, sanitized tyranny. A third modernity that solves our problems at the price of a human future is a cruel perversion of capitalism and of the digital capabilities it commands. It is also an unacceptable affront to democracy. I repeat Thomas Piketty's warning: "A market economy... if left to itself... contains powerful forces of divergence, which are potentially threatening to democratic societies and to the values of social justice on which they are based."[68] This is precisely the whirlwind that we will reap at the hands of surveillance capitalism, an unprecedented form of raw capitalism that is surely contributing to the tempering of commitment to the democratic prospect as it successfully bends populations to its soft-spoken will. It gives so much, but it takes even more.

Surveillance capitalism arrived on the scene with democracy already on the ropes, its early life sheltered and nourished by neoliberalism's claims to freedom that set it at a distance from the lives of people. Surveillance capitalists quickly learned to exploit the gathering momentum aimed at hollowing out democracy's meaning and muscle. Despite the democratic promise of its rhetoric and capabilities, it contributed to a new Gilded Age of extreme wealth inequality, as well as to once-unimaginable new forms of economic exclusivity and new sources of social inequality that separate the tuners from the tuned. Among the many insults to democracy and democratic institutions imposed by this *coup des gens,* I count the unauthorized expropriation of human experience; the hijack of the division of learning in society; the structural independence from people; the stealthy imposition of the hive collective; the rise of instrumentarian power and the radical indifference that sustains its extractive logic; the construction, ownership, and operation of the means of behavior modification that is Big Other; the abrogation of the elemental right to the future tense and the elemental right to sanctuary; the degradation of the self-determining individual as the fulcrum of democratic life; and the insistence on psychic numbing as the answer to its illegitimate quid pro quo. We can now see that surveillance capitalism takes an even more expansive turn toward domination than its neoliberal source code would predict, claiming its right to freedom *and* knowledge, while setting its sights on a collectivist vision that claims the totality of society. Though still sounding like Hayek, and even Smith, its antidemocratic collectivist ambitions reveal it as an insatiable child devouring its aging fathers.

Cynicism is seductive and can blind us to the enduring fact that democracy remains our only channel for reformation. It is the one idea to have emerged from the long story of human oppression that insists upon a people's inalienable right to rule themselves. Democracy may be under siege, but we cannot allow its many injuries to deflect us from allegiance to its promise. It is precisely in recognition of this dilemma that Piketty refuses to concede defeat, arguing that even "abnormal" dynamics of accumulation have been— and can again be—mitigated by democratic institutions that produce durable and effective countermeasures: "If we are to regain control of capital, we must bet everything on democracy...."[69]

Democracy is vulnerable to the unprecedented, but the strength of democratic institutions is the clock that determines the duration and destructiveness of that vulnerability. In a democratic society the debate and contest afforded by still-healthy institutions can shift the tide of public opinion against unexpected sources of oppression and injustice, with legislation and jurisprudence eventually to follow.

VI. Be the Friction

This promise of democracy reflects an enduring lesson that I absorbed from Milton Friedman at the University of Chicago as a nineteen-year-old undergraduate wedged in the back of a seminar room and straining to hear his instruction of the Chilean doctoral candidates who would soon lead their country to cataclysm, marching under the Friedman-Hayek flag. The professor was an optimist and a tireless educator who believed that legislative and judicial action invariably reflect the public opinion of twenty to thirty years earlier. It was an insight that he and Hayek—the two have been described as "soul mates and adversaries"—had crafted and transformed into systematic strategies and tactics.[70] As Hayek told Robert Bork in a 1978 interview, "I'm operating on public opinion. I don't even believe that before public opinion has changed, a change in the law will do any good...the primary thing is to change opinion...."[71] Friedman's conviction oriented him toward the long game as he threw himself into the distinctly nonacademic project of neoliberal evangelism with a steady stream of popular articles, books, and television programs. He was always sensitive to the impact of local experience, from school textbooks to grassroots political campaigns.

The critical role of public opinion explains why even the most destructive "ages" do not last forever. I echo here what Edison said a century ago: that capitalism is "all wrong, out of gear." The instability of Edison's day threatened every promise of industrial civilization. It had to give way, he insisted, to a new synthesis that reunited capitalism and its populations. Edison was prophetic. Capitalism has survived the *longue durée* less because of any specific capability and more because of its plasticity. It survives and thrives by periodically renewing its roots in the social, finding new ways to generate new wealth by meeting new needs. Its evolution has been marked by a convergence of basic principles—private property, the profit motive, and growth—but with new forms, norms, and practices in each era.[72] This is precisely the lesson of Ford's discovery and the logic behind successive episodes of revitalization over many centuries. "There is no single variety of capitalism or organization of production," Piketty writes. "This will continue to be true in the future, no doubt more than ever: New forms of organization and ownership remain to be invented."[73] Harvard philosopher Roberto Unger enlarges on this point, arguing that market forms can take any number of distinct legal and institutional directions, *"each with dramatic consequences for every aspect of social life"* and *"immense importance for the future of humanity."*[74]

When I speak to my children or an audience of young people, I try to alert them to the historically contingent nature of "the thing that has us" by calling attention to ordinary values and expectations before surveillance capitalism began its campaign of psychic numbing. "It is not OK to have to hide in your own life; it is not normal," I tell them. "It is not OK to spend your lunchtime conversations comparing software that will camouflage you and protect you from continuous unwanted invasion." *Five trackers blocked. Four trackers blocked. Fifty-nine trackers blocked, facial features scrambled, voice disguised...*

I tell them that the word "search" has meant a daring existential journey, not a finger tap to already existing answers; that "friend" is an embodied mystery that can be forged only face-to-face and heart-to-heart; and that "recognition" is the glimmer of homecoming we experience in our beloved's face, not "facial recognition." I say that it is not OK to have our best instincts for connection, empathy, and information exploited by a draconian quid pro quo that holds these goods hostage to the pervasive strip search of our lives. It is not OK for every move, emotion, utterance, and desire to be catalogued, manipulated, and then used to surreptitiously herd us through the future tense for the sake of someone else's profit. "These things are brand-new," I tell them. "They are unprecedented. You should not take them for granted because they are not OK."

If democracy is to be replenished in the coming decades, it is up to us to rekindle the sense of outrage and loss over what is being taken from us. In this I do not mean only our "personal information." What is at stake here is the human expectation of sovereignty over one's own life and authorship of one's own experience. What is at stake is the inward experience from which we form the will to will and the public spaces to act on that will. What is at stake is the dominant principle of social ordering in an information civilization and our rights as individuals and societies to answer the questions *Who knows? Who decides? Who decides who decides?* That surveillance capitalism has usurped so many of our rights in these domains is a scandalous abuse of digital capabilities and their once grand promise to democratize knowledge and meet our thwarted needs for effective life. Let there be a digital future, but let it be a human future first.

I reject inevitability, and it is my hope that as a result of our journey together, you will too. We are at the beginning of this story, not the end. If we engage the oldest questions now, there is still time to take the reins and redirect the action toward a human future that we can call home. I turn once more to Tom Paine, who called upon each generation to assert its will when illegitimate forces hijack the future and we find ourselves hurled toward a destiny that we did not choose: "The rights of men in society are neither devisable, nor transferable, nor annihilable, but are descendible only; and it is not in the power of any generation to intercept finally and cut off the descent. If the present generation or any other, are disposed to be slaves, it does not lessen the right of the succeeding generation to be free: wrongs cannot have a legal descent."[75]

Whatever has gone wrong, the responsibility to right it is renewed with each generation. Pity us and those who come next if we forfeit a human future to powerful companies and a rogue capitalism that fail to honor our needs or serve our genuine interests. Worse still would be our own voiceless capitulation to the message of inevitability that is power's velvet-gloved right hand. Hannah Arendt, referring to her work on the origins of totalitarianism, wrote that "the natural human reaction to such conditions is one of anger and indignation because these conditions are against the dignity of man. If I describe these conditions without permitting my indignation to interfere, then I have lifted this particular phenomenon out of its context in human society and have thereby robbed it of part of its nature, deprived it of one of its important inherent qualities."[76]

So it is for me and perhaps for you: the bare facts of surveillance capitalism necessarily arouse my indignation because they demean human dignity. The future of this narrative will depend upon the indignant citizens, journalists, and scholars drawn to this frontier project; indignant elected officials and policy makers who understand that their authority originates in the foundational values of democratic communities; and, especially, indignant young people who act in the knowledge that effectiveness without autonomy is not effective, dependency-induced compliance is no social contract, a hive with no exit can never be a home, experience without sanctuary is but a shadow, a life that requires hiding is no life, touch without feel reveals no truth, and freedom from uncertainty is no freedom.

We return here to George Orwell, but perhaps not in the way you might imagine. In an indignant 1946 review of James Burnham's best seller, *The Managerial Revolution*, Orwell takes aim at Burnham for his cowardly attachment to power. The thesis of Burnham's 1940 book was that capitalism, democracy, and socialism would not survive World War II. All would be replaced by a new planned centralized society modeled on totalitarianism. A new "managerial" class composed of executives, technicians, bureaucrats, and soldiers would concentrate all power and privilege in their own hands: an aristocracy of talent built on a semi-slave society. Throughout the book, Burnham insisted on the "inevitability" of this future and extolled the managerial capabilities evident in German and Russian political leadership. Writing in 1940, Burnham prophesied a Germany victory and the "managed" society to follow. Later, as the war still raged and the Red Army scored key successes, Burnham wrote a series of supplemental notes to later editions of the book in which he asserted with equal certainty that Russia would dominate the world.

Orwell's disgust is palpable: "It will be seen that at each point Burnham is predicting *a continuation of the thing that is happening.* Now, the tendency to do this is not simply a bad habit, like inaccuracy or exaggeration, which one can correct by taking thought. It is a major mental disease, and its roots lie partly in cowardice and partly in the worship of power, which is not fully separable from cowardice." Burnham's "sensational" contradictions revealed his own enthrallment with power and a complete failure to ascertain the creative principle in human history. "In each case," Orwell thundered, "he was obeying the same instinct: the instinct to bow down before the conqueror of the moment, to accept the existing trend as irreversible."[77]

Orwell reviled Burnham for his absolute failure of "moral effort," expressed in his profound loss of bearings. Under these conditions, "literally anything can become right or wrong if the dominant class of the moment so wills it." Burnham's loss of bearings allowed him "to think of Nazism as something rather admirable, something that could and probably would build up a workable and durable social order."[78]

Burnham's cowardice is a cautionary tale. We are living in a moment when surveillance capitalism and its instrumentarian power appear to be invincible. Orwell's courage demands that we refuse to cede the future to illegitimate power. He asks us to break the spell of enthrallment, helplessness, resignation, and numbing. We answer his call when we bend ourselves toward friction, rejecting the smooth flows of coercive confluence. Orwell's courage sets us against the relentless tides of dispossession that demean all human experience. Friction, courage, and bearings are the resources we require to begin the shared work of synthetic declarations that claim the digital future as a human place, demand that digital capitalism operate as an inclusive force bound to the people it must serve, and defend the division of learning in society as a source of genuine democratic renewal.

Arendt, like Orwell, asserts the possibility of new beginnings that do not cleave to already visible lines of power. She reminds us that every beginning, seen from the perspective of the framework that it interrupts, is a miracle. The capacity for performing such miracles is entirely human, she argues, because it is the source of all freedom: "What usually remains intact in the epochs of petrification and foreordained doom is the faculty of freedom itself, the sheer capacity to begin, which animates and inspires all human activities and is the hidden source...of all great and beautiful things."[79]

The decades of economic injustice and immense concentrations of wealth that we call the Gilded Age succeeded in teaching people how they did not want to live. That knowledge empowered them to bring the Gilded Age to an end, wielding the armaments of progressive legislation and the New Deal. Even now, when we recall the lordly "barons" of the late nineteenth century, we call them "robbers."

Surely the Age of Surveillance Capitalism will meet the same fate as it teaches us how *we* do not want to live. It instructs us in the irreplaceable value of our greatest moral and political achievements by threatening to destroy them. It reminds us that shared trust is the only real protection from uncertainty. It demonstrates that power untamed by democracy can only lead to exile and despair. Friedman's cycle of public opinion and durable law now reverts to us: it is up to us to use our knowledge, to regain our bearings, to stir others to do the same, and to found a new beginning. In the conquest of nature, industrial capitalism's victims were mute. Those who would try to conquer human nature will find their intended victims full of voice, ready to name danger and defeat it. This book is intended as a contribution to that collective effort.

The Berlin Wall fell for many reasons, but above all it was because the people of East Berlin said, "No more!" We too can be the authors of many "great and beautiful" new facts that reclaim the digital future as humanity's home. No more! Let this be *our* declaration.

DETAILED TABLE OF CONTENTS

ACKNOWLEDGMENTS

The two people who contributed the most to this work at the beginning no longer stand beside me, that I might shower them with gratitude at the end. When a lightning fire destroyed our home in 2009, thousands of books along with every trace of my scholarly career and new work in progress vanished in a few hours. I thought that I would never write again, but my brilliant and beloved husband, Jim Maxmin, insisted that time would bring rebirth. It did. For nearly thirty years, Jim was my first, last, and most important reader and interlocutor. He patiently absorbed early drafts of these chapters, as we excitedly argued our way through the new ideas. It remains incredible to me that we cannot share the fruit of this long labor. Jim's great love and boundless enthusiasm fortified me for the long road, in work and life. His spirit lives through these pages in ways too numerous to reckon.

Frank Schirrmacher, Germany's courageous public intellectual and a publisher of the *Frankfurter Allgemeine Zeitung*, was an extraordinary source of support and inspiration as I began to piece together my theories of surveillance capitalism and instrumentarian power. Frank urged me to write for *FAZ* while I pursued the longer work, insisting on publishing essays that, in my monkish way, I would have incubated for many more months or years. I learned so much from our endless discussions, and it is because of Frank that my work on Big Other and surveillance capitalism became useful public frameworks long before this book was finalized. Though he left us four years ago, I still reach for my phone to share a new thought with him. Frank! I am also grateful to current and former colleagues at *FAZ*, especially Edo Reents and Jordan Mejias who helped bring my essays there to fruition.

My deepest gratitude goes to the technology and data science professionals who gave so generously of their time, knowledge, and reflections over several years of interviews, as I pieced together my understanding of surveillance capitalism, its mechanisms and imperatives. I would gladly thank each one of these insightful talented teachers by name, but for my promises of confidentiality and anonymity. I also owe a debt of thanks to the families in the UK and Spain whom I interviewed during the worst years of the Great Recession. They taught me so much about "the collision" and how it set the stage upon which surveillance capitalism flourished. Special thanks to the "Montes" family, featured in Chapter 2, and to "David," the attorney featured in the Pokémon Go narrative, for allowing me to elaborate their stories, though their names and details have been changed to protect their anonymity.

Several colleagues made invaluable contributions to this work, each one exemplifying the big-hearted selflessness of true scholarship when we gather in the service of ideas. I shall never forget these gifts. Privacy advocate Marc Rotenberg has been an extraordinary colleague who read and commented on several drafts and helped me hone my understanding of key themes as they bear on privacy law. Berkeley legal scholar Chris Jay Hoofnagle graciously read a version of the manuscript in its entirety at a crucial stage. His generous comments made an important contribution to this work. The erudite Frank Pasquale read parts of the manuscript at an early stage, offering sage advice, insights, and enthusiasm. David Lidsky brought his immense editorial talents to an earlier draft of the manuscript. His collegiality, conceptual grasp, and unparalleled mastery of craft helped me to discover the final structure of this work. My deepest thanks also go to several other colleagues who read drafts of individual chapters at various junctures during the last five years and shared their discerning comments: Paul Schwartz, Artemi Rallo, Mikkel Flyverbom, David Lyon, Martha Poon, Mathias Doepfner, Karyn Allen, and Peter van den Heuvel. Both Chris Soghian and Bruce Schneier tolerated my questions on encryption and data processing as I worked out my understanding of the foundational mechanisms of surveillance capitalism.

Thinking and writing are solitary projects, and I am thankful to colleagues who offered opportunities to share ideas with scholars and students. Jonathan Zittrain invited my participation as a Faculty Associate at the Harvard Law School's Berkman Klein Center for Internet and Society at an important time in the development of my thinking. David Lyon and David Murakami Wood hosted me at Queens University toward the end of this project, where discussions with faculty, graduate students, and undergraduates energized the final stages of writing. Special thanks go to Queens undergraduates Helen Kosc and Qianli Chen for their insightful comments. There were many other wonderful invitations that I forfeited to my obligation to these pages, and I remain deeply appreciative of my colleagues around the world who voiced their interest in this work. Though they could not know it, their enthusiasm kept me going.

Thanks to Leslie Willcocks at the London School of Economics and Chris Sauer at Oxford for their early intellectual support. As co-editors of a special issue of the *Journal of Information Technology* devoted to the theme of "big data," they enthusiastically embraced my first academic paper on surveillance capitalism, "Big Other: Surveillance Capitalism and the Prospects of an Information Civilization," and helped speed its way to publication. Special thanks to the Senior Scholars of the International Conference on Information Systems for their recognition of "Big Other" with the ICIS's 2016 Best Paper Award, which also fortified my commitment to this larger work.

My literary agent Wayne Kabak resonated with this project from the start and has been tireless in his encouragement and support. I treasure his friendship and wise counsel. My editor John Mahaney has been a passionate advocate of this undertaking, bringing his years of editorial wisdom to every draft and keeping me focused on the road ahead. Thanks to Kristina Fazzalaro, Jaime Leifer, Collin Tracy, Stephanie Summerhays, and the entire Public-Affairs team for their enduring commitment and support.

My ambitions for this book could not have been fulfilled without my cherished home team. My citations manager, William Dickie, joined me in 2014, when he couldn't have had any idea of what he was getting into. Instead of running away as the project grew, he rolled up his sleeves and learned to master the citations process with a determination that evolved into mastery. I am so grateful for his patient spirit, kind heart, thoughtful contributions, and friendship as he quietly and diligently pursued his responsibilities. My research assistant Jordan Keenan joined this project in early 2015 and hit the ground running with outstanding contributions as he mastered the professional art of the research deep dive. I have relied on him through the twists and turns of this intellectual journey, and I watched him rise to each new challenge with grace and gusto as he trained his lively intelligence on new research territories. As if that were not enough, his unflappable good cheer, laid-back humor, and quiet sensitivity have made him an invaluable companion on this adventure.

My children selflessly supported my work on this book for these many years. They listened to my ideas, nurtured me through my frustrations, and celebrated my successes with so much love, patience, and devotion. When forks in the road required urgent decisions, they gathered round to help. My daughter Chloe Maxmin has been a steadfast and wise counselor, reading each draft of each chapter and offering the kind of unvarnished incisive commentary that all writers need but few receive. I learned that chapters were complete only when Chloe signed off on them. My son, Jake Maxmin, endured this book project throughout his college and now graduate school careers. Jake has been a fierce advocate, showering me with texts and phone calls, "You've got this mama," whenever I faced a difficult stretch or deadline and stepping in to provide crucial help with final edits. Chloe and Jake's astute advice and inspiration never failed to get me over the mountain. Thank you *mis vidas, mis corazones.* I would not be here without you.

I offer loving thanks to my friends who shared holidays and celebrations when I was lost to writing and every table, chair, and floor was covered with research materials: Minda Gold, Jacques Vesery, Isaac Vesery, Jonah Vesery, Lisa Katz, Ed, Theo, and Toby Seidel, Mary Dee Choate Grant, Garret Grant, Kathy Leeman, Kerry Altiero. Kathy Leeman read the book in its entirety before I began final revisions. Her perceptiveness and fervor were immensely helpful during the last year of writing. Virginia Alicia Hasenbalg-Corabianu cheered me on from Paris ("please finish this book already!"), hosted my children, and heroically translated one of my published lectures on surveillance capitalism into French. Susan Tross was an unconditional wellspring of love and support. My friend and "adopted son" Canyon Woodward has been a faithful cheerleader, always galvanizing my spirit and my stories. Finally, these declarations of acknowledgment are not complete without mention of Pachi Maxmin, my stalwart loving companion.

Every author acknowledges what I am about to write, because it is true: In the end, one faces the page in solitude. Anything in this book that falls short of the trust invested in me is my responsibility alone.

NOTES

CHAPTER ONE

1. Martin Hilbert, "Technological Information Inequality as an Incessantly Moving Target: The Redistribution of Information and Communication Capacities Between 1986 and 2010," *Journal of the American Society for Information Science and Technology* 65, no. 4 (2013): 821–35, https://doi.org/10.1002/asi.23020.
2. By 2014, about twenty years after the invention of the world wide web, an extensive survey by Pew Research found 87 percent of Americans using the internet. Among those, 76 percent regarded it as "a good thing for society" and 90 percent as "a good thing for me." Indeed, people routinely call 9-1-1 when Facebook is down. In less than two decades after the Mosaic browser was released to the public, enabling easy access to the world wide web, a 2010 BBC poll found that 79 percent of people in twenty-six countries considered internet access to be a fundamental human right. Six years later, the United Nations adopted specific language on internet access: "Everyone has the right to freedom of opinion and expression; this right includes freedom to hold opinions without interference and to seek, receive and impart information and ideas through any media and regardless of frontiers." See Susannah Fox and Lee Rainie, "The web at 25 in the U.S.," *PewResearchCenter*, February 27, 2014, http://www.pewinternet.org/2014/02/27/the-web-at-25-in-the-u-s; "911 Calls About Facebook Outage Angers L.A. County Sheriff's Officials," *Los Angeles Times*, August 1, 2014, http://www.latimes.com/local/lanow/la-me-ln-911-calls-about-facebook-outage-angers-la-sheriffs-officials-20140801-htmlstory.html; "Internet Access 'a Human Right,'" *BBC News*, March 8, 2010, http://news.bbc.co.uk/2/hi/8548190.stm; "The Promotion, Protection and Enjoyment of Human Rights on the Internet," United Nations Human Rights Council, June 27, 2016, https://www.article19.org/data/files/Internet_Statement_Adopted.pdf.
3. João Leal, *The Making of Saudade: National Identity and Ethnic Psychology in Portugal* (Amsterdam: Het Spinhuis, 2000), https://run.unl.pt/handle/10362/4386.
4. Cory D. Kidd et al., "The Aware Home: A Living Laboratory for Ubiquitous Computing Research," in *Proceedings of the Second International Workshop on*

Cooperative Buildings, Integrating Information, Organization, and Architecture, CoBuild '99 (London: Springer-Verlag, 1999), 191–98, http://dl.acm.org /citation.cfm?id=645969.674887.

5. "Global Smart Homes Market 2018 by Evolving Technology, Projections & Estimations, Business Competitors, Cost Structure, Key Companies and Forecast to 2023," *Reuters,* February 19, 2018, https://www.reuters.com /brandfeatures/venture-capital/article?id=28096.

6. Ron Amadeo, "Nest Is Done as a Standalone Alphabet Company, Merges with Google," *Ars Technica,* February 7, 2018, https://arstechnica.com /gadgets/2018/02/nest-is-done-as-a-standalone-alphabet-company-merges -with-google; Leo Kelion, "Google-Nest Merger Raises Privacy Issues," *BBC News,* February 8, 2018, http://www.bbc.com/news/technology-42989073.

7. Kelion, "Google-Nest Merger Raises Privacy Issues."

8. Rick Osterloh and Marwan Fawaz, "Nest to Join Forces with Google's Hardware Team," Google, February 7, 2018, https://www.blog.google/inside-google /company-annoucements/nest-join-forces-googles-hardware-team.

9. Grant Hernandez, Orlando Arias, Daniel Buentello, and Yier Jin, "Smart Nest Thermostat: A Smart Spy in Your Home," *Black Hat USA,* 2014, https://www .blackhat.com/docs/us-14/materials/us-14-Jin-Smart-Nest-Thermostat-A -Smart-Spy-In-Your-Home-WP.pdf.

10. Guido Noto La Diega, "Contracting for the 'Internet of Things': Looking into the Nest" (research paper, Queen Mary University of London, School of Law, 2016); Robin Kar and Margaret Radin, "Pseudo-Contract & Shared Meaning Analysis" (legal studies research paper, University of Illinois College of Law, November 16, 2017), https://papers.ssrn.com/abstract=3083129.

11. Hernandez, Arias, Buentello, and Jin, "Smart Nest Thermostat."

12. For a prescient early treatment of these issues, see Langdon Winner, "A Victory for Computer Populism," *Technology Review* 94, no. 4 (1991): 66. See also Chris Jay Hoofnagle, Jennifer M. Urban, and Su Li, "Privacy and Modern Advertising: Most US Internet Users Want 'Do Not Track' to Stop Collection of Data About Their Online Activities" (BCLT Research Paper, Rochester, NY: Social Science Research Network, October 8, 2012), https://papers.ssrn.com /abstract=2152135; Joseph Turow et al., "Americans Reject Tailored Advertising and Three Activities That Enable It," Annenberg School for Communication, September 29, 2009, http://papers.ssrn.com/abstract=1478214; Chris Jay Hoofnagle and Jan Whittington, "Free: Accounting for the Costs of the Internet's Most Popular Price," *UCLA Law Review* 61 (February 28, 2014): 606; Jan Whittington and Chris Hoofnagle, "Unpacking Privacy's Price," *North Carolina Law Review* 90 (January 1, 2011): 1327; Chris Jay Hoofnagle, Jennifer King, Su Li, and Joseph Turow, "How Different Are Young Adults from Older Adults When It Comes to Information Privacy Attitudes & Policies?" April 14, 2010, http://repository.upenn.edu/asc_papers/399.

13. The phrase is from Roberto Mangabeira Unger, "The Dictatorship of No Alternatives," in *What Should the Left Propose?* (London: Verso, 2006), 1–11.

14. Jared Newman, "Google's Schmidt Roasted for Privacy Comments," *PCWorld*, December 11, 2009, http://www.pcworld.com/article/184446/googles_schmidt _roasted_for_privacy_comments.html.

15. Max Weber, *Economy and Society: An Outline of Interpretive Sociology* (Berkeley, CA: University of California Press, 1978), 1:67.

CHAPTER TWO

1. Roben Farzad, "Apple's Earnings Power Befuddles Wall Street," *Bloomberg Businessweek*, August 7, 2011, https://www.bloomberg.com/news /articles/2011-07-28/apple-s-earnings-power-befuddles-wall-street.

2. "iTunes Music Store Sells Over One Million Songs in First Week," *Apple Newsroom*, March 9, 2018, https://www.apple.com/newsroom/2003/05 /05iTunes-Music-Store-Sells-Over-One-Million-Songs-in-First-Week.

3. Jeff Sommer, "The Best Investment Since 1926? Apple," *New York Times*, September 22, 2017, https://www.nytimes.com/2017/09/22/business/apple -investment.html.

4. See Shoshana Zuboff and James Maxmin, *The Support Economy: How Corporations Are Failing Individuals and the Next Episode of Capitalism* (New York: Penguin, 2002), 230.

5. Henry Ford, "Mass Production," *Encyclopedia Britannica* (New York: Encyclopedia Britannica, 1926), 821, http://memory.loc.gov/cgi-bin/query /h?ammem/coolbib:@field(NUMBER+@band(amrlg+lg48)).

6. Lizabeth Cohen, *A Consumers' Republic: The Politics of Mass Consumption in Postwar America* (New York: Knopf, 2003); Martin J. Sklar, *The Corporate Reconstruction of American Capitalism: 1890–1916: The Market, the Law, and Politics* (New York: Cambridge University Press, 1988).

7. Emile Durkheim, *The Division of Labor in Society* (New York: Free Press, 1964), 275 (italics mine).

8. Durkheim, *The Division of Labor in Society,* 266.

9. Ulrich Beck and Mark Ritter, *Risk Society: Towards a New Modernity* (Thousand Oaks, CA: Sage, 1992).

10. For readers interested in a more detailed analysis of the rise of this phenomenon, I recommend the extended discussion in Zuboff and Maxmin, *The Support Economy.* See also Ulrich Beck and Elisabeth Beck-Gernsheim, *Individualization: Institutionalized Individualism and Its Social and Political Consequences* (London: Sage, 2002); Ulrich Beck, "Why 'Class' Is Too Soft a Category to Capture the Explosiveness of Social Inequality at the Beginning of the Twenty-First Century," *British Journal of Sociology* 64, no. 1 (2013): 63–74; Ulrich Beck and Edgar Grande, "Varieties of Second Modernity: The

Cosmopolitan Turn in Social and Political Theory and Research," *British Journal of Sociology* 61, no. 3 (2010): 409–43.

11. Beck and Ritter, *Risk Society*.

12. Talcott Parsons, *Social Structure and Personality* (New York: Free Press, 1964).

13. Beck and Beck-Gernsheim, *Individualization*.

14. Erik Erikson, *Childhood and Society* (New York: W. W. Norton, 1993), 279.

15. Ronald Inglehart, *Culture Shift in Advanced Industrial Society* (Princeton, NJ: Princeton University Press, 1990); Ronald F. Inglehart, "Changing Values Among Western Publics from 1970 to 2006," *West European Politics* 31, nos. 1–2 (2008): 130–46; Ronald Inglehart and Christian Welzel, "How We Got Here: How Development Leads to Democracy," *Foreign Affairs* 88, no. 2 (2012): 48–50; Ronald Inglehart and Wayne E. Baker, "Modernization, Cultural Change, and the Persistence of Traditional Values," *American Sociological Review* 65, no. 1 (2000): 19; Mette Halskov Hansen, *iChina: The Rise of the Individual in Modern Chinese Society*, ed. Rune Svarverud (Copenhagen: Nordic Institute of Asian Studies, 2010); Yunxiang Yan, *The Individualization of Chinese Society* (Oxford: Bloomsbury Academic, 2009); Arthur Kleinman et al., *Deep China: The Moral Life of the Person* (Berkeley: University of California Press, 2011); Chang Kyung-Sup and Song Min-Young, "The Stranded Individual Under Compressed Modernity: South Korean Women in Individualization Without Individualism," *British Journal of Sociology* 61, no. 3 (2010); Chang Kyung-Sup, "The Second Modern Condition? Compressed Modernity as Internalized Reflexive Cosmopolitization," *British Journal of Sociology* 61, no. 3 (2010); Munenori Suzuki et al., "Individualizing Japan: Searching for Its Origin in First Modernity," *British Journal of Sociology* 61, no. 3 (2010); Anthony Elliott, Masataka Katagiri, and Atsushi Sawai, "The New Individualism and Contemporary Japan: Theoretical Avenues and the Japanese New Individualist Path," *Journal for the Theory of Social Behavior* 42, no. 4 (2012); Mitsunori Ishida et al., "The Individualization of Relationships in Japan," *Soziale Welt* 61 (2010): 217–35; David Tyfield and John Urry, "Cosmopolitan China?" *Soziale Welt* 61 (2010): 277–93.

16. Beck and Beck-Gernsheim, *Individualization*; Ulrich Beck, *A God of One's Own: Religion's Capacity for Peace and Potential for Violence*, trans. Rodney Livingstone (Cambridge, UK: Polity, 2010).

17. Thomas M. Franck, *The Empowered Self: Law and Society in an Age of Individualism* (Oxford: Oxford University Press, 2000).

18. Beck and Beck-Gernsheim, *Individualization*, xxii.

19. Daniel Stedman Jones, *Masters of the Universe: Hayek, Friedman, and the Birth of Neoliberal Politics* (Princeton, NJ: Princeton University Press, 2012); T. Flew, "Michel Foucault's *The Birth of Biopolitics* and Contemporary Neo-Liberalism Debates," *Thesis Eleven* 108, no. 1 (2012): 44–65, https://doi.org/10.1177/0725513611421481; Philip Mirowski, *Never Let a Serious Crisis Go to Waste: How Neoliberalism Survived the Financial Meltdown* (London:

Verso, 2013); Gérard Duménil and Dominique Lévy, *The Crisis of Neoliberalism* (Cambridge, MA: Harvard University Press, 2013); Pierre Dardot and Christian Laval, *The New Way of the World: On Neoliberal Society* (Brooklyn: Verso, 2013); António Ferreira, "The Politics of Austerity as Politics of Law," *Oñati Socio-Legal Series* 6, no. 3 (2016): 496–519; David M. Kotz, *The Rise and Fall of Neoliberal Capitalism* (Cambridge, MA: Harvard University Press, 2017); Philip Mirowski and Dieter Plehwe, eds., *The Road from Mont Pelerin: The Making of the Neoliberal Thought Collective* (Cambridge, MA: Harvard University Press, 2009); Wendy Brown, *Undoing the Demos: Neoliberalism's Stealth Revolution* (New York: Zone, 2015); David Jacobs and Lindsey Myers, "Union Strength, Neoliberalism, and Inequality: Contingent Political Analyses of US Income Differences Since 1950," *American Sociological Review* 79 (2014): 752–74; Angus Burgin, *The Great Persuasion: Reinventing Free Markets Since the Depression* (Cambridge, MA: Harvard University Press, 2012); Greta R. Krippner, *Capitalizing on Crisis: The Political Origins of the Rise of Finance* (Cambridge, MA: Harvard University Press, 2011).

20. Jones, *Masters of the Universe*, 215. See also Krippner, *Capitalizing on Crisis*.

21. Mirowski, Dardot and Laval, and Jones provide detailed accounts of these developments.

22. Friedrich August von Hayek, *The Fatal Conceit: The Errors of Socialism*, ed. William Warren Bartley, vol. 1, *The Collected Works of Friedrich August Hayek* (Chicago: University of Chicago Press, 1988), 14–15.

23. Mirowski, *Never Let a Serious Crisis Go to Waste*, 53–67.

24. Michael C. Jensen and William H. Meckling, "Theory of the Firm: Managerial Behavior, Agency Costs and Ownership Structure," *Journal of Financial Economics* 3, no. 4 (1976): 12.

25. Krippner, *Capitalizing on Crisis*.

26. Karl Polanyi, *The Great Transformation: The Political and Economic Origins of Our Time* (Boston: Beacon, 2001), 79.

27. Martin J. Sklar, *The United States as a Developing Country: Studies in U.S. History in the Progressive Era and the 1920s* (Cambridge: Cambridge University Press, 1992); Sanford M. Jacoby, *Modern Manors: Welfare Capitalism Since the New Deal* (Princeton, NJ: Princeton University Press, 1998); Michael Alan Bernstein, *The Great Depression: Delayed Recovery and Economic Change in America, 1929–1939*, Studies in Economic History and Policy (Cambridge, MA: Cambridge University Press, 1987); C. Goldin and R. A. Margo, "The Great Compression: The Wage Structure in the United States at Mid-century," *Quarterly Journal of Economics* 107, no. 1 (1992): 1–34; Edwin Amenta, "Redefining the New Deal," in *The Politics of Social Policy in the United States*, ed. Theda Skocpol, Margaret Weir, and Ann Shola Orloff (Princeton, NJ: Princeton University Press, 1988), 81–122.

28. Ian Gough, Anis Ahmad Dani, and Harjan de Haan, "European Welfare States: Explanations and Lessons for Developing Countries," in *Inclusive States: Social*

Policies and Structural Inequalities (Washington, DC: World Bank, 2008); Peter Baldwin, *The Politics of Social Solidarity: Class Bases of the European Welfare State, 1875–1975* (Cambridge: Cambridge University Press, 1990); John Kenneth Galbraith, Sean Wilentz, and James K. Galbraith, *The New Industrial State* (Princeton, NJ: Princeton University Press, 1967); Gerald Davis, "The Twilight of the Berle and Means Corporation," *Seattle University Law Review* 34, no. 4 (2011): 1121–38; Alfred Dupont Chandler, *Essential Alfred Chandler: Essays Toward a Historical Theory of Big Business,* ed. Thomas K. McCraw (Boston: Harvard Business School Press, 1988).

29. Jones, *Masters of the Universe,* 217.
30. See, for example, Vivien A. Schmidt and Mark Thatcher, eds., *Resilient Liberalism in Europe's Political Economy* (Cambridge: Cambridge University Press, 2013); Kathleen Thelen, *Varieties of Liberalization and the New Politics of Social Solidarity* (Cambridge: Cambridge University Press, 2014); Peter Kingstone, *The Political Economy of Latin America: Reflections on Neoliberalism and Development* (New York: Routledge, 2010); Jeffry Frieden, Manuel Pastor, Jr., and Michael Tomz, *Modern Political Economy and Latin America: Theory and Policy* (Boulder, CO: Routledge, 2000); Giuliano Bonoli and David Natali, *The Politics of the New Welfare State* (Oxford: Oxford University Press, 2012); Richard Münch, *Inclusion and Exclusion in the Liberal Competition State: The Cult of the Individual* (New York: Routledge, 2012), http://site.ebrary.com /id/10589064; Kyung-Sup Chang, *Developmental Politics in Transition: The Neoliberal Era and Beyond* (Basingstoke, UK: Palgrave Macmillan, 2012); Zsuzsa Ferge, "The Changed Welfare Paradigm: The Individualization of the Social," *Social Policy & Administration* 31, no. 1 (1997): 20–44.
31. Gerald F. Davis, *Managed by the Markets: How Finance Re-shaped America* (Oxford: Oxford University Press, 2011); Davis, "The Twilight of the Berle and Means Corporation"; Özgür Orhangazi, "Financialisation and Capital Accumulation in the Non-financial Corporate Sector: A Theoretical and Empirical Investigation on the US Economy: 1973–2003," *Cambridge Journal of Economics* 32, no. 6 (2008): 863–86; William Lazonick, "The Financialization of the U.S. Corporation: What Has Been Lost, and How It Can Be Regained," in *The Future of Financial and Securities Markets* (Fourth Annual Symposium of the Adolf A. Berle, Jr. Center for Corporations, Law and Society of the Seattle School of Law, London, 2012); Yuri Biondi, "The Governance and Disclosure of the Firm as an Enterprise Entity," *Seattle University Law Review* 36, no. 2 (2013): 391–416; Robert Reich, "Obama's Transition Economic Advisory Board: The Full List," *US News & World Report,* November 7, 2008, http://www.usnews .com/news/campaign-2008/articles/2008/11/07/obamas-transition-economic -advisory-board-the-full-listn; Robert B. Reich, *Beyond Outrage: What Has Gone Wrong with Our Economy and Our Democracy, and How to Fix It,* rev. ed. (New York: Vintage, 2012).

32. Michael Jensen, "Eclipse of the Public Corporation," *Harvard Business Review,* September–October, 1989.

33. Michael C. Jensen, "Value Maximization, Stakeholder Theory, and the Corporate Objective Function," *Business Ethics Quarterly* 12, no. 2 (2002): 235–56.

34. Thomas I. Palley, "Financialization: What It Is and Why It Matters" (white paper, Levy Economics Institute of Bard College, 2007), http://www .levyinstitute.org/pubs/wp_525.pdf; Jon Hanson and Ronald Chen, "The Illusion of Law: The Legitimating Schemas of Modern Policy and Corporate Law," *Michigan Law Review* 103, no. 1 (2004): 1–149; Henry Hansmann and Reinier Kraakman, "The End of History for Corporate Law" (working paper, Discussion Paper Series, Harvard Law School's John M. Olin Center for Law, Economics and Business, 2000), http://lsr.nellco.org/cgi/viewcontent .cgi?article=1068&context=harvard_olin.

35. Davis, "The Twilight of the Berle and Means Corporation," 1131.

36. Gerald F. Davis, "After the Corporation," *Politics & Society* 41, no. 2 (2013): 41.

37. Juta Kawalerowicz and Michael Biggs, "Anarchy in the UK: Economic Deprivation, Social Disorganization, and Political Grievances in the London Riot of 2011," *Social Forces* 94, no. 2 (2015): 673–98, https://doi.org/10.1093/sf/sov052.

38. Paul Lewis et al., "Reading the Riots: Investigating England's Summer of Disorder," *London School of Economics and Political Science,* 2011, 17, http:// eprints.lse.ac.uk/46297.

39. Saskia Sassen, "Why Riot Now? Malaise Among Britain's Urban Poor Is Nothing New. So Why Did It Finally Tip into Widespread, Terrifying Violence?" *Daily Beast,* August 15, 2011, http://www.donestech.net/ca /why_riot_now_by_saskia_sassen_newsweek.

40. Lewis et al., "Reading the Riots," 25.

41. In addition to Lewis et al., "Reading the Riots," see also Kawalerowicz and Biggs, "Anarchy in the UK"; James Treadwell et al., "Shopocalypse Now: Consumer Culture and the English Riots of 2011," *British Journal of Criminology* 53, no. 1 (2013): 1–17, https://doi.org/10.1093/bjc/azs054; Tom Slater, "From 'Criminality' to Marginality: Rioting Against a Broken State," *Human Geography* 4, no. 3 (2011): 106–15.

42. Thomas Piketty, *Capital in the Twenty-First Century* (Cambridge, MA: Belknap Press, 2014). Piketty integrated years of income data to conclude that income inequality in the US and the UK has reached levels not seen since the nineteenth century. The top decile of US wage earners steadily increased its share of national income from 35 percent in the 1980s to over 46 percent in 2010. The bulk of this increase comes from the top 1 percent, whose share rose from 9 percent to 20 percent, about half of which went to the 0.1 percent. Piketty calculates that 60–70 percent of the top 0.1 percent of the income hierarchy is composed of managers who have succeeded in obtaining "historically unprecedented" compensation thanks to new value-maximizing incentive structures.

43. On the general theme of the salience of democratically oriented social, political, and economic institutions in the mitigation of economic outcomes, see Daron Acemoglu and James Robinson's monumental *Why Nations Fail: The Origins of Power, Prosperity, and Poverty* (New York: Crown Business, 2012). It is also the focus of Robert Reich's work on inequality and regressive economic policy: Robert B. Reich, *Aftershock: The Next Economy and America's Future* (New York: Vintage, 2011). See also Michael Stolleis, *History of Social Law in Germany* (Heidelberg: Springer, 2014), www.springer.com/us/book/9783642384530; Mark Hendrickson, *American Labor and Economic Citizenship: New Capitalism from World War I to the Great Depression* (Cambridge: Cambridge University Press, 2013); Swank, "The Political Sources of Labor Market Dualism in Postindustrial Democracies, 1975–2011"; Emin Dinlersoz and Jeremy Greenwood, "The Rise and Fall of Unions in the U.S." (NBER working paper, US Census Bureau, 2012), http://www.nber.org/papers/w18079; Basak Kus, "Financialization and Income Inequality in OECD Nations: 1995–2007," *Economic and Social Review* 43, no. 4 (2012): 477–95; Viki Nellas and Elisabetta Olivieri, "The Change of Job Opportunities: The Role of Computerization and Institutions" (Quaderni DSE working paper, University of Bologna & Bank of Italy, 2012), http://papers .ssrn.com/sol3/papers.cfm?abstract_id=1983214; Gough, Dani, and de Haan, "European Welfare States"; Landon R. Y. Storrs, *Civilizing Capitalism: The National Consumers' League, Women's Activism, and Labor Standards in the New Deal Era*, rev. ed. (Chapel Hill: University of North Carolina Press, 2000); Ferge, "The Changed Welfare Paradigm"; Jacoby, *Modern Manors*; Sklar, *The United States as a Developing Country*; J. Bradford De Long and Barry Eichengreen, "The Marshall Plan: History's Most Successful Structural Adjustment Program," in *Post–World War II Economic Reconstruction and Its Lessons for Eastern Europe Today*, ed. Rudiger Dornbusch (Cambridge, MA: MIT Press, 1991); Baldwin, *The Politics of Social Solidarity*; Amenta, "Redefining the New Deal"; Robert H. Wiebe, *The Search for Order: 1877–1920* (New York: Hill and Wang, 1967); John Maynard Keynes, "Economic Possibilities for Our Grandchildren," in *Essays in Persuasion* (New York: W. W. Norton, 1930).

   By 2014, a Standard and Poor's report concluded that income inequality impedes economic growth and destabilizes the social fabric, a fact that Henry Ford had long ago acknowledged with his five-dollar day. See "How Increasing Income Inequality Is Dampening US Economic Growth, and Possible Ways to Change the Tide," *S&P Capital IQ*, Global Credit Portal Report, August 5, 2014, https://www.globalcreditportal.com/ratingsdirect /renderArticle.do?articleId=1351366&SctArtId=255732&from=CM&nsl _code=LIME&sourceObjectId=8741033&sourceRevId=1&fee_ind=N&exp _date=20240804-19:41:13.

44. Tcherneva, "Reorienting Fiscal Policy: A Bottom-Up Approach," 57. See also Francisco Rodriguez and Arjun Jayadev, "The Declining Labor Share of

Income," *Journal of Globalization and Development* 3, no. 2 (2013): 1–18; Oliver
Giovannoni, "What Do We Know About the Labor Share and the Profit Share?
Part III: Measures and Structural Factors" (working paper, Levy Economics
Institute at Bard College, 2014), http://www.levyinstitute.org/publications
/what-do-we-know-about-the-labor-share-and-the-profit-share-part-3
-measures-and-structural-factors; Dirk Antonczyk, Thomas DeLeire, and
Bernd Fitzenberger, "Polarization and Rising Wage Inequality: Comparing the
U.S. and Germany" (IZA discussion papers, Institute for the Study of Labor,
March 2010), https://ideas.repec.org/p/iza/izadps/dp4842.html; Duane Swank,
"The Political Sources of Labor Market Dualism in Postindustrial Democracies,
1975–2001," conference paper presented at the American Political Science
Association Annual Meeting, Chicago, 2013; David Jacobs and Lindsey Myers,
"Union Strength, Neoliberalism, and Inequality: Contingent Political Analyses
of US Income Differences Since 1950," *American Sociological Review* 79 (2014):
752–74; Viki Nellas and Elisabetta Olivieri, "The Change of Job Opportunities:
The Role of Computerization and Institutions" (Quaderni DSE working paper,
University of Bologna & Bank of Italy, 2012), http://papers.ssrn.com/sol3
/papers.cfm?abstract_id=1983214; Gough, Dani, and de Haan, "European
Welfare States: Explanations and Lessons for Developing Countries."

45. Jonathan D. Ostry, "Neoliberalism: Oversold?" *Finance & Development* 53, no. 2
(2016): 38–41; as another U.S. economist concluded, "The Great Recession of
2008 finally stripped away the illusion of economic expansion, revealing instead
the bare bones of financial capitalism's achievements: income stagnation since
the mid-1970s for the majority pitted against immense concentrations of wealth
for a tiny minority." See Josh Bivens, "In 2013, Workers' Share of Income in the
Corporate Sector Fell to Its Lowest Point Since 1950," *Economic Policy Institute*
(blog), September 4, 2014, http://www.epi.org/publication/2013-workers-share
-income-corporate-sector.

Studies of financial deepening—liberalization, financialization—in both
developed and less developed economies have shown that it is linked to new
instabilities, including bankruptcies, bank failures, extreme asset volatility, and
recession in the real sectors of the economy. See, for example, Malcolm Sawyer,
"Financial Development, Financialisation and Economic Growth" (working
paper, Financialisation, Economy, Society & Sustainable Development Project,
2014), http://fessud.eu/wpcontent/uploads/2013/04/Financialisation-and
-growth-Sawyer-working-paper-21.pdf. See also William A. Galston, "The New
Challenge to Market Democracies: The Political and Social Costs of Economic
Stagnation" (research report, Brookings Institution, 2014), http://www
.brookings.edu/research/reports2/2014/10/new-challenge-market-democracies;
Joseph E. Stiglitz, *The Price of Inequality: How Today's Divided Society Endangers
Our Future* (New York: W. W. Norton, 2012); James K. Galbraith, *Inequality
and Instability: A Study of the World Economy Just Before the Great Crisis* (New

York: Oxford University Press, 2012); Ronald Dore, "Financialization of the Global Economy," *Industrial and Corporate Change* 17, no. 6 (2008): 1097–1112; Philip Arestis and Howard Stein, "An Institutional Perspective to Finance and Development as an Alternative to Financial Liberalisation," *International Review of Applied Economics* 19, no. 4 (2005): 381–98; Asil Demirguc-Kunt and Enrica Detragiache, "The Determinants of Banking Crises in Developing and Developed Countries," *Staff Papers—International Monetary Fund* 45, no. 1 (1998): 81–109.

46. Emanuele Ferragina, Mark Tomlinson, and Robert Walker, "Poverty, Participation and Choice," *JRF*, May 28, 2013, https://www.jrf.org.uk /report/poverty-participation-and-choice.

47. Helen Kersley et al., "Raising the Benchmark: The Role of Public Services in Tackling the Squeeze on Pay," *New Economics Foundation,* https://www.unison .org.uk/content/uploads/2013/12/On-line-Catalogue219732.pdf.

48. Sally Gainsbury and Sarah Neville, "Austerity's £18bn Impact on Local Services," *Financial Times,* July 19, 2015, http://www.ft.com/intl/cms/s/2 /5fcbd0c4-2948-11e5-8db8-c033edba8a6e.html?ftcamp=crm/email/2015719 /nbe/InTodaysFT/product#axzz3gRAfXkt4.

49. Carmen DeNavas-Walt and Bernadette D. Proctor, "Income and Poverty in the United States: 2014," US Census Bureau, September 2015, http://www .census.gov/content/dam/Census/library/publications/2014/demo/p60 -249.pdf; Thomas Gabe, "Poverty in the United States: 2013," *Congressional Research Service*, September 25, 2014, http://digitalcommons.ilr.cornell.edu /key_workplace/1329.

50. Alisha Coleman-Jensen, Mark Nord, and Anita Singh, "Household Food Security in the United States in 2012" (economic research report, US Department of Agriculture, September 2013), https://www.ers.usda.gov /webdocs/publications/45129/39937_err-155.pdf?v=42199.

51. Piketty, *Capital in the Twenty-First Century,* 334–35. See also Theda Skocpol and Vanessa Williamson, *The Tea Party and the Remaking of Republican Conservatism,* rev. ed. (New York: Oxford University Press, 2016); Naomi Oreskes and Erik M. Conway, *Merchants of Doubt: How a Handful of Scientists Obscured the Truth on Issues from Tobacco Smoke to Global Warming* (London: Bloomsbury, 2010).

52. Nicholas Confessore, "The Families Funding the 2016 Presidential Election," *New York Times,* October 10, 2015, https://www.nytimes.com /interactive/2015/10/11/us/politics/2016-presidential-election-super-pac -donors.html.

53. Historian Nancy MacLean and journalist Jane Mayer document hidden operations of radical-right ideologues and their billionaire backers who command unlimited funds for the purposes of political and public manipulation, relying on clandestine networks of think tanks, donor

organizations, and media outlets to skillfully exploit citizen unrest and drive it toward extremist views. See Nancy MacLean, *Democracy in Chains: The Deep History of the Radical Right's Stealth Plan for America* (New York: Viking, 2017); Jane Mayer, *Dark Money: The Hidden History of the Billionaires Behind the Rise of the Radical Right* (New York: Anchor, 2017).

54. Piketty, *Capital in the Twenty-First Century*, 571.

55. Milan Zafirovski, "'Neo-Feudalism' in America? Conservatism in Relation to European Feudalism," *International Review of Sociology* 17, no. 3 (2007): 393–427, https://doi.org/10.1080/03906700701574323; Alain Supiot, "The Public-Private Relation in the Context of Today's Refeudalization," *International Journal of Constitutional Law* 11, no. 1 (2013): 129–45, https://doi.org/10.1093/icon/mos050; Daniel J. H. Greenwood, "Neofeudalism: The Surprising Foundations of Corporate Constitutional Rights," *University of Illinois Law Review* 163 (2017).

56. Piketty, *Capital in the Twenty-First Century*, 237–70.

57. For a poignant and powerful exploration of these themes, see Carol Graham, *Happiness for All? Unequal Hopes and Lives in Pursuit of the American Dream* (Princeton, NJ: Princeton University Press, 2017); David G. Blanchflower and Andrew Oswald, "Unhappiness and Pain in Modern America: A Review Essay, and Further Evidence, on Carol Graham's 'Happiness for All?'" (NBER working paper, November 2017).

58. See Tim Newburn et al., "David Cameron, the Queen and the Rioters' Sense of Injustice," *Guardian*, December 5, 2011, http://www.theguardian.com/uk/2011/dec/05/cameron-queen-injustice-english-rioters.

59. Slater, "From 'Criminality' to Marginality."

60. Todd Gitlin, *Occupy Nation: The Roots, the Spirit, and the Promise of Occupy Wall Street* (New York: Harper Collins, 2012); Zeynep Tufekci, *Twitter and Tear Gas: The Power and Fragility of Networked Protest* (New Haven, CT: Yale University Press, 2017). See also Andrew Gavin Marshall, "World of Resistance Report: Davos Class Jittery amid Growing Warnings of Global Unrest," *Occupy .com*, July 4, 2014, http://www.occupy.com/article/world-resistance-report -davos-class-jittery-amid-growing-warnings-global-unrest.

61. Todd Gitlin, "Occupy's Predicament: The Moment and the Prospects for the Movement," *British Journal of Sociology* 64, no. 1 (2013): 3–25, https://doi .org/10.1111/1468-4446.12001.

62. Anthony Barnett, "The Long and Quick of Revolution," *Open Democracy*, February 2, 2015, https://www.opendemocracy.net/anthony-barnett /long-and-quick-of-revolution.

63. Peter Wells and Paul Nieuwenhuis, "Transition Failure: Understanding Continuity in the Automotive Industry," *Technological Forecasting and Social Change* 79, no. 9 (2012): 1681–92, https://doi.org/10.1016/j .techfore.2012.06.008.

64. Steven Levy, *In the Plex: How Google Thinks, Works, and Shapes Our Lives* (New York: Simon & Schuster, 2011), 172–73.

65. Bobbie Johnson, "Privacy No Longer a Social Norm, Says Facebook Founder," *Guardian*, January 10, 2010, https://www.theguardian.com/technology/2010/jan/11/facebook-privacy.

66. See Charlene Li, "Close Encounter with Facebook Beacon," *Forrester*, November 23, 2007, https://web.archive.org/web/20071123023712/http://blogs.forrester.com/charleneli/2007/11/close-encounter.html.

67. Peter Linzer, "Contract as Evil," *Hastings Law Journal* 66 (2015): 971; Paul M. Schwartz, "Internet Privacy and the State," *Connecticut Law Review* 32 (1999): 815–59; Daniel J. Solove, "Privacy Self-Management and the Consent Dilemma," *Harvard Law Review* 126, no. 7 (2013): 1880–1904.

68. Yannis Bakos, Florencia Marotta-Wurgler, and David R. Trossen, "Does Anyone Read the Fine Print? Consumer Attention to Standard-Form Contracts," *Journal of Legal Studies* 43, no. 1 (2014): 1–35, https://doi.org/10.1086/674424; Tess Wilkinson-Ryan, "A Psychological Account of Consent to Fine Print," *Iowa Law Review* 99 (2014): 1745; Thomas J. Maronick, "Do Consumers Read Terms of Service Agreements When Installing Software? A Two-Study Empirical Analysis," *International Journal of Business and Social Research* 4, no. 6 (2014): 137–45; Mark A. Lemley, "Terms of Use," *Minnesota Law Review* 91 (2006), https://papers.ssrn.com/abstract=917926; Nili Steinfeld, "'I Agree to the Terms and Conditions': (How) Do Users Read Privacy Policies Online? An Eye-Tracking Experiment," *Computers in Human Behavior* 55 (2016): 992–1000, https://doi.org/10.1016/j.chb.2015.09.038; Victoria C. Plaut and Robert P. Bartlett, "Blind Consent? A Social Psychological Investigation of Non-readership of Click-Through Agreements," *Law and Human Behavior*, June 16, 2011, 1–23, https://doi.org/10.1007/s10979-011-9288-y.

69. Ewa Luger, Stuart Moran, and Tom Rodden, "Consent for All: Revealing the Hidden Complexity of Terms and Conditions," in *Proceedings of the SIGCHI Conference on Human Factors in Computing Systems*, CHI '13 (New York: ACM, 2013), 2687–96, https://doi.org/10.1145/2470654.2481371.

70. Debra Cassens Weiss, "Chief Justice Roberts Admits He Doesn't Read the Computer Fine Print," *ABA Journal*, October 20, 2010, http://www.abajournal.com/news/article/chief_justice_roberts_admits_he_doesnt_read_the_computer_fine_print.

71. Margaret Jane Radin, *Boilerplate: The Fine Print, Vanishing Rights, and the Rule of Law* (Princeton, NJ: Princeton University Press, 2012), 14.

72. Radin, *Boilerplate,* 16–17.

73. Nancy S. Kim, *Wrap Contracts: Foundations and Ramifications* (Oxford: Oxford University Press, 2013), 50–69.

74. Jon Leibowitz, "Introductory Remarks at the FTC Privacy Roundtable," *FTC,* December 7, 2009, http://www.ftc.gov/speeches/leibowitz/091207.pdf.

75. Aleecia M. McDonald and Lorrie Faith Cranor, "The Cost of Reading Privacy Policies," *Journal of Policy for the Information Society*, 4, no. 3 (2008), http://hdl .handle.net/1811/72839.

76. Kim, *Wrap Contracts*, 70–72.

77. For an example of this rhetoric, see Tom Hayes, "America Needs a Department of 'Creative Destruction,'" *Huffington Post*, October 27, 2011, https://www .huffingtonpost.com/tom-hayes/america-needs-a-departmen_b_1033573.html.

78. Joseph A. Schumpeter, *Capitalism, Socialism, and Democracy* (New York: Harper Perennial Modern Classics, 2008), 68.

79. Schumpeter, *Capitalism*, 83.

80. Joseph A. Schumpeter, *The Economics and Sociology of Capitalism*, ed. Richard Swedberg (Princeton, NJ: Princeton University Press, 1991), 412 (italics mine).

81. Schumpeter, *Capitalism*, 83.

82. Yochai Benkler, *The Wealth of Networks: How Social Production Transforms Markets and Freedom* (New Haven, CT: Yale University Press, 2006).

83. Tom Worden, "Spain's Economic Woes Force a Change in Traditional Holiday Habits," *Guardian*, August 8, 2011, http://www.theguardian.com/world/2011 /aug/08/spain-debt-crisis-economy-august-economy.

84. Suzanne Daley, "On Its Own, Europe Backs Web Privacy Fights," *New York Times*, August 9, 2011, http://www.nytimes.com/2011/08/10/world /europe/10spain.html.

85. Ankit Singla et al., "The Internet at the Speed of Light" (ACM Press, 2014), https://doi.org/10.1145/2670518.2673876; Taylor Hatmaker, "There Could Soon Be Wi-Fi That Moves at the Speed of Light," *Daily Dot*, July 14, 2014, https:// www.dailydot.com/debug/sisoft-li-fi-vlc-10gbps.

86. "Google Spain SL v. Agencia Española de Protección de Datos (Case C-131/12 (May 13, 2014)," *Harvard Law Review* 128, no. 2 (2014): 735.

87. *Google Spain*, 2014 E.C.R. 317, 80–81.

88. Paul M. Schwartz and Karl-Nikolaus Peifer, "Transatlantic Data Privacy," *Georgetown Law Journal* 106, no. 115 (2017): 131, https://papers.ssrn.com /abstract=3066971. A few of the many excellent analyses of the right to be forgotten include Dawn Nunziato, "Forget About It? Harmonizing European and American Protections for Privacy, Free Speech, and Due Process" (GWU Law School Public Law Research Paper, George Washington University, January 1, 2015), http://scholarship.law.gwu.edu/faculty_publications/1295; Jeffrey Rosen, "The Right to Be Forgotten," *Stanford Law Review Online* 64 (2012): 88; "The Right to Be Forgotten (Google v. Spain)," *EPIC.org*, October 30, 2016, https://epic.org/privacy/right-to-be-forgotten; Ambrose Jones, Meg Leta, and Jef Ausloos, "The Right to Be Forgotten Across the Pond," *Journal of Information Policy* 3 (2012): 1–23; Hans Graux, Jef Ausloos, and Peggy Valcke, "The Right to Be Forgotten in the Internet Era," Interdisciplinary Centre for Law and ICT, November 12, 2012, http://www.researchgate.net

/publication/256039959_The_Right_to_Be_Forgotten_in_the_Internet_Era; Franz Werro, "The Right to Inform v. the Right to Be Forgotten: A Transatlantic Clash," *Liability in the Third Millennium*, May 2009, 285–300; "Google Spain SL v. Agencia Española de Protección de Datos." For a comprehensive review, see Anita L. Allen and Marc Rotenberg, *Privacy Law and Society*, 3rd ed. (St. Paul: West, 2016), 1520–52.

89. "Judgement in Case C-131/12: Google Spain SL, Google Inc. v Agencia Española de Protección de Datos, Mario Costeja González" (Court of Justice of the European Union, May 13, 2014), https://curia.europa.eu/jcms/upload/docs /application/pdf/2014-05/cp140070en.pdf.

90. Federico Fabbrini, "The EU Charter of Fundamental Rights and the Rights to Data Privacy: The EU Court of Justice as a Human Rights Court," in *The EU Charter of Fundamental Rights as a Binding Instrument: Five Years Old and Growing*, ed. Sybe de Vries, Ulf Burnitz, and Stephen Weatherill (Oxford: Hart, 2015), 21–22.

91. For excellent background on "free speech" and the First Amendment in cyberlaw, see Anupam Chander and Uyên Lê, "The Free Speech Foundations of Cyberlaw" (UC Davis Legal Studies Research Paper 351, September 2013, School of Law, University of California, Davis).

92. Henry Blodget, "Hey, Europe, Forget the 'Right to Be Forgotten'—Your New Google Ruling Is Nuts!" *Business Insider*, May 14, 2014, http://www .businessinsider.com/europe-google-ruling-2014-5.

93. Greg Sterling, "Google Co-Founder Sergey Brin: I Wish I Could Forget the 'Right to Be Forgotten,'" *Search Engine Land*, May 28, 2014, http:// searchengineland.com/google-co-founder-brin-wish-forget-right-forgotten -192648.

94. Richard Waters, "Google's Larry Page Resists Secrecy but Accepts Privacy Concerns," *Financial Times*, May 30, 2014, http://www.ft.com/cms/s/f3b127ea -e708-11e3-88be-00144feabdc0,Authorised=false.html?_i_location=http %3A%2F%2Fwww.ft.com%2Fcms%2Fs%2F0%2Ff3b127ea-e708-11e3-88be -00144feabdc0.html%3Fsiteedition%3Duk&siteedition=uk&_i_referer=https %3A%2F%2Fduckduckgo.com.

95. James Vincent, "Google Chief Eric Schmidt Says 'Right to Be Forgotten' Ruling Has Got the Balance 'Wrong,'" *Independent*, May 15, 2014, http://www .independent.co.uk/life-style/gadgets-and-tech/google-chief-eric-schmidt-says -right-to-be-forgotten-ruling-has-got-the-balance-wrong-9377231.html.

96. Pete Brodnitz et al., "Beyond the Beltway February 26–27 Voter Poll," *Beyond the Beltway Insights Initiative*, February 27, 2015, http://web.archive .org/web/20160326035834/http://beltway.bsgco.com/about; Mary Madden and Lee Rainie, "Americans' Attitudes About Privacy, Security and Surveillance," *PewResearchCenter* (blog), May 20, 2015, http://www.pewinternet.org/2015 /05/20/americans-attitudes-about-privacy-security-and-surveillance. A national poll conducted by Software Advice found that 61 percent of Americans believe

that some version of the right to be forgotten is necessary, 39 percent want a European-style blanket right to be forgotten, and nearly half were concerned that "irrelevant" search results can harm a person's reputation. A survey by YouGov found that 55 percent of Americans would support legislation similar to the right to be forgotten, compared to only 14 percent who would not. A US survey by Benenson Strategy Group and SKDKnickerbocker published nearly a year after the EU decision found that 88 percent of respondents somewhat (36 percent) or strongly (52 percent) supported a US law that would let them petition companies such as Google, Yahoo!, and Bing to remove certain personal information that appears in search results. See Daniel Humphries, "U.S. Attitudes Toward the 'Right to Be Forgotten,'" *Software Advice,* September 5, 2014, https://www.softwareadvice.com/security/industryview/right-to-be-forgotten-2014; Jake Gammon, "Americans Would Support 'Right to Be Forgotten,'" *YouGov,* December 6, 2017, https://today.yougov.com/news/2014/06/02/americans-would -support-right-be-forgotten; Mario Trujillo, "Public Wants 'Right to Be Forgotten' Online," *Hill,* March 19, 2015, http://thehill.com/policy/technology /236246-poll-public-wants-right-to-be-forgotten-online.

97. Francis Collins, "Vaccine Research: New Tactics for Tackling HIV," *NIH Director's Blog,* June 30, 2015, https://directorsblog.nih.gov/2015/06/30/vaccine -research-new-tactics-for-tackling-hiv; Liz Szabo, "Scientists Making Progress on AIDS Vaccine, but Slowly," *USAToday.com,* August 8, 2012, http://www .usatoday.com/news/health/story/2012-07-25/aids-vaccine/56485460/1.

98. Collins, "Vaccine Research."

99. Szabo, "Scientists Making Progress on AIDS Vaccine."

100. See Mary Madden and Lee Rainie, "Americans' Attitudes About Privacy, Security and Surveillance," *PewResearchCenter* (blog), May 20, 2015, http:// www.pewinternet.org/2015/05/20/americans-attitudes-about-privacy-security -and-surveillance.

CHAPTER THREE

1. See the discussion in David A. Hounshell, *From the American System to Mass Production, 1800–1932: The Development of Manufacturing Technology in the United States,* 7th ed., Studies in Industry and Society 4 (Baltimore: Johns Hopkins University Press, 1997).

2. See Reinhard Bendix, *Work and Authority in Industry: Ideologies of Management in the Course of Industrialization* (Berkeley: University of California Press, 1974).

3. David Farber, *Sloan Rules: Alfred P. Sloan and the Triumph of General Motors* (Chicago: University of Chicago Press, 2005); Henry Ford, *My Life and Work* (Garden City, NY: Ayer, 1922).

4. Chris Jay Hoofnagle, "Beyond Google and Evil: How Policy-Makers, Journalists, and Consumers Should Talk Differently About Google and Privacy," *First Monday,* April 6, 2009.

5. Reed Albergotti et al., "Employee Lawsuit Accuses Google of 'Spying Program,'" *Information,* December 20, 2016, https://www.theinformation .com/employee-lawsuit-accuses-google-of-spying-program.

6. See Steven Levy, *In the Plex: How Google Thinks, Works, and Shapes Our Lives* (New York: Simon & Schuster, 2011), 116; Hal R. Varian, "Biography of Hal R. Varian," UC Berkeley School of Information Management & Systems, October 3, 2017, http://people.ischool.berkeley.edu/~hal/people/hal/biography.html; "Economics According to Google," *Wall Street Journal,* July 19, 2007, http:// blogs.wsj.com/economics/2007/07/19/economics-according-to-google; Steven Levy, "Secret of Googlenomics: Data-Fueled Recipe Brews Profitability," *Wired,* May 22, 2009, http://archive.wired.com/culture/culturereviews/magazine /17-06/nep_googlenomics; Hal R. Varian, "Beyond Big Data," *Business Economics* 49, no. 1 (2014): 27–31.

Although it's important to note that Hal Varian is not an executive leader at Google, there is a great deal in the public record to suggest that he has played a leading role in helping Google's leaders grasp the operations and implications of their own commercial logic as well as its extension and elaboration. I compare Varian's insights to those of James Couzens at Ford. Couzens was an investor and businessman—he would later become a US senator—who served as general manager at Ford. He helped shepherd Ford's spectacular success with his clear grasp of the new logic of mass production and its economic significance. He was not a theoretician or a prolific writer like Varian, but his correspondence and articles were graced with unusual insight and have remained a vital source for scholars of mass production.

Varian spent several years as a consultant to Google before becoming the firm's chief economist in 2007. He notes in his biographical material that "since 2002 he has been involved in many aspects of the company, including auction design, econometric analysis, finance, corporate strategy, and public policy." When the *Wall Street Journal* reported on Varian's new position at Google in 2007, it noted that the position entailed building "a team of economists, statisticians, and analysts to assist the company in 'marketing, in human resources, in strategy, in policy related stuff.'" In his book on Google, Steven Levy quotes Eric Schmidt reflecting on how the firm learned to exploit its new "click economy": "We have Hal Varian and we have the physicists." In Levy's 2009 *Wired* article on "Googlenomics," Schmidt credits Varian's early examination of the firm's ad auctions with providing the eureka moment that clarified the true nature of Google's business: "All of a sudden, we realized we were in the auction business."

In the work that I cite here, Varian frequently illustrates his points with examples from Google. He often uses the first-person plural in these instances, such as "Google has been so successful with our own experiments that we have made them available to our advertisers and publishers in two programs." Therefore, it seems fair to assume that Varian's perspectives provide critical insights into the premises and objectives that define this new market form.

7. Hal R. Varian, "Computer Mediated Transactions," *American Economic Review* 100, no. 2 (2010): 1–10, https://doi.org/10.1257/aer.100.2.1; Varian, "Beyond Big Data." The first article, published in 2010, is the text of Varian's Richard T. Ely lecture. The second is also about computer-mediated transactions and overlaps substantially with the material in the Ely lecture.

8. Varian, "Beyond Big Data," 27.

9. "Machine Intelligence," *Research at Google,* 2018, https://web.archive.org /web/20180427114330/https://research.google.com/pubs/MachineIntelligence .html.

10. Ellen Meiksins Wood, *The Origin of Capitalism: A Longer View* (London: Verso, 2002), 125.

11. Wood, *The Origin of Capitalism,* 76, 93.

12. Levy, *In the Plex,* 46; Jennifer Lee, "Postcards from Planet Google," *New York Times,* November 28, 2002, http://www.nytimes.com/2002/11/28/technology /circuits/28goog.html.

13. Kenneth Cukier, "Data, Data Everywhere," *Economist,* February 25, 2010, http:// www.economist.com/node/15557443.

14. Levy, *In the Plex,* 46–48.

15. "Google Receives $25 Million in Equity Funding," *Google News,* July 7, 1999, http://googlepress.blogspot.com/1999/06/google-receives-25-million-in-equity .html.

16. Hal R. Varian, "Big Data: New Tricks for Econometrics," *Journal of Economic Perspectives* 28, no. 2 (2014): 113.

17. Sergey Brin and Lawrence Page, "The Anatomy of a Large-Scale Hypertextual Web Search Engine," *Computer Networks and ISDN Systems* 30, nos. 1–7 (1998): 18, https://doi.org/10.1016/S0169-7552(98)00110-X.

18. "NEC Selects Google to Provide Search Services on Japan's Leading BIGLOBE Portal Site," *Google Press,* December 18, 2000, http://googlepress.blogspot.com /2000/12/nec-selects-google-to-provide-search.html; "Yahoo! Selects Google as Its Default Search Engine Provider," *Google Press,* June 26, 2000, http:// googlepress.blogspot.com/2000/06/yahoo-selects-google-as-its-default.html.

19. Wood, *The Origin of Capitalism,* 125. Conflicts were already emerging between serving the interests of an expanding user base and the needs of the portals.

20. Scarlet Pruitt, "Search Engines Sued Over 'Pay-for-Placement,'" *CNN.com,* February 4, 2002, http://edition.cnn.com/2002/TECH/internet/02/04/search .engine.lawsuit.idg/index.html.

21. Saul Hansell, "Google's Toughest Search Is for a Business Model," *New York Times,* April 8, 2002, http://www.nytimes.com/2002/04/08/business/google-s -toughest-search-is-for-a-business-model.html.

22. Elliot Zaret, "Can Google's Search Engine Find Profits?" *ZDNet,* June 14, 1999, http://www.zdnet.com/article/can-googles-search-engine-find-profits.

23. John Greenwald, "Doom Stalks the Dotcoms," *Time,* April 17, 2000.

24. Alex Berenson and Patrick McGeehan, "Amid the Stock Market's Losses, a Sense the Game Has Changed," *New York Times,* April 16, 2000, http://www.nytimes.com/2000/04/16/business/amid-the-stock-market-s-losses-a-sense-the-game-has-changed.html; Laura Holson and Saul Hansell, "The Maniac Markets: The Making of a Market Bubble," *New York Times,* April 23, 2000.

25. Ken Auletta, *Googled: The End of the World as We Know It* (New York: Penguin, 2010).

26. Levy, *In the Plex,* 83.

27. Michel Ferrary and Mark Granovetter, "The Role of Venture Capital Firms in Silicon Valley's Complex Innovation Network," *Economy and Society* 38, no. 2 (2009): 347–48, https://doi.org/10.1080/03085140902786827.

28. Dave Valliere and Rein Peterson, "Inflating the Bubble: Examining Dot-Com Investor Behaviour," *Venture Capital* 6, no. 1 (2004): 1–22, https://doi.org/10.1080/1369106032000152452.

29. Valliere and Peterson, "Inflating the Bubble," 17–18. See also Udayan Gupta, ed., *Done Deals: Venture Capitalists Tell Their Stories* (Boston: Harvard Business School Press, 2000), 170–71, 190. Junfu Zhang, "Access to Venture Capital and the Performance of Venture-Backed Startups in Silicon Valley," *Economic Development Quarterly* 21, no. 2 (2007): 124–47.

30. Among the first generation of Silicon Valley venture-backed internet startups, 12.5 percent had completed IPOs by the end of 2001, compared to 7.3 percent in the rest of the country, while only 4.2 percent of valley startups attained profitability, a significantly lower proportion than in the rest of the country.

31. Zhang, 124–47.

32. Patricia Leigh Brown, "Teaching Johnny Values Where Money Is King," *New York Times,* March 10, 2000, http://www.nytimes.com/2000/03/10/us/teaching-johnny-values-where-money-is-king.html.

33. Kara Swisher, "Dot-Com Bubble Has Burst; Will Things Worsen in 2001?" *Wall Street Journal,* December 19, 2000, http://www.wsj.com/articles/SB9770911 8336535099.

34. S. Humphreys, "Legalizing Lawlessness: On Giorgio Agamben's State of Exception," *European Journal of International Law* 17, no. 3 (2006): 677–87, https://doi.org/10.1093/ejil/chl020.

35. Levy, *In the Plex,* 83–85.

36. Levy, 86–87 (italics mine).

37. See Lee, "Postcards."

38. Lee.

39. Lee.

40. Auletta, *Googled.*

41. John Markoff and G. Pascal Zachary, "In Searching the Web, Google Finds Riches," *New York Times,* April 13, 2003, http://www.nytimes.com/2003/04/13 /business/in-searching-the-web-google-finds-riches.html.

42. Peter Coy, "The Secret to Google's Success," *Bloomberg.com,* March 6, 2006, http://www.bloomberg.com/news/articles/2006-03-05/the-secret-to-googles -success (italics mine).

43. For example, consider this exemplary sample of Google patents filed during this general time frame: Krishna Bharat, Stephen Lawrence, and Mehran Sahami, Generating user information for use in targeted advertising, US9235849 B2, filed December 31, 2003, and issued January 12, 2016, http://www.google.com /patents/US9235849; Jacob Samuels Burnim, System and method for targeting advertisements or other information using user geographical information, US7949714 B1, filed December 5, 2005, and issued May 24, 2011, http://www .google.com/patents/US7949714; Alexander P. Carobus et al., Content-targeted advertising using collected user behavior data, US20140337128 A1, filed July 25, 2014, and issued November 13, 2014, http://www.google.com/patents /US20140337128; Jeffrey Dean, Georges Harik, and Paul Buchheit, Methods and apparatus for serving relevant advertisements, US20040059708 A1, filed December 6, 2002, and issued March 25, 2004, http://www.google.com/patents /US20040059708; Jeffrey Dean, Georges Harik, and Paul Buchheit, Serving advertisements using information associated with e-mail, US20040059712 A1, filed June 2, 2003, and issued March 25, 2004, http://www.google.com/patents /US20040059712; Andrew Fikes, Ross Koningstein, and John Bauer, System and method for automatically targeting web-based advertisements, US8041601 B2, issued October 18, 2011, http://www.google.com/patents/US8041601; Georges R. Harik, Generating information for online advertisements from internet data and traditional media data, US8438154 B2, filed September 29, 2003, and issued May 7, 2013, http://www.google.com/patents/US8438154; Georges R. Harik, Serving advertisements using a search of advertiser web information, US7647299 B2, filed June 30, 2003, and issued January 12, 2010, http://www .google.com/patents/US7647299; Rob Kniaz, Abhinay Sharma, and Kai Chen, Syndicated trackable ad content, US7996777 B2, issued August 9, 2011, http:// www.google.com/patents/US7996777; Method of delivery, targeting, and measuring advertising over networks, USRE44724 E1, filed May 24, 2000, and issued January 21, 2014, http://www.google.com/patents/USRE44724.

44. Three distinguished computer scientists, Krishna Bharat, Stephen Lawrence, and Meham Sahami, invented the technologies and techniques described in this patent (Generating user information for use in targeted advertising).

45. Bharat, Lawrence, and Sahami, Generating user information.

46. Bharat, Lawrence, and Sahami, 11.

47. Bharat, Lawrence, and Sahami, 11–12.

48. Bharat, Lawrence, and Sahami, 15 (italics mine).

49. Bharat, Lawrence, and Sahami, 15.
50. Bharat, Lawrence, and Sahami, 18.
51. Bharat, Lawrence, and Sahami, 12.
52. Bharat, Lawrence, and Sahami, 12 (italics mine).
53. Empirical work suggests the primacy of decision rights in users' privacy assessments. See Laura Brandimarte, Alessandro Acquisti, and George Loewenstein, "Misplaced Confidences: Privacy and the Control Paradox," *Social Psychological and Personality Science* 4, no. 3 (2010): 340–47.
54. Bharat, Lawrence, and Sahami, Generating user information, 17 (italics mine).
55. Bharat, Lawrence, and Sahami, 16–17. The list of attributes included the content (e.g., words, Anchortext, etc.) of websites that the user has visited (or visited in a certain time period); demographic information; geographic information; psychographic information; previous queries (and/or associated information) that the user has made; information about previous advertisements that the user has been shown, has selected, and/or has made purchases after viewing; information about documents (e.g., word processor files) viewed/requested and/or edited by the user; user interests; browsing activity; and previous purchasing behavior.
56. Bharat, Lawrence, and Sahami, Generating user information, 13.
57. Douglas Edwards, *I'm Feeling Lucky* (Boston: Houghton Mifflin Harcourt, 2011), 268.
58. Levy, *In the Plex,* 101.
59. The term is discussed in a video interview with Eric Schmidt and his colleague/coauthor Jonathan Rosenberg. See Eric Schmidt and Jonathan Rosenberg, "How Google Works," interview by Computer History Museum, October 15, 2014, https://youtu.be/3tNpYpcU5s4?t=3287.
60. See, for example, Edwards, *I'm Feeling Lucky,* 264–70.
61. See Levy, *In the Plex,* 13, 32, 35, 105–6 (quotation from 13); John Battelle, *The Search: How Google and Its Rivals Rewrote the Rules of Business and Transformed Our Culture* (New York: Portfolio, 2006), 65–66, 74, 82; Auletta, *Googled.*
62. See Levy, *In the Plex,* 94.
63. Humphreys, "Legalizing Lawlessness."
64. Michael Moritz, "Much Ventured, Much Gained," interview, *Foreign Affairs,* February 2015, https://www.foreignaffairs.com/interviews/2014-12-15/much-ventured-much-gained.
65. Hounshell, *From the American System,* 247–48.
66. Hounshell, 10.
67. Richard S. Tedlow, *Giants of Enterprise: Seven Business Innovators and the Empires They Built* (New York: HarperBusiness, 2003), 159–60; Donald Finlay Davis, *Conspicuous Production: Automobiles and Elites in Detroit 1899–1933* (Philadelphia, PA: Temple University Press, 1989), 122.

68. David M. Kristol, "HTTP Cookies: Standards, Privacy, and Politics," *ArXiv:Cs/0105018,* May 9, 2001, http://arxiv.org/abs/cs/0105018.

69. Richard M. Smith, "The Web Bug FAQ," *Electronic Frontier Foundation,* November 11, 1999, https://w2.eff.org/Privacy/Marketing/web_bug.html.

70. Kristol, "HTTP Cookies," 9–16; Richard Thieme, "Uncompromising Position: An Interview About Privacy with Richard Smith," *Thiemeworks,* January 2, 2000, http://www.thiemeworks.com/an-interview-with-richard-smith.

71. Kristol, "HTTP Cookies," 13–15.

72. "Amendment No. 9 to Form S-1 Registration Statement Under the Securities Act of 1933 for Google Inc.," Securities and Exchange Commission, August 18, 2004, https://www.sec.gov/Archives/edgar/data/1288776/000119312512025336/d260164d10k.htm.

73. Henry Ford, "Mass Production," *Encyclopedia Britannica* (New York: Encyclopedia Britannica, 1926), 821, http://memory.loc.gov/cgi-bin/query/h?ammem/coolbib:@field(NUMBER+@band(amrlg+lg48)).

74. See Levy, *In the Plex,* 69.

75. Edwards, *I'm Feeling Lucky,* 340–45.

76. Battelle, *The Search.*

77. Levy, *In the Plex,* 69.

78. See Hansell, "Google's Toughest Search."

79. See Markoff and Zachary, "In Searching the Web."

80. William O. Douglas, "Dissenting Statement of Justice Douglas, Regarding Warden v. Hayden, 387 U.S. 294" (US Supreme Court, April 12, 1967), https://www.law.cornell.edu/supremecourt/text/387/294; Nita A. Farahany, "Searching Secrets," *University of Pennsylvania Law Review* 160, no. 5 (2012): 1271.

81. George Orwell, *Politics and the English Language* (Peterborough: Broadview, 2006).

82. A typical example is this statement from the *Economist:* "Google exploits information that is a by-product of user interactions, or data exhaust, which is automatically recycled to improve the service or create an entirely new product." "Clicking for Gold," *Economist,* February 25, 2010, http://www.economist.com/node/15557431.

83. Valliere and Peterson, "Inflating the Bubble," 1–22.

84. See Lev Grossman, "Exclusive: Inside Facebook's Plan to Wire the World," *Time.com* (blog), December 2015, http://time.com/facebook-world-plan.

85. David Kirkpatrick, *The Facebook Effect: The Inside Story of the Company That Is Connecting the World* (New York: Simon & Schuster, 2011), 257.

86. Kirkpatrick, *The Facebook Effect,* 80; Auletta, *Googled.*

87. See Auletta, *Googled.*

88. Kirkpatrick, *The Facebook Effect,* 266.

89. "Selected Financial Data for Alphabet Inc.," Form 10-K, Commission File, United States Securities and Exchange Commission, December 31, 2016,

https://www.sec.gov/Archives/edgar/data/1652044/000165204417000008
/goog10-kq42016.htm#s58C60B74D56A630AD6EA2B64F53BD90C.
According to Alphabet's account in 2016, its revenues were $90,272,000,000.
This included "Google segment revenues of $89.5 billion with revenue growth
of 20% year over year and Other Bets revenues of $0.8 billion with revenue
growth of 82% year over year." Google Segment advertising revenues were
$79,383,000,000, or 88.73 percent of Google Segment revenues.

90. "Google Search Statistics—Internet Live Stats," Internet Live Stats, September
20, 2017, http://www.internetlivestats.com/google-search-statistics; Greg
Sterling, "Data: Google Monthly Search Volume Dwarfs Rivals Because
of Mobile Advantage," *Search Engine Land,* February 9, 2017, http://
searchengineland.com/data-google-monthly-search-volume-dwarfs-rivals
-mobile-advantage-269120. This translated into 76 percent of all desktop
searches and 96 percent of mobile searches in the US and corresponding shares
of 87 percent and 95 percent worldwide.

91. Roben Farzad, "Google at $400 Billion: A New No. 2 in Market Cap,"
BusinessWeek, February 12, 2014, http://www.businessweek.com
/articles/2014-02-12/google-at-400-billion-a-new-no-dot-2-in-market-cap.

92. "Largest Companies by Market Cap Today," *Dogs of the Dow,* 2017, https://
web.archive.org/web/20180701094340/http://dogsofthedow.com/largest
-companies-by-market-cap.htm.

93. Jean-Charles Rochet and Jean Tirole, "Two-Sided Markets: A Progress Report,"
RAND Journal of Economics 37, no. 3 (2006): 645–67.

94. For a discussion on this point and its relation to online target advertising, see
Katherine J. Strandburg, "Free Fall: The Online Market's Consumer Preference
Disconnect" (working paper, New York University Law and Economics,
October 1, 2013).

95. Kevin Kelly, "The Three Breakthroughs That Have Finally Unleashed AI on the
World," *Wired,* October 27, 2014, https://www.wired.com/2014/10/future-of
-artificial-intelligence.

96. Xiaoliang Ling et al., "Model Ensemble for Click Prediction in Bing Search
Ads," in *Proceedings of the 26th International Conference on World Wide Web
Companion,* 689–98, https://doi.org/10.1145/3041021.3054192.

97. Ruoxi Wang et al., "Deep & Cross Network for Ad Click Predictions,"
ArXiv:1708.05123 [Computer Science. Learning], August 16, 2017, http://arxiv
.org/abs/1708.05123.

CHAPTER FOUR

1. See Steven Levy, "Secret of Googlenomics: Data-Fueled Recipe Brews
Profitability," *Wired,* May 22, 2009, http://archive.wired.com/culture/culture
reviews/magazine/17-06/nep_googlenomics.

2. Douglas Edwards, *I'm Feeling Lucky* (Boston: Houghton Mifflin Harcourt,
2011), 291.

3. Karl Polanyi, *The Great Transformation: The Political and Economic Origins of Our Time*, 2nd ed. (Boston: Beacon, 2001), 75–76.

4. Karl Marx, *Capital*, trans. David Fernbach, 3rd ed. (Penguin, 1992), Chapter 6.

5. Hannah Arendt, *The Origins of Totalitarianism* (New York: Schocken, 2004), 198.

6. Michael J. Sandel, *What Money Can't Buy: The Moral Limits of Markets* (New York: Farrar, Straus and Giroux, 2013).

7. David Harvey, *The New Imperialism* (New York: Oxford University Press, 2005), 153.

8. Sergey Brin, "2004 Founders' IPO Letter," Google, https://abc.xyz/investor/founders-letters/2004.

9. Cato Institute, *Eric Schmidt Google/Cato Interview*, YouTube, 2014, https://www.youtube.com/watch?v=BH3vjTz8OII.

10. Nick Summers, "Why Google Is Issuing a New Kind of Toothless Stock," *Bloomberg.com*, April 3, 2014, https://www.bloomberg.com/news/articles/2014-04-03/why-google-is-issuing-c-shares-a-new-kind-of-powerless-stock. When shareholders objected to the system at the company's annual general meeting by casting 180 million votes for an equal-voting-rights resolution, they were swamped by the 551 million votes owned by the founders.

11. Eric Lam, "New Google Share Classes Issued as Founders Cement Grip," *Bloomberg.com*, April 3, 2014, https://www.bloomberg.com/news/articles/2014-04-03/new-google-shares-hit-market-as-founders-cement-grip-with-split.

12. Tess Townsend, "Alphabet Shareholders Want More Voting Rights but Larry and Sergey Don't Want It That Way," *Recode*, June 13, 2017, https://www.recode.net/2017/6/13/15788892/alphabet-shareholder-proposals-fair-shares-counted-equally-no-supervote.

13. Ronald W. Masulis, Cong Wang, and Fei Xie, "Agency Problems at Dual-Class Companies," *Journal of Finance* 64, no. 4 (2009): 1697–1727, https://doi.org/10.1111/j.1540-6261.2009.01477.x; Randall Smith, "One Share, One Vote?" *Wall Street Journal*, October 28, 2011, https://www.wsj.com/articles/SB10001424052970203911804576653591322367506. In 2017 Snap's IPO offered only nonvoting shares, leaving its founders with 70 percent control of all votes and the remainder in the hand of pre-IPO investors. See Maureen Farrell, "In Snap IPO, New Investors to Get Zero Votes, While Founders Keep Control," *Wall Street Journal*, January 17, 2017, http://www.wsj.com/articles/in-snap-ipo-new-investors-to-get-zero-votes-while-founders-keep-control-1484568034. Other IPOs featured super-voting shares with as many as thirty to ten thousand times as many votes per share as the ordinary class of stock. See Alfred Lee, "Where Supervoting Rights Go to the Extreme," *Information*, March 22, 2016.

14. "Power Play: How Zuckerberg Wrested Control of Facebook from His Shareholders," *VentureBeat* (blog), February 2, 2012, https://venturebeat.com/2012/02/01/zuck-power-play. https://www.nasdaq.com/articles/facebook-fb-decides-against-creation-class-c-shares-2017-09-25.

15. Spencer Feldman, "IPOs in 2016 Increasingly Include Dual-Class Shareholder Voting Rights," *Securities Regulation & Law Report*, 47 SRLR 1342, July 4, 2016; R. C. Anderson, E. Ottolenghi, and D. M. Reeb, "The Extreme Control Choice," paper presented at the Research Workshop on Family Business, Lehigh University, 2017.

16. Adam Hayes, "Facebook's Most Important Acquisitions," *Investopedia*, February 11, 2015, http://www.investopedia.com/articles/investing/021115 /facebooks-most-important-acquisitions.asp; Rani Molla, "Google Parent Company Alphabet Has Made the Most AI Acquisitions," *Recode*, May 19, 2017, https://www.recode.net/2017/5/19/15657758/google-artificial -intelligence-ai-investments; "The Race for AI: Google, Baidu, Intel, Apple in a Rush to Grab Artificial Intelligence Startups," *CB Insights Research*, July 21, 2017, http://www.cbinsights.com/research/top-acquirers-ai-startups -ma-timeline.

17. "Schmidt: We Paid $1 Billion Premium for YouTube," *CNET*, March 27, 2018, https://www.cnet.com/news/schmidt-we-paid-1-billion-premium-for -youtube.

18. Adrian Covert, "Facebook Buys WhatsApp for $19 Billion," *CNNMoney*, February 19, 2014, http://money.cnn.com/2014/02/19/technology/social /facebook-whatsapp/index.html.

19. Tim Fernholz, "How Mark Zuckerberg's Control of Facebook Lets Him Print Money," *Quartz* (blog), March 27, 2014, https://qz.com/192779/how-mark -zuckerbergs-control-of-facebook-lets-him-print-money.

20. Duncan Robinson, "Facebook Faces EU Fine Over WhatsApp Data-Sharing," *Financial Times*, December 20, 2016, https://www.ft.com/content/f652746c -c6a4-11e6-9043-7e34c07b46ef; Tim Adams, "Margrethe Vestager: 'We Are Doing This Because People Are Angry,'" *Observer*, September 17, 2017, http:// www.theguardian.com/world/2017/sep/17/margrethe-vestager-people-feel -angry-about-tax-avoidance-european-competition-commissioner; "WhatsApp FAQ—How Do I Choose Not to Share My Account Information with Facebook to Improve My Facebook Ads and Products Experiences?" *WhatsApp.com*, August 28, 2016, https://www.whatsapp.com/faq/general/26000016.

21. Eric Schmidt and Jared Cohen, *The New Digital Age: Transforming Nations, Businesses, and Our Lives* (New York: Vintage, 2014).

22. Arendt, *The Origins of Totalitarianism*, 183.

23. Vinod Khosla, "Fireside Chat with Google Co-Founders, Larry Page and Sergey Brin," *Khosla Ventures*, July 3, 2014, http://www.khoslaventures .com/fireside-chat-with-google-co-founders-larry-page-and-sergey-brin.

24. Holman W. Jenkins, "Google and the Search for the Future," *Wall Street Journal*, August 14, 2010, http://www.wsj.com/articles/SB10001424052748704901104575 5423294099527212.

25. See Lillian Cunningham, "Google's Eric Schmidt Expounds on His Senate Testimony," *Washington Post*, September 30, 2011, http://www.washingtonpost

.com/national/on-leadership/googles-eric-schmidt-expounds-on-his-senate
-testimony/2011/09/30/gIQAPyVgCL_story.html.

26. Pascal-Emmanuel Gobry, "Eric Schmidt to World Leaders at EG8: Don't
Regulate Us, or Else," *Business Insider,* May 24, 2011, http://www
.businessinsider.com/eric-schmidt-google-eg8-2011-5.

27. See Jay Yarow, "Google CEO Larry Page Wants a Totally Separate World Where
Tech Companies Can Conduct Experiments on People," *Business Insider,* May
16, 2013, http://www.businessinsider.com/google-ceo-larry-page-wants-a-place
-for-experiments-2013-5.

28. Conor Dougherty, "Tech Companies Take Their Legislative Concerns to the
States," *New York Times,* May 27, 2016, http://www.nytimes.com/2016/05/28
/technology/tech-companies-take-their-legislative-concerns-to-the-states.html;
Tim Bradshaw, "Google Hits Out at Self-Driving Car Rules," *Financial Times,*
December 18, 2015, http://www.ft.com/intl/cms/s/0/d4afee02-a517
-11e5-97e1-a754d5d9538c.html?ftcamp=crm/email/20151217/nbe
/InTodaysFT/product#axzz3ufyqWRo2; Jon Brodkin, "Google and Facebook
Lobbyists Try to Stop New Online Privacy Protections," *Ars Technica,* May 24,
2017, https://arstechnica.com/tech-policy/2017/05/google-and-facebook
-lobbyists-try-to-stop-new-online-privacy-protections.

29. Robert H. Wiebe, *The Search for Order: 1877–1920* (New York: Hill and
Wang, 1967), 135–37. Wiebe summarizes the worldview put forth by the
millionaires as they coordinated to repulse the electoral threat to industrial
capital, and it will sound familiar to anyone who has read the justifications of
the Silicon Valley tycoons and their adulation of all things "destructive" and
"entrepreneurial." According to the nineteenth-century catechism, only the
"highest types of their race discovered more effective ways to combine land,
labor, and capital, and drew society upward as the rest reorganized behind their
leaders." The majority of "ordinary talent" was left to divide what remained after
the requirements of capital, and "the weakest simply disappeared." The result
was to be "an ever improving race winnowed by competition." Any violation
of these "natural laws" would only benefit "the survival of the unfittest" and
reverse the evolution of the race.

30. David Nasaw, "Gilded Age Gospels," in *Ruling America: A History of Wealth
and Power in a Democracy,* ed. Steve Fraser and Gary Gerstle (Cambridge, MA:
Harvard University Press, 2005), 124–25.

31. Nasaw, "Gilded Age," 132.

32. Nasaw, 146.

33. Lawrence M. Friedman, *American Law in the 20th Century* (New Haven, CT:
Yale University Press, 2004), 15–28.

34. Nasaw, "Gilded Age," 148.

35. For two outstanding discussions, see Chris Jay Hoofnagle, *Federal Trade
Commission: Privacy Law and Policy* (New York: Cambridge University Press,
2016); Julie E. Cohen, "The Regulatory State in the Information Age," *Theoretical*

Inquiries in Law 17, no. 2 (2016), http://www7.tau.ac.il/ojs/index.php/til/article /view/1425.

36. Jodi L. Short, "The Paranoid Style in Regulatory Reform," *Hastings Law Journal* 63 (January 12, 2011): 633.

37. For a wonderful collection of essays on this subject, see Steve Fraser and Gary Gerstle, eds., *The Rise and Fall of the New Deal Order 1930–1980* (Princeton, NJ: Princeton University Press, 1989).

38. Alan Brinkley, *Liberalism and Its Discontents* (Cambridge, MA: Harvard University Press, 2000).

39. Short, "The Paranoid Style," 44–46.

40. Short, 52–53. Economic historian Philip Mirowski summarizes the "meta-theses" that since the 1980s helped to constitute neoliberalism as a loose "paradigm," despite its amorphous, multifaceted, and sometimes contradictory theories and practices. Among these, several became essential shelter for the surveillance capitalists' bold actions, secret operations, and rhetorical misdirection: (1) democracy was to be constrained in favor of actively reconstructing the state as the agent of a stable market society; (2) the entrepreneur and the corporation were conflated, enshrining "corporate personhood," rather than the rights of citizens, as the focus of legal protections; (3) freedom was defined negatively, as "freedom from" interference in the natural laws of competition, and all control was understood as coercive, except for market control; and (4) inequality of wealth and rights was accepted and even celebrated as a necessary feature of a successful market system and a force for progress. Later, surveillance capitalism's success, its aggressive rhetoric, and its leaders' willingness to fight every challenge, both in the courts and in the court of public opinion, further cemented these guiding principles in US politics, economic policies, and regulatory approaches. See Philip Mirowski, *Never Let a Serious Crisis Go to Waste: How Neoliberalism Survived the Financial Meltdown* (London: Verso, 2013). See also Wendy Brown, *Undoing the Demos: Neoliberalism's Stealth Revolution* (New York: Zone Books, 2015); David M. Kotz, *The Rise and Fall of Neoliberal Capitalism* (Cambridge, MA: Harvard University Press, 2015), 166–75.

41. Frank A. Pasquale, "Privacy, Antitrust, and Power," *George Mason Law Review* 20, no. 4 (2013): 1009–24.

42. There is wide-ranging scholarship on the internet companies' claims to First Amendment protections as a defense against regulation. Here are a few among many important contributions: Andrew Tutt, "The New Speech," *Hastings Constitutional Law Quarterly*, 41 (July 17, 2013): 235; Richard Hasen, "Cheap Speech and What It Has Done (to American Democracy)," *First Amendment Law Review* 16 (January 1, 2017), http://scholarship.law.uci.edu /faculty_scholarship/660; Dawn Nunziato, "With Great Power Comes Great Responsibility: Proposed Principles of Digital Due Process for ICT Companies" (GWU Law School Public Law research paper, George Washington University,

January 1, 2013), http://scholarship.law.gwu.edu/faculty_publications/1293; Tim Wu, "Machine Speech," *University of Pennsylvania Law Review* 161, no. 6 (2013): 1495; Dawn Nunziato, "Forget About It? Harmonizing European and American Protections for Privacy, Free Speech, and Due Process" (GWU Law School Public Law research paper, George Washington University, January 1, 2015), http://scholarship.law.gwu.edu/faculty_publications/1295; Marvin Ammori, "The 'New' *New York Times*: Free Speech Lawyering in the Age of Google and Twitter," *Harvard Law Review* 127 (June 20, 2014): 2259–95; Jon Hanson and Ronald Chen, "The Illusion of Law: The Legitimating Schemas of Modern Policy and Corporate Law," *Legitimating Schemas of Modern Policy and Corporate Law* 103, no. 1 (2004): 1–149.

43. Steven J. Heyman, "The Third Annual C. Edwin Baker Lecture for Liberty, Equality, and Democracy: The Conservative-Libertarian Turn in First Amendment Jurisprudence" (SSRN Scholarly Paper, Rochester, NY: Social Science Research Network, October 8, 2014), 300, https://papers.ssrn.com/abstract=2497190.

44. Heyman, "The Third Annual C. Edwin Baker Lecture," 277; Andrew Tutt, "The New Speech."

45. Daniel J. H. Greenwood, "Neofederalism: The Surprising Foundations of Corporate Constitutional Rights," *University of Illinois Law Review* 163 (2017): 166, 221.

46. Frank A. Pasquale, "The Automated Public Sphere" (Legal Studies research paper, University of Maryland, November 10, 2017).

47. Ammori, "The 'New' *New York Times*," 2259–60.

48. Adam Winkler, *We the Corporations* (New York: W. W. Norton, 2018), xxi.

49. "Section 230 of the Communications Decency Act," Electronic Frontier Foundation, n.d., https://www.eff.org/issues/cda230.

50. Christopher Zara, "The Most Important Law in Tech Has a Problem," *Wired,* January 3, 2017.

51. David S. Ardia, "Free Speech Savior or Shield for Scoundrels: An Empirical Study of Intermediary Immunity Under Section 230 of the Communications Decency Act" (SSRN Scholarly Paper, Rochester, NY: Social Science Research Network, June 16, 2010), https://papers.ssrn.com/abstract=1625820.

52. Paul Ehrlich, "Communications Decency Act 230," *Berkeley Technology Law Journal* 17 (2002): 404.

53. Ardia, "Free Speech Savior or Shield for Scoundrels."

54. See Zara, "The Most Important Law in Tech."

55. Zara.

56. David Lyon, *Surveillance After September 11,* Themes for the 21st Century (Malden, MA: Polity, 2003), 7; Jennifer Evans, "Hijacking Civil Liberties: The USA Patriot Act of 2001," *Loyola University Chicago Law Journal* 33, no. 4 (2002): 933; Paul T. Jaeger, John Carlo Bertot, and Charles R. McClure, "The Impact of the USA Patriot Act on Collection and Analysis of Personal

Information Under the Foreign Intelligence Surveillance Act," *Government Information Quarterly* 20, no. 3 (2003): 295–314, https://doi.org/10.1016/S0740-624X(03)00057-1.

57. The first wave of consumer-oriented privacy legislation in the US dates from the 1970s, with important landmark bills in the US Congress such as the Fair Credit Reporting Act in 1970 and the Fair Information Practices Principles in 1973. The OECD adopted a strong set of privacy guidelines in 1980, and the EU's first Data Protection Directive took effect in 1998. See Peter Swire, "The Second Wave of Global Privacy Protection: Symposium Introduction," *Ohio State Law Journal* 74, no. 6 (2013): 842–43; Peter P. Swire, "Privacy and Information Sharing in the War on Terrorism," *Villanova Law Review* 51, no. 4 (2006): 951; Ibrahim Altaweel, Nathaniel Good, and Chris Jay Hoofnagle, "Web Privacy Census," *Technology Science*, December 15, 2015, https://techscience.org/a/2015121502.

58. Swire, "Privacy and Information Sharing," 951; Swire, "The Second Wave"; Hoofnagle, *Federal Trade Commission;* Brody Mullins, Rolfe Winkler, and Brent Kendall, "FTC Staff Wanted to Sue Google," *Wall Street Journal,* March 20, 2015; Daniel J. Solove and Woodrow Hartzog, "The FTC and the New Common Law of Privacy," *Columbia Law Review* 114, no. 3 (2014): 583–676; Brian Fung, "The FTC Was Built 100 Years Ago to Fight Monopolists. Now, It's Washington's Most Powerful Technology Cop," *Washington Post,* September 25, 2014, https://www.washingtonpost.com/blogs/the-switch/wp/2014/09/25/the-ftc-was-built-100-years-ago-to-fight-monopolists-now-its-washingtons-most-powerful-technology-cop; Stephen Labaton, "The Regulatory Signals Shift; F.T.C. Serves as Case Study of Differences Under Bush," *New York Times,* June 12, 2001, http://www.nytimes.com/2001/06/12/business/the-regulatory-signals-shift-ftc-serves-as-case-study-of-differences-under-bush.html; Tanzina Vega and Edward Wyatt, "U.S. Agency Seeks Tougher Consumer Privacy Rules," *New York Times,* March 26, 2012, http://www.nytimes.com/2012/03/27/business/ftc-seeks-privacy-legislation.html.

59. Robert Pitofsky et al., "Privacy Online: Fair Information Practices in the Electronic Marketplace: A Federal Trade Commission Report to Congress," Federal Trade Commission, May 1, 2000, 35, https://www.ftc.gov/reports/privacy-online-fair-information-practices-electronic-marketplace-federal-trade-commission.

60. Pitofsky et al., "Privacy Online," 36–37. The proposed legislation would set forth a basic level of privacy protection for all visits to consumer-oriented commercial websites to the extent not already provided by the Children's Online Privacy Protection Act (COPPA). Such legislation would set out the basic standards of practice governing the collection of information online and provide an implementing agency with the authority to promulgate more-detailed standards pursuant to the Administrative Procedure Act, including

authority to enforce those standards. All consumer-oriented commercial websites that collect personal identifying information from or about consumers online, to the extent not covered by the COPPA, would be required to comply with the four widely accepted fair information practices:

(1) Notice. Websites would be required to provide consumers clear and conspicuous notice of their information practices, including what information they collect, how they collect it (e.g., directly or through nonobvious means such as cookies), how they use it, how they provide Choice, Access, and Security to consumers, whether they disclose the information collected to other entities, and whether other entities are collecting information through the site.

(2) Choice. Websites would be required to offer consumers choices as to how their personal identifying information is used beyond the use for which the information was provided (e.g., to consummate a transaction). Such choice would encompass both internal secondary uses (such as marketing back to consumers) and external secondary uses (such as disclosing data to other entities).

(3) Access. Websites would be required to offer consumers reasonable access to the information a website has collected about them, including a reasonable opportunity to review the information and to correct inaccuracies or delete information.

(4) Security. Websites would be required to take reasonable steps to protect the security of the information they collect from consumers.

61. Swire, "The Second Wave," 845.
62. Paul M. Schwartz, "Systematic Government Access to Private-Sector Data in Germany," *International Data Privacy Law* 2, no. 4 (2012): 289, 296; Ian Brown, "Government Access to Private-Sector Data in the United Kingdom," *International Data Privacy Law* 2, no. 4 (2012): 230, 235; W. Gregory Voss, "After Google Spain and Charlie Hebdo: The Continuing Evolution of European Union Data Privacy Law in a Time of Change," *Business Lawyer* 71, no. 1 (2015): 281; Mark Scott, "Europe, Shaken by Paris Attacks, Weighs Security with Privacy Rights," *New York Times—Bits Blog,* September 18, 2015; Frank A. Pasquale, "Privacy, Antitrust, and Power," *George Mason Law Review* 20, no. 4 (2013): 1009–24; Alissa J. Rubin, "Lawmakers in France Move to Vastly Expand Surveillance," *New York Times,* May 5, 2015, http://www.nytimes.com/2015/05/06/world/europe/french-legislators-approve-sweeping-intelligence-bill.html; Georgina Prodham and Michael Nienaber, "Merkel Urges Germans to Put Aside Fear of Big Data," *Reuters,* June 9, 2015, https://www.reuters.com/article/us-germany-technology-merkel/merkel-urges-germans-to-put-aside-fear-of-big-data-idUSKBN0OP2EM 20150609.

63. Richard A. Clarke et al., *The NSA Report: Liberty and Security in a Changing World* (Princeton, NJ: Princeton University Press, 2014), 27, 29; Declan McCullagh, "How 9/11 Attacks Reshaped U.S. Privacy Debate," *CNET*, September 9, 2011, http://www.cnet.com/news/how-911-attacks-reshaped-u-s-privacy -debate. *The NSA Report*, compiled by the president's review panel in 2013, describes the intelligence mandate that made this possible: "The September 11 attacks were a vivid demonstration of the need for detailed information about the activities of potential terrorists...some information, which could have been useful, was not collected and other information, which could have helped to prevent the attacks, was not shared among departments....One thing seemed clear: If the government was overly cautious in its efforts to detect and prevent terrorist attacks, the consequences for the nation could be disastrous."

64. Hoofnagle, *Federal Trade Commission*, 158.

65. See Andrea Peterson, "Former NSA and CIA Director Says Terrorists Love Using Gmail," *Washington Post*, September 15, 2013, https://www .washingtonpost.com/news/the-switch/wp/2013/09/15/former-nsa-and-cia -director-says-terrorists-love-using-gmail.

66. Marc Rotenberg, "Security and Liberty: Protecting Privacy, Preventing Terrorism," testimony before the National Commission on Terrorist Attacks upon the United States, 2003.

67. Swire, "The Second Wave," 846.

68. Hoofnagle, *Federal Trade Commission*, Chapter 6.

69. Lyon, *Surveillance After September 11*, 15.

70. Richard Herbert Howe, "Max Weber's Elective Affinities: Sociology Within the Bounds of Pure Reason," *American Journal of Sociology* 84, no. 2 (1978): 366–85.

71. See Joe Feuerherd, "'Total Information Awareness' Imperils Civil Rights, Critics Say," *National Catholic Reporter*, November 29, 2002, http://natcath.org/NCR _Online/archives2/2002d/112902/112902d.htm.

72. See Matt Marshall, "Spying on Startups," *Mercury News*, November 17, 2002.

73. Marshall, "Spying on Startups."

74. Mark Williams Pontin, "The Total Information Awareness Project Lives On," *MIT Technology Review*, April 26, 2006, https://www.technologyreview .com/s/405707/the-total-information-awareness-project-lives-on.

75. John Markoff, "Taking Spying to Higher Level, Agencies Look for More Ways to Mine Data," *New York Times*, February 25, 2006, http://www.nytimes .com/2006/02/25/technology/25data.html.

76. Inside Google, "Lost in the Cloud: Google and the US Government," Consumer Watchdog, January 2011, insidegoogle.com/wp-content/uploads/2011/01 /GOOGGovfinal012411.pdf.

77. Nafeez Ahmed, "How the CIA Made Google," *Medium* (blog), January 22, 2015, https://medium.com/insurge-intelligence/how-the-cia-made -google-e836451a959e.

78. Verne Kopytoff, "Google Has Lots to Do with Intelligence," *SFGate*, March 30, 2008, http://www.sfgate.com/business/article/Google-has-lots-to-do-with -intelligence-3221500.php.

79. Noah Shachtman, "Exclusive: Google, CIA Invest in 'Future' of Web Monitoring," *Wired*, July 28, 2010, http://www.wired.com/2010/07/exclusive -google-cia.

80. Ryan Gallagher, "The Surveillance Engine: How the NSA Built Its Own Secret Google," *Intercept* (blog), August 25, 2014, https://firstlook.org/theintercept /2014/08/25/icreach-nsa-cia-secret-google-crisscross-proton.

81. Robyn Winder and Charlie Speight, "Untangling the Web: An Introduction to Internet Research," National Security Agency Center for Digital Content, March 2013, http://www.governmentattic.org/8docs/UntanglingTheWeb-NSA_2007 .pdf.

82. Richard O'Neill, *Seminar on Intelligence, Command, and Control* (Cambridge, MA: Highlands Forums Press, 2001), http://www.pirp.harvard.edu/pubs_pdf /o'neill/o'neill-i01-3.pdf; Richard P. O'Neill, "The Highlands Forum Process," interview by Oettinger, April 5, 2001.

83. Mary Anne Franks, "Democratic Surveillance" (SSRN Scholarly Paper, Rochester, NY: Social Science Research Network, November 2, 2016), https:// papers.ssrn.com/abstract=2863343.

84. Ahmed, "How the CIA Made Google."

85. Stephanie A. DeVos, "The Google-NSA Alliance: Developing Cybersecurity Policy at Internet Speed," *Fordham Intellectual Property, Media and Entertainment Law Journal* 21, no. 1 (2011): 173–227.

86. The elective affinities that linked government operations to Google and the wider commercial surveillance project are evident in the decade that followed 9/11 as the NSA strived to become more like Google, analyzing and integrating Google's capabilities in a variety of domains. For detailed insight, see "Lost in the Cloud: Google and the US Government," *Inside Google*, January 2011, insidegoogle.com/wp-content/uploads/2011/01/GOOGGovfinal012411.pdf; Ahmed, "How the CIA Made Google"; Verne Kopytoff, "Google Has Lots to Do with Intelligence," *SFGate*, March 30, 2008, http://www.sfgate.com/business /article/Google-has-lots-to-do-with-intelligence-3221500.php; "Google Acquires Keyhole Corp—News Announcements," *Google Press*, October 27, 2004, http://googlepress.blogspot.com/2004/10/google-acquires-keyhole-corp .html; Josh G. Lerner et al., "In-Q-Tel: Case 804-146," *Harvard Business School Publishing*, February 2004, 1–20; Winder and Speight, "Untangling the Web"; Gallagher, "The Surveillance Engine"; Ellen Nakashima, "Google to Enlist NSA to Help It Ward Off Cyberattacks," *Washington Post*, February 4, 2010, http:// www.washingtonpost.com/wp-dyn/content/article/2010/02/03/AR20100 20304057.html; Mike Scarcella, "DOJ Asks Court to Keep Secret Any Partnership Between Google, NSA," *BLT: Blog of Legal Times*, March 9, 2012, 202, http://legaltimes.typepad.com/blt/2012/03/doj-asks-court-to-keep-secret

-any-partnership-between-google-nsa.html; Shane Harris, *@WAR: The Rise of the Military-Internet Complex* (Boston: Houghton Mifflin Harcourt, 2014), 175.

87. Jack Balkin, "The Constitution in the National Surveillance State," *Minnesota Law Review* 93, no. 1 (2008), http://digitalcommons.law.yale.edu /fss_papers/225.

88. Jon D. Michaels, "All the President's Spies: Private-Public Intelligence Partnerships in the War on Terror," *California Law Review* 96, no. 4 (2008): 901–66.

89. Michaels, "All the President's Spies," 908; Chris Hoofnagle, "Big Brother's Little Helpers: How ChoicePoint and Other Commercial Data Brokers Collect and Package Your Data for Law Enforcement," *North Carolina Journal of International Law and Commercial Regulation* 29 (January 1, 2003): 595; Junichi P. Semitsu, "From Facebook to Mug Shot: How the Dearth of Social Networking Privacy Rights Revolutionized Online Government Surveillance," *Pace Law Review* 31, no. 1 (2011).

90. Mike McConnell, "Mike McConnell on How to Win the Cyber-War We're Losing," *Washington Post,* February 28, 2010, http://www.washingtonpost.com /wp-dyn/content/story/2010/02/25/ST2010022502680.html.

91. Davey Alba, "Pentagon Taps Eric Schmidt to Make Itself More Google-ish," *Wired,* March 2, 2016, https://www.wired.com/2016/03/ex-google -ceo-eric-schmidt-head-pentagon-innovation-board; Lee Fang, "The CIA Is Investing in Firms That Mine Your Tweets and Instagram Photos," *Intercept,* April 14, 2016, https://theintercept.com/2016/04/14/in-undisclosed-cia -investments-social-media-mining-looms-large.

92. Fred H. Cate and James X. Dempsey, eds., *Bulk Collection: Systematic Government Access to Private-Sector Data* (New York: Oxford University Press, 2017), xxv–xxvi.

93. Michael Alan Bernstein, *The Great Depression: Delayed Recovery and Economic Change in America, 1929–1939,* Studies in Economic History and Policy (Cambridge, MA: Cambridge University Press, 1987), chapters 1, 8.

94. http://bits.blogs.nytimes.com/2008/11/07/how-obamas-internet-campaign -changed-politics/?_r=0.

95. Sasha Issenberg, "The Romney Campaign's Data Strategy," *Slate,* July 17, 2012, http://www.slate.com/articles/news_and_politics/victory_lab/2012/07/the _romney_campaign_s_data_strategy_they_re_outsourcing_.single.html. See also Joe Lertola and Bryan Christie Design, "A Short History of the Next Campaign," *Politico,* February 27, 2014, http://www.politico.com/magazine /story/2014/02/a-short-history-of-the-next-campaign-103672.html.

96. Daniel Kreiss and Philip N. Howard, "New Challenges to Political Privacy: Lessons from the First U.S. Presidential Race in the Web 2.0 Era," *International Journal of Communication* 4 (2010): 1032–50.

97. Sasha Issenberg, *The Victory Lab: The Secret Science of Winning Campaigns* (New York: Crown, 2012), 271.

98. "I'm doing this personally," Schmidt told journalists. "Google is officially neutral" in the campaign. His first appearance with Obama was at an event in Florida where they moderated a panel on the economy. Schmidt told the *Wall Street Journal* that his planned endorsement of Obama was a "natural evolution" of his role as an informal advisor. See Monica Langley and Jessica E. Vascellaro, "Google CEO Backs Obama," *Wall Street Journal*, October 20, 2008, http://www.wsj.com/articles/SB122446734650049199; and Jeff Zeleny and Elisabeth Bumiller, "Candidates Face Off Over Economic Plans," *New York Times*, October 21, 2008, http://www.nytimes.com/2008/10/22/us/politics/22campaign.html.

99. Robert Reich, "Obama's Transition Economic Advisory Board: The Full List," *US News & World Report*, November 7, 2008, http://www.usnews.com/news/campaign-2008/articles/2008/11/07/obamas-transition-economic-advisory-board-the-full-listn.

100. Eamon Javers, "Obama–Google Connection Scares Competitors," *Politico*, November 10, 2008, http://www.politico.com/news/stories/1108/15487.html.

101. "Diary of a Love Affair: Obama and Google (Obama@Google)," *Fortune*, November 14, 2007, http://archive.fortune.com/galleries/2009/fortune/0910/gallery.obama_google.fortune/2.html.

102. Brody Mullins, "Google Makes Most of Close Ties with White House," *Wall Street Journal*, March 24, 2015, https://www.wsj.com/articles/google-makes-most-of-close-ties-to-white-house-1427242076.

103. Jim Rutenberg, "Data You Can Believe In: The Obama Campaign's Digital Masterminds Cash In," *New York Times*, June 20, 2013, https://www.nytimes.com/2013/06/23/magazine/the-obama-campaigns-digital-masterminds-cash-in.html.

104. Rutenberg, "Data You Can Believe In."

105. Lillian Cunningham, "Google's Eric Schmidt Expounds on His Senate Testimony," *Washington Post*, September 30, 2011, http://www.washingtonpost.com/national/on-leadership/googles-eric-schmidt-expounds-on-his-senate-testimony/2011/09/30/gIQAPyVgCL_story.html.

106. "Google's Revolving Door Explorer (US)," *Google Transparency Project*, April 15, 2016, http://www.googletransparencyproject.org/googles-revolving-door-explorer-us; Tess VandenDolder, "Is Google the New Revolving Door?" *DC Inno*, September 9, 2014, http://dcinno.streetwise.co/2014/09/09/is-google-the-new-revolving-door; "Revolving Door | OpenSecrets—Employer Search: Google Inc.," *OpenSecrets.org*, February 23, 2017, https://www.opensecrets.org/revolving/search_result.php?priv=Google+Inc; Yasha Levine, "The Revolving Door Between Google and the Department of Defense," *PandoDaily* (blog), April 23, 2014, http://pando.com/2014/04/23/the-revolving-door-between-google-and-the-department-of-defense; Cecilia Kang and Juliet Eilperin, "Why Silicon Valley Is the New Revolving Door for Obama Staffers," *Washington Post*, February 27, 2015, http://www.washingtonpost.com/business

/economy/as-obama-nears-close-of-his-tenure-commitment-to-silicon-valley
-is-clear/2015/02/27/3bee8088-bc8e-11e4-bdfa-b8e8f594e6ee_story.html.

107. Eric Schmidt and Jonathan Rosenberg, *How Google Works* (New York: Grand Central, 2014), 255.

108. Deborah D'Souza, "Big Tech Spent Record Amounts on Lobbying Under Trump," *Investopedia*, July 11, 2017, https://www.investopedia.com/tech /what-are-tech-giants-lobbying-trump-era; Brodkin, "Google and Facebook Lobbyists"; Natasha Lomas, "Google Among Top Lobbyists of Senior EC Officials," *TechCrunch* (blog), June 24, 2015, http://social.techcrunch .com/2015/06/24/google-among-top-lobbyists-of-senior-ec-officials; "Google's European Revolving Door," *Google Transparency Project*, September 25, 2017, http://googletransparencyproject.org/articles/googles-european-revolving-door.

109. "Google Enlisted Obama Officials to Lobby States on Driverless Cars," *Google Transparency Project*, March 29, 2018, https://googletransparencyproject.org /articles/google-enlisted-obama-officials-lobby-states-driverless-cars.

110. "Tech Companies Are Pushing Back Against Biometric Privacy Laws," *Bloomberg.com*, July 20, 2017, https://www.bloomberg.com/news/articles/2017 -07-20/tech-companies-are-pushing-back-against-biometric-privacy-laws; "Biometric Privacy Laws: Illinois and the Fight Against Intrusive Tech," March 29, 2018, https://news.law.fordham.edu/jcfl/2018/03/20/biometric-privacy -laws-illinois-and-the-fight-against-intrusive-tech; April Glaser, "Facebook Is Using an 'NRA Approach' to Defend Its Creepy Facial Recognition Programs," *Slate*, August 4, 2017, http://www.slate.com/blogs/future_tense/2017/08/04 /facebook_is_fighting_biometric_facial_recognition_privacy_laws.html; Conor Dougherty, "Tech Companies Take Their Legislative Concerns to the States," *New York Times*, May 27, 2016, http://www.nytimes.com/2016/05/28/tech nology/tech-companies-take-their-legislative-concerns-to-the-states.html.

111. Schmidt had joined the board of the New America Foundation in 1999. As of 2013, he was chairman of its Board of Directors, and his financial contribution remained in the top tier, matched by only three other donors: the US State Department, the Lumina Foundation, and the Bill and Melinda Gates Foundation. The secondary level of contributors includes Google. See http:// newamerica.net/about/funding. The foundation is a fulcrum in Washington policy discourse, and its board members constitute a who's who of the policy establishment. See http://newamerica.net/about/board.

112. Tom Hamburger and Matea Gold, "Google, Once Disdainful of Lobbying, Now a Master of Washington Influence," *Washington Post*, April 12, 2014, http:// www.washingtonpost.com/politics/how-google-is-transforming-power-and -politicsgoogle-once-disdainful-of-lobbying-now-a-master-of-washington -influence/2014/04/12/51648b92-b4d3-11e3-8cb6-284052554d74_story.html.

113. For additional sources, see David Dayen, "Google's Insidious Shadow Lobbying: How the Internet Giant Is Bankrolling Friendly Academics—and Skirting

Federal Investigations," *Salon.com,* November 24, 2015, https://www.salon.com
/2015/11/24/googles_insidious_shadow_lobbying_how_the_internet_giant_is
_bankrolling_friendly_academics_and_skirting_federal_investigations.

114. Nick Surgey, "The Googlization of the Far Right: Why Is Google Funding
Grover Norquist, Heritage Action and ALEC?" *PR Watch,* November 27, 2013,
http://www.prwatch.org/news/2013/11/12319/google-funding-grover-norquist
-heritage-action-alec-and-more. *PR Watch* is a publication of the Center for
Media and Democracy. I encourage the interested reader to access this article
for a full list of Google's antigovernment funding recipients and an analysis of
their positions and research agendas.

115. Mike McIntire, "ALEC, a Tax-Exempt Group, Mixes Legislators and Lobbyists,"
New York Times, April 21, 2012, https://www.nytimes.com/2012/04/22/us/alec
-a-tax-exempt-group-mixes-legislators-and-lobbyists.html; Nick Surgey, "The
Googlization of the Far Right: Why Is Google Funding Grover Norquist,
Heritage Action and ALEC?" *PR Watch,* November 27, 2013, http://www
.prwatch.org/news/2013/11/12319/google-funding-grover-norquist-heritage
-action-alec-and-more; "What Is ALEC?—ALEC Exposed," Center for Media
and Democracy, February 22, 2017, http://www.alecexposed.org/wiki/What_is
_ALEC%3F; Katie Rucke, "Why Are Tech Companies Partnering with ALEC?"
Mint Press News (blog), December 13, 2013, http://www.mintpressnews
.com/tech-companies-partnering-alec/175074.

116. "2014 Fellows—Policy Fellowship—Google," https://www.google.com/policy
fellowship/2014fellows.html.

117. Brody Mullins and Jack Nicas, "Paying Professors: Inside Google's Academic
Influence Campaign," *Wall Street Journal,* July 14, 2017, https://www.wsj.com
/articles/paying-professors-inside-googles-academic-influence-campaign
-1499785286.

118. Kenneth P. Vogel, "Google Critic Ousted from Think Tank Funded by
the Tech Giant," *New York Times,* August 30, 2017, https://www.nytimes
.com/2017/08/30/us/politics/eric-schmidt-google-new-america.html;
Hope Reese, "The Latest Google Controversy Shows How Corporate
Funding Stifles Criticism," *Vox,* September 5, 2017, https://www.vox.com
/conversations/2017/9/5/16254910/google-controversy-new-america-barry
-lynn.

CHAPTER FIVE

1. "Google Management Discusses Q3 2011 Results—Earnings Call Transcript
About Alphabet Inc. (GOOG)," *Seeking Alpha,* October 14, 2011, http://
seekingalpha.com/article/299518-google-management-discusses-q3-2011
-results-earnings-call-transcript (italics mine).

2. See Ken Auletta, *Googled: The End of the World as We Know It* (New York:
Penguin, 2010).

3. Here are some of Edelman's articles and works: Benjamin Edelman, "Bias in Search Results? Diagnosis and Response," *Indian Journal of Law and Technology* 7 (2011): 16–32; Benjamin Edelman and Zhenyu Lai, "Design of Search Engine Services: Channel Interdependence in Search Engine Results" (working paper, Working Knowledge, Harvard Business School, March 9, 2015), *Journal of Marketing Reseach* 53, no. 6 (2016): 881–900; Benjamin Edelman, "Leveraging Market Power Through Tying and Bundling: Does Google Behave Anti-competitively?" (working paper, no. 14–112, Harvard, May 28, 2014), http://www.benedelman.org/publications/google-tying-2014-05-12.pdf; Benjamin Edelman et al., *Exclusive Preferential Placement as Search Diversion: Evidence from Flight Search* (Social Science Research Network, 2013); Benjamin Edelman, "Google Tying Google Plus and Many More," Benedelman.org, January 12, 2012, http://www.benedelman.org/news/011212-1.html; Benjamin Edelman, "Hard-Coding Bias in Google 'Algorithmic' Search Results," Benedelman.org, November 15, 2010, http://www.benedelman.org/hardcoding.

4. Ashkan Soltani, Andrea Peterson, and Barton Gellman, "NSA Uses Google Cookies to Pinpoint Targets for Hacking," *Washington Post,* December 10, 2013, https://www.washingtonpost.com/news/the-switch/wp/2013/12/10/nsa-uses -google-cookies-to-pinpoint-targets-for-hacking.

5. Michael Luca et al., "Does Google Content Degrade Google Search? Experimental Evidence" (working paper, NOM Unit, Harvard Business School, August 2016), http://papers.ssrn.com/abstract=2667143.

6. Alistair Barr, "How Google Aims to Delve Deeper into Users' Lives," *Wall Street Journal,* May 28, 2015, http://www.wsj.com/articles/how-google-aims-to-delve -deeper-into-users-lives-1432856623.

7. See Erick Schonfeld, "Schmidt: 'Android Adoption Is About to Explode,'" *TechCrunch* (blog), October 15, 2009, http://social.techcrunch .com/2009/10/15/schmidt-android-adoption-is-about-to-explode.

8. Bill Gurley, "The Freight Train That Is Android," *Above the Crowd,* March 25, 2011, http://abovethecrowd.com/2011/03/24/freight-train-that-is-android.

9. Steve Kovach, "Eric Schmidt: We'll Have 2 Billion People Using Android Thanks to Cheap Phones," *Business Insider,* April 16, 2013, http://www .businessinsider.com/eric-schmidt-on-global-android-growth-2013-4 (italics mine); Ina Fried, "Eric Schmidt on the Future of Android, Motorola, Cars and Humanity (Video)," *AllThingsD* (blog), May 8, 2013, http://allthingsd .com/20130508/eric-schmidt-on-the-future-of-android-motorola-cars-and -humanity-video.

10. See Ameet Sachdev, "Skyhook Sues Google After Motorola Stops Using Its Location-Based Software," *Chicago Tribune,* August 19, 2011, http://articles .chicagotribune.com/2011-08-19/business/ct-biz-0819-chicago-law -20110819_1_google-s-android-google-risks-google-spokesperson. See also the May 2011 in-depth analysis of more than 750 pages of unsealed court

documents: Nilay Patel, "How Google Controls Android: Digging Deep into the Skyhook Filings," *Verge,* May 12, 2011, http://www.theverge .com/2011/05/12/google-android-skyhook-lawsuit-motorola-samsung.

11. "Complaint of Disconnect, Inc.—Regarding Google's Infringement of Article 102 TFEU Through Bundling into the Android Platform and the Related Exclusion of Competing Privacy and Security Technology, No. COMP/40099," June 2015, https://assets.documentcloud.org/documents/2109044/disconnect -google-antitrust-complaint.pdf.

12. Among other noteworthy studies, a 2015 study by Harvard researcher Jinyan Zang tested 110 of the most popular free apps in both Android (Google) and iOS (Apple) environments. Researchers found that 73 percent of Android apps compared to 16 percent of iOS apps share personally identifying information (PII) with third parties. The researchers also found that many mobile apps share sensitive user information with third parties "and that *they do not need visible permission requests* to access the data" (italics mine). See Jinyan Zang et al., "Who Knows What About Me? A Survey of Behind the Scenes Personal Data Sharing to Third Parties by Mobile Apps," *Journal of Technology Science,* October 30, 2015, http://techscience.org/a/2015103001.

 Another detailed study, by Luigi Vigneri and his colleagues at EURECOM in 2015, looked closely at the 5,000 newest and most popular applications in the Google Play store. Researchers found that 500 of these apps connect to more than 500 distinct URLs, and 25 connect to more than 1,000 URLs. Multiple URLs can connect to the same "domain." So the researchers also examined the domains that were most often the source of these connections. Nine of the top twenty domains behind these hidden connections are web services run by Google. Of the remaining eleven, three are owned or affiliated with Google. The other eight are Google competitors in behavioral futures markets, including Facebook, Samsung, and Scorecard Research, a data broker that sells surplus behavior to its customers.

 The researchers then took another valuable step. They characterized each of the URLs visited by these apps as either "ad-related" or "user-tracking related" and found that 66 percent of the apps contact an average of 40 ad-related URLs, although in some cases it's more than 1,000. Of the five top domains represented by these URLs, three belong to Google. When it comes to tracking, the data suggest that the competition for behavioral surplus is even more contested. Among the apps in the study, 73 percent did not connect to tracking sites, but 16 percent connected to 100 or more tracking sites. Google remains the dominant force here, with 44 percent of the tracker-related domains, followed by 32 percent operated by AT Internet, a privately held "digital intelligence" firm that specializes in "behavior analysis." Four of the ten most-intensive tracking apps in Google Play were also awarded Google's "Top Developer Badge." See Luigi Vigneri et al., "Taming the Android AppStore:

Lightweight Characterization of Android Applications," *ArXiv:1504.06093 [Computer Science]*, April 23, 2015, http://arxiv.org/abs/1504.06093.

A team of researchers from the University of Washington led by Adam Lerner and Anna Simpson studied the growth of web trackers from 1996 to 2016. Not surprisingly from our point of view, web tracking increased in tandem with the rise and institutionalization of surveillance capitalism, as did third-party connections. The researchers note that although earlier trackers recorded routine data oriented toward product stability, the more recent rise in trackers is of those that capture and analyze personal information. In 2000 only about 5 percent of sites contacted at least five third parties, but by 2016, 40 percent of sites sent data to third parties. Among trackers with the "most power to capture profiles of user behavior across many sites," google-analytics .com is cited as a "remarkable outlier," gathering more data from more sites than any other entity does. The researchers conclude that despite the privacy concerns that have received so much attention in recent years, tracking has substantially expanded in "scope and complexity" on a clear upward trend line. In other words, there is more tracking now than at any time since the launch of the internet, even as citizens and governments try to protect individual privacy. See Adam Lerner et al., "Internet Jones and the Raiders of the Lost Trackers: An Archeological Study of Web Tracking from 1996–2016," in *Proceedings of the Workshop on End-to-End, Sense-and-Respond Systems, Applications, and Services: (EESR '05), June 5, 2005, Seattle* (Berkeley, CA: USENIX Association, 2005), http://portal.acm.org/toc.cfm?id=1072530.

13. Ibrahim Altaweel, Nathan Good, and Chris Jay Hoofnagle, "Web Privacy Census" (SSRN scholarly paper, Social Science Research Network, December 15, 2015), https://papers.ssrn.com/abstract=2703814.

14. Timothy Libert, "Exposing the Invisible Web: An Analysis of Third-Party HTTP Requests on 1 Million Websites," *International Journal of Communication* 9 (October 28, 2015): 18.

15. Altaweel, Good, and Hoofnagle, "Web Privacy Census."

16. Mengwei Xu et al., "AppHolmes: Detecting and Characterizing App Collusion Among Third-Party Android Markets," *Association for Computing Machinery,* 2017, https://www.microsoft.com/en-us/research/publication/appholmes -detecting-characterizing-app-collusion-among-third-party-android-markets.

17. See "Press | Yale Privacy Lab," https://privacylab.yale.edu/press.html; and "Exodus Privacy," Exodus-Privacy, https://exodus-privacy.eu.org/. See also Yael Grauer, "Staggering Variety of Clandestine Trackers Found In Popular Android Apps," *Intercept,* November 24, 2017, https://theintercept.com/2017/11/24 /staggering-variety-of-clandestine-trackers-found-in-popular-android-apps/.

18. "Complaint of Disconnect, Inc.," 2.

19. "Complaint of Disconnect, Inc.," 3.

20. Vigneri et al., "Taming the Android AppStore"; "Antitrust/Cartel Cases—40099 Google Android," European Union Commission on Competition, April 15,

2015, http://ec.europa.eu/competition/elojade/isef/case_details.cfm?proc
_code=1_40099.

21. "European Commission—Press Release—Antitrust: Commission Sends
Statement of Objections to Google on Android Operating System and
Applications," European Commission, April 20, 2016, http://europa.eu/rapid
/press-release_IP-16-1492_en.htm.

22. "Complaint of Disconnect, Inc.," 40.

23. Marc Rotenberg, phone interview with author, June 2014.

24. Jennifer Howard, "Publishers Settle Long-Running Lawsuit Over Google's
Book-Scanning Project," *Chronicle of Higher Education,* October 4, 2012,
https://chronicle.com/article/Publishers-Settle-Long-Running/134854;
"Google Books Settlement and Privacy," *EPIC.org,* October 30, 2016, https://
epic.org/privacy/googlebooks; Juan Carlos Perez, "Google Books Settlement
Proposal Rejected," *PCWorld,* March 22, 2011, http://www.pcworld.com
/article/222882/article.html; Eliot Van Buskirk, "Justice Dept. to Google
Books: Close, but No Cigar," *Wired,* February 5, 2010, http://www.wired
.com/2010/02/justice-dept-to-google-books-close-but-no-cigar; Miguel Helft,
"Opposition to Google Books Settlement Jells," *New York Times—Bits Blog,*
April 17, 2009, https://bits.blogs.nytimes.com/2009/04/17/opposition-to
-google-books-settlement; Brandon Butler, "The Google Books Settlement:
Who Is Filing and What Are They Saying?" *Association of Research Libraries*
28 (2009): 9; Ian Chant, "Authors Guild Appeals Dismissal of Google Books
Lawsuit," *Library Journal,* April 16, 2014, http://lj.libraryjournal.com/2014/04
/litigation/authors-guild-appeals-dismissal-of-google-books-lawsuit.

25. "Investigations of Google Street View," *EPIC.org,* 2014, https://epic.org/privacy
/streetview; David Kravets, "An Intentional Mistake: The Anatomy of Google's
Wi-Fi Sniffing Debacle," *Wired,* May 2, 2012, https://www.wired.com/2012/05
/google-wifi-fcc-investigation; Clint Boulton, "Google WiFi Privacy Breach
Challenged by 38 States," *eWeek,* July 21, 2010, http://www.eweek.com/c/a
/Search-Engines/Google-WiFi-Privacy-Breach-Challenged-by-38-States-
196191; Alastair Jamieson, "Google Will Carry On with Camera Cars Despite
Privacy Complaints Over Street Views," *Telegraph,* April 9, 2009, http://www
.telegraph.co.uk/technology/google/5130068/Google-will-carry-on-with
-camera-cars-despite-privacy-complaints-over-street-views.html; Gareth
Corfield, "'At Least I Can Walk Away with My Dignity'—Streetmap Founder
After Google Lawsuit Loss," *Register,* February 20, 2017, https://www.the
register.co.uk/2017/02/20/streetmap_founder_kate_sutton_google_lawsuit.

26. Joseph Menn, Daniel Schäfer, and Tim Bradshaw, "Google Set for Probes on
Data Harvesting," *Financial Times,* May 17, 2010, http://www.ft.com/intl/cms
/s/2/254ff5b6-61e2-11df-998c-00144feab49a.html#axzz3JjXPNno5.

27. Julia Angwin, "Google in New Privacy Probes," *Wall Street Journal,* March 16,
2012, http://online.wsj.com/articles/SB10001424052702304692804577283382
1586827892; Julia Angwin, "Google, FTC Near Settlement on Privacy," *Wall*

Street Journal, July 10, 2012, http://www.wsj.com/articles/SB100014240527023
0356770457751708117855304‌6; Jonathan Owen, "Google in Court Again Over
'Right to Be Above British Law' on Alleged Secret Monitoring," *Independent,*
December 8, 2014, http://www.independent.co.uk/news/uk/crime/google
-challenges-high-court-decision-on-alleged-secret-monitoring-9911411.html.

28. "Testimony of Benjamin Edelman Presented Before the United States House of
Representatives Committee on the Judiciary Task Force on Competition Policy
and Antitrust Laws," June 27, 2008; Brody Mullins, Rolfe Winkler, and Brent
Kendall, "Inside the U.S. Antitrust Probe of Google," *Wall Street Journal,* March
19, 2015, http://www.wsj.com/articles/inside-the-u-s-antitrust-probe-of-google
-1426793274.

29. Nate Anderson, "Why Google Keeps Your Data Forever, Tracks You with Ads,"
Ars Technica, March 8, 2010, http://arstechnica.com/tech-policy/news/2010/03
/google-keeps-your-data-to-learn-from-good-guys-fight-off-bad-guys.ars;
Kevin J. O'Brien and Thomas Crampton, "E.U. Probes Google Over Data
Retention Policy," *New York Times,* May 26, 2007, http://www.nytimes
.com/2007/05/26/business/26google.html; Mark Bergen, "Google Manipulates
Search Results, According to Study from Yelp and Legal Star Tim Wu," *Recode,*
June 29, 2015, http://www.recode.net/2015/6/29/11563936/yelp-teams-with
-legal-star-tim-wu-to-trounce-google-in-new-study.

30. David Snelling, "Google Maps Is Tracking You! How Your Smartphone
Knows Your Every Move," *Express,* August 18, 2014, http://www.express
.co.uk/life-style/science-technology/500811/Google-Maps-is-tracking-your
-every-move; Jason Mick, "ACLU Fights for Answers on Police Phone Location
Data Tracking," *Daily Tech,* August 4, 2011, http://www.dailytech.com
/ACLU+Fights+for+Answers+on+Police+Phone+Location+Data+Tracking
/article22352.htm.

31. "Google Glass and Privacy," *EPIC.org,* October 6, 2017, https://epic.org/privacy
/google/glass.

32. Benjamin Herold, "Google Under Fire for Data-Mining Student Email
Messages," *Education Week,* March 26, 2014, http://www.edweek.org/ew
/articles/2014/03/13/26google.h33.html; Quinten Plummer, "Google Email Tip
-Off Draws Privacy Concerns," *Tech Times,* August 5, 2014, http://www
.techtimes.com/articles/12194/20140805/google-email-tip-off-draws
-privacy-concerns.htm.

33. Grant Gross, "French Fine Google Over Change in Privacy Policy," *PCWorld,*
January 8, 2014, http://www.pcworld.com/article/2085920/french-fine-google
-over-change-in-privacy-policy.html; Dheepthika Laurent, "Facebook, Twitter
and Google Targeted in French Lawsuit," *CNN.com,* March 26, 2014, http://
www.cnn.com/2014/03/25/world/europe/france-social-media-lawsuit/index
.html; Mark Milian, "Google to Merge User Data Across Its Services," *CNN.com,*
January 25, 2012, http://www.cnn.com/2012/01/24/tech/web/google-privacy
-policy/index.html; Martin Gijzemijter, "Google's Privacy Policy Merger

'Against Dutch Law,'" *ZDNet,* November 29, 2013, http://www.zdnet.com
/article/googles-privacy-policy-merger-against-dutch-law; Zack Whittaker,
"Google Faces EU State Fines Over Privacy Policy Merger," *ZDNet,* April 2,
2013, http://www.zdnet.com/article/google-faces-eu-state-fines-over-privacy
-policy-merger.

34. Peter Fleischer, "Street View and Privacy," *Google Lat Long,* September 24, 2007,
https://maps.googleblog.com/2007/09/street-view-and-privacy.html.

35. See Stephen Hutcheon, "We're Not Stalking You or Helping Terrorists, Says
Google Earth Boss," *Sydney Morning Herald,* January 30, 2009, http://www.smh
.com.au/news/technology/biztech/were-not-stalking-you-or-helping-terrorists
-says-google-earthboss/2009/01/30/1232818692103.html.

36. See Jamieson, "Google Will Carry On with Camera Cars."

37. Kevin J. O'Brien and Claire Cain Miller, "Germany's Complicated Relationship
with Google Street View," *New York Times—Bits Blog,* April 23, 2013,
http://bits.blogs.nytimes.com/2013/04/23/germanys-complicated
-relationship-with-google-street-view.

38. Peter Fleischer, "Data Collected by Google Cars," *Google Europe,* April 27, 2010,
https://europe.googleblog.com/2010/04/data-collected-by-google-cars.html.

39. "In the Matter of Google, Inc.: Notice of Apparent Liability for Forfeiture,
File No.: EB-10-IH-4055, NAL/Acct. No.: 201232080020, FRNs: 0010119691,
0014720239," Federal Communications Commission, April 13, 2012, 12–13.

40. Kevin J. O'Brien, "Google's Data Collection Angers European Officials,"
New York Times, May 15, 2010, http://www.nytimes.com/2010/05/16/
technology/16google.html; "Commissioner's Findings—PIPEDA Report of
Findings #2011-001: Report of Findings: Google Inc. WiFi Data Collection—
Office of the Privacy Commissioner of Canada," June 6, 2011, https://www
.priv.gc.ca/en/opc-actions-and-decisions/investigations/investigations-into
-businesses/2011/pipeda-2011-001; CNIL, "Délibération de La Commission
Nationale de l'Informatique et Des Libertés Decision No. 2011-035 of the
Restricted Committee Imposing a Financial Penalty on the Company Google
Inc.," 2011-035 § (2011), https://www.legifrance.gouv.fr/affichCnil.do?&id
=CNILTEXT000023733987; "Final Findings, Dutch Data Protection Authority
Investigation into the Collection of Wifi Data by Google Using Street View
Cars—Z2010-00582—DDPA Decision," December 7, 2010, https://web
.archive.org/web/20130508060039/http://www.dutchdpa.nl/downloads_overig
/en_pb_20110811_google_final_findings.pdf; "Investigations of Google Street
View"; Kevin J. O'Brien, "Europe Pushes Google to Turn Over Wi-Fi Data,"
New York Times, June 27, 2010, http://www.nytimes.com/2010/06/28
/technology/28google.html.

41. "In the Matter of Google, Inc.: Notice of Apparent Liability for Forfeiture";
O'Brien, "Google's Data Collection."

42. EPIC maintained a website charting the time line of Street View events and
the outcomes of legal challenges both domestically and internationally. See

"Investigations of Google Street View"; "Ben Joffe v. Google," *EPIC.org,* 2017, https://epic.org/amicus/google-street-view; "FCC Investigation of Google Street View," *EPIC.org,* 2017, https://www.epic.org/privacy/google/fcc_investi gation_of_google_st.html; Mark A. Chavez and Marc Rotenberg, "Brief for Amicus Curiae: Electronic Privacy Information Center in Support of Plaintiffs—In Re Google Street View Electronic Communications Litigation, Case No. 5:10-Md-02184-JW," US District Court for Northern District of California San Jose Division, April 11, 2011, https://epic.org/privacy/streetview /EPICStreetviewAmicus.pdf.

43. See Maija Palmer and Lionel Barber, "Google to Hand Over Intercepted Data," *Financial Times,* June 3, 2010, http://www.ft.com/cms/s/2/db664044-6f43-11df -9f43-00144feabdc0.html.

44. "In the Matter of Google, Inc.: Notice of Apparent Liability for Forfeiture."

45. Denis Howe, "Wardriving," *Dictionary.com,* http://www.dictionary.com/browse /wardriving.

46. "In the Matter of Google, Inc.: Notice of Apparent Liability for Forfeiture," 11.

47. "In the Matter of Google, Inc.: Notice of Apparent Liability for Forfeiture," 11–12.

48. David Streitfeld, "Google Concedes That Drive-By Prying Violated Privacy," *New York Times,* March 13, 2013, http://bits.blogs.nytimes.com/2013/03/13 /daily-report-google-concedes-that-drive-by-prying-violated-privacy.

49. David Streitfeld, "Google Admits Street View Project Violated Privacy," *New York Times,* March 12, 2013, http://www.nytimes.com/2013/03/13/technology /google-pays-fine-over-street-view-privacy-breach.html.

50. "In the Matter of Google, Inc.: Notice of Apparent Liability for Forfeiture," 11; "Google to Give Governments Street View Data," *New York Times,* June 3, 2010, https://www.nytimes.com/2010/06/04/business/global/04google.html.

51. Alan Eustace, "Creating Stronger Privacy Controls Inside Google," *Google Public Policy Blog,* October 22, 2010, https://publicpolicy.googleblog .com/2010/10/creating-stronger-privacy-controls.html.

52. "Measures (Guidance) Concerning Protection of 'Secrecy of Communication' to Google Inc.," Ministry of Internal Affairs and Communications, November 11, 2011, http://www.soumu.go.jp/menu_news/s-news/01kiban08_02000056 .html; "Navigating Controversy: Google Launches Street View Germany," *Spiegel Online,* November 18, 2010, http://www.spiegel.de/international/business /navigating-controversy-google-launches-street-view-germany-a-729793 .html; Matt McGee, "Google Street View Debuts in Germany, Blurry Houses Included," *Search Engine Land,* November 1, 2010, http://searchengineland .com/google-street-view-germany-blurry-houses-included-54632.

53. Arne Gerhards, "Fine Imposed upon Google," Hamburg Commissioner for Data Protection and Freedom of Information, April 22, 2013, https://www .datenschutz-hamburg.de/fileadmin/user_upload/documents/PressRelease _2013-04-22_Google-Wifi-Scanning.pdf.

54. Matt McGee, "Google Has Stopped Street View Photography in Germany," *Search Engine Land,* April 10, 2011, http://searchengineland .com/google-has-stopped-street-view-photography-germany-72368.

55. Peter Fleischer, "Street View in Switzerland," *Google Europe,* November 13, 2009, https://europe.googleblog.com/2009/11/street-view-in-switzerland .html; Scott Capper, "Google Faces Court Action Over Street View," *Swissinfo. ch,* November 16, 2009, http://www.swissinfo.ch/eng/business/google -faces-court-action-over-street-view/7656246; Anita Greil and Katharina Bart, "Swiss Court to Rule on Google Street View," *Wall Street Journal,* February 24, 2011, http://www.wsj.com/articles/SB10001424052748703 4086045761637707589841 78; Frank Jordans, "Google Threatens to Shut Down Swiss Street View," *Phys.org,* May 11, 2011, http://phys.org/news /2011-05-google-appeals-street-view-switzerland.html; Kevin J. O'Brien and David Streitfeld, "Swiss Court Orders Modifications to Google Street View," *New York Times,* June 8, 2012, http://www.nytimes.com/2012/06/09 /technology/09iht-google09.html; "Google Beefs Up Restricted Swiss Street View," *swissinfo.ch,* May 19, 2015, http://www.swissinfo.ch/eng /society/eagle-eye_google-beefs-up-restricted-swiss-street-view/41439290.

56. However, India continues to block Street View operations, and the corporation considered it to be too costly to comply with requirements imposed by Austria and Germany after bans there were lifted. For details of bans in key countries, see "New Developments Regarding Google Street View," Austrian Data Protection Agency, April 4, 2016, http://web.archive.org/web/20160404072538 /https://www.dsb.gv.at/site/6733/default.aspx; Helena Smith Athens, "Google Street View Banned from Greece," *Guardian,* May 12, 2009, https://www.the guardian.com/technology/2009/may/12/google-street-view-banned-greece; John Ribeiro, "Google Street View in India Faces Challenges," *PCWorld,* May 26, 2011, http://www.pcworld.com/article/228713/article.html; Danuta Pavilenene, "Google Street View Banned from Lithuanian Streets," *Baltic Course,* May 23, 2012, http://www.baltic-course.com/eng/Technology/?doc =57649.

57. Liz Gannes, "Ten Years of Google Maps, from Slashdot to Ground Truth," *Recode,* February 8, 2015, http://www.recode.net/2015/2/8/11558788/ten -years-of-google-maps-from-slashdot-to-ground-truth.

58. Kashmir Hill, "Google's Privacy Director Is Stepping Down," *Forbes,* April 1, 2013, http://www.forbes.com/sites/kashmirhill/2013/04/01/googles-privacy -director-is-stepping-down/print; ScroogledTruth, *Scroogled—Dr. Alma Whitten (Google's Privacy Engineering Lead) Before Congress,* 2013, https://www .youtube.com/watch?v=vTLEJsunCfI.

59. Steve Lohr and David Streitfeld, "Engineer in Google's Street View Is Identified," *New York Times,* April 30, 2012, http://www.nytimes.com/2012/05/01/ technology/engineer-in-googles-street-view-is-identified.html; Farhad Manjoo, "Is It Time to Stop Trusting Google?" *Slate,* May 1, 2012, http://www

.slate.com/articles/technology/technology/2012/05/marius_milner_google
_wi_fi_snooping_assessing_the_disturbing_fcc_report_on_the_company_s
_street_view_program_.html; John V. Hanke et al., A system and method for
transporting virtual objects in a parallel reality game, United States US8968099
B1, filed November 1, 2012, and issued March 3, 2015, https://patents.google
.com/patent/US8968099B1/en.

60. See Alexis C. Madrigal, "How Google Builds Its Maps—and What It Means for
the Future of Everything," *Atlantic,* September 6, 2012, http://www.theatlantic
.com/technology/archive/2012/09/how-google-builds-its-maps-and-what-it
-means-for-the-future-of-everything/261913.

61. Brian McClendon, "Building a Better Map of Europe," *Google Maps,* December
5, 2012, https://maps.googleblog.com/2012/12/building-better-map-of-europe
.html.

62. "TIGER Geodatabases," US Census Bureau, 2016, https://www.census.gov/geo
/maps-data/data/tiger-geodatabases.html.

63. Madrigal, "How Google Builds Its Maps" (italics mine).

64. See Gannes, "Ten Years of Google Maps."

65. Soufi Esmaeilzadeh, "'See Inside' with Google Maps Business View," *Google Lat
Long,* December 17, 2014, https://maps.googleblog.com/2014/12/see-inside
-with-google-maps-business.html; "Google Street View—What It Takes to Be
Trusted," Google Street View, November 10, 2016, https://www.google.com
/streetview/earn; "About—Google Maps," Google Maps, November 10, 2016,
https://www.google.com/maps/about/partners.

66. James Vincent, "Skybox: Google Maps Goes Real-Time—but Would You Want
a Spy in the Sky Staring into Your Letter Box?" *Independent,* June 21, 2014,
http://www.independent.co.uk/life-style/gadgets-and-tech/news/skybox
-google-maps-goes-real-time-but-would-you-want-a-spy-in-the-sky-staring
-into-your-letter-box-9553934.html; "DigitalGlobe Hosts U.S. Secretary of
Commerce Pritzker for a Discussion on Commerce in Colorado | Seeing a
Better World™," *DigitalGlobe Blog,* June 25, 2014, http://blog.digitalglobe
.com/2014/06/25/digitalglobehostsussecretarypritzker; Ellen Huet, "Google
Buys Skybox Imaging—Not Just for Its Satellites," *Forbes,* June 10, 2014, http://
www.forbes.com/sites/ellenhuet/2014/06/10/google-buys-skybox-imaging
-not-just-for-its-satellites.

67. Tom Warren, "Google Just Showed Me the Future of Indoor Navigation," *Verge,*
February 23, 2016, http://www.theverge.com/2016/2/23/11094020/google
-lenovo-project-tango-indoor-navigation.

68. Sophia Lin, "Making of Maps: The Cornerstones," *Google Maps,* September 4,
2014, https://maps.googleblog.com/2014/09/making-of-maps-cornerstones
.html.

69. Alistair Barr, "Google Maps Guesses Where You're Headed Now," *Wall Street
Journal* (blog), January 13, 2016, http://blogs.wsj.com/digits/2016/01/13
/google-maps-guesses-where-youre-headed-now.

70. Akshay Kannan, "Introducing Nearby: A New Way to Discover the Things Around You," *Official Android Blog,* June 9, 2016, https://android.googleblog .com/2016/06/introducing-nearby-new-way-to-discover.html.

71. Kieren McCarthy, "Delete Google Maps? Go Ahead, Says Google, We'll Still Track You," *Register,* September 12, 2016, http://www.theregister.co.uk /2016/09/12/turn_off_location_services_go_ahead_says_google_well_still _track_you.

72. John B. Harley, *The New Nature of Maps: Essays in the History of Cartography,* ed. Paul Laxton (Baltimore, MD: Johns Hopkins University Press, 2001), 58–59.

73. See Stephen Graves, "Niantic Labs' John Hanke on Alternate Reality Games and the Future of Storytelling," *PC&Tech Authority,* October 13, 2014.

74. David DiSalvo, "The Banning of Google Glass Begins (and They Aren't Even Available Yet)," *Forbes,* March 10, 2013, http://www.forbes.com/sites /daviddisalvo/2013/03/10/the-ban-on-google-glass-begins-and-they-arent -even-available-yet; David Streitfeld, "Google Glass Picks Up Early Signal: Keep Out," *New York Times,* May 6, 2013, http://www.nytimes.com/2013/05/07 /technology/personaltech/google-glass-picks-up-early-signal-keep-out.html.

75. Aaron Smith, "U.S. Views of Technology and the Future," *Pew Research Center: Internet, Science & Tech* (blog), April 17, 2014, http://www.pewinternet .org/2014/04/17/us-views-of-technology-and-the-future.

76. Drew FitzGerald, "Now Google Glass Can Turn You into a Live Broadcast," *Wall Street Journal,* June 24, 2014, http://www.wsj.com/articles/now-google -glass-can-turn-you-into-a-live-broadcast-1403653079.

77. See Amir Efrati, "Google Glass Privacy Worries Lawmakers," *Wall Street Journal,* May 17, 2013, http://www.wsj.com/articles/SB100014241278873247 67004578487661143483672.

78. "We're Graduating from Google[x] Labs," Google, January 15, 2015, https:// plus.google.com/app/basic/stream/z124trxirsruxvcdp23otv4qerfwghdhv04.

79. Alistair Barr, "Google Glass Gets a New Name and Hires from Amazon," *Wall Street Journal,* September 16, 2015.

80. Fred O'Connor, "Google is Making Glass 'Ready for Users,' Says Schmidt," *PCWorld,* March 23, 2015, http://www.pcworld.com/article/2900632/google-is -making-glass-ready-for-users-says-schmidt.html; "Looking Ahead for WhatsApp," *WhatsApp* (blog), August 25, 2016, https://blog.whatsapp.com /10000627/Looking-ahead-for-WhatsApp.

81. Alistair Barr, "Google's Tough Search for New Platforms on Display at I/O," *Wall Street Journal,* May 27, 2015, http://www.wsj.com/articles/googles-tough -search-for-new-platforms-on-display-at-i-o-1432748457.

82. Jay Kothari, "A New Chapter for Glass," *Team at X* (blog), July 18, 2017, https:// blog.x.company/a-new-chapter-for-glass-c7875d40bf24.

83. See, for example, Darrell Etherington, "Google Glass Is Back with Hardware Focused on the Enterprise," *TechCrunch* (blog), July 18, 2017, http://social .techcrunch.com/2017/07/18/google-glass-is-back-with-hardware-focused-on

-the-enterprise; Hayley Tsukayama, "Google Will Stop Selling Glass to the General Public, but Google Says the Device Is Not Dead Yet," *Washington Post*, January 15, 2015, https://www.washingtonpost.com/news/the-switch /wp/2015/01/15/google-will-stop-selling-glass-to-the-general-public-but -google-says-the-device-is-not-dead-yet; Brid-Aine Parnell, "NYPD Dons Google Tech Specs: Part Man. Part Machine. All Glasshole," *Register*, February 10, 2014, http://www.theregister.co.uk/2014/02/10/nypd_tests_google_glass.

84. Arnold Roosendaal, "Facebook Tracks and Traces Everyone: Like This!" (SSRN Scholarly Paper, Rochester, NY: Social Science Research Network, November 30, 2010), https://papers.ssrn.com/abstract=1717563.

85. Jose Antonio Vargas, "The Face of Facebook," *New Yorker*, September 13, 2010, https://www.newyorker.com/magazine/2010/09/20/the-face-of-facebook.

86. Cynthia Ghazali, "Facebook Keeps Tabs on Users Even After They Sign Off: Report," *NY Daily News*, November 18, 2011, http://www.nydailynews.com /news/money/facebook-tabs-users-sign-report-article-1.979848.

87. Amir Efrati, "'Like' Button Follows Web Users," *Wall Street Journal*, May 18, 2011, https://www.wsj.com/articles/SB10001424052748704281504576329441 432995616. See also, Emil Protalinski, "Facebook Denies Cookie Tracking Allegations," *ZDNet*, October 3, 2011, https://www.zdnet.com/article/facebook -denies-cookie-tracking-allegations/; Riva Richmond, "As 'Like' Buttons Spread, So Do Facebook's Tentacles," *New York Times—Bits Blog*, September 27, 2011, https://bits.blogs.nytimes.com/2011/09/27/as-like-buttons-spread-so-do -facebooks-tentacles/; Julia Angwin, "It's Complicated: Facebook's History of Tracking You," *ProPublica*, June 17, 2014, https://www.propublica.org/article /its-complicated-facebooks-history-of-tracking-you; Rainey Reitman, "Facebook's Hotel California: Cross-Site Tracking and the Potential Impact on Digital Privacy Legislation," Electronic Frontier Foundation, October 10, 2011, https://www.eff.org/deeplinks/2011/10/facebook%E2%80%99s-hotel-california -cross-site-tracking-and-potential-impact-digital-privacy.

88. Asher Moses, "Facebook's Privacy Lie: Aussie Exposes 'Tracking' as New Patent Uncovered," *The Sydney Morning Herald*, October 4, 2011, https://www.smh .com.au/technology/facebooks-privacy-lie-aussie-exposes-tracking-as-new -patent-uncovered-20111004-1l61i.html.

89. Moses; Emil Protalinski, "Facebook Denies Cookie Tracking Allegations;" Emil Protalinski, "Facebook Fixes Cookie Behavior After Logging Out," *ZDNet*, September 27, 2011, https://www.zdnet.com/article/facebook -fixes-cookie-behavior-after-logging-out/; Nik Cubrilovic, "Facebook Fixes Logout Issue, Explains Cookies," *New Web Order*, September 27, 2011, https:// web.archive.org/web/20140701103652/https://www.nikcub.com/posts/face book-fixes-logout-issue-explains-cookies-2/.

90. Kent Matthew Schoen, Gregory Luc Dingle, and Timothy Kendall, "Communicating information in a social network system about activities

from another domain," WO2011097624 A3, filed February 8, 2011, and issued September 22, 2011, http://www.google.com/patents/WO2011097624A3.

91. Emil Protalinski, "Facebook Denies Patent Is Used for Tracking Logged-out Users," *ZDNet*, October 3, 2011, https://www.zdnet.com/article/facebook -denies-patent-is-used-for-tracking-logged-out-users/. See also Michael Arrington, "Facebook: Brutal Dishonesty," *Uncrunched* (blog), October 2, 2011, https://uncrunched.com/2011/10/01/brutal-dishonesty/.

92. Just a day after the Cubrilovic post, the *Hill* confirmed that Facebook had filed to start its own political action committee, which aimed to support candidates who "share our goals of promoting the value of innovation" and making the world "more open and connected." Gautham Nagesh, "Facebook to Form Its Own PAC to Back Political Candidates," *Hill*, September 26, 2011, http://thehill .com/policy/technology/183951-facebook-forming-own-pac-to-back -candidates.

93. "Facebook Settles FTC Charges That It Deceived Consumers by Failing to Keep Privacy Promises," Federal Trade Commission, November 29, 2011, https:// www.ftc.gov/news-events/press-releases/2011/11/facebook-settles-ftc -charges-it-deceived-consumers-failing-keep.

94. "FTC Facebook Settlement," Electronic Privacy Information Center, December 2009, https://epic.org/privacy/ftc/facebook/.

95. "Facebook Settles FTC Charges That It Deceived Consumers." See also Emily Steel and April Dembosky, "Facebook Raises Fears with Ad Tracking," *Financial Times*, September 23, 2012, https://www.ft.com/ content/6cc4cf0a-0584-11e2-9ebd-00144feabdc0.

96. "Facebook Custom Audiences: Target Facebook Ads by Email List," *Jon Loomer Digital*, September 24, 2012, https://www.jonloomer.com/2012/09/24 /facebook-custom-audiences/.

97. Tom Simonite, "Facebook Will Now Target Ads Based on What Its Like Buttons Saw You Do," *MIT Technology Review*, September 16, 2015, https://www .technologyreview.com/s/541351/facebooks-like-buttons-will-soon -track-your-web-browsing-to-target-ads; Cotton Delo, "Facebook to Use Web Browsing History for Ad Targeting," *AdAge*, June 12, 2014, http://adage.com /article/digital/facebook-web-browsing-history-ad-targeting/293656; Violet Blue, "Facebook Turns User Tracking 'Bug' into Data Mining 'Feature' for Advertisers," *ZDNet*, https://www.zdnet.com/article/facebook-turns-user -tracking-bug-into-data-mining-feature-for-advertisers/.

98. Julia Angwin, "Google Has Quietly Dropped Ban on Personally Identifiable Web Tracking," *ProPublica*, October 21, 2016, https://www.propublica.org/article /google-has-quietly-dropped-ban-on-personally-identifiable-web-tracking; Jack Nicas, "Privacy Groups Seek Regulatory Review of Google Privacy Policy," *Wall Street Journal*, December 19, 2016, http://www.wsj.com/articles/privacy -groups-seek-regulatory-review-of-google-privacy-policy-1482190366.

99. Ross Hunter, Farhad Zaman, and Kennedy Liu, "Global Top 100 Companies by Market Capitalisation," IPO Center, Price Waterhouse Coopers, March 31, 2017, http://www.pwc.com/gx/en/audit-services/assets/pdf/global-top -100-companies-2017-final.pdf; Deborah Crawford et al., "Facebook, Inc. (FB)—Fourth Quarter and Full Year 2016 Results Conference Call," February 1, 2017, https://s21.q4cdn.com/399680738/files/doc_financials/2016/Q4/Q4'16 -Earnings-Transcript.pdf.

100. Julia Kollewe, "Google and Facebook Bring in One-Fifth of Global Ad Revenue," *Guardian,* May 1, 2017, http://www.theguardian.com/media/2017 /may/02/google-and-facebook-bring-in-one-fifth-of-global-ad-revenue; Paul Murphy, "It Seems Google and Facebook Really Are Taking ALL the Growth in Ad Revenue," *Financial Times,* April 26, 2017, http://ftalphaville .ft.com/2017/04/26/2187891/it-seems-google-and-facebook-really-are-taking -all-the-growth-in-ad-revenue; Mathew Ingram, "Google and Facebook Have Taken Over the Digital Ad Industry," *Fortune,* January 4, 2017, http://fortune .com/2017/01/04/google-facebook-ad-industry.

101. Kara Swisher, "Microsoft's Point Man on Search—Satya Nadella—Speaks: 'It's a Game of Scale,'" *AllThingsD* (blog), August 4, 2009, http://allthingsd.com/2009 0804/microsofts-point-man-on-search-satya-nadella-speaks-its-a-game-of-scale.

102. Julie Bort, "Satya Nadella Just Launched Microsoft into a New $1.6 Trillion Market," *Business Insider,* April 15, 2014, http://www.businessinsider.com/micro soft-launches-iot-cloud-2014-4.

103. Satya Nadella, "A Data Culture for Everyone," *Official Microsoft Blog,* April 15, 2014, https://blogs.microsoft.com/blog/2014/04/15/a-data-culture-for -everyone.

104. Richard Qian, "Understand Your World with Bing," *Bing Blogs,* March 21, 2013, http://blogs.bing.com/search/2013/03/21/understand-your-world-with-bing.

105. See Dan Farber, "Microsoft's Bing Seeks Enlightenment with Satori," *CNET,* July 30, 2013, https://www.cnet.com/news/microsofts-bing-seeks -enlightenment-with-satori.

106. Greg Sterling, "Milestone: Bing Now Profitable as Windows 10 Success Boosts Usage," *Search Engine Land,* October 23, 2015, http://searchengineland.com /milestone-bing-now-profitable-as-windows-10-success-boosts-usage-234285.

107. See Ginny Marvin, "After a Year of Transition, Microsoft Execs Say, 'We're All in on Search,'" *Search Engine Land,* November 23, 2015, http://searchengineland .com/microsoft-execs-all-in-on-search-bing-ads-next-236746.

108. "Cortana and Privacy," Microsoft, November 11, 2016, https://privacy .microsoft.com/en-US/windows-10-cortana-and-privacy.

109. See Dan Kedmey, "Here's What Really Makes Microsoft's Cortana So Amazing," *Time,* July 20, 2015, http://time.com/3960670/windows-10-cortana.

110. See "Artificial Intelligence: A Virtual Assistant for Life," *Financial Times,* February 23, 2017, https://www.ft.com/content/4f2f97ea-b8ec-11e4-b8e6 -00144feab7de.

111. "Microsoft Outlines Intelligence Vision and Announces New Innovations for Windows 10," *Microsoft News Center* (blog), March 30, 2016, https://news.microsoft.com/2016/03/30/microsoft-outlines-intelligence-vision-and-announces-new-innovations-for-windows-10.

112. Chris Messina, "Conversational Commerce: Messaging Apps Bring the Point of Sale to You," *Medium,* January 16, 2015, https://medium.com/chris-messina/conversational-commerce-92e0bccfc3ff#.sdpy3xp3b.

113. Shish Shridhar, "We Don't Need Yet Another App, Conversations Are the New App," *Microsoft Developer Blogs—the ShiSh List,* May 21, 2016, https://blogs.msdn.microsoft.com/shishirs/2016/05/21/we-dont-need-yet-another-app-conversations-are-the-new-app.

114. Terry Myerson, "Hello World: Windows 10 Available on July 29," *Windows Experience Blog,* June 1, 2015, https://blogs.windows.com/windowsexperience/2015/06/01/hello-world-windows-10-available-on-july-29.

115. David Auerbach, "Broken Windows Theory," *Slate,* August 3, 2015, http://www.slate.com/articles/technology/bitwise/2015/08/windows_10_privacy_problems_here_s_how_bad_they_are_and_how_to_plug_them.html.

116. Peter Bright, "Even When Told Not to, Windows 10 Just Can't Stop Talking to Microsoft," *Ars Technica,* August 13, 2015, https://arstechnica.com/information-technology/2015/08/even-when-told-not-to-windows-10-just-cant-stop-talking-to-microsoft.

117. Amul Kalia, "With Windows 10, Microsoft Blatantly Disregards User Choice and Privacy: A Deep Dive," *Electronic Frontier Foundation,* August 17, 2016, https://www.eff.org/deeplinks/2016/08/windows-10-microsoft-blatantly-disregards-user-choice-and-privacy-deep-dive; Conner Forrest, "Windows 10 Violates Your Privacy by Default, Here's How You Can Protect Yourself," *TechRepublic,* August 4, 2015, http://www.techrepublic.com/article/windows-10-violates-your-privacy-by-default-heres-how-you-can-protect-yourself; Alec Meer, "Windows 10 Is Spying on You: Here's How to Stop It," *Rock, Paper, Shotgun* (blog), July 30, 2015, https://www.rockpapershotgun.com/2015/07/30/windows-10-privacy-settings.

118. "About Us—LinkedIn," LinkedIn, November 11, 2016, https://press.linkedin.com/about-linkedin; Satya Nadella et al., "Slides from Microsoft Investors Call Announcing LinkedIn Acquisition—World's Leading Professional Cloud + Network—Microsoft's and LinkedIn's Vision for the Opportunity Ahead," June 13, 2016, https://ncmedia.azureedge.net/ncmedia/2016/06/msft_announce_160613.pdf.

119. Nadella et al., "Slides from Microsoft."

120. Supantha Mukherjee, "Microsoft's Market Value Tops $500 Billion Again After 17 Years," *Reuters,* January 27, 2015, https://www.reuters.com/article/us-microsoft-results-research/microsofts-market-value-tops-500-billion-again-after-17-years-idUSKBN15B1L6.

121. Brian Fung, "Internet Providers Want to Know More About You Than Google Does, Privacy Groups Say," *Washington Post*, January 20, 2016, https://www.washingtonpost.com/news/the-switch/wp/2016/01/20/your-internet-provider-is-turning-into-a-data-hungry-tech-company-consumer-groups-warn.

122. Melissa Parietti, "The World's Top 10 Telecommunications Companies," *Investopedia*, March 2, 2016, http://www.investopedia.com/articles/markets/030216/worlds-top-10-telecommunications-companies.asp; Eric Griffith, "The Fastest ISPs of 2016," *PCMAG*, August 31, 2016, http://www.pcmag.com/article/346232/the-fastest-isps-of-2016.

123. See Mark Bergen and Alex Kantrowitz, "Verizon Looks to Target Its Mobile Subscribers with Ads," *Advertising Age*, May 21, 2014, http://adage.com/article/digital/verizon-target-mobile-subscribers-ads/293356.

124. See Julia Angwin and Mike Tigas, "How This Company Is Using Zombie Cookies to Track Verizon Customers," *ProPublica*, January 14, 2015, https://www.propublica.org/article/zombie-cookie-the-tracking-cookie-that-you-cant-kill.

125. Robert McMillan, "Verizon's 'Perma-Cookie' Is a Privacy-Killing Machine," *Wired*, October 2014, http://www.wired.com/2014/10/verizons-perma-cookie.

126. Jacob Hoffman-Andrews, "Verizon Injecting Perma-Cookies to Track Mobile Customers, Bypassing Privacy Controls," *Electronic Frontier Foundation*, November 3, 2014, https://www.eff.org/deeplinks/2014/11/verizon-x-uidh.

127. Julia Angwin and Jeff Larson, "Somebody's Already Using Verizon's ID to Track Users," *ProPublica*, October 30, 2014, http://www.propublica.org/article/somebodys-already-using-verizons-id-to-track-users.

128. Jacob Hoffman-Andrews, "How Verizon and Turn Defeat Browser Privacy Protections," *Electronic Frontier Foundation*, January 14, 2015, https://www.eff.org/deeplinks/2015/01/verizon-and-turn-break-browser-privacy-protections.

129. Julia Angwin, "AT&T Stops Using Undeletable Phone Tracking IDs," *ProPublica*, November 14, 2014, http://www.propublica.org/article/att-stops-using-undeletable-phone-tracking-ids; Angwin and Tigas, "How This Company Is Using Zombie Cookies."

130. Angwin and Larson, "Somebody's Already Using Verizon's ID."

131. Jonathan Mayer, "The Turn-Verizon Zombie Cookie," *Web Policy* (blog), January 14, 2015, http://webpolicy.org/2015/01/14/turn-verizon-zombie-cookie; Allison Schiff, "Can You Identify Me Now? A Deep Dive on Verizon's Data Practices," *AdExchanger*, October 9, 2014, http://adexchanger.com/data-exchanges/can-you-identify-me-now-a-deep-dive-on-verizons-data-practices.

132. Jacob Hoffman-Andrews, "Under Senate Pressure, Verizon Plans Supercookie Opt-Out," *Electronic Frontier Foundation*, February 2, 2015, https://www.eff.org/deeplinks/2015/02/under-senate-pressure-verizon-improves-its-supercookie-opt-out.

133. Bill Nelson et al., "Letter to Mr. Lowell C. McAdam, Chairman and CEO of Verizon Communications from United States Senate Committee on Commerce, Science, and Transportation," January 29, 2015, http://thehill.com/sites/default /files/nelson-blumenthal-schatz-markey_letter_to_verizon_re_supercookies .pdf.

134. Brian X. Chen and Natasha Singer, "Verizon Wireless to Allow Complete Opt Out of Mobile 'Supercookies,'" *New York Times—Bits Blog,* January 30, 2015, http://bits.blogs.nytimes.com/2015/01/30/verizon-wireless-to-allow-complete -opt-out-of-mobile-supercookies.

135. See Edmund Ingham, "Verizon Had One Thing on Its Mind When It Agreed to Buy AOL: CEO Tim Armstrong," *Forbes,* May 13, 2015, http://www.forbes.com /sites/edmundingham/2015/05/13/verizon-had-one-thing-on-its-mind -when-it-agreed-to-buy-aol-ceo-tim-armstrong. See also Alexander Nazaryan, "How Tim Armstrong Bested Marissa Mayer," *Newsweek,* July 25, 2016, http:// www.newsweek.com/marissa-mayer-tim-armstrong-nerd-prom-483539.

136. "Advertising Programs Privacy Notice—October 2015," Verizon, December 7, 2015, http://www.verizon.com/about/privacy/advertising-programs-privacy -notice; Julia Angwin and Jeff Larson, "Verizon's Zombie Cookie Gets New Life," *ProPublica,* October 6, 2015, https://www.propublica.org /article/verizons-zombie-cookie-gets-new-life.

137. Julia Angwin, "Verizon to Pay $1.35 Million to Settle Zombie Cookie Privacy Charges," *ProPublica,* March 7, 2016, https://www.propublica.org/article /verizon-to-pay-1.35-million-to-settle-zombie-cookie-privacy-charges.

138. Mike Shields and Ryan Knutson, "AOL's Tim Armstrong Aims to Build Digital-Ad Empire at Verizon," *Wall Street Journal,* March 30, 2016, http://www .wsj.com/articles/aols-tim-armstrong-aims-to-build-digital-ad-empire-at -verizon-1459330200.

139. Tom Wheeler, "Statement of Chairman Tom Wheeler in Reply to WC Docket No. 16-106—Protecting the Privacy of Customers of Broadband and Other Telecommunications Services," Federal Communications Commission, 2016.

140. Alina Selyukh, "FCC Votes to Propose New Privacy Rules for Internet Service Providers," *NPR.org,* March 31, 2016, http://www.npr.org/sections/thetwo -way/2016/03/31/472528382/fcc-votes-to-propose-new-privacy-rules-for -internet-service-providers.

141. "FCC Adopts Privacy Rules to Give Broadband Consumers Increased Choice, Transparency and Security for Their Personal Data," Federal Communications Commission, October 27, 2016, https://www.fcc.gov/document/fcc-adopts -broadband-consumer-privacy-rules; Wendy Davis, "Broadband Providers Push Back Against Tough Privacy Proposal," *MediaPost,* March 10, 2016, http:// www.mediapost.com/publications/article/270983/broadband-providers-push -back-against-tough-privac.html; Brian Fung and Craig Timberg, "The FCC Just Passed Sweeping New Rules to Protect Your Online Privacy," *Washington*

Post, October 27, 2016, https://www.washingtonpost.com/news/the-switch
/wp/2016/10/27/the-fcc-just-passed-sweeping-new-rules-to-protect-your
-online-privacy.

142. Michelle Castillo, "AOL's Tim Armstrong: Yahoo Helps Verizon Compete
Against Facebook, Google," *CNBC,* July 25, 2016, http://www.cnbc
.com/2016/07/25/aol-ceo-tim-armstrong-yahoo-deal-helps-verizon-against
-facebook-google.html.

143. See Kara Swisher, "AOL's Tim Armstrong Says 'Scale Is Imperative' in
the Verizon-Yahoo Deal," *Recode,* July 25, 2016, http://www.recode.net
/2016/7/25/12269980/aol-tim-armstrong-scale-imperative-yahoo-deal.

144. See Ingrid Lunden, "AOL CEO on Yahoo Deal: 'We Want to Get to 2B Users,'"
TechCrunch (blog), July 25, 2016, http://social.techcrunch.com/2016/07/25/aol
-ceo-armstrongs-yahoo-memo-well-work-closely-with-marissa.

145. Tom Wheeler, "How the Republicans Sold Your Privacy to Internet Providers,"
New York Times, March 29, 2017, https://www.nytimes.com/2017/03/29
/opinion/how-the-republicans-sold-your-privacy-to-internet-providers.html;
"Republicans Attack Internet Privacy," *New York Times,* March 29, 2017, https://
www.nytimes.com/2017/03/29/opinion/republicans-attack-internet-privacy
.html; Cecilia Kang, "Congress Moves to Overturn Obama-Era Online Privacy
Rules," *New York Times,* March 28, 2017, https://www.nytimes.com/2017/03/28
/technology/congress-votes-to-overturn-obama-era-online-privacy-rules
.html; "The House Just Voted to Wipe Out the FCC's Landmark Internet
Privacy Protections," *Washington Post,* March 28, 2017, https://www.washington
post.com/news/the-switch/wp/2017/03/28/the-house-just-voted-to-wipe-out
-the-fccs-landmark-internet-privacy-protections.

146. Brian Fung, "It's Begun: Internet Providers Are Pushing to Repeal Obama-Era
Privacy Rules," *Washington Post,* January 4, 2017, https://www.washingtonpost
.com/news/the-switch/wp/2017/01/04/its-begun-cable-companies-are-pushing
-to-repeal-obama-era-internet-privacy-rules.

147. Wheeler, "How the Republicans Sold Your Privacy."

148. Jack Marshall, "With Washington's Blessing, Telecom Giants Can Mine Your
Web History," *Wall Street Journal,* March 30, 2017, https://www.wsj.com/articles
/with-washingtons-blessing-telecom-giants-can-mine-your-web-history
-1490869801; Olivia Solon, "What to Know Before Lawmakers Decide if ISPs
Can Sell Your Browsing History," *Guardian,* March 28, 2017, https://www
.theguardian.com/technology/2017/mar/28/internet-service-providers-sell
-browsing-history-house-vote; Bruce Schneier, "Snoops May Soon Be Able to
Buy Your Browsing History. Thank the US Congress," *Guardian,* March 30,
2017, http://www.theguardian.com/commentisfree/2017/mar/30/snoops-buy
-your-browsing-history-us-congress; Jeremy Gillula, "Five Creepy Things
Your ISP Could Do if Congress Repeals the FCC's Privacy Protections,"
Electronic Frontier Foundation, March 19, 2017, https://www.eff.org/deep

links/2017/03/five-creepy-things-your-isp-could-do-if-congress-repeals-fccs -privacy-protections.

149. Elizabeth Dwoskin, "Lending Startups Look at Borrowers' Phone Usage to Assess Creditworthiness," *Wall Street Journal,* December 1, 2015, http://www .wsj.com/articles/lending-startups-look-at-borrowers-phone-usage-to-assess -creditworthiness-1448933308.

150. Daniel Bjorkegren and Darrell Grissen, "Behavior Revealed in Mobile Phone Usage Predicts Loan Repayment" (SSRN scholarly paper, Social Science Research Network, July 13, 2015), https://papers.ssrn.com/abstract =2611775.

151. See Dwoskin, "Lending Startups Look at Borrowers' Phone Usage."

152. Dwoskin.

153. See Caitlin Dewey, "Creepy Startup Will Help Landlords, Employers and Online Dates Strip-Mine Intimate Data from Your Facebook Page," *Washington Post,* June 9, 2016, https://www.washingtonpost.com/news/the-intersect/ wp/2016/06/09/creepy-startup-will-help-landlords-employers-and-online -dates-strip-mine-intimate-data-from-your-facebook-page.

154. Frank Pasquale, "The Dark Market for Personal Data," *New York Times,* October 16, 2014, http://www.nytimes.com/2014/10/17/opinion/the-dark-market-for -personal-data.html.

155. "hiQ Labs—Home," hiQ Labs, August 26, 2017, https://www.hiqlabs.com.

156. Christopher Ingraham, "Analysis: Politics Really Is Ruining Thanksgiving, According to Data from 10 Million Cellphones," *Washington Post,* November 15, 2017, https://www.washingtonpost.com/news/wonk/wp/2017/11/15 /politics-really-is-ruining-thanksgiving-according-to-data-from-10-million -cellphones. For the research working paper, see M. Keith Chen and Ryne Rohla, "Politics Gets Personal: Effects of Political Partisanship and Advertising on Family Ties," *ArXiv:1711.10602 [Economics],* November 28, 2017, http:// arxiv.org/abs/1711.10602.

CHAPTER SIX

1. Matthew Restall, *Seven Myths of the Spanish Conquest* (Oxford: Oxford University Press, 2004), 19.

2. See Felipe Fernández-Armesto, *1492: The Year the World Began* (New York: HarperOne, 2010), 196.

3. John R. Searle, *Making the Social World: The Structure of Human Civilization* (Oxford: Oxford University Press, 2010), 85–86.

4. Searle, *Making the Social World,* 13.

5. Restall, *Seven Myths,* 65.

6. Restall, 19.

7. For a fascinating analysis of the *Requirimiento,* see Paja Faudree, "How to Say Things with Wars: Performativity and Discursive Rupture in the *Requerimiento*

of the Spanish Conquest," *Journal of Linguistic Anthropology* 22, no. 3 (2012): 182–200.

8. Bartolomé de las Casas, *A Brief Account of the Destruction of the Indies* (Penguin Classics), Kindle 334–38.

9. de las Casas, *A Brief Account,* 329–33.

10. David Hart, "On the Origins of Google," National Science Foundation, August 17, 2004, http://www.nsf.gov/discoveries/disc_summ.jsp?cntn_id=100 660&org=NSF.

11. Eric Schmidt and Jared Cohen, *The New Digital Age: Transforming Nations, Businesses, and Our Lives* (New York: Vintage, 2014), 9–10.

12. Mark Muro et al., "Digitalization and the American Workforce," Metropolitan Policy Program, Brookings Institution, November 15, 2017, https://www .brookings.edu/research/digitalization-and-the-american-workforce. As the report observes,

> In 2002, a one-point increase in digitalization score predicted a $166.20 (in 2016 dollars) increase in real annual average wages for occupations with the same education requirements. By 2016 this wage premium had almost doubled to $292.80. In sum, workers with superior digital skills are more and more earning higher wages (all other things being equal) than similarly educated workers with fewer digital skills.... Thus, a sizable portion of the nation's critical middle-skill employment now requires dexterity with basic IT tools, standard health monitoring technology, computer numerical control equipment, basic enterprise management software, customer relationship management software like Salesforce or SAP, or spreadsheet programs like Microsoft Excel.... In sum, tens of millions of jobs that provide the best routes toward economic inclusion for workers without a college degree turn out to be less and less accessible to workers who lack basic digital skills. (23–24, 33)

13. Philipp Brandes, Roger Wattenhofer, and Stefan Schmid, "Which Tasks of a Job Are Susceptible to Computerization?" *Bulletin of EATCS* 3, no. 120 (2016), http://bulletin.eatcs.org/index.php/beatcs/article/view/467; Carl Benedikt Frey and Michael Osborne, "The Future of Employment: How Susceptible Are Jobs to Computerisation?" *Technological Forecasting and Social Change* 114 (September 17, 2013): 254–80; Seth G. Benzell et al., "Robots Are Us: Some Economics of Human Replacement" (National Bureau of Economic Research, 2015), http://www.nber.org/papers/w20941; Carl Benedikt Frey, "Doing Capitalism in the Digital Age," *Financial Times,* October 1, 2014, https://www .ft.com/content/293780fc-4245-11e4-9818-00144feabdc0.

14. Frey and Osborne, "The Future of Employment"; Martin Krzywdzinski, "Automation, Skill Requirements and Labour-Use Strategies: High-Wage and Low-Wage Approaches to High-Tech Manufacturing in the Automotive Industry," *New Technology, Work and Employment* 32, no. 3 (2017): 247–67, https://doi.org/10.1111/ntwe.12100; Frey, "Doing Capitalism"; William Lazonick, "Labor in the Twenty-First Century: The Top 0.1% and the

Disappearing Middle-Class" (working paper, Institute for New Economic Thinking, February 2015), https://www.ineteconomics.org/research/research -papers/labor-in-the-twenty-first-century-the-top-0-1-and-the-disappearing -middle-class; Dirk Antonczyk, Thomas DeLeire, and Bernd Fitzenberger, "Polarization and Rising Wage Inequality: Comparing the U.S. and Germany" (IZA Discussion Paper, Institute for the Study of Labor, March 2010), https:// ideas.repec.org/p/iza/izadps/dp4842.html; Erik Brynjolfsson and Andrew McAfee, *The Second Machine Age: Work, Progress, and Prosperity in a Time of Brilliant Technologies* (New York: W. W. Norton, 2016); Daron Acemoglu and David Autor, "What Does Human Capital Do? A Review of Goldin and Katz's 'The Race Between Education and Technology,'" *Journal of Economic Literature* 50, no. 2 (2012): 426–63; Sang Yoon Lee and Yongseok Shin, "Horizontal and Vertical Polarization: Task-Specific Technological Change in a Multi-Sector Economy" (SSRN Scholarly Paper, Rochester, NY: Social Science Research Network, March 1, 2017), https://papers.ssrn.com /abstract=2941261.

15. Kathleen Thelen, *Varieties of Liberalization and the New Politics of Social Solidarity* (Cambridge: Cambridge University Press, 2014); Olivier Giovannoni, "What Do We Know About the Labor Share and the Profit Share? Part III: Measures and Structural Factors" (working paper, Levy Economics Institute at Bard College, 2014), http://www.levyinstitute.org/publications/what-do-we -know-about-the-labor-share-and-the-profit-share-part-3-measures-and -structural-factors; Francisco Rodriguez and Arjun Jayadev, "The Declining Labor Share of Income," *Journal of Globalization and Development* 3, no. 2 (2013): 1–18; Antonczyk, DeLeire, and Fitzenberger, "Polarization and Rising Wage Inequality"; Duane Swank, "The Political Sources of Labor Market Dualism in Postindustrial Democracies, 1975–2011" (American Political Science Association Annual Meeting, Chicago: Social Science Research Network, 2013), https://papers.ssrn.com/sol3/papers.cfm?abstract_id=2299566; David Jacobs and Lindsey Myers, "Union Strength, Neoliberalism, and Inequality: Contingent Political Analyses of US Income Differences Since 1950," *American Sociological Review* 79 (2014): 752–74; Viki Nellas and Elisabetta Olivieri, "The Change of Job Opportunities: The Role of Computerization and Institutions" (Quaderni DSE Working Paper, University of Bologna & Bank of Italy, 2012), http://papers.ssrn.com/sol3/papers.cfm?abstract_id=1983214; Ian Gough, Anis Ahmad Dani, and Harjan de Haan, "European Welfare States: Explanations and Lessons for Developing Countries," in *Inclusive States: Social Policies and Structural Inequalities* (Washington, DC: World Bank, 2008).

16. Martin R. Gillings, Martin Hilbert, and Darrell J. Kemp, "Information in the Biosphere: Biological and Digital Worlds," *Trends in Ecology and Evolution* 31, no. 3 (2016).

17. Emile Durkheim, *The Division of Labor in Society* (New York: Free Press, 1964), 41.

18. Durkheim, *The Division of Labor,* 60–61.

19. Harvard legal scholar John Palfrey observed the "read-only" nature of electronic surveillance in his wonderful 2008 essay, "The Public and the Private at the United States Border with Cyberspace," *Mississippi Law Journal* 78 (2008): 241–94, especially 249.

20. Frank Pasquale, *The Black Box Society* (Cambridge, MA: Harvard University Press, 2015), 60–61.

21. Martin Hilbert, "Toward a Synthesis of Cognitive Biases: How Noisy Information Processing Can Bias Human Decision Making," *Psychological Bulletin* 138, no. 2 (2012): 211–37, https://doi.org/10.1037/a0025940; Martin Hilbert, "Big Data for Development: From Information- to Knowledge Societies" (United Nations ECLAC Report, Social Science Research Network, 2013), 4, http://papers.ssrn.com/sol3/papers.cfm?abstract_id=2205145; Viktor Mayer-Schönberger and Kenneth Cukier, *Big Data: A Revolution That Will Transform How We Live, Work, and Think* (Boston: Houghton Mifflin Harcourt, 2013), 9.

22. Hilbert, "Toward a Synthesis of Cognitive Biases."

23. Paul Borker, "What Is Hyperscale?" *Digital Realty,* February 2, 2018, https://www.digitalrealty.com/blog/what-is-hyperscale; Paul McNamara, "What Is Hyperscale and Why Is It so Important to Enterprises?" http://cloudblog.ericsson.com/digital-services/what-is-hyperscale-and-why-is-it-so-important-to-enterprises; James Manyika and Michael Chui, "Digital Era Brings Hyperscale Challenges," *Financial Times,* August 13, 2014, http://www.ft.com/intl/cms/s/0/f30051b2-1e36-11e4-bb68-00144feabdc0.html?siteedition=intl#axzz3JjXPNno5; Cade Metz, "Building an AI Chip Saved Google from Building a Dozen New Data Centers," *Wired,* April 5, 2017, https://www.wired.com/2017/04/building-ai-chip-saved-google-building-dozen-new-data-centers.

24. Smaller firms without hyperscale revenues can leverage some of these capabilities with cloud computing services.

25. Catherine Dong, "The Evolution of Machine Learning," *TechCrunch,* August 8, 2017, http://social.techcrunch.com/2017/08/08/the-evolution-of-machine-learning; Metz, "Building an AI Chip"; "Google Data Center FAQ," *Data Center Knowledge,* March 16, 2017, http://www.datacenterknowledge.com/archives/2017/03/16/google-data-center-faq.

26. ARK Investment Management, "Google: The Full Stack AI Company," *Seeking Alpha,* May 25, 2017, https://seekingalpha.com/article/4076671-google-full-stack-ai-company; Alon Halevy, Peter Norvig, and Fernando Pereira, "The Unreasonable Effectiveness of Data," *Intelligent Systems, IEEE* 24 (2009): 8–12, https://doi.org/10.1109/MIS.2009.36.

27. See Tom Krazit, "Google's Urs Hölzle Still Thinks Its Cloud Revenue Will Catch Its Ad Revenue, but Maybe Not by 2020," *GeekWire,* November 15, 2017,

https://www.geekwire.com/2017/googles-urs-holzle-still-thinks-cloud-revenue
-will-catch-ad-revenue-maybe-not-2020.

28. Norm Jouppi, "Google Supercharges Machine Learning Tasks with TPU
Custom Chip," *Google Cloud Platform Blog,* May 18, 2016, https://cloud
platform.googleblog.com/2016/05/Google-supercharges-machine-learning
-tasks-with-custom-chip.html; Jeff Dean and Urs Hölzle, "Build and Train
Machine Learning Models on Our New Google Cloud TPUs," Google,
May 17, 2017, https://blog.google/topics/google-cloud/google-cloud-offer
-tpus-machine-learning; Yevgeniy Sverdlik, "Google Ramped Up Data Center
Spend in 2016," *Data Center Knowledge,* February 1, 2017, http://www.data
centerknowledge.com/archives/2017/02/01/google-ramped-data-center
-spend-2016; Courtney Flatt, "Google's All-Renewable Energy Plan to Include
Data Center in Oregon," *Oregon Public Broadcasting,* December 6, 2016, http://
www.opb.org/news/article/google-says-it-will-consume-only-renewable
-energy.

29. Michael Feldman, "Market for Artificial Intelligence Projected to Hit $36
Billion by 2025," *Top500,* August 30, 2016, https://www.top500.org
/news/market-for-artificial-intelligence-projected-to-hit-36-billion-by
-2025.

30. Kevin McLaughlin and Mike Sullivan, "Google's Relentless AI Appetite,"
Information, January 10, 2017, https://www.theinformation.com/googles
-relentless-ai-appetite.

31. Cade Metz, "Tech Giants Are Paying Huge Salaries for Scarce A.I. Talent," *New
York Times,* October 22, 2017, https://www.nytimes.com/2017/10/22
/technology/artificial-intelligence-experts-salaries.html; "Artificial Intelligence
Is the New Black," *Paysa Blog,* April 18, 2017, https://www.paysa.com/blog/2017
/04/17/artificial-intelligence-is-the-new-black.

32. Ian Sample, "Big Tech Firms' AI Hiring Frenzy Leads to Brain Drain at UK
Universities," *Guardian,* November 2, 2017, http://www.theguardian.com
/science/2017/nov/02/big-tech-firms-google-ai-hiring-frenzy-brain-drain-uk
-universities.

33. Pedro Domingos, *The Master Algorithm: How the Quest for the Ultimate
Learning Machine Will Remake Our World* (New York: Basic, 2015), 12–13;
Cade Metz, "Why A.I. Researchers at Google Got Desks Next to the Boss," *New
York Times,* February 19, 2018, https://www.nytimes.com/2018/02/19
/technology/ai-researchers-desks-boss.html.

34. Spiros Simitis, "Reviewing Privacy in an Information Society," *University of
Pennsylvania Law Review* 135, no. 3 (1987): 710, https://doi.org/10.2307
/3312079.

35. Paul M. Schwartz, "The Computer in German and American Constitutional
Law: Towards an American Right of Informational Self-Determination,"
American Journal of Comparative Law 37 (1989): 676.

CHAPTER SEVEN

1. Chris Matyszczyk, "The Internet Will Vanish, Says Google's Eric Schmidt," *CNET,* January 22, 2015, http://www.cnet.com/news/the-internet-will -vanish-says-googles-schmidt.

2. Mark Weiser, "The Computer for the 21st Century," *Scientific American,* September 1991.

3. Mark Weiser and John Seely Brown, "The Coming Age of Calm Technology," in *Beyond Calculation* (New York: Springer, 1997), 75–85, https://doi.org/10.1007 /978-1-4612-0685-9_6; Weiser, "The Computer for the 21st Century."

4. Janina Bartje, "IoT Analytics the Top 10 IoT Application Areas—Based on Real IoT Projects," *IOT Analytics,* August 16, 2016, https://iot-analytics.com/top -10-iot-project-application-areas-q3-2016.

5. Kevin D. Werbach and Nicolas Cornell, "Contracts Ex Machina" (SSRN Scholarly Paper, Rochester, NY: Social Science Research Network, March 18, 2017), https://papers.ssrn.com/abstract=2936294.

6. Christy Pettey, "Treating Information as an Asset," *Smarter with Gartner,* February 17, 2016, www.gartner.com/smarterwithgartner/treating-informa tion-as-an-asset (italics mine).

7. R. Stuart MacKay, "Bio-medical Telemetry: Sensing and Transmitting Biological Information from Animals and Man," *Quarterly Review of Biology* 44, no. 4 (1969): 18–23.

8. MacKay, "Bio-medical Telemetry."

9. MacKay.

10. Roland Kays et al., "Terrestrial Animal Tracking as an Eye on Life and Planet," *Science* 348, no. 6240 (2015), https://doi.org/10.1126/science.aaa2478.

11. P. Ramesh Kumar, Ch. Srikanth, and K. L. Sailaja, "Location Identification of the Individual Based on Image Metadata," *Procedia Computer Science* 85 (2016): 451–54, https://doi.org/10.1016/j.procs.2016.05.191; Anuradha Vishwakarma et al., "GPS and RFID Based Intelligent Bus Tracking and Management System," *International Research Journal of Engineering and Technology* 3, no. 3 (2016); Nirali Panchal, "GPS Based Vehicle Tracking System and Using Analytics to Improve the Performance," *ResearchGate,* June 2016, https://www.researchgate .net/publication/304129283_GPS_Based_Vehicle_Tracking_System_and _Using_Analytics_to_Improve_The_Performance.

12. Mark Prigg, "Software That Can Track People as They Walk from Camera to Camera," *Mail Online,* November 18, 2014, http://www.dailymail.co.uk /sciencetech/article-2838633/Software-track-people-walk-camera-camera-say -tracked-Boston-bombers-hours.html.

13. Joseph A. Paradiso, "Our Extended Sensoria: How Humans Will Connect with the Internet of Things," *MIT Technology Review,* August 1, 2017, https://www .technologyreview.com/s/608566/our-extended-sensoria-how-humans-will -connect-with-the-internet-of-things.

14. Gershon Dublon and Edwina Portocarrerro, "ListenTree: Audio-Haptic Display in the Natural Environment," 2014, https://smartech.gatech.edu/handle/1853/52083.

15. Gershon Dublon et al., "DoppelLab: Tools for Exploring and Harnessing Multimodal Sensor Network Data," in *IEEE Sensors Proceedings,* 2011, 1612–15, http://dspace.mit.edu/handle/1721.1/80402.

16. Gershon Dublon and Joseph A. Paradiso, "Extra Sensory Perception," *Scientific American,* June 17, 2014.

17. Paradiso, "Our Extended Sensoria" (italics mine).

18. Dublon and Paradiso, "Extra Sensory Perception."

19. Kevin Ashton, a former Procter and Gamble brand manager who pioneered the marriage of radio-enabled microchips and physical products, birthed the term "internet of things," and helped drive RFID innovation at MIT's Media Lab, criticizes the US government for its lack of a comprehensive vision for the "internet of things" and the leadership of private firms in this domain. See Kevin Ashton, "America Last?" *Politico,* June 29, 2015, http://www.politico.com /agenda/story/2015/06/kevin-ashton-internet-of-things-in-the-us-000102.

20. See Nick Statt, "What the Volkswagen Scandal Means for the Future of Connected Devices," *Verge,* October 21, 2015, http://www.theverge.com/2015 /10/21/9556153/internet-of-things-privacy-paranoia-data-volkswagen-scandal.

21. Matt Weinberger, "Companies Stand to Make a Lot of Money Selling Data from Smart Devices, Says Microsoft," *Business Insider,* December 6, 2015, http://www .businessinsider.com/microsoft-azure-internet-of-things-boss-sam-george -interview-2015-12; "Live on a Screen Near You: IoT Slam, a New Virtual Conference for All Things IoT," *Microsoft IoT Blog,* December 9, 2015, https:// blogs.microsoft.com/iot/2015/12/09/live-on-a-screen-near-you-iot-slam-a -new-virtual-conference-for-all-things-iot.

22. "The Economy of Things: Extracting New Value from the Internet of Things," *IBM Institute for Business Value,* 2014, http://www-935.ibm.com/services/us /gbs/thoughtleadership/economyofthings.

23. For a compelling discussion of unstructured data and their roots in everyday life, see Ioanna D. Constantiou and Jannis Kallinikos, "New Games, New Rules: Big Data and the Changing Context of Strategy," *Journal of Information Technology,* September 9, 2014, 1–14.

24. Bryan Glick, "Executive Interview: Harriet Green, IBM's Internet of Things Chief," *ComputerWeekly,* April 7, 2016, http://www.computerweekly.com /news/450280673/Executive-interview-Harriet-Green-IBMs-internet-of -things-chief.

25. "Dark Data," *Gartner IT Glossary,* May 7, 2013, http://www.gartner.com /it-glossary/dark-data; Isaac Sacolick, "Dark Data—a Business Definition," *Social, Agile, and Transformation,* April 10, 2013, http://blogs.starcio.com /2013/04/dark-data-business-definition.html; Heather Johnson, "Digging Up Dark Data: What Puts IBM at the Forefront of Insight Economy," *SiliconANGLE,* October 30, 2015, https://siliconangle.com/blog/2015/10/30

/ibm-is-at-the-forefront-of-insight-economy-ibminsight; Ed Tittel, "The Dangers of Dark Data and How to Minimize Your Exposure," *CIO,* September 24, 2014, https://www.cio.com/article/2686755/data-analytics/the-dangers-of -dark-data-and-how-to-minimize-your-exposure.html; Derek Gascon, "Thwart 'Dark Data' Risk with Data Classification Tools," *SearchCompliance,* July 2014, http://searchcompliance.techtarget.com/answer/Thwart-dark-data-risk-with -data-classification-tools.

26. Glick, "Executive Interview."

27. Hal R. Varian, "Computer Mediated Transactions," *American Economic Review* 100, no. 2 (2010): 1–10, https://doi.org/10.1257/aer.100.2.1.

28. Hal R. Varian, "Beyond Big Data," *Business Economics* 49, no. 1 (2014): 27–31.

29. Varian, "Beyond Big Data," 30.

30. See Dan Kraut, "Allstate Mulls Selling Driver Data," *Bloomberg.com,* May 28, 2015, http://www.bloomberg.com/news/articles/2015-05-28/allstate-seeks-to -follow-google-as-ceo-mulls-selling-driver-data.

31. Rachel Ward and Rebecca Lancaster, "The Contribution of Individual Factors to Driving Behaviour: Implications for Managing Work-Related Road Safety" (research report, Doherty Innovation Centre, Midlothian, UK, 2002), http:// www.hse.gov.uk/research/rrhtm/rr020.htm.

32. "Insurers Need to Plug into the Internet of Things—or Risk Falling Behind," McKinsey, January 8, 2017, http://www.mckinsey.com/industries/financial -services/our-insights/insurers-need-to-plug-into-the-internet-of-things-or -risk-falling-behind.

33. "Overcoming Speed Bumps on the Road to Telematics," Deloitte University Press, April 21, 2014, https://dupress.deloitte.com/dup-us-en/industry /insurance/telematics-in-auto-insurance.html.

34. "Overcoming Speed Bumps on the Road to Telematics."

35. Leslie Scism, "State Farm Is There: As You Drive," *Wall Street Journal,* August 5, 2013.

36. "Insurers Need to Plug into the Internet of Things."

37. Joseph Reifel, Alyssa Pei, Neeti Bhardwaj, and Shamik Lala, "The Internet of Things: Opportunity for Insurers," ATKearney, 2014, https://www.atkearney .co.uk/documents/10192/5320720/internet+of+Things+-+Opportunity+for +Insurers.pdf/4654e400-958a-40d5-bb65-1cc7ae64bc72.

38. Steve Johansson, "Spireon Reaches 2.4 Million Subscribers, Becoming Industry's Largest Aftermarket Vehicle Telematics Company," *BusinessWire,* August 17, 2015, http://www.businesswire.com/news/home/20150817005365 /en/Spireon-Reaches-2.4-Million-Subscribers-Industry%E2%80%99s-Largest.

39. Rebecca Kates, "Geotab Launches a World Leading Driver Safety Telematics Solution That Speaks to the Driver Inside the Vehicle," Geotab, September 10, 2015, https://www.geotab.com/press-release/geotab-launches-driver-safety -telematics-solution-that-speaks-to-the-driver-inside-the-vehicle.

40. Brad Jarvis et al., Insurance rate optimization through driver behavior monitoring, US20150006207 A1, published January 2015, https://patents.google.com/patent/US20150006207A1/en.
41. Brad Jarvis et al., Operator benefits and rewards through sensory tracking of a vehicle, US20150019270 A1, published January 2015, 2015, http://www.google.com/patents/US20150019270.
42. Joao Lima, "Insurers Look Beyond Connected Cars for IOT Driven Business Boom," *Computer Business Review*, December 9, 2015, http://www.cbronline.com/news/internet-of-things/insurers-look-beyond-connected-cars-for-iot-driven-business-boom-4748866.
43. Sam Ramji, "Looking Beyond the Internet of Things Hype: Here's What's in Store," *VentureBeat*, March 28, 2014, http://venturebeat.com/2014/03/28/looking-beyond-the-internet-of-things-hype-heres-whats-in-store.
44. "Overcoming Speed Bumps on the Road to Telematics."
45. Corin Nat, "Think Outside the Box—Motivate Drivers Through Gamification," Spireon, August 11, 2015, https://web.archive.org/web/20150811014300/spireon.com/motivate-drivers-through-gamification; "Triad Isotopes," 2017, http://www.triadisotopes.com.
46. "Overcoming Speed Bumps on the Road to Telematics."
47. See Byron Pope, "Experts Examine Auto Telematics' Pitfalls, Potential," *Ward's Auto*, June 20, 2013, http://wardsauto.com/technology/experts-examine-auto-telematics-pitfalls-potential.
48. "Analytics Trends 2016, the Next Evolution," Deloitte, 2016, https://www2.deloitte.com/us/en/pages/deloitte-analytics/articles/analytics-trends.html.
49. "Insurers Need to Plug into the Internet of Things"; "Navigating the Challenges and Opportunities in Financial Services," Deloitte Touche, 2015, https://www2.deloitte.com/content/dam/Deloitte/au/Documents/financial-services/deloitte-au-fs-fsi-outlook-focus-2015-090215.pdf.
50. "Dell Services Have Launched a New Internet of Things Insurance Accelerator," Dell, November 17, 2015, http://www.dell.com/learn/al/en/alcorp1/press-releases/2015-11-17-dell-services-launch-of-internet-of-things; "Microsoft and American Family Insurance Launch Startup Accelerator Focused on Home Automation," *Microsoft News Center*, June 17, 2014, https://news.microsoft.com/2014/06/17/microsoft-and-american-family-insurance-launch-startup-accelerator-focused-on-home-automation.
51. Gabe Nelson, "Who Owns the Dashboard? Apple, Google or the Automakers?" *Advertising Age*, December 15, 2014, http://adage.com/article/digital/owns-dashboard-apple-google-automakers/296200.
52. "Google Looks to Partner with Insurance Companies in France," *Fortune*, September 13, 2016, http://fortune.com/2016/09/13/google-france-insurance-partners.
53. Varian, "Beyond Big Data," 30 (italics mine).
54. Varian.

55. Herman Kahn and Anthony J. Wiener, *The Year 2000* (New York: Macmillan, 1967).
56. Kahn and Wiener, *The Year 2000*, 97–98.
57. Kahn and Wiener, 410–11.
58. Paul T. David and William R. Ewald, "The Study of the Future," *Public Administration Review* 28, no. 2 (1968): 187–93, https://doi.org/10.2307/974089.
59. Neil P. Hurley, "The Future and the Nearsighted Faust," *Review of Politics* 30, no. 4 (1968): 521–23.
60. Langdon Winner, *Autonomous Technology: Technics-out-of-Control as a Theme in Political Thought* (Cambridge, MA: MIT Press, 1978), 7–8.
61. Dublon and Paradiso, "Extra Sensory Perception," 37–44.
62. Frank E. Manuel and Fritzie P. Manuel, *Utopian Thought in the Western World* (Cambridge, MA: Belknap Press, 1979), 6.
63. Krishan Kumar, *Utopia and Anti-utopia in Modern Times* (Oxford: Blackwell, 1991); Andrzej Walicki, *Marxism and the Leap to the Kingdom of Freedom: The Rise and Fall of the Communist Utopia* (Stanford, CA: Stanford University Press, 1997); Gregory Claeys, *Searching for Utopia: The History of an Idea* (New York: Thames & Hudson, 2011); Roland Schaer, Gregory Claeys, and Lyman Tower Sargent, *Utopia: The Search for Ideal Society in the Western World* (New York: Oxford University Press, 2000); Perry Anderson, *Arguments Within English Marxism*, 2nd ed. (London: Verso, 1980).
64. In the 1867 preface to the first edition of Marx's *Capital*, one finds the following:
     Intrinsically, it is not a question of the higher or lower degree of development of the social antagonisms that result from the natural laws of capitalist production. It is a question of these laws themselves, of these tendencies working with iron necessity towards inevitable results. The country that is more developed industrially only shows, to the less developed, the image of its own future. See Karl Marx and Ernest Mandel, *Capital*, vol. 1, trans. Ben Fowkes (New York: Penguin, 1990), 91.
65. Manuel and Manuel, *Utopian Thought*, 3–4.
66. Eric Schmidt and Jared Cohen, *The New Digital Age: Transforming Nations, Businesses, and Our Lives* (New York: Vintage, 2014), 5.
67. Schmidt and Cohen, *The New Digital Age*, 253–54.
68. Dublon and Paradiso, "Extra Sensory Perception."
69. John Steinbeck, *The Grapes of Wrath* (New York: Viking, 1939).
70. Winner, *Autonomous Technology*, 6.
71. Langdon Winner, "Do Artifacts Have Politics?" *Daedalus* 109, no. 1 (1980): 99.
72. For a wonderful related discussion, see Alasdair Morrison, "Uses of Utopia," in *Utopias*, ed. Peter Alexander and Roger Gill (La Salle, IL: Open Court, 1983), 149–50.
73. "Digital Transformation Map," Cisco, August 3, 2018, https://www.cisco.com/c/m/en_us/solutions/industries/smart-connected-communities/digital-transformation-map.html; and Anil Menon, "Announcing Cisco Kinetic for Cities," *Cisco Blogs*, October 4, 2017, https://blogs.cisco.com/news/announcing-kinetic-for-cities (italics mine).

74. "Titan and Control Group Become Intersection," *PR Newswire*, September 16, 2015, http://www.prnewswire.com/news-releases/titan-and-control-group -become-intersection-300144002.html. Sidewalk acquired a company called Intersection to build and administer the kiosks. Intersection was formed in the merger of Control, an urban-focused-technology company, and Titan, an ad media firm. Intersection describes itself as "an urban experience, technology and media company [that] will work with cities to solve modern challenges and reinvent the urban experience, creating more connected, livable and prosperous cities. The company will leverage the groundbreaking engineering and design capabilities of Control Group, which has more than a decade of experience with merging digital technologies with real-world user experiences in urban environments, with Titan, one of the country's largest municipal and transit media companies and a leader in digital out-of-home advertising."

75. Conor Dougherty, "Cities to Untangle Traffic Snarls, with Help from Alphabet Unit," *New York Times*, March 17, 2016, http://www.nytimes.com/2016/03/18 /technology/cities-to-untangle-traffic-snarls-with-help-from-alphabet-unit .html.

76. "Sidewalk Labs | Team—Alphabet," Sidewalk Labs, October 2, 2017, https:// www.sidewalklabs.com/team.

77. Dougherty, "Cities to Untangle."

78. See Dougherty.

79. See Diana Budds, "How Google Is Turning Cities into R&D Labs," *Co.Design*, February 22, 2016, https://www.fastcodesign.com/3056964/design-moves /how-google-is-turning-cities-into-rd-labs.

80. Mark Harris, "Secretive Alphabet Division Aims to Fix Public Transit in US by Shifting Control to Google," *Guardian*, June 27, 2016, http://www.theguardian .com/technology/2016/jun/27/google-flow-sidewalk-labs-columbus -ohio-parking-transit.

81. *Google City: How the Tech Juggernaut Is Reimagining Cities—Faster Than You Realize*, 2016, https://www.youtube.com/watch?v=JXN9QHHD8eA.

82. *Google City.*

83. *Google City.*

84. See Budds, "How Google Is Turning Cities into R&D Labs."

85. Jessica E. Lessin, "Alphabet's Sidewalk Preps Proposal for Digital District," *I nformation*, April 14, 2016, https://www.theinformation.com/sidewalk-labs -preps-proposal-for-digital-district.

86. Eliot Brown, "Alphabet's Next Big Thing: Building a 'Smart' City," *Wall Street Journal*, April 27, 2016, http://www.wsj.com/articles/alphabets-next-big -thing-building-a-smart-city-1461688156.

87. Shane Dingman, "With Toronto, Alphabet Looks to Revolutionize City-Building," *Globe and Mail*, October 17, 2017, https://beta.theglobeandmail.com /report-on-business/with-toronto-alphabet-looks-to-revolutionize-city-build ing/article36634779.

CHAPTER EIGHT

1. Jan Wolfe, "Roomba Vacuum Maker iRobot Betting Big on the 'Smart' Home," *Reuters*, July 28, 2017, https://www.reuters.com/article/irobot-strategy/roomba -vacuum-maker-irobot-betting-big-on-the-smart-home-idUSL1N1KJ1BA; Melissa Wen, "iRobot Shares Surge on Strong Sales of Roomba Vacuum Cleaners," *Reuters*, July 26, 2017, https://www.reuters.com/article/us-irobot -stocks/irobot-shares-surge-on-strong-sales-of-roomba-vacuum-cleaners -idUSKBN1AB2QW.
2. Wolfe, "Roomba Vacuum Maker iRobot Betting Big."
3. Lance Ulanoff, "iRobot CEO Says the Company Won't Share Your Roomba Home Mapping Data Without Your OK," *Mashable*, July 25, 2017, http:// mashable.com/2017/07/25/irobot-wants-to-sell-home-mapping-data.
4. "iRobot HOME," Google Play Store, August 12, 2018, https://play.google.com /store/apps/details?id=com.irobot.home&hl=en. See also Alex Hern, "Roomba Maker May Share Maps of Users' Homes with Google, Amazon or Apple," *Guardian*, July 25, 2017, http://www.theguardian.com/technology/2017/jul/25 /roomba-maker-could-share-maps-users-homes-google-amazon-apple-irobot -robot-vacuum.
5. "How It Works | Smart Bed Technology & Sleep Tracking | It Bed," SleepNumber.com, October 6, 2017, https://itbed.sleepnumber.com/how-it -works.
6. "Sleep Number Privacy Policy," SleepNumber.com, September 18, 2017, https:// www.sleepnumber.com/sn/en/privacy-policy.
7. Guido Noto La Diega and Ian Walden, "Contracting for the 'Internet of Things': Looking into the Nest" (research paper, Queen Mary University of London, School of Law, 2016).
8. Jonathan A. Obar and Anne Oeldorf-Hirsch, "The Biggest Lie on the Internet: Ignoring the Privacy Policies and Terms of Service Policies of Social Networking Services," in *Facebook/Social Media 2* (TPRC 44: The 44th Research Conference on Communication, Information and Internet Policy, Arlington, VA: Social Science Research Network, 2016), https://papers.ssrn.com/abstract =2757465.
9. For an important discussion of this larger theme as it applies to digital products, see Aaron Perzanowski and Chris Hoofnagle, "What We Buy When We 'Buy Now'" (SSRN scholarly paper, Social Science Research Network, May 13, 2016), https://papers.ssrn.com/abstract=2778072.
10. Michelle Locke, "Ready for Liquor Bottles Smart Enough to Talk Smart Phones?" *Phys.org*, May 21, 2015, http://phys.org/news/2015-05-ready-liquor -bottles-smart.html; Joseph Cox, "This Rectal Thermometer Is the Logical Conclusion of the Internet of Things," *Motherboard*, January 14, 2016, http:// motherboard.vice.com/read/this-rectal-thermometer-is-the-logical-conclusion -of-the-internet-of-things.

11. Shona Ghosh, "How Absolut Vodka Will Use the Internet of Things to Sell More Than 'Static Pieces of Glass,'" *Campaign US,* August 6, 2015, http://www .campaignlive.com/article/absolut-vodka-will-use-internet-things-sell-static -pieces-glass/1359074.

12. See Locke, "Ready for Liquor Bottles Smart Enough to Talk?"

13. "Global Smart Homes Market 2018 by Evolving Technology, Projections & Estimations, Business Competitors, Cost Structure, Key Companies and Forecast to 2023," *Reuters,* February 19, 2018, https://www.reuters.com /brandfeatures/venture-capital/article?id=28096.

14. "Sproutling Wearable Baby Monitor," Mattel, December 8, 2017, http://fisher -price.mattel.com/shop/en-us/fp/sproutling-sleep-wearable-fnf59.

15. See Perzanowski and Hoofnagle, "What We Buy."

16. Amie Thuener, "Letter to SEC from Google Finance Director Re: Google Inc," Securities and Exchange Commission, January 29, 2013, https://www.sec.gov /Archives/edgar/data/1288776/000128877613000074/filename1.htm.

17. See Stacey Higginbotham, "Qualcomm Has Devised New Technology That Can Help Unlock Your Smartphone Using Iris Scans," *MIT Technology Review,* March 29, 2017, https://www.technologyreview.com/s/603964/qualcomm -wants-your-smartphone-to-have-energy-efficient-eyes.

18. Ben S. Cook et al., "Only Skin Deep," *IEEE Microwave Magazine,* May 2013. Smart skin is thus imagined, an innovation that "could set the foundation for the emergence of the first generation of truly long-range, fully-printable, chipless, flexible, low cost, wireless sensors for ubiquitous applications in Smart Skins and the Internet of Things...."

19. J. G. D. Hester and M. M. Tentzeris, "Inkjet-Printed Van-Atta Reflectarray Sensors: A New Paradigm for Long-Range Chipless Low Cost Ubiquitous Smart Skin Sensors of the Internet of Things," in *2016 IEEE MTT-S International Microwave Symposium (IMS),* 2016, 1–4, https://doi.org/10.1109/MWSYM .2016.7540412.

20. Cook et al., "Only Skin Deep."

21. Michael Galvin, "Attract Customers with Beacons, Geotagging & Geofencing," *New Perspective,* March 21, 2016, http://www.npws.net/blog/attract-customers -with-beacons-geotagging-geofencing.

22. For the most comprehensive description of these and other surveillance processes now at use in retailing, see Joseph Turow, *The Aisles Have Eyes: How Retailers Track Your Shopping, Strip Your Privacy, and Define Your Power* (New Haven, CT: Yale University Press, 2017).

23. Jimm Fox, "Life-Pattern Marketing and Geo Targeting," *One Market Media,* March 26, 2009, http://onemarketmedia.com/2009/03/26/life-pattern-market ing-and-geo-targetting.

24. Galvin, "Attract Customers with Beacons."

25. Monte Zweben, "Life-Pattern Marketing: Intercept People in Their Daily Routines," *SeeSaw Networks,* March 2009.

26. Monica Anderson, "6 Facts About Americans and Their Smartphones," *PewResearchCenter,* April 1, 2015, http://www.pewresearch.org/fact -tank/2015/04/01/6-facts-about-americans-and-their-smartphones; "Most Smartphone Owners Use Location-Based Services—EMarketer," *eMarketer,* April 22, 2016, https://www.emarketer.com/Article/Most-Smartphone-Owners -Use-Location-Based-Services/1013863; Chris Smith, "Why Location Data Is One of the Most Coveted Details Apps Collect About You," *BGR,* March 25, 2015, http://bgr.com/2015/03/25/smartphone-app-location-data.

27. Hazim Almuhimedi et al., "Your Location Has Been Shared 5,398 Times! A Field Study on Mobile App Privacy Nudging," in *Proceedings of the 33rd Annual ACM Conference on Human Factors in Computing Systems,* CHI '15 (New York: ACM, 2015), 787–96, https://doi.org/10.1145/2702123.2702210.

28. Tim Moynihan, "Apps Snoop on Your Location Way More Than You Think," *Wired,* March 25, 2015, https://www.wired.com/2015/03/apps-snoop-loca tion-way-think.

29. Byron Spice, "Study Shows People Act to Protect Privacy When Told How Often Phone Apps Share Personal Information," *Carnegie Mellon University News,* March 23, 2015, https://www.cmu.edu/news/stories/archives/2015 /march/privacy-nudge.html.

30. Russell Brandom, "Police Are Filing Warrants for Android's Vast Store of Location Data," *Verge,* June 1, 2016, http://www.theverge.com/2016/6/1 /11824118/google-android-location-data-police-warrants.

31. Keith Collins, "Google Collects Android Users' Locations Even When Location Services Are Disabled," *Quartz,* November 21, 2017, https://qz.com/1131515 /google-collects-android-users-locations-even-when-location-services-are -disabled.

32. Gerard Sans, "Your Timeline: Revisiting the World That You've Explored," *Google Lat Long,* July 21, 2015, https://maps.googleblog.com/2015/07/your -timeline-revisiting-world-that.html; Nathan Ingraham, "Google Knows Where You've Been, and Your Timeline for Maps Shows You," *Verge,* July 21, 2015, http://www.theverge.com/2015/7/21/9012035/google-your-timeline -location-history.

33. Here are just a few examples of the de-anonymization literature. In 1997 data privacy researcher Latanya Sweeney famously demonstrated that using publicly available population registers (e.g., a voter list), she could identify the medical records of Massachusetts Governor William Weld from medical information that had already been stripped of all explicit identifiers, such as name, address, and Social Security number. See Latanya Sweeney, "Statement of Latanya Sweeney, PhD Before the Privacy and Integrity Advisory Committee of the Department of Homeland Security—'Privacy Technologies for Homeland Security,'" US Department of Homeland Security, June 15, 2005, https://www .dhs.gov/xlibrary/assets/privacy/privacy_advcom_06-2005_testimony_sweeney

.pdf; Latanya Sweeney, "Only You, Your Doctor, and Many Others May Know," *Technology Science,* September 29, 2015, https://techscience.org/a/2015092903; Latanya Sweeney, "Matching a Person to a Social Security Number," *Data Privacy Lab,* October 13, 2017, https://dataprivacylab.org/dataprivacy /projects/ssnwatch/index.html; Sean Hooley and Latanya Sweeney, "Survey of Publicly Available State Health Databases—Data Privacy Lab, IQSS," Harvard University, 2013; Yves-Alexandre de Montjoye et al., "Unique in the Shopping Mall: On the Reidentifiability of Credit Card Metadata," *Science* 347, no. 6221 (2015): 536–39, https://doi.org/10.1126/science.1256297; Jessica Su, Ansh Shukla, Sharad Goel, and Arvind Narayanan, "De-Anonymizing Web Browsing Data with Social Networks," in *26th International Conference on World Wide Web Pages* (Perth, Australia: ACM, 2017), 1261–69, https://doi.org/10.1145 /3038912.3052714.

34. Paul Ohm, "Broken Promises of Privacy: Responding to the Surprising Failure of Anonymization," *UCLA Law Review* 57 (August 2010): 1701.
35. de Montjoye et al., "Unique in the Shopping Mall." See also Yves-Alexandre de Montjoye, "Computational Privacy: Towards Privacy-Conscientious Uses of Metadata," Massachusetts Institute of Technology, 2015, http://dspace.mit .edu/handle/1721.1/101850; Nicholas D. Lane et al., "On the Feasibility of User De-Anonymization from Shared Mobile Sensor Data," in *Proceedings of the Third International Workshop on Sensing Applications on Mobile Phones: PhoneSense '12,* 2012, http://dl.acm.org/citation.cfm?id=2389148.
36. Christina DesMarais, "This Smartphone Tracking Tech Will Give You the Creeps," *PCWorld,* May 22, 2012, http://www.pcworld.com/article/255802 /new_ways_to_track_you_via_your_mobile_devices_big_brother_or_good _business_.html. See also "Smartphones—Diagram of Sensors," *Broadcom.com,* February 22, 2018, https://www.broadcom.com/applications/wireless/smart -phones.
37. Arvind Narayanan and Edward W. Felten, "No Silver Bullet: De-identification Still Doesn't Work," July 9, 2014, http://randomwalker.info/publications/no -silver-bullet-de-identification.pdf.
38. Hal Hodson, "Baidu Uses Millions of Users' Location Data to Make Predictions," *New Scientist,* July 20, 2016, https://www.newscientist.com /article/2098206-baidu-uses-millions-of-users-location-data-to-make-predic tions.
39. In 2015, 29.5 million US adults used wearable devices—mostly fitness trackers like Under Armour's and smart watches—representing an increase of 57.7 percent over 2014. See Mary Ellen Berglund, Julia Duvall, and Lucy E. Dunne, "A Survey of the Historical Scope and Current Trends of Wearable Technology Applications," in *Proceedings of the 2016 ACM International Symposium on Wearable Computers,* ISWC '16 (New York: ACM, 2016), 40–43, https://doi .org/10.1145/2971763.2971796; Kate Kaye, "FTC: Fitness Apps Can Help You

Shred Calories—and Privacy," *Advertising Age,* May 7, 2014, http://adage
.com/article/privacy-and-regulation/ftc-signals-focus-health-fitness-data
-privacy/293080.

40. Michelle De Mooy and Shelten Yuen, "Towards Privacy-Aware Research and
Development in Wearable Health," *Hawaii International Conference on System
Sciences 2017 (HICSS-50),* January 4, 2017, http://aisel.aisnet.org/hicss-50/hc
/security_for_healthcare/4.

41. Sarah Perez, "Google and Levi's 'Connected' Jacket That Lets You Answer Calls,
Use Maps and More Is Going on Sale," *TechCrunch,* September 25, 2017, http://
social.techcrunch.com/2017/09/25/google-and-levis-connected-jacket-that
-lets-you-answer-calls-use-maps-and-more-goes-on-sale.

42. Just a few illustrations from recent literature include Ya-Li Zheng et al.,
"Unobtrusive Sensing and Wearable Devices for Health Informatics," *IEEE
Transactions on Biomedical Engineering* 61, no. 5 (2014): 1538–54, https://doi
.org/10.1109/TBME.2014.2309951; Claire Furino et al., "Synthetic Skin-Like
Sensing in Wearable Garments," *Rutgers Governor's School of Engineering and
Technology Research Journal,* July 16, 2016, http://www.soe.rutgers.edu
/sites/default/files/imce/pdfs/gset-2016/synth%20skin.pdf; Preeti Kumari, Lini
Mathew, and Poonam Syal, "Increasing Trend of Wearables and Multimodal
Interface for Human Activity Monitoring: A Review," *Biosensors and
Bioelectronics* 90, Supplement C (April 15, 2017): 298–307, https://doi.org/10
.1016/j.bios.2016.12.001; Arpan Pal, Arijit Mukherjee, and Swarnava Dey,
"Future of Healthcare—Sensor Data-Driven Prognosis," in *Wireless World
in 2050 and Beyond: A Window into the Future!* Springer Series in Wireless
Technology (Cham, Switzerland: Springer, 2016), 93–109, https://doi
.org/10.1007/978-3-319-42141-4_9.

43. "Ovum Report: The Future of E-Commerce—the Road to 2026," *Criteo,* 2015,
http://www.criteo.com/resources/ovum-future-ecommerce.

44. C. S. Pattichis et al., "Wireless Telemedicine Systems: An Overview," *IEEE
Antennas and Propagation Magazine* 44, no. 2 (2002): 143–53.

45. A. Solanas et al., "Smart Health: A Context-Aware Health Paradigm Within
Smart Cities," *IEEE Communications Magazine* 52, no. 8 (2014): 74–81, https://
doi.org/10.1109/MCOM.2014.6871673.

46. Subhas Chandra Mukhopadhyay, "Wearable Sensors for Human Activity
Monitoring: A Review," *IEEE Sensors Journal* 15, no. 3 (2015): 1321–30, https://
doi.org/10.1109/JSEN.2014.2370945; Stephen S. Intille, Jonathan Lester, James
F. Sallis, and Glen Duncan, "New Horizons in Sensor Development," *Medicine
& Science in Sports & Exercise* 44 (January 2012): S24–31, https://doi
.org/10.1249/MSS.0b013e3182399c7d; P. Castillejo, J. F. Martínez, J. Rodríguez
-Molina, and A. Cuerva, "Integration of Wearable Devices in a Wireless Sensor
Network for an E-health Application," *IEEE Wireless Communications* 20, no.
4 (2013): 38–49; J. Cheng, O. Amft, G. Bahle, and P. Lukowicz, "Designing

Sensitive Wearable Capacitive Sensors for Activity Recognition," *IEEE Sensors Journal* 13, no. 10 (2013): 3935–47; D. De Rossi and P. Veltink, "Wearable Technology for Biomechanics: E-Textile or Micromechanical Sensors?" *IEEE Engineering in Medicine and Biology Magazine,* May 20, 2010, 37–43.

47. In 2012 Pew Research reported that 53 percent of Americans owned smartphones and, of those, 20 percent had downloaded at least one health-related app. By 2015, a national survey found that 71 percent of Americans owned a smartphone or other wireless device and, of those, 32 percent had downloaded at least one health-related app, fulfilling "the public's desire for 'anywhere, anytime' monitoring, diagnosis, and treatment." See Susannah Fox and Maeve Duggan, "Mobile Health 2012," *Pew Research Center: Internet, Science & Tech,* November 8, 2012, http://www.pewinternet.org/2012/11/08/mobile-health-2012; Mark Brohan, "Mobile Will Be a Top Health Industry Trend in 2016," *MobileStrategies360,* December 11, 2015, https://web-beta.archive.org/web/20160403231014/https://www.mobilestrategies360.com/2015/12/11/mobile-will-be-top-health-industry-trend-2016. *Forbes* reported on the excitement with which industry met this news: "Big pharm companies are launching hundreds of mobile apps...while corporations take extra mobile strides to keep employees' health insurance costs down...." See Jennifer Elias, "In 2016, Users Will Trust Health Apps More Than Their Doctors," *Forbes,* December 31, 2015, http://www.forbes.com/sites/jenniferelias/2015/12/31/in-2016-users-will-trust-health-apps-more-than-their-doctors.

48. Gabrielle Addonizio, "The Privacy Risks Surrounding Consumer Health and Fitness Apps with HIPAA's Limitations and the FTC's Guidance," *Health Law Outlook* 9, no. 1 (2016), http://scholarship.shu.edu/health-law-outlook/vol9/iss1/1.

49. "Mobile Health App Developers: FTC Best Practices," Federal Trade Commission, April 2016, https://www.ftc.gov/tips-advice/business-center/guidance/mobile-health-app-developers-ftc-best-practices; "Mobile Privacy Disclosures: Building Trust Through Transparency," Federal Trade Commission, February 2013, https://www.ftc.gov/sites/default/files/documents/reports/mobile-privacy-disclosures-building-trust-through-transparency-federal-trade-commission-staff-report/130201mobileprivacyreport.pdf; Harrison Kaminsky, "FDA States It Will Not Regulate Fitness Trackers and Wellness Apps," *Digital Trends,* July 31, 2016, http://www.digitaltrends.com/health-fitness/fda-will-not-regulate-fitness-wellness-apps.

50. Tobias Dehling et al., "Exploring the Far Side of Mobile Health: Information Security and Privacy of Mobile Health Apps on iOS and Android," *JMIR MHealth and UHealth* 3, no. 1 (2015): 1–26, https://doi.org/10.2196/mhealth.3672. In 2013 an analysis by the Privacy Rights Clearinghouse evaluated a range of health and fitness apps according to their level of privacy risk, including the expropriation of personal information, the sensitivity of that

information, and its degree of dissemination. See "Mobile Health and Fitness Apps: What Are the Privacy Risks?" *Privacy Rights Clearinghouse,* July 1, 2013, https://www.privacyrights.org/consumer-guides/mobile-health-and-fitness -apps-what-are-privacy-risks; Bruno M. Silva et al., "A Data Encryption Solution for Mobile Health Apps in Cooperation Environments," *Journal of Medical Internet Research* 15, no. 4 (2013): e66, https://doi.org/10.2196 /jmir.2498; Miloslava Plachkinova, Steven Andres, and Samir Chatterjee, "A Taxonomy of MHealth Apps—Security and Privacy Concerns," 48th Hawaii International Conference on System Sciences, 2015, 3187–96, https://doi.org /10.1109/HICSS.2015.385; Soumitra S. Bhuyan et al., "Privacy and Security Issues in Mobile Health: Current Research and Future Directions," *Health Policy and Technology,* January 2017, https://doi.org/10.1016/j.hlpt.2017.01 .004; Borja Martínez-Pérez, Isabel de la Torre-Díez, and Miguel López-Coronado, "Privacy and Security in Mobile Health Apps: A Review and Recommendations," *Journal of Medical Systems* 39, no. 1 (2015), https://doi .org/10.1007/s10916-014-0181-3.

51. Andrew Hilts, Christopher Parsons, and Jeffrey Knockel, "Every Step You Fake: A Comparative Analysis of Fitness Tracker Privacy and Security," *Open Effect,* 2016, https://openeffect.ca/fitness-trackers.

52. Sarah R. Blenner et al., "Privacy Policies of Android Diabetes Apps and Sharing of Health Information," *JAMA* 315, no. 10 (2016): 1051–52, https://doi.org/10 .1001/jama.2015.19426 (italics mine).

53. Erin Marine, "Biometric Privacy Laws: Illinois and the Fight Against Intrusive Tech," Fordham Law School, March 29, 2018, https://news.law.fordham.edu /jcfl/2018/03/20/biometric-privacy-laws-illinois-and-the-fight-against-intru sive-tech.

54. Jared Bennett, "Saving Face: Facebook Wants Access Without Limits," *Center for Public Integrity,* July 31, 2017, https://www.publicintegrity.org/2017/07/31 /21027/saving-face-facebook-wants-access-without-limits.

55. Allan Holmes and Jared Bennett, "Why Mark Zuckerberg's Senate Hearing Could Mean Little for Facebook's Privacy Reform," *Center for Public Integrity,* April 10, 2018, https://www.publicintegrity.org/2018/04/10/21665/why-mark -zuckerbergs-senate-hearing-could-mean-little-facebooks-privacy-reform.

56. Bennett, "Saving Face."

57. Yaniv Taigman et al., "DeepFace: Closing the Gap to Human-Level Performance in Face Verification," *Facebook Research,* April 14, 2018, https://research.fb.com /publications/deepface-closing-the-gap-to-human-level-performance-in-face -verification.

58. Aviva Rutkin, "Facebook Can Recognise You in Photos Even If You're Not Looking," *New Scientist,* April 14, 2018, https://www.newscientist.com /article/dn27761-facebook-can-recognise-you-in-photos-even-if-youre-not -looking.

59. Bennett, "Saving Face," 13; April Glaser, "Facebook Is Using an 'NRA Approach' to Defend Its Creepy Facial Recognition Programs," *Slate*, August 4, 2017, http://www.slate.com/blogs/future_tense/2017/08/04/facebook_is_fighting _biometric_facial_recognition_privacy_laws.html; Kartikay Mehrotra, "Tech Companies Are Pushing Back Against Biometric Privacy Laws," *Bloomberg.com*, July 20, 2017, https://www.bloomberg.com/news/articles/2017-07-20/tech -companies-are-pushing-back-against-biometric-privacy-laws; Ally Marotti, "Proposed Changes to Illinois' Biometric Law Concern Privacy Advocates," *Chicago Tribune*, April 13, 2018, http://www.chicagotribune.com/business/ct -biz-illinois-biometrics-bills-20180409-story.html.

60. Kashmir Hill, "You're Being Secretly Tracked with Facial Recognition, Even in Church," *Splinter*, April 14, 2018, https://splinternews.com/youre-being -secretly-tracked-with-facial-recognition-e-1793848585; Robinson Meyer, "Who Owns Your Face?" *Atlantic*, July 2, 2015.

61. "Privacy Best Practice Recommendations for Commercial Facial Recognition Use," NTIA, https://www.ntia.doc.gov/files/ntia/publications/privacy_best _practices_recommendations_for_commercial_use_of_facial_recogntion.pdf.

62. Alvaro Bedoya et al., "Statement on NTIA Privacy Best Practice Recommendations for Commercial Facial Recognition Use," Consumer Federation of America, April 14, 2018, https://consumerfed.org/press_release /statement-ntia-privacy-best-practice-recommendations-commercial-facial -recognition-use.

CHAPTER NINE

1. Satya Nadella et al., "Satya Nadella: Microsoft Ignite 2016," September 26, 2016, https://news.microsoft.com/speeches/satya-nadella-microsoft-ignite-2016.

2. Hal R. Varian, "Beyond Big Data," *Business Economics* 49, no. 1 (2014): 28–29.

3. Neil McKendrick, "The Consumer Revolution of Eighteenth-Century England," in *Birth of a Consumer Society: The Commercialization of Eighteenth-Century England*, ed. John Brewer and J. H. Plumb (Bloomington: Indiana University Press, 1982), 11.

4. Nathaniel Forster, *An Enquiry into the Causes of the Present High Price of Provisions* (London: J. Fletcher, 1767), 41.

5. Adam Smith, *The Wealth of Nations*, ed. Edwin Cannan (New York: Modern Library, 1994).

6. Lee Rainie and Janna Anderson, "The Future of Privacy: Above-and -Beyond Responses: Part 1," Pew Research Center: Internet, Science & Tech, December 18, 2014, http://www.pewinternet.org/2014/12/18 /above-and-beyond-responses-part-1-2/.

7. Tom Simonite, "Google's Answer to Siri Thinks Ahead," *MIT Technology Review*, September 28, 2012, https://www.technologyreview.com/s/429345

/googles-answer-to-siri-thinks-ahead; Dieter Bohn, "Google Now: Behind the Predictive Future of Search," *Verge*, October 29, 2012, http://www.theverge .com/2012/10/29/3569684/google-now-android-4-2-knowledge-graph-neural -networks.

8. *Introducing Google Now*, 2012, https://www.youtube.com/watch?v=pPqliPz HYyc; Simonite, "Google's Answer to Siri."

9. Bohn, "Google Now."

10. Drew Olanoff and Josh Constine, "Facebook Is Adding a Personal Assistant Called 'M' to Your Messenger App," *TechCrunch*, August 26, 2015, http://social .techcrunch.com/2015/08/26/facebook-is-adding-a-personal-assistant-called -m-to-your-messenger-app; Amir Efrati, "Facebook Preps 'Moneypenny' Assistant," *Information*, July 13, 2015, https://www.theinformation.com/coming -soon-to-facebook-messenger-moneypenny-assistant.

11. Jessi Hempel, "Facebook Launches M, Its Bold Answer to Siri and Cortana," *Wired*, August 2015, https://www.wired.com/2015/08/facebook-launches-m -new-kind-virtual-assistant.

12. Andrew Orlowski, "Facebook Scales Back AI Flagship after Chatbots Hit 70% F-AI-Lure Rate," *Register*, February 22, 2017, https://www.theregister.co.uk /2017/02/22/facebook_ai_fail/.

13. Cory Weinberg, "How Messenger and 'M' Are Shifting Gears," *Information*, February 22, 2017, https://www.theinformation.com/how-messenger-and-m -are-shifting-gears.

14. For example, a 2004 white paper from the Kansas City Federal Reserve singles out "voice recognition" as a significant threat to future employment rates: "Advances in voice recognition technology, expert systems, and artificial intelligence may eventually allow computers to handle many customer service jobs and perhaps even routine x-ray screening." See C. Alan Garner, "Offshoring in the Service Sector: Economic Impact and Policy Issues," *Economic Review* 89, no. 3 (2004): 5–37. Frey and Osborne's much-cited 2013 study of technological unemployment sounded the same theme: "Moreover, a company called SmartAction now provides call computerisation solutions that use ML technology and advanced speech recognition to improve upon conventional interactive voice response systems, realising cost savings of 60 to 80 percent over an outsourced call center consisting of human labour." See Carl Benedikt Frey and Michael Osborne, "The Future of Employment: How Susceptible Are Jobs to Computerisation?" *Technological Forecasting and Social Change* 114 (2013): 254–80. See also a follow-up study: Philipp Brandes, Roger Wattenhofer, and Stefan Schmid, "Which Tasks of a Job Are Susceptible to Computerization?" *Bulletin of EATCS* 3, no. 120 (2016), http://bulletin.eatcs .org/index.php/beatcs/article/view/467.

15. "Dave Limp, Exec Behind Amazon's Alexa: Full Transcript of Interview," *Fortune*, July 14, 2016, http://fortune.com/2016/07/14/amazon-alexa-david -limp-transcript.

16. See Matthew Lynley, "Google Unveils Google Assistant, a Virtual Assistant That's a Big Upgrade to Google Now," *TechCrunch*, May 18, 2016, http:// social.techcrunch.com/2016/05/18/google-unveils-google-assistant-a-big -upgrade-to-google-now. See also Minda Smiley, "Google I/O Conference: Three Takeaways for Marketers," *Drum*, May 19, 2016, http://www.thedrum .com/news/2016/05/19/google-io-conference-three-takeaways-marketers. For Pichai's speech, see Sundar Pichai, "Google I/O 2016 Keynote," *Singju Post*, May 20, 2016, http://singjupost.com/google-io-2016-keynote-full-transcript.

17. Pichai, "Google I/O 2016 Keynote."

18. Pichai.

19. Jing Cao and Dina Bass, "Why Google, Microsoft and Amazon Love the Sound of Your Voice," *Bloomberg Businessweek*, December 13, 2016, https://www .bloomberg.com/news/articles/2016-12-13/why-google-microsoft-and-amazon -love-the-sound-of-your-voice.

20. A. J. Dellinger, "I Took a Job Listening to Your Siri Conversations," *Daily Dot*, March 2, 2015, https://www.dailydot.com/debug/siri-google-now-cortana -conversations.

21. "Global Smart Appliances Market 2016–2020," *Technavio*, April 10, 2017, https://www.technavio.com/report/global-home-kitchen-and-large-appliances -global-smart-appliances-market-2016-2020; Adi Narayan, "Samsung Wants to Put Your Home on a Remote," *BusinessWeek: Technology*, December 11, 2014.

22. Alex Hern, "Samsung Rejects Concern over 'Orwellian' Privacy Policy," Guardian, February 9, 2015, http://www.theguardian.com/technology/2015 /feb/09/samsung-rejects-concern-over-orwellian-privacy-policy.

23. Hern, "Samsung Rejects Concern."

24. Electronic Privacy Information Center, "EPIC—Samsung 'SmartTV' Complaint," *EPIC.org*, May 9, 2017, https://epic.org/privacy/internet/ftc /samsung/.

25. "EPIC—Samsung 'SmartTV' Complaint," *EPIC.org*, May 9, 2017, https:// epic.org/privacy/internet/ftc/samsung; "Samsung Privacy Policy," Samsung, February 10, 2015, http://www.samsung.com/us/common/privacy.html; "Nuance Communications, Inc. Privacy Policy General Information," Nuance, December 2015, https://www.nuance.com/about-us/company-policies/privacy -policies.html.

26. Committee on Privacy and Consumer Protection, "Connected Televisions," Pub. L. No. 1116, § 35, 22948.20-2298.25, 2015, https://leginfo.legislature.ca .gov/faces/billTextClient.xhtml?bill_id=201520160AB1116.

27. Megan Wollerton, "Voice Control Comes to the Forefront of the Smart Home," *CNET*, December 1, 2014, https://www.cnet.com/news/voice-control-roundup; David Katzmaier, "Think Smart TV Is Dumb? Samsung Aims to Change Your Mind by Controlling Your Gear," *CNET*, April 14, 2016, https://www.cnet.com /news/think-smart-tv-is-dumb-samsung-aims-to-change-your-mind-by -controlling-your-gear; David Pierce, "Soon, You'll Be Able to Control Every

Corner of Your Smart Home with a Single Universal Remote," *Wired,* January 7, 2016, https://www.wired.com/2016/01/smart-home-universal-remote.

28. "VIZIO to Pay $2.2 Million to FTC, State of New Jersey to Settle Charges It Collected Viewing Histories on 11 Million Smart Televisions Without Users' Consent," Federal Trade Commission, February 6, 2017, https://www.ftc.gov /news-events/press-releases/2017/02/vizio-pay-22-million-ftc-state-new -jersey-settle-charges-it; Nick Visser, "Vizio to Pay Millions After Secretly Spying on Customers, Selling Viewer Data," *Huffington Post,* February 7, 2017, https://www.huffingtonpost.com/entry/vizio-settlement_us_589962 dee4b0c1284f27e534.

29. Lesley Fair, "What Vizio Was Doing Behind the TV Screen," Federal Trade Commission, February 6, 2017, https://www.ftc.gov/news-events /blogs/business-blog/2017/02/what-vizio-was-doing-behind-tv-screen.

30. Maureen K. Ohlhausen, "Concurring Statement of Acting Chairman Maureen K. Ohlhausen—In the Matter of Vizio, Inc.—Matter No. 1623024," Federal Trade Commission, February 6, 2017, https://www.ftc.gov/system/files /documents/public_statements/1070773/vizio_concurring_statement_of _chairman_ohlhausen_2-6-17.pdf.

31. EPIC.org, "Federal Trade Commission—In the Matter of Genesis Toys and Nuance Communications—Complaint and Request for Investigation, Injunction and Other Relief," December 6, 2016, https://epic.org/privacy /kids/EPIC-IPR-FTC-Genesis-Complaint.pdf; Kate Cox, "These Toys Don't Just Listen to Your Kid; They Send What They Hear to a Defense Contractor," *Consumerist,* December 6, 2016, https://consumerist.com/2016/12/06/these -toys-dont-just-listen-to-your-kid-they-send-what-they-hear-to-a-defense -contractor.

32. "Federal Trade Commission—In the Matter of Genesis Toys."

33. "Federal Trade Commission—In the Matter of Genesis Toys." See also Jiri Havelka and Raimo Bakis, Systems and methods for facilitating communication using an interactive communication system, US20170013124 A1, filed July 6, 2015, and issued January 12, 2017, http://www.google.com/patents /US20170013124.

34. James Vlahos, "Barbie Wants to Get to Know Your Child," *New York Times,* September 16, 2015, http://www.nytimes.com/2015/09/20/magazine/barbie -wants-to-get-to-know-your-child.html?action=click&pgtype=Homepage&m odule=photo-spot-region&region=top-news&WT.nav=top-news; Evie Nagy, "After the Fracas Over Hello Barbie, ToyTalk Responds to Its Critics," *Fast Company,* May 23, 2015, https://www.fastcompany.com/3045676/after-the -fracas-over-hello-barbie-toytalk-responds-to-its-critics; Mike Krieger, "Big Barbie Is Watching You: Meet the WiFi-Connected Doll That Talks to Your Kids & Records Them," *Zero Hedge,* January 7, 2013, http://www.zerohedge.com /news/2015-02-23/big-barbie-watching-you-meet-wifi-connected-doll -talks-your-kids-records-them; Issie Lapowsky, "Pixar Vets Reinvent Speech

Recognition So It Works for Kids," *Wired,* September 25, 2014, https://www
.wired.com/2014/09/toytalk; Tim Moynihan, "Barbie Has a New Super-Dope
Dreamhouse That's Voice-Activated and Connected to the Internet," *Wired,*
September 15, 2016, https://www.wired.com/2016/09/barbies-new-smart
-home-crushing-hard; Irina D. Manta and David S. Olson, "Hello Barbie: First
They Will Monitor You, Then They Will Discriminate Against You. Perfectly,"
Alabama Law Review 135, no. 67 (2015), http://papers.ssrn.com/sol3/papers
.cfm?abstract_id=2578815. See also "ToyTalk | Legal | Hello Barbie/Barbie
Hello Dreamhouse Privacy Policy," *ToyTalk,* March 30, 2017, https://www.toy
talk.com/hellobarbie/privacy.

35. Moynihan, "Barbie Has a New Super-Dope Dreamhouse."

36. Paul Ziobro and Joann S. Lublin, "Mattel Finds Its New CEO at Google," *Wall
Street Journal,* January 18, 2017.

37. James Vincent, "German Watchdog Tells Parents to Destroy Wi-Fi–Connected
Doll Over Surveillance Fears," *Verge,* February 17, 2017, http://www.theverge
.com/2017/2/17/14647280/talking-doll-hack-cayla-german-government
-ban; Thomas Claburn, "Smash Up Your Kid's Bluetooth-Connected Cayla
'Surveillance' Doll, Germany Urges Parents," *Register,* February 17, 2017,
https://www.theregister.co.uk/2017/02/17/cayla_doll_banned_in_germany;
Hayley Tsukayama, "Mattel Has Canceled Plans for a Kid-Focused AI
Device That Drew Privacy Concerns," *Washington Post,* October 4, 2017,
https://www.washingtonpost.com/news/the-switch/wp/2017/10/04/mattel
-has-an-ai-device-to-soothe-babies-experts-are-begging-them-not-to-sell-it.

38. Frank Pasquale, "Will Amazon Take Over the World?" *Boston Review,* July 20,
2017, https://bostonreview.net/class-inequality/frank-pasquale-will-amazon
-take-over-world; Lina M. Khan, "Amazon's Antitrust Paradox," April 16, 2018,
https://www.yalelawjournal.org/note/amazons-antitrust-paradox.

39. Kevin McLaughlin et al., "Bezos Ordered Alexa App Push," *Information,*
November 16, 2016, https://www.theinformation.com/bezos-ordered-alexa
-app-push; "The Real Reasons That Amazon's Alexa May Become the Go-To AI
for the Home," *Fast Company,* April 8, 2016, https://www.fastcompany.com
/3058721/app-economy/the-real-reasons-that-amazons-alexa-may-become
-the-go-to-ai-for-the-home; "Amazon Lex—Build Conversation Bots," Amazon
Web Services, February 24, 2017, https//aws.amazon.com/lex.

40. "Dave Limp, Exec Behind Amazon's Alexa."

41. Aaron Tilley and Priya Anand, "Apple Loses Ground to Amazon in Smart Home
Deals with Builders," *Information,* April 16, 2018, https://www.theinformation
.com/articles/apple-loses-ground-to-amazon-in-smart-home-deals-with
-builders.

42. Sapna Maheshwari, "Hey, Alexa, What Can You Hear? And What Will You
Do with It?" *New York Times,* March 31, 2018, https://www.nytimes
.com/2018/03/31/business/media/amazon-google-privacy-digital-assistants
.html.

43. Alex Hern, "Amazon to Release Alexa-Powered Smartglasses, Reports Say," *Guardian,* September 20, 2017, http://www.theguardian.com /technology/2017/sep/20/amazon-alexa-smartglasses-google-glass-snapchat -spectacles-voice-assistant; Scott Gillum, "Why Amazon Is the New Google for Buying," *MediaInsider,* September 14, 2017, https://www.mediapost.com /publications/article/307348/why-amazon-is-the-new-google-for-buying.html; Mike Shields, "Amazon Looms Quietly in Digital Ad Landscape," *Wall Street Journal,* October 6, 2016, http://www.wsj.com/articles/amazon-looms-quietly -in-digital-ad-landscape-1475782113;

44. Keith Naughton and Spencer Soper, "Alexa, Take the Wheel: Ford Models to Put Amazon in Driver Seat," *Bloomberg.com,* January 5, 2017, https://www .bloomberg.com/news/articles/2017-01-05/steering-wheel-shopping-arrives-as -alexa-hitches-ride-with-ford; Ryan Knutson and Laura Stevens, "Amazon and Google Consider Turning Smart Speakers into Home Phones," *Wall Street Journal,* February 15, 2017, https://www.wsj.com/articles/amazon-google -dial-up-plans-to-turn-smart-speakers-into-home-phones-1487154781; Kevin McLaughlin, "AWS Takes Aim at Call Center Industry," *Information,* February 28, 2017, https://www.theinformation.com/aws-takes-aim-at-call -center-industry.

45. Lucas Matney, "Siri-Creator Shows Off First Public Demo of Viv, 'The Intelligent Interface for Everything,'" *TechCrunch,* http://social.techcrunch.com /2016/05/09/siri-creator-shows-off-first-public-demo-of-viv-the-intelligent -interface-for-everything.

46. Shoshana Zuboff, *In the Age of the Smart Machine: The Future of Work and Power* (New York: Basic, 1988), 381.

47. Zuboff, *In the Age of the Smart Machine,* 362–86.

48. Zuboff, 383.

49. The five-factor model has become the standard since the 1980s because it lends itself easily to computational analysis. The model is based on a taxonomy of personality traits along five dimensions: extraversion (the tendency to be outgoing and energetic while seeking stimulation in the company of others), agreeableness (warmth, compassion, and cooperativeness), conscientiousness (the tendency to exhibit self-discipline, organization, and achievement orientation), neuroticism (the susceptibility to unpleasant emotions), and openness to experience (the tendency to be intellectually curious, creative, and open to feelings).

50. Jennifer Golbeck, Cristina Robles, and Karen Turner, "Predicting Personality with Social Media," in *CHI '11 Extended Abstracts on Human Factors in Computing Systems,* CHI EA '11 (New York: ACM, 2011), 253–62, https://doi .org/10.1145/1979742.1979614.

51. Jennifer Golbeck, Cristina Robles, Michon Edmondson, and Karen Turner, "Predicting Personality from Twitter," in *2011 IEEE Third International*

Conference on Privacy, Security, Risk and Trust and 2011 IEEE Third International Conference on Social Computing (PASSAT-SocialCom 2011), ed. Institute of Electrical and Electronics Engineers and Computer Society (Boston: IEEE, 2011).

52. Golbeck, Robles, and Turner, "Predicting Personality with Social Media."

53. Daniele Quercia et al., "Our Twitter Profiles, Our Selves: Predicting Personality with Twitter," IEEE, 2011, 180–85, https://doi.org/10.1109/PASSAT/Social Com.2011.26.

54. From 2010, the Psychometrics Centre was a "strategic research network" within Cambridge University, with research alliances across the university's varied disciplines. As the results of this growing body of work were pulled into the orbit of surveillance capitalism's supply operations, it is worth noting, the Psychometrics Centre was invited to relocate to the University's Judge School of Business. The Judge School's embrace of the centre was explicitly linked to its commercial prospects and above all to the prospects of depth surplus and personality prediction as they could be applied to surveillance capitalism's prediction requirements. For example, in announcing the integration of the centre into the Judge School's campus and research program, its director noted, "Today, the digital traces we leave behind allow machines to treat all of our online activity as a 'test.' Our Facebook Likes, the words we use in tweets and emails and the images we upload all provide 'items' from which the machine can learn who we are, what drives and motivates us and how we differ from each other. Psychometrics is at the forefront of developments in ambient intelligence and the Internet of Things, powering connected environments that are sensitive and responsive to our needs." The school's director of business development shared even more pointed observations regarding the immediate commercial utility of data once produced solely for the promise of personal feedback: "The Centre's expertise in assessment, measurement, and prediction will enhance the Cambridge Judge Business School's ability to push forward the boundaries of value creation for its global client network. We will... deliver world-leading support in some of the fascinating challenges of modern commerce...." See "Psychometrics Centre Moves to Cambridge Judge Business School—the Psychometrics Centre," University of Cambridge, July 19, 2016, http://www.psychometrics.cam.ac.uk/news/Move_to_JBS; "Dr David Stillwell, Deputy Director—the Psychometrics Centre," March 9, 2017, http://www.psychometrics.cam.ac.uk/about-us/directory/david-stillwell.

55. Bobbie Johnson, "Privacy No Longer a Social Norm, Says Facebook Founder," *Guardian,* January 10, 2010, https://www.theguardian.com/technology/2010/jan/11/facebook-privacy.

56. Yoram Bachrach et al., "Personality and Patterns of Facebook Usage," *Microsoft Research,* January 1, 2012, https://www.microsoft.com/en-us/research/publication/personality-and-patterns-of-facebook-usage.

57. Michal Kosinski, David Stillwell, and Thore Graepel, "Private Traits and Attributes Are Predictable from Digital Records of Human Behavior," *Proceedings of the National Academy of Sciences of the United States of America* 110, no. 15 (2013): 5802–5.

58. G. Park et al., "Automatic Personality Assessment Through Social Media Language," *Journal of Personality and Social Psychology* 108, no. 6 (2015): 934–52.

59. Michal Kosinski et al., "Mining Big Data to Extract Patterns and Predict Real-Life Outcomes," *Stanford Graduate School of Business* 21, no. 4 (2016): 1; Michal Kosinski, "Dr Michal Kosinski," February 28, 2018, http://www.michalkosinski .com.

60. Park et al., "Automatic Personality Assessment." See also Peter J. Rentfrow et al., "Divided We Stand: Three Psychological Regions of the United States and Their Political, Economic, Social, and Health Correlates," *Journal of Personality and Social Psychology* 105, no. 6 (2013): 996–1012; Dejan Markovikj, Sonja Gievska, Michal Kosinski, and David Stillwell, "Mining Facebook Data for Predictive Personality Modeling," *Association for the Advancement of Artificial Intelligence,* 2013, https://www.gsb.stanford.edu /sites/gsb/files/conf-presentations/miningfacebook.pdf; H. Andrew Schwartz et al., "Predicting Individual Well-Being Through the Language of Social Media," in *Biocomputing 2016: Proceedings of the Pacific Symposium,* 2016, 516–27, https://doi.org/10.1142/9789814749411_0047; H. Andrew Schwartz et al., "Extracting Human Temporal Orientation from Facebook Language," *Proceedings of the 2015 Conference of the North American Chapter of the Association for Computational Linguistics: Human Language Technologies,* 2015, http://www.academia.edu/15692796/Extracting_Human_Temporal _Orientation_from_Facebook_Language; David M. Greenberg et al., "The Song Is You: Preferences for Musical Attribute Dimensions Reflect Personality," *Social Psychological and Personality Science* 7, no. 6 (2016): 597–605, https:// doi.org/10.1177/1948550616641473; Michal Kosinski, David Stillwell, and Thore Graepel, "Private Traits and Attributes Are Predictable from Digital Records of Human Behavior," *Proceedings of the National Academy of Sciences of the United States of America* 110, no. 15 (2013): 5802–5.

61. Wu Youyou, Michal Kosinski, and David Stillwell, "Computer-Based Personality Judgments Are More Accurate Than Those Made by Humans," *Proceedings of the National Academy of Sciences* 112, no. 4 (2015): 1036–40, https://doi.org/10.1073/pnas.1418680112.

62. Tsung-Yi Chen, Meng-Che Tsai, and Yuh-Min Chen, "A User's Personality Prediction Approach by Mining Network Interaction Behaviors on Facebook," *Online Information Review* 40, no. 7 (2016): 913–37.

63. Sam Biddle, "Facebook Uses Artificial Intelligence to Predict Your Future Actions for Advertisers, Says Confidential Document," *Intercept,* April 13, 2018, https://theintercept.com/2018/04/13/facebook-advertising-data-artificial -intelligence-ai.

64. The five-factor model is popular in part because it lends itself to easy assessment with simple protocols. The traits that it describes enjoy face validity: commonsensical and readily observed. For example, someone who is well organized is likely to score high on conscientiousness. A person who prefers to be surrounded by groups of friends is likely to score high on extraversion, and so on. Similarly, Kosinski and his coauthors point out the close association between Facebook "likes" and the five trait dimensions: "Participants with high openness to experience tend to 'like' Salvador Dali, meditation, or TED talks…." These correlations are obvious and thus easy to score, program, and scale. Human judges cannot compete on scale, but they exceed the machines in scope. Kosinski and his colleagues know this, acknowledging that human perception is "flexible" and "able to capture many subconscious cues unavailable to machines." See Youyou, Kosinski, and Stillwell, "Computer-Based Personality Judgments Are More Accurate."

65. CaPPr, *Interview with Michal Kosinski on Personality and Facebook Likes,* May 20, 2015, https://www.youtube.com/watch?v=pJGuWKqwYRk.

66. Leqi Liu et al., "Analyzing Personality Through Social Media Profile Picture Choice," *Association for the Advancement of Artificial Intelligence,* 2016, https://sites.sas .upenn.edu/sites/default/files/danielpr/files/persimages16icwsm.pdf; Sharath Chandra Guntuku et al., "Do Others Perceive You as You Want Them To? Modeling Personality Based on Selfies," in *Proceedings of the 1st International Workshop on Affect & Sentiment in Multimedia,* ASM '15 (New York: ACM, 2015), 21–26, https://doi.org/10.1145/2813524.2813528; Bruce Ferwerda, Markus Schedl, and Marko Tkalcic, "Using Instagram Picture Features to Predict Users' Personality," in *MultiMedia Modeling* (Cham, Switzerland: Springer, 2016), 850–61, https://doi .org/10.1007/978-3-319-27671-7_71; Golbeck, Robles, and Turner, "Predicting Personality with Social Media"; Chen, Tsai, and Chen, "A User's Personality Prediction Approach"; Schwartz et al., "Predicting Individual Well-Being."

67. CaPPr, *Interview with Michal Kosinski.*

68. "IBM Cloud Makes Hybrid a Reality for the Enterprise," IBM, February 23, 2015, https://www-03.ibm.com/press/us/en/pressrelease/46136.wss.

69. "IBM Watson Personality Insights," IBM Watson Developer Cloud, October 14, 2017, https://personality-insights-livedemo.mybluemix.net; "IBM Personality Insights—Needs," IBM Watson Developer Cloud, October 14, 2017, https:// console.bluemix.net/docs/services/personality-insights/needs.html#needs; "IBM Personality Insights—Values," IBM Watson Developer Cloud, October 14, 2017, https://console.bluemix.net/docs/services/personality-insights/values .html#values.

70. "IBM Personality Insights—Use Cases," IBM Cloud Docs, November 8, 2017, https://console.bluemix.net/docs/services/personality-insights/usecases.html #usecases.

71. See Vibha Sinha, "Personality of Your Agent Matters—an Empirical Study on Twitter Conversations—Watson Dev," *Watson,* November 3, 2016, https://

developer.ibm.com/watson/blog/2016/11/03/personality-of-your-agent
-matters-an-empirical-study-on-twitter-conversations. In a study undertaken
with data broker Acxiom, the two giant corporations set out to determine
whether IBM's personality insights more accurately predict consumption
preferences than the more standard demographic information amassed by the
data brokers. The findings were affirmative. After examining 133 consumption
preferences of about 785,000 US individuals, the addition of personality data
improved prediction accuracy for 115 preferences (86.5 percent). Personality
data alone provided better prediction accuracy than demographic data for
23 of those preferences. The researchers note with some enthusiasm that in
61 percent of the cases, Watson's "personality insights" can accurately predict
certain preference categories, such as "camping/hiking," "without collecting any
data from the user." They concede that an individual's income is also a powerful
predictor of consumption, but they complain that income information is "very
sensitive" and "hard to collect" compared to personality data, which "can be
directly derived from the people's social media profile." See IBM-Acxiom,
"Improving Consumer Consumption Preference Prediction Accuracy with
Personality Insights," March 2016, https://www.ibm.com/watson/developer
cloud/doc/personality-insights/applied.shtml.

72. IBM-Acxiom, "Improving Consumer Consumption Preference Prediction
Accuracy."

73. "Social Media Analytics," Xerox Research Center Europe, April 3, 2017, http://
www.xrce.xerox.com/Our-Research/Natural-Language-Processing
/Social-Media-Analytics; Amy Webb, "8 Tech Trends to Watch in 2016,"
Harvard Business Review, December 8, 2015, https://hbr.org/2015/12/8-tech
-trends-to-watch-in-2016; Christina Crowell, "Machines That Talk to Us May
Soon Sense Our Feelings, Too," *Scientific American,* June 24, 2016, https://www
.scientificamerican.com/article/machines-that-talk-to-us-may-soon-sense-our
-feelings-too; R. G. Conlee, "How Automation and Analytics Are Changing
Customer Care," *Conduent Blog,* July 18, 2016, https://www.blogs.conduent
.com/2016/07/18/how-automation-and-analytics-are-changing-customer-care;
Ryan Knutson, "Call Centers May Know a Surprising Amount About You," *Wall
Street Journal,* January 6, 2017, http://www.wsj.com/articles/that-anonymous
-voice-at-the-call-center-they-may-know-a-lot-about-you-1483698608.

74. Nicholas Confessore and Danny Hakim, "Bold Promises Fade to Doubts for a
Trump-Linked Data Firm," *New York Times,* March 6, 2017, https://www
.nytimes.com/2017/03/06/us/politics/cambridge-analytica.html; Mary
-Ann Russon, "Political Revolution: How Big Data Won the US Presidency
for Donald Trump," *International Business Times UK,* January 20, 2017, http://
www.ibtimes.co.uk/political-revolution-how-big-data-won-us
-presidency-donald-trump-1602269; Grassegger and Krogerus, "The Data
That Turned the World Upside Down"; Carole Cadwalladr, "Revealed: How US
Billionaire Helped to Back Brexit," *Guardian,* February 25, 2017, https://www

.theguardian.com/politics/2017/feb/26/us-billionaire-mercer-helped
-back-brexit; Paul-Olivier Dehaye, "The (Dis)Information Mercenaries Now
Controlling Trump's Databases," *Medium*, January 3, 2017, https://medium
.com/personaldata-io/the-dis-information-mercenaries-now-controlling
-trumps-databases-4f6a20d4f3e7; Harry Davies, "Ted Cruz Using Firm
That Harvested Data on Millions of Unwitting Facebook Users," *Guardian*,
December 11, 2015, https://www.theguardian.com/us-news/2015
/dec/11/senator-ted-cruz-president-campaign-facebook-user-data.

75. Concordia, *The Power of Big Data and Psychographics*, 2016, https://www
.youtube.com/watch?v=n8Dd5aVXLCc.

76. See "Speak the Customer's Language with Behavioral Microtargeting," *Dealer
Marketing*, December 1, 2016, http://www.dealermarketing.com/speak-the
-customers-language-with-behavioral-microtargeting.

77. Sam Biddle, "Facebook Uses Artificial Intelligence to Predict Your Future
Actions for Advertisers, Says Confidential Document," *Intercept*, April 13, 2018,
https://theintercept.com/2018/04/13/facebook-advertising-data-artificial-
intelligence-ai/.

78. "Introducing FB Flow: Facebook's AI Backbone," Jeffrey Dunn, Facebook
Code, May 9, 2016, https://code.facebook.com/posts/1072626246134461/
introducing-fblearner-flow-facebook-s-ai-backbone.

79. Andy Kroll, "Cloak and Data: The Real Story Behind Cambridge Analytica's
Rise and Fall," *Mother Jones*, March 24, 2018, https://www.motherjones.com
/politics/2018/03/cloak-and-data-cambridge-analytica-robert-mercer.

80. Carole Cadwalladr, "'I Made Steve Bannon's Psychological Warfare Tool': Meet
the Data War Whistleblower," *Guardian*, March 18, 2018, http://www.the
guardian.com/news/2018/mar/17/data-war-whistleblower-christopher-wylie
-faceook-nix-bannon-trump; Kroll, "Cloak and Data."

81. Matthew Rosenberg, Nicholas Confessore, and Carole Cadwalladr, "How
Trump Consultants Exploited the Facebook Data of Millions," *New York Times*,
March 17, 2018, https://www.nytimes.com/2018/03/17/us/politics
/cambridge-analytica-trump-campaign.html; Emma Graham-Harrison and
Carole Cadwalladr, "Revealed: 50 Million Facebook Profiles Harvested for
Cambridge Analytica in Major Data Breach," *Guardian*, March 17, 2018, http://
www.theguardian.com/news/2018/mar/17/cambridge-analytica-facebook
-influence-us-election; Julia Carrie Wong and Paul Lewis, "Facebook Gave Data
About 57bn Friendships to Academic," *Guardian*, March 22, 2018, http://www
.theguardian.com/news/2018/mar/22/facebook-gave-data-about-57bn
-friendships-to-academic-aleksandr-kogan; Olivia Solon, "Facebook Says
Cambridge Analytica May Have Gained 37m More Users' Data," *Guardian*,
April 4, 2018, http://www.theguardian.com/technology/2018/apr/04/facebook
-cambridge-analytica-user-data-latest-more-than-thought.

82. Paul Lewis and Julia Carrie Wong, "Facebook Employs Psychologist Whose
Firm Sold Data to Cambridge Analytica," *Guardian*, March 18, 2018, http://

www.theguardian.com/news/2018/mar/18/facebook-cambridge-analytica
-joseph-chancellor-gsr.

83. Kroll, "Cloak and Data."
84. Frederik Zuiderveen Borgesius et al., "Online Political Microtargeting: Promises and Threats for Democracy" (SSRN Scholarly Paper, Rochester, NY: Social Science Research Network, February 9, 2018), https://papers.ssrn.com /abstract=3128787.
85. See Cadwalladr, "'I Made Steve Bannon's Psychological Warfare Tool.'"
86. Charlotte McEleny, "European Commission Issues €3.6m Grant for Tech That Measures Content 'Likeability,'" *CampaignLive.co.uk*, April 20, 2015, http:// www.campaignlive.co.uk/article/european-commission-issues-€36m-grant -tech-measures-content-likeability/1343366.
87. "2016 Innovation Radar Prize Winners," *Digital Single Market*, September 26, 2016, https://ec.europa.eu/digital-single-market/en/news/2016-innovation -radar-prize-winners.
88. "Affective Computing Market—Global Industry Analysis, Size, Share, Growth, Trends and Forecast 2015–2023," *Transparency Market Research*, 2017, http://www.transparencymarketresearch.com/affective-computing-market .html.
89. Patrick Mannion, "Facial-Recognition Sensors Adapt to Track Emotions, Mood, and Stress," *EDN*, March 3, 2016, http://www.edn.com/electronics-blogs /sensor-ee-perception/4441565/Facial-recognition-sensors-adapt-to-track -emotions—mood—and-stress; "Marketers, Welcome to the World of Emotional Analytics," *MarTech Today*, January 12, 2016, https://martechtoday .com/marketers-welcome-to-the-world-of-emotional-analytics-159152; Ben Virdee-Chapman, "5 Companies Using Facial Recognition to Change the World," *Kairos*, May 26, 2016, https://www.kairos.com/blog/5-companies -using-facial-recognition-to-change-the-world; "Affectiva Announces New Facial Coding Solution for Qualitative Research," Affectiva, May 7, 2014, https://web-beta.archive.org/web/20160625173829/http://www.affectiva.com /news/affectiva-announces-new-facial-coding-solution-for-qualitative -research; Ahmad Jalal, Shaharyar Kamal, and Daijin Kim, "Human Depth Sensors-Based Activity Recognition Using Spatiotemporal Features and Hidden Markov Model for Smart Environments," *Journal of Computer Networks and Communications* (2016), https://doi.org/10.1155/2016/8087545; M. Kakarla and G. R. M. Reddy, "A Real Time Facial Emotion Recognition Using Depth Sensor and Interfacing with Second Life Based Virtual 3D Avatar," in *International Conference on Recent Advances and Innovations in Engineering (ICRAIE-2014)*, 2014, 1–7, https://doi.org/10.1109/ICRAIE.2014.6909153.
90. "Sewa Project: Automatic Sentiment Analysis in the Wild," SEWA, April 25, 2017, https://sewaproject.eu/description.

91. Mihkel Jäätma, "Realeyes—Emotion Measurement," Realeyes Data Services, 2016, https://www.realeyesit.com/Media/Default/Whitepaper/Realeyes_White paper.pdf.

92. Mihkel Jäätma, "Realeyes—Emotion Measurement."

93. Alex Browne, "Realeyes—Play Your Audience Emotions to Stay on Top of the Game," Realeyes, February 21, 2017, https://www.realeyesit.com/blog/play -your-audience-emotions.

94. "Realeyes—Emotions," Realeyes, April 2, 2017, https://www.realeyesit.com /emotions.

95. "See What Industrial Advisors Think About SEWA," SEWA, April 24, 2017, https://sewaproject.eu/qa#ElissaMoses.

96. Roland Marchand, *Advertising the American Dream: Making Way for Modernity, 1920–1940* (Berkeley: University of California Press, 1985).

97. Some key early papers include Paul Ekman and Wallace V. Friesen, "The Repertoire of Nonverbal Behavior: Categories, Origins, Usage and Coding," *Semiotica* 1, no. 1 (1969): 49–98; Paul Ekman and Wallace V. Friesen, "Constants Across Cultures in the Face and Emotion," *Journal of Personality and Social Psychology* 17, no. 2 (1971): 124–29; P. Ekman and W. V. Friesen, "Nonverbal Leakage and Clues to Deception," *Psychiatry* 32, no. 1 (1969): 88–106; Paul Ekman, E. Richard Sorenson, and Wallace V. Friesen, "Pan-Cultural Elements in Facial Displays of Emotion," *Science* 164, no. 3875 (1969): 86–88, https://doi.org/10.1126/science.164.3875.86; Paul Ekman, Wallace V. Friesen, and Silvan S. Tomkins, "Facial Affect Scoring Technique: A First Validity Study," *Semiotica* 3, no. 1 (1971), https://doi.org/10.1515/semi.1971.3.1.37.

98. Ekman and Friesen, "Nonverbal Leakage."

99. Ekman and Friesen, "The Repertoire of Nonverbal Behavior."

100. Paul Ekman, "An Argument for Basic Emotions," *Cognition and Emotion* 6, nos. 3–4 (1992): 169–200, https://doi.org/10.1080/02699939208411068.

101. Ekman and colleagues published an article describing their own approach to "automatic facial expression measurement" in 1997, the same year as Rosalind W. Picard's book *Affective Computing* (Cambridge, MA: MIT Press, 2000).

102. Rosalind W. Picard, *Affective Computing*, Chapter 3.

103. Picard, *Affective Computing*, 244.

104. Picard, Chapter 4, especially 123–24, 136–37.

105. Barak Reuven Naveh, Techniques for emotion detection and content delivery, US20150242679 A1, filed February 25, 2014, and issued August 27, 2015, http:// www.google.com/patents/US20150242679.

106. Naveh, Techniques for emotion detection and content delivery, paragraph 32.

107. "Affective Computing Market by Technology (Touch-Based and Touchless), Software (Speech Recognition, Gesture Recognition, Facial Feature Extraction, Analytics Software, & Enterprise Software), Hardware, Vertical, and Region—Forecast to 2021," *MarketsandMarkets*, March 2017, http://

www.marketsandmarkets.com/Market-Reports/affective-computing-market
-130730395.html.

108. Raffi Khatchadourian, "We Know How You Feel," *New Yorker,* January 19, 2015, http://www.newyorker.com/magazine/2015/01/19/know-feel.

109. Khatchadourian, "We Know How You Feel."

110. Khatchadourian.

111. "Affectiva," *Crunchbase,* October 22, 2017, https://www.crunchbase.com /organization/affectiva.

112. Lora Kolodny, "Affectiva Raises $14 Million to Bring Apps, Robots Emotional Intelligence," *TechCrunch,* May 25, 2016, http://social.techcrunch.com/2016 /05/25/affectiva-raises-14-million-to-bring-apps-robots-emotional-intelli gence; Rana el Kaliouby, "Emotion Technology Year in Review: Affectiva in 2016," Affectiva, December 29, 2016, http://blog.affectiva.com/emotion -technology-year-in-review-affectiva-in-2016.

113. Matthew Hutson, "Our Bots, Ourselves," *Atlantic,* March 2017, https://www .theatlantic.com/magazine/archive/2017/03/our-bots-ourselves/513839.

114. Patrick Levy-Rosenthal, "Emoshape Announces Production of the Emotions Processing Unit II," *Emoshape | Emotions Synthesis,* January 18, 2016, http:// emoshape.com/emoshape-announces-production-of-the-emotions-process ing-unit-ii.

115. Tom Foster, "Ready or Not, Companies Will Soon Be Tracking Your Emotions," *Inc.com,* June 21, 2016, https://www.inc.com/magazine/201607/tom-foster /lightwave-monitor-customer-emotions.html; "Emotion as a Service," Affectiva, March 30, 2017, http://www.affectiva.com/product/emotion-as-a-service; "Affectiva Announces Availability of Emotion as a Service, a New Data Solution, and Version 2.0 of Its Emotion-Sensing SDK," *PR Newswire,* September 8, 2015, http://www.prnewswire.com/news-releases/affectiva-announces-availability-of -emotion-as-a-service-a-new-data-solution-and-version-20-of-its-emotion -sensing-sdk-300139001.html.

116. See Khatchadourian, "We Know How You Feel."

117. Jean-Paul Sartre, *Being and Nothingness,* trans. Hazel E. Barnes (New York: Washington Square, 1993), 573.

118. Jean-Paul Sartre, *Situations* (New York: George Braziller, 1965), 333.

119. "Kairos for Market Researchers," Kairos, March 9, 2017, https://www.kairos .com/human-analytics/market-researchers.

120. Picard, *Affective Computing,* 119, 123, 244, 123–24, 136–37. See also Chapter 4.

121. Rosalind Picard, "Towards Machines That Deny Their Maker—Lecture with Rosalind Picard," *VBG,* April 22, 2016, http://www.vbg.net/ueber-uns/agenda /termin/3075.html.

122. Joseph Weizenbaum, "Not Without Us," *SIGCAS Computers and Society* 16, nos. 2–3 (1986): 2–7, https://doi.org/10.1145/15483.15484.

CHAPTER TEN

1. Richard H. Thaler and Cass R. Sunstein, *Nudge: Improving Decisions About Health, Wealth, and Happiness,* rev. ed. (New York: Penguin, 2009).
2. Elizabeth J. Lyons et al., "Behavior Change Techniques Implemented in Electronic Lifestyle Activity Monitors: A Systematic Content Analysis," *Journal of Medical Internet Research* 16, no. 8 (2014), e192, https://doi.org/10.2196/jmir.3469.

The commercial theory and practice of behavior modification assume an inescapable networked presence and its cornucopia of digital tools. A team of British researchers surveyed fifty-five behavioral experts in order to compile "a consensually agreed hierarchically structured taxonomy of techniques used in behavior change interventions." This exercise identified ninety-three distinct behavior-change techniques, which were grouped into sixteen methodological clusters: "scheduled consequences," "reward and threat," "repetition and substitution," "antecedents," "associations," "feedback and monitoring," "goals and planning," "social support," "comparison of behavior," "communication of natural consequences," "self-belief," "comparison of outcomes," "shaping knowledge," "regulation," "identity," and "covert learning."

The researchers warn that behavior modification is a "fast-moving field." As illustration, they note that the first such taxonomy, published only four years earlier, had identified just twenty-two behavior-change techniques, many of which were individually oriented and required face-to-face interaction and relationship building. In contrast, the newer techniques are aimed at "community and population-level interventions," a fact that speaks to the migration of behavior-change operations to the novel capabilities (contextual control, automated digital nudges, operant conditioning at scale) of the internet-enabled tools upon which economies of action depend. See Susan Michie et al., "The Behavior Change Technique Taxonomy (v1) of 93 Hierarchically Clustered Techniques: Building an International Consensus for the Reporting of Behavior Change Interventions," *Annals of Behavioral Medicine* 46, no. 1 (2013): 81–95, https://doi.org/10.1007/s12160-013-9486-6.
3. Hal R. Varian, "Beyond Big Data," *Business Economics* 49, no. 1 (2014): 6.
4. Varian, "Beyond Big Data," 7.
5. Robert M. Bond et al., "A 61-Million-Person Experiment in Social Influence and Political Mobilization," *Nature* 489, no. 7415 (2012): 295–98, https://doi.org/10.1038/nature11421.
6. Bond et al., "A 61-Million-Person Experiment."
7. Andrew Ledvina, "10 Ways Facebook Is Actually the Devil," AndrewLedvina.com, July 4, 2017, http://andrewledvina.com/code/2014/07/04/10-ways-facebook-is-the-devil.html.

8. Jonathan Zittrain, "Facebook Could Decide an Election Without Anyone Ever Finding Out," *New Republic*, June 1, 2014, http://www.newrepublic.com/article /117878/information-fiduciary-solution-facebook-digital-gerrymandering; Jonathan Zittrain, "Engineering an Election," *Harvard Law Review* 127 (June 20, 2014): 335; Reed Albergotti, "Facebook Experiments Had Few Limits," *Wall Street Journal*, July 2, 2014, http://www.wsj.com/articles/facebook-experiments -had-few-limits-1404344378; Charles Arthur, "If Facebook Can Tweak Our Emotions and Make Us Vote, What Else Can It Do?" *Guardian*, June 30, 2014, https://www.theguardian.com/technology/2014/jun/30/if-facebook-can-tweak -our-emotions-and-make-us-vote-what-else-can-it-do; Sam Byford, "Facebook Offers Explanation for Controversial News Feed Psychology Experiment," *Verge*, June 29, 2014, https://www.theverge.com/2014/6/29/5855710/facebook -responds-to-psychology-research-controversy; Chris Chambers, "Facebook Fiasco: Was Cornell's Study of 'Emotional Contagion' an Ethics Breach?" *Guardian*, July 1, 2014, https://www.theguardian.com/science/head -quarters/2014/jul/01/facebook-cornell-study-emotional-contagion-ethics -breach.

9. Adam D. I. Kramer, Jamie E. Guillory, and Jeffrey T. Hancock, "Experimental Evidence of Massive-Scale Emotional Contagion Through Social Networks," *Proceedings of the National Academy of Sciences* 111, no. 24 (2014): 8788–90, https://doi.org/10.1073/pnas.1320040111.

10. Kramer, Guillory, and Hancock, "Experimental Evidence of Massive-Scale Emotional Contagion."

11. Matthew R. Jordan, Dorsa Amir, and Paul Bloom, "Are Empathy and Concern Psychologically Distinct?" *Emotion* 16, no. 8 (2016): 1107–16, https://doi .org/10.1037/emo0000228; Marianne Sonnby-Borgström, "Automatic Mimicry Reactions as Related to Differences in Emotional Empathy," *Scandinavian Journal of Psychology* 43, no. 5 (2002): 433–43, https://doi.org/10.1111/1467 -9450.00312; Rami Tolmacz, "Concern and Empathy: Two Concepts or One?" *American Journal of Psychoanalysis* 68, no. 3 (2008): 257–75, https://doi .org/10.1057/ajp.2008.22; Ian E. Wickramasekera and Janet P. Szlyk, "Could Empathy Be a Predictor of Hypnotic Ability?" *International Journal of Clinical and Experimental Hypnosis* 51, no. 4 (2003): 390–99, https://doi.org/10.1076 /iceh.51.4.390.16413; E. B. Tone and E. C. Tully, "Empathy as a 'Risky Strength': A Multilevel Examination of Empathy and Risk for Internalizing Disorders," *Development and Psychopathology* 26, no. 4 (2014): 1547–65, https://doi .org/10.1017/S0954579414001199; Ulf Dimberg and Monika Thunberg, "Empathy, Emotional Contagion, and Rapid Facial Reactions to Angry and Happy Facial Expressions: Empathy and Rapid Facial Reactions," *PsyCh Journal* 1, no. 2 (2012): 118–27, https://doi.org/10.1002/pchj.4; Tania Singer and Claus Lamm, "The Social Neuroscience of Empathy," *Annals of the New York Academy of Sciences* 1156 (April 1, 2009): 81–96, https://doi.org/10.1111/j.1749-

6632.2009.04418.x; Douglas F. Watt, "Social Bonds and the Nature of Empathy," *Journal of Consciousness Studies* 12, nos. 8–9 (2005): 185–209.

12. Jocelyn Shu et al., "The Role of Empathy in Experiencing Vicarious Anxiety," *Journal of Experimental Psychology: General* 146, no. 8 (2017): 1164–88, https://doi.org/10.1037/xge0000335; Tone and Tully, "Empathy as a 'Risky Strength.'"

13. Chambers, "Facebook Fiasco"; Adrienne LaFrance, "Even the Editor of Facebook's Mood Study Thought It Was Creepy," *Atlantic*, June 28, 2014, https://www.theatlantic.com/technology/archive/2014/06/even-the-editor-of-facebooks-mood-study-thought-it-was-creepy/373649.

14. See LaFrance, "Even the Editor."

15. Vindu Goel, "Facebook Tinkers with Users' Emotions in News Feed Experiment, Stirring Outcry," *New York Times*, June 29, 2014, https://www.nytimes.com/2014/06/30/technology/facebook-tinkers-with-users-emotions-in-news-feed-experiment-stirring-outcry.html.

16. Albergotti, "Facebook Experiments Had Few Limits"; Chambers, "Facebook Fiasco."

17. Inder M. Verma, "Editorial Expression of Concern and Correction Regarding 'Experimental Evidence of Massive-Scale Emotional Contagion Through Social Networks,'" *Proceedings of the National Academy of Sciences* 111, no. 29 (2014): 8788–90.

18. James Grimmelmann, "Law and Ethics of Experiments on Social Media Users," *Colorado Technology Law Journal* 13 (January 1, 2015): 255.

19. Michelle N. Meyer et al., "Misjudgements Will Drive Social Trials Underground," *Nature* 511 (July 11, 2014): 265; Michelle Meyer, "Two Cheers for Corporate Experimentation," *Colorado Technology Law Journal* 13 (May 7, 2015): 273.

20. Darren Davidson, "Facebook Targets 'Insecure' to Sell Ads," *Australian*, May 1, 2017.

21. Antonio Garcia-Martinez, "I'm an Ex-Facebook Exec: Don't Believe What They Tell You About Ads," *Guardian*, May 2, 2017, https://www.theguardian.com/technology/2017/may/02/facebook-executive-advertising-data-comment.

22. Dylan D. Wagner and Todd F. Heatherton, "Self-Regulation and Its Failure: The Seven Deadly Threats to Self-Regulation," in *APA Handbook of Personality and Social Psychology* (Washington, DC: American Psychological Association, 2015), 805–42, https://pdfs.semanticscholar.org/2e62/15047e3a296184c3698f3553255ffabd46c7.pdf (italics mine); William M. Kelley, Dylan D. Wagner, and Todd F. Heatherton, "In Search of a Human Self-Regulation System," *Annual Review of Neuroscience* 38, no. 1 (2015): 389–411, https://doi.org/10.1146/annurev-neuro-071013-014243.

23. David Modic and Ross J. Anderson, "We Will Make You Like Our Research: The Development of a Susceptibility-to-Persuasion Scale" (SSRN scholarly paper, Social Science Research Network, April 28, 2014), https://papers.ssrn

.com/abstract=2446971. See also Mahesh Gopinath and Prashanth U. Nyer, "The Influence of Public Commitment on the Attitude Change Process: The Effects of Attitude Certainty, PFC and SNI" (SSRN scholarly paper, Social Science Research Network, August 29, 2007), https://papers.ssrn.com /abstract=1010562.

24. See Dyani Sabin, "The Secret History of 'Pokémon Go' as Told by the Game's Creator," *Inverse,* February 28, 2017, https://www.inverse.com/article/28485 -pokemon-go-secret-history-google-maps-ingress-john-hanke-updates.

25. Tim Bradshaw, "The Man Who Put 'Pokémon Go' on the Map," *Financial Times,* July 27, 2016, https://www.ft.com/content/7209d7ca-49d3-11e6 -8d68-72e9211e86ab.

26. Sebastian Weber and John Hanke, "Reality as a Virtual Playground," *Making Games,* January 22, 2015, http://www.makinggames.biz/feature/reality-as-a -virtual-playground,7286.html.

27. "John Hanke at SXSW 2017: We'll Announce Some New Products at the Next Event!" Pokemon GO Hub, March 10, 2017, http://web.archive.org /web/20170330220737/https://pokemongohub.net/john-hanke-sxsw-2017 -well-announce-new-products-next-event.

28. Sabin, "The Secret History of 'Pokémon Go.'"

29. Weber and Hanke, "Reality as a Virtual Playground."

30. See Hal Hodson, "Why Google's Ingress Game Is a Data Gold Mine," *New Scientist,* September 28, 2012, https://www.newscientist.com/article/mg 21628936-200-why-googles-ingress-game-is-a-data-gold-mine.

31. Sabin, "The Secret History of 'Pokémon Go.'"

32. Ryan Wynia, "Behavior Design Bootcamp with Stanford's Dr. BJ Fogg," *Technori,* October 19, 2012, http://technori.com/2012/10/2612-behavior-design -bootcamp; Ryan Wynia, "BJ Fogg's Behavior Design Bootcamp: Day 2," *Technori,* October 22, 2012, http://technori.com/2012/10/2613-behavior-de sign-bootcamp-day-2. The Stanford researcher B. J. Fogg in his 2003 book *Persuasive Technology* recognized that computer game designers seek to change people's behaviors with Skinnerian-style conditioning, concluding that "good game play and effective operant conditioning go hand in hand."

33. Kevin Werbach, "(Re)Defining Gamification: A Process Approach," in *Persuasive Technology,* Lecture Notes in Computer Science, International Conference on Persuasive Technology (Cham, Switzerland: Springer, 2014), 266–72, https://doi.org/10.1007/978-3-319-07127-5_23; Kevin Werbach and Dan Hunter, *For the Win: How Game Thinking Can Revolutionize Your Business* (Philadelphia: Wharton Digital Press, 2012).

34. Michael Sailer et al., "How Gamification Motivates: An Experimental Study of the Effects of Specific Game Design Elements on Psychological Need Satisfaction," *Computers in Human Behavior* 69 (April 2017): 371–80, https:// doi.org/10.1016/j.chb.2016.12.033; J. Hamari, J. Koivisto, and H. Sarsa, "Does Gamification Work?—a Literature Review of Empirical Studies on

Gamification," in *47th Hawaii International Conference on System Sciences,* 2014, 3025–34, https://doi.org/10.1109/HICSS.2014.377; Carina Soledad González and Alberto Mora Carreño, "Methodological Proposal for Gamification in the Computer Engineering Teaching," *2014 International Symposium on Computers in Education (SIIE),* 1–34; Dick Schoech et al., "Gamification for Behavior Change: Lessons from Developing a Social, Multiuser, Web-Tablet Based Prevention Game for Youths," *Journal of Technology in Human Services* 31, no. 3 (2013): 197–217, https://doi.org/10.108 0/15228835.2013.812512.

35. Yu-kai Chou, "A Comprehensive List of 90+ Gamification Cases with ROI Stats," *Yu-Kai Chou: Gamification & Behavioral Design,* January 23, 2017, http://yukaichou.com/gamification-examples/gamification-stats-figures.

36. Ian Bogost, "Persuasive Games: Exploitationware," *Gamasutra,* May 3, 2011, http://www.gamasutra.com/view/feature/134735/persuasive_games_exploita tionware.php; Adam Alter, *Irresistible: The Rise of Addictive Technology and the Business of Keeping Us Hooked* (New York: Penguin, 2017).

37. Jessica Conditt, "The Pokémon Go Plus Bracelet Is Great for Grinding," *Engadget,* September 17, 2016, https://www.engadget.com/2016/09/17/pokemon-go-plus -hands-on; Sarah E. Needleman, "'Pokémon Go' Wants to Take Monster Battles to the Street," *Wall Street Journal,* September 10, 2015, https://blogs.wsj.com/digits /2015/09/10/pokemon-go-wants-to-take-monster-battles-to-the-street; Patience Haggin, "Alphabet Spinout Scores Funding for Augmented Reality Pokémon Game," *Wall Street Journal,* February 26, 2016, https://blogs.wsj.com/venture capital/2016/02/26/alphabet-spinout-scores-funding-for-augmented-reality -pokemon-game.

38. Joseph Schwartz, "5 Charts That Show Pokémon GO's Growth in the US," *Similarweb Blog,* July 10, 2016, https://www.similarweb.com/blog/pokemon-go.

39. Nick Wingfield and Mike Isaac, "Pokémon Go Brings Augmented Reality to a Mass Audience," *New York Times,* July 11, 2016, https://www.nytimes.com/2016/07/12 /technology/pokemon-go-brings-augmented-reality-to-a-mass-audience.html.

40. Polly Mosendz and Luke Kawa, "Pokémon Go Brings Real Money to Random Bars and Pizzerias," *Bloomberg.com,* July 11, 2016, https://www.bloomberg.com /news/articles/2016-07-11/pok-mon-go-brings-real-money-to-random-bars -and-pizzerias; Abigail Gepner, Jazmin Rosa, and Sophia Rosenbaum, "There's a Pokémon in My Restaurant, and Business Is Booming," *New York Post,* July 12, 2016, http://nypost.com/2016/07/12/pokemania-runs-wild-through-city -causing-crime-accidents; Jake Whittenberg, "Pokemon GO Saves Struggling Wash. Ice Cream Shop," *KSDK,* August 9, 2016, http://www.ksdk.com/news /pokemon-go-saves-struggling-business/292596081.

41. Wingfield and Isaac, "Pokémon Go Brings Augmented Reality."

42. Sabin, "The Secret History of 'Pokémon Go.'"

43. Tim Bradshaw and Leo Lewis, "Advertisers Set for a Piece of 'Pokémon Go' Action," *Financial Times,* July 13, 2016; Jacky Wong, "Pokémon Mania Makes

Mint for Bank of Kyoto," *Wall Street Journal*, July 12, 2016, https://blogs.wsj .com/moneybeat/2016/07/12/pokemon-mania-makes-mint-for-bank-of-kyoto.

44. See Bradshaw and Lewis, "Advertisers Set for a Piece" (italics mine).

45. Jon Russell, "Pokémon Go Will Launch in Japan Tomorrow with Game's First Sponsored Location," *TechCrunch*, July 19, 2016, http://social.techcrunch .com/2016/07/19/pokemon-go-is-finally-launching-in-japan-tomorrow; Takashi Mochizuki, "McDonald's Unit to Sponsor 'Pokémon Go' in Japan," *Wall Street Journal*, July 19, 2016, http://www.wsj.com/articles/mcdonalds-unit-to-sponsor -pokemon-go-in-japan-1468936459; Stephen Wilmot, "An Alternative Way to Monetize Pokémon Go," *Wall Street Journal*, July 29, 2016, https://blogs.wsj.com /moneybeat/2016/07/29/an-alternative-way-to-monetize-pokemon-go; "Pokémon GO Frappuccino at Starbucks," *Starbucks Newsroom*, December 8, 2016, https://news.starbucks.com/news/starbucks-pokemon-go; Megan Farokhmanesh, "Pokémon Go Is Adding 10.5K Gym and Pokéstop Locations at Sprint Stores," *Verge*, December 7, 2016, http://www.theverge.com/2016/12 /7/13868086/pokemon-go-sprint-store-new-gyms-pokestops; Mike Ayers, "Pokémon Tracks Get a Pokémon Go Bump on Spotify," *Wall Street Journal*, July 12, 2016, https://blogs.wsj.com/speakeasy/2016/07/12/pokemon-tracks -get-a-pokemon-go-bump-on-spotify; Josie Cox, "Insurer Offers Pokémon Go Protection (But It's Really Just Coverage for Your Phone)," *Wall Street Journal* (blog), July 22, 2016, https://blogs.wsj.com/moneybeat/2016/07/22/insurer-offers -pokemon-go-protection-but-its-really-just-coverage-for-your-phone; Ben Fritz, "Disney Looks to Tech Behind Pokemon Go," *Wall Street Journal*, August 5, 2016.

46. See Adam Sherrill, "Niantic Believes Pokémon GO Has 'Only Just Scratched the Surface' of AR Gameplay Mechanics," *Gamnesia*, May 5, 2017, https:// www.gamnesia.com/news/niantic-believes-pokemon-go-has-only-just-scrat ched-the-surface-of-ar.

47. Joseph Bernstein, "You Should Probably Check Your Pokémon Go Privacy Settings," *BuzzFeed*, July 11, 2016, https://www.buzzfeed.com/josephbernstein /heres-all-the-data-pokemon-go-is-collecting-from-your-phone.

48. Natasha Lomas, "Pokémon Go Wants to Catch (Almost) All Your App Permissions," *TechCrunch*, July 16, 2016, http://social.techcrunch.com/2016 /07/11/pokemon-go-wants-to-catch-almost-all-your-permissions.

49. Marc Rotenberg, Claire Gartland, and Natashi Amlani, "EPIC Letter to FTC Chair Edith Ramirez," July 22, 2016, 4, https://epic.org/privacy/ftc/FTC-letter -Pokemon-GO-07-22-2016.pdf.

50. Al Franken, "Letter to John Hanke, CEO of Niantic, Inc. from U.S. Senator Al Franken," July 12, 2016, http://www.businessinsider.com/us-senator-al-franken -writes-to-pokmon-go-developers-niantic-privacy-full-letter2016-7.

51. Courtney Greene Power, "Letter to U.S. Senator Al Franken from General Counsel for Niantic, Inc. Courtney Greene Power," August 26, 2016.

52. Rebecca Lemov, *World as Laboratory: Experiments with Mice, Mazes, and Men* (New York: Hill and Wang, 2005), 189.

53. H. Keith Melton and Robert Wallace, *The Official CIA Manual of Trickery and Deception* (New York: William Morrow, 2010), 4.

54. Lemov, *World as Laboratory*, 189; Ellen Herman, *The Romance of American Psychology: Political Culture in the Age of Experts* (Berkeley: University of California Press, 1995), 129.

55. Melton and Wallace, *The Official CIA Manual of Trickery and Deception.*

56. "Church Committee: Book I—Foreign and Military Intelligence," Mary Ferrell Foundation, 1975, 390, https://www.maryferrell.org/php/showlist.php?docset =1014.

57. Lemov, *World as Laboratory*, 200.

58. Alexandra Rutherford, "The Social Control of Behavior Control: Behavior Modification, Individual Rights, and Research Ethics in America, 1971–1979," *Journal of the History of the Behavioral Sciences* 42, no. 3 (2006): 206, https://doi .org/10.1002/jhbs.20169.

59. Noam Chomsky, "The Case Against B. F. Skinner," *New York Review of Books*, December 30, 1971.

60. John L. McClellan et al., "Individual Rights and the Federal Role in Behavior Modification; A Study Prepared by the Staff of the Subcommittee on Constitutional Rights of the Committee on the Judiciary, United States Senate, Ninety-Third Congress, Second Session," November 1974, iii–iv, https://eric .ed.gov/?id=ED103726.

61. McClellan et al., "Individual Rights," IV, 21.

62. McClellan et al., 13–14.

63. P. London, "Behavior Technology and Social Control—Turning the Tables," *APA Monitor* (April 1974): 2 (italics mine); Rutherford, "The Social Control of Behavior Control."

64. Rutherford, "The Social Control of Behavior Control," 213.

65. "The Belmont Report—Office of the Secretary—Ethical Principles and Guidelines for the Protection of Human Subjects of Research—the National Commission for the Protection of Human Subjects of Biomedical and Behavioral Research," Regulations & Policy, Office for Human Research Protections, US Department of Health, Education and Welfare, January 28, 2010, https://www.hhs.gov/ohrp/regulations-and-policy/belmont-report /index.html; Rutherford, "The Social Control of Behavior Control," 215.

66. See Rutherford, "The Social Control of Behavior Control," 217.

67. Daniel W. Bjork, *B. F. Skinner: A Life* (New York: Basic, 1993), 220.

68. "Anthropotelemetry: Dr. Schwitzgebel's Machine." *Harvard Law Review* 80, no. 2 (1966): 403–21, https://doi.org/10.2307/1339322 (italics mine).

CHAPTER ELEVEN

1. Hannah Arendt, *The Life of the Mind*, vol. 2, *Willing* (New York: Harcourt Brace Jovanovich, 1978), 13–14.

2. Hannah Arendt, *The Human Condition* (Chicago: University of Chicago Press, 1998), 244.
3. See also the discussion in John R. Searle, *Making the Social World: The Structure of Human Civilization* (Oxford: Oxford University Press, 2010), 133.
4. Searle, *Making the Social World,* 133, 136.
5. Searle, 194–95. See also Harvard Law School professor Alan Dershowitz, who offers a pragmatic theory of human rights that is relevant to my analysis. He argues that "rights are those fundamental preferences that experience and history—especially of great injustices—have taught are so essential that the citizenry should be persuaded to entrench them and not make them subject to easy change by shifting majorities." Rights, in this way, are derived from wrongs. His is a "bottom-up" approach because there is typically far more consensus on what constitutes a terrible injustice than there is agreement on the conditions for perfect justice. Alan M. Dershowitz, *Rights from Wrongs: A Secular Theory of the Origins of Rights* (New York: Basic, 2004), 81–96.
6. Sir Henry Maine, *Ancient Law* (New York: E. P. Dutton & Co. Inc., 1861).
7. Liam B. Murphy, "The Practice of Promise and Contract" (working paper, New York University Public Law and Legal Theory, 2014), 2069; Avery W. Katz, "Contract Authority—Who Needs It?" *University of Chicago Law Review* 81, no. 4 (2014): 27; Robin Bradley Kar, "Contract as Empowerment," *University of Chicago Law Review* 83, no. 2 (2016): 1.
8. Hal Varian, "Beyond Big Data," *Business and Economics* 49, no. 1 (January 2014). This perfect information is what behavioral economists call "unbounded rationality" or "unrestricted cognitive competence." See Oliver E. Williamson, *The Economic Institutions of Capitalism* (New York: Free Press, 1998), 30.
9. For a powerful examination of this problem in relation to click-wrap and other forms of boilerplate, see Robin Kar and Margaret Radin, "Pseudo-contract & Shared Meaning Analysis" (Legal Studies Research Paper, University of Illinois College of Law, November 16, 2017), https://papers.ssrn.com /abstract=3083129.
10. Weber argued that the "decentralization of lawmaking" expressed in the "private ordering of contracts" does not necessarily produce "a decrease in the degree of coercion." He warned that when a legal order imposes few "mandatory and prohibitory norms and ever so many 'freedoms' and 'empowerments,' [it] can nonetheless...facilitate a quantitative and qualitative increase not only of coercion in general but quite specifically of authoritarian coercion." This is precisely how early-twentieth-century industrial employers used their rights of freedom of contract to employ child labor, demand twelve-hour workdays, and impose dangerous working conditions, and it is precisely how we have been saddled with illegitimate and audacious click -wrap agreements whose authors have found similar shelter in their claims

to freedom of contract. See Max Weber, *Economy and Society: An Outline of Interpretive Sociology,* vol. 2 (Berkeley: University of California Press, 1978), 668–81.

11. Hal R. Varian, "Economic Scene; If There Was a New Economy, Why Wasn't There a New Economics?" *New York Times,* January 17, 2002, http://www .nytimes.com/2002/01/17/business/economic-scene-if-there-was-a-new -economy-why-wasn-t-there-a-new-economics.html.

12. Williamson, *The Economic Institutions of Capitalism.*

13. Oliver E. Williamson, "The Theory of the Firm as Governance Structure: From Choice to Contract," *Journal of Economic Perspectives* 16, no. 3 (2002): 174.

14. Williamson, *The Economic Institutions of Capitalism,* 30–31, 52.With his usual insight, Evgeny Morozov made this connection in a prescient 2014 discussion of the origins of "Big Data" analytics in the ambitions of socialist planners. Evgeny Morozov, "The Planning Machine," *The New Yorker,* October 6, 2014, https://www.newyorker.com/magazine/2014/10/13/planning-machine.

15. "Repo Man Helps Pay Off Bill for Elderly Couple's Car," *ABC News,* November 23, 2016, http://abcnews.go.com/US/repo-man-helps-pays-off -bill-elderly-couples/story?id=43738753; Sarah Larimer, "A Repo Man Didn't Want to Seize an Elderly Couple's Car, So He Helped Pay It Off for Them Instead," *Washington Post,* November 24, 2016, https://www.washingtonpost .com/news/inspired-life/wp/2016/11/24/a-repo-man-didnt-want-to-seize-an -elderly-couples-car-so-he-helped-pay-it-off-for-them-instead/?utm_term =.5ab21c4510ab.

16. Timothy D. Smith, Jeffrey T. Laitman, and Kunwar P. Bhatnagar, "The Shrinking Anthropoid Nose, the Human Vomeronasal Organ, and the Language of Anatomical Reduction," *Anatomical Record* 297, no. 11 (2014): 2196–2204, https://doi.org/10.1002/ar.23035.

17. Chris Jay Hoofnagle and Jennifer King, "Research Report: What Californians Understand About Privacy Offline" (SSRN Scholarly Paper, Rochester, NY: Social Science Research Network, May 15, 2008), http://papers.ssrn.com /abstract=1133075.

18. Joseph Turow et al., "Americans Reject Tailored Advertising and Three Activities That Enable It," Annenberg School for Communication, September 29, 2009, http://papers.ssrn.com/abstract=1478214; Joseph Turow, Michael Hennessy, and Nora Draper, "The Tradeoff Fallacy: How Marketers Are Misrepresenting American Consumers and Opening Them Up to Exploitation," Annenberg School for Communication, June 2015, https://www.asc.upenn.edu /news-events/publications/tradeoff-fallacy-how-marketers-are-misrepresenting -american-consumers-and; Lee Rainie, "Americans' Complicated Feelings About Social Media in an Era of Privacy Concerns," *Pew Research Center,*

March 27, 2018, http://www.pewresearch.org/fact-tank/2018/03/27/americans
-complicated-feelings-about-social-media-in-an-era-of-privacy-concerns.

19. Filippo Tommaso Marinetti, *The Futurist Manifesto* (Paris, France: Le Figaro,
1909); F. T. Marinetti and R. W. Flint, *Marinetti: Selected Writings* (New York:
Farrar, Straus and Giroux, 1972); Harlan K. Ullman and James P. Wade, *Shock
and Awe: Achieving Rapid Dominance* (Forgotten Books, 2008).

20. Greg Mitchell, *The Tunnels: Escapes Under the Berlin Wall and the Historic Films
the JFK White House Tried to Kill* (New York: Crown, 2016); Kristen Greishaber,
"Secret Tunnels That Brought Freedom from Berlin's Wall," *Independent,*
October 18, 2009, http://www.independent.co.uk/news/world/europe/secret
-tunnels-that-brought-freedom-from-berlins-wall-1804765.html.

21. Mary Elise Sarotte, *The Collapse: The Accidental Opening of the Berlin Wall*
(New York: Basic, 2014), 181.

22. Karl Polanyi, *The Great Transformation: The Political and Economic Origins of
Our Time,* 2nd ed. (Boston: Beacon, 2001), 137.

23. Ellen Meiksins Wood, *The Origin of Capitalism: A Longer View* (London: Verso,
2002).

CHAPTER TWELVE

1. Peter S. Menell, "2014: Brand Totalitarianism" (UC Berkeley Public Law
Research Paper, University of California, September 4, 2013), http://papers
.ssrn.com/abstract=2318492; "Move Over, Big Brother," *Economist,* December
2, 2004, http://www.economist.com/node/3422918; Wojciech Borowicz,
"Privacy in the Internet of Things Era," *Next Web,* October 18, 2014, http://
thenextweb.com/dd/2014/10/18/privacy-internet-things-era-will-nsa-know
-whats-fridge; Tom Sorell and Heather Draper, "Telecare, Surveillance, and the
Welfare State," *American Journal of Bioethics* 12, no. 9 (2012): 36–44, https://doi
.org/10.1080/15265161.2012.699137; Christina DesMarais, "This Smartphone
Tracking Tech Will Give You the Creeps," *PCWorld,* May 22, 2012, http://www
.pcworld.com/article/255802/new_ways_to_track_you_via_your_mobile
_devices_big_brother_or_good_business_.html; Rhys Blakely, "'We Thought
Google Was the Future but It's Becoming Big Brother,'" *Times,* September 19,
2014, http://www.thetimes.co.uk/tto/technology/internet/article4271776.ece;
CPDP Conferences, *Technological Totalitarianism, Politics and Democracy,*
2016, http://www.internet-history.info/media-library/mediaitem/2389
-technological-totalitarianism-politics-and-democracy.html; Julian Assange,
"The Banality of 'Don't Be Evil,'" *New York Times,* June 1, 2013, https://www
.nytimes.com/2013/06/02/opinion/sunday/the-banality-of-googles-dont-be
-evil.html; Julian Assange, "Julian Assange on Living in a Surveillance Society,"
New York Times, December 4, 2014, https://www.nytimes.com/2014/12/04
/opinion/julian-assange-on-living-in-a-surveillance-society.html; Michael

Hirsh, "We Are All Big Brother Now," *Politico,* July 23, 2015, https://www
.politico.com/magazine/story/2015/07/big-brother-technology-trial-120477
.html; "Apple CEO Tim Cook: Apple Pay Is Number One," *CBS News,* October
28, 2014, http://www.cbsnews.com/news/apple-ceo-tim-cook-apple-pay-is
-number-one; Mathias Döpfner, "An Open Letter to Eric Schmidt: Why We
Fear Google," *FAZ.net,* April 17, 2014, http://www.faz.net/1.2900860; Sigmar
Gabriel, "Sigmar Gabriel: Political Consequences of the Google Debate,"
Frankfurter Allgemeine Zeitung, May 20, 2014, http://www.faz.net/aktuell
/feuilleton/debatten/the-digital-debate/sigmar-gabriel-consequences-of-the
-google-debate-12948701-p6.html; Cory Doctorow, "Unchecked Surveillance
Technology Is Leading Us Towards Totalitarianism," *International Business
Times,* May 5, 2017, http://www.ibtimes.com/unchecked-surveillance
-technology-leading-us-towards-totalitarianism-opinion-2535230; Martin
Schulz, "Transcript of Keynote Speech at Cpdp2016 on Technological,
Totalitarianism, Politics and Democracy," *Scribd,* 2016, https://www.scribd
.com/document/305093114/Keynote-Speech-at-Cpdp2016-on-Technological
-Totalitarianism-Politics-and-Democracy.

2. Mussolini appointed Gentile to his cabinet as minister for Public Instruction
when he first assumed power in 1922, describing Gentile as his "teacher." See
A. James Gregor, *Giovanni Gentile: Philosopher of Fascism* (New Brunswick, NJ:
Routledge, 2004), 60.

3. Gregor, *Giovanni Gentile,* 30.

4. Gregor, 62–63.

5. Benito Mussolini, *The Doctrine of Fascism* (Hawaii: Haole Church Library, 2015), 4.

6. Frank Westerman, *Engineers of the Soul: The Grandiose Propaganda of Stalin's
Russia,* trans. Sam Garrett (New York: Overlook, 2012), 32–34 (italics mine);
Robert Conquest, *Stalin: Breaker of Nations* (New York: Penguin, 1992).

7. Westerman, *Engineers of the Soul,* 22–29.

8. Waldemar Gurian, "Totalitarianism as Political Religion," in *Totalitarianism,* ed.
Carl J. Friedrich (New York: Grosset & Dunlap, 1964), 120. Ironically, perhaps,
many scholars concluded that Italy never in fact did become a truly totalitarian
state, citing the continuity of institutions—such as the Catholic Church—and
the absence of mass murder. Some argue that theory and practice were more
fully elaborated in Germany and even more extensively—and over a longer
period—in the Soviet Union, despite Soviet elites' rejection of the term in their
reluctance to be identified with fascism.

9. Claude Lefort, "The Concept of Totalitarianism," *Democratiya* 9 (2007): 183–84.

10. On Duranty as a Soviet apologist, see Westerman, *Engineers of the Soul,* 188;
Robert Conquest, *The Great Terror: A Reassessment* (Oxford: Oxford University
Press, 2007), 468.

11. Conquest, *The Great Terror,* 485.

12. Conquest, 447.

13. Conquest, 405.

14. Conquest, *Stalin*, 222, 228.

15. Conquest, 229.

16. Walter Duranty, "What's Going On in Russia?" *Look*, August 15, 1939, 21.

17. Duranty won the Pulitzer Prize in 1932 for *New York Times* articles written from Moscow in 1931. Subsequently, Duranty's reporting was challenged by anti-Stalinist groups who regarded Duranty as a tool for Stalinist propaganda. The Pulitzer committee investigated the allegations for six months and in the end decided not to revoke the award. Years later, the *New York Times* would conclude that Duranty's work was some of the worst reporting in the history of that newspaper. None of this prevented *Look* and other periodicals from continuing to rely on Duranty's accounts of life in the Soviet Union, certainly a significant contribution to the slow pace of public recognition of the unique features of totalitarian power.

18. Conquest, *The Great Terror*, 467–68.

19. Conquest, 486.

20. Carl J. Friedrich, "The Problem of Totalitarianism—an Introduction," in *Totalitarianism*, ed. Carl J. Friedrich (New York: Grosset & Dunlap, 1964), 1.

21. Friedrich, "The Problem of Totalitarianism," 1–2. Friedrich was born and educated in Germany and worked as an advisor to the US military governor of Germany from 1946 to 1949.

22. Hannah Arendt, *The Origins of Totalitarianism* (New York: Schocken, 2004), 387.

23. Arendt, *The Origins*, 431.

24. Arendt, xxvii.

25. Arendt, 429.

26. Carl J. Friedrich, ed., *Totalitarianism* (New York: Grosset & Dunlap, 1954); Carl J. Friedrich and Zbigniew Brzezinski, *Totalitarian Dictatorship and Autocracy* (Cambridge, MA: Harvard University Press, 1956); Theodor Adorno, "Education After Auschwitz," in *Critical Models: Interventions and Catchwords* (New York: Columbia University Press, 1966); Theodor W. Adorno, "The Schema of Mass Culture," in *Culture Industry: Selected Essays on Mass Culture* (New York: Routledge, 1991); Theodor W. Adorno, "On the Question: 'What Is German?'" *New German Critique* no. 36 (Autumn, 1985): 121–31; Gurian, "Totalitarianism as Political Religion"; Raymond Aron, *Democracy and Totalitarianism*, Nature of Human Society Series (London: Weidenfeld & Nicolson, 1968).

27. See Raul Hilberg's monumental account of the destruction of European Jews, which conveys the complexity of the Nazis' mass mobilizations, including systems of transportation and production, military operations, ministerial hierarchies, orchestrated secrecy, and the recruitment of friends and neighbors to terrorize and murder their friends and neighbors: *The Destruction of the European Jews* (New York: Holmes & Meier, 1985). See also Daniel

Jonah Goldhagen, *Hitler's Willing Executioners: Ordinary Germans and the Holocaust* (New York: Vintage, 1997); Jan T. Gross, *Neighbors: The Destruction of the Jewish Community in Jedwabne, Poland* (New York: Penguin, 2002); Christopher R. Browning, *Ordinary Men: Reserve Police Battalion 101 and the Final Solution in Poland* (New York: Harper Perennial, 1998); Norman M. Naimark, *Stalin's Genocides (Human Rights and Crimes Against Humanity)* (Princeton, NJ: Princeton University Press, 2012). In Russia, Solzhenitsyn's account of the concentration camp system reveals the corps of party officials and ordinary citizens required to "feed" the constant flow of terror. See Aleksandr Solzhenitsyn, *The Gulag Archipelago* (New York: Harper & Row, 1973).

28. Richard Shorten, *Modernism and Totalitarianism—Rethinking the Intellectual Sources of Nazism and Stalinism, 1945 to the Present* (New York: Palgrave Macmillan, 2012), 50.

29. Shorten, *Modernism,* Chapter 1.

30. Claude Lefort, *The Political Forms of Modern Society: Bureaucracy, Democracy, Totalitarianism,* ed. John B. Thompson (Cambridge, MA: MIT Press, 1986), 297–98.

31. Hannah Arendt, *Essays in Understanding* (New York: Schocken, 1994), 343.

32. "Behavior: Skinner's Utopia: Panacea, or Path to Hell?" *Time,* September 20, 1971, http://content.time.com/time/magazine/article/0,9171,909994,00.html.

33. B. F. Skinner, "Current Trends in Experimental Psychology," in *Cumulative Record,* 319, www.bfskinner.org (italics mine).

34. See Ludy T. Benjamin Jr. and Elizabeth Nielsen-Gammon, "B. F. Skinner and Psychotechnology: The Case of the Heir Conditioner," *Review of General Psychology* 3, no. 3 (1999): 155–67, https://doi.org/10.1037/1089-2680.3.3.155.

35. B. F. Skinner, *About Behaviorism* (New York: Vintage, 1976), 1–9.

36. John B. Watson, "Psychology as the Behaviorist Views It," *Psychological Review* 20 (1913): 158–77.

37. Lefort, *The Political Forms of Modern Society;* "Max Planck," *Complete Dictionary of Scientific Biography* (Detroit, MI: Scribner's, 2008), http://www.encyclopedia.com/people/science-and-technology/physics-biographies/max-planck.

38. "Max Karl Ernst Ludwig Planck," Nobel-winners.com, December 16, 2017, http://www.nobel-winners.com/Physics/max_karl_ernst_ludwig_planck.html. As a biographer of Max Planck writes, "Planck recalled that his 'original decision to devote myself to science was a direct result of the discovery... that the laws of human reasoning coincide with the laws governing the sequences of the impressions we receive from the world about us; that, therefore, pure reasoning can enable man to gain an insight into the mechanism of the [world]....' He deliberately decided, in other words, to become a theoretical physicist at a time when theoretical physics was not yet recognized as a discipline in its own right. But he went further: he concluded that the existence

of physical laws presupposes that the 'outside world is something independent from man, something absolute, and the quest for the laws which apply to this absolute appeared…as the most sublime scientific pursuit in life.'"

39. Erwin Esper examines the factors that isolated Meyer within American psychology and therefore deprived much of his work of the recognition it should have enjoyed. See, especially, Erwin A. Esper, "Max Meyer in America," *Journal of the History of the Behavioral Sciences* 3, no. 2 (1967): 107–31, https://doi.org/10.1002/1520-6696(196704)3:2<107:AID-JHBS2300030202>3.0.CO;2-F.

40. Esper, "Max Meyer."

41. Skinner, *About Behaviorism,* 14.

42. Skinner reiterated Meyer's definitive importance to the radical behaviorist point of view. See, for example, his 1967 debate with the philosopher Brand Blanshard, where he stated the following: "A special problem arises from the inescapable fact that a small part of the universe is enclosed within the skin of each of us. It is not different in kind from the rest of the universe, but because our contact with it is intimate and in some ways exclusive, it receives special consideration. It is said to be known in a special way, to contain the immediately given, to be the first thing a man knows and according to some the only thing he can really know. Philosophers, following Descartes, begin with it in their analysis of mind. Almost everyone seems to begin with it in explaining his own behavior. There is, however, another possible starting point—the behavior of what Max Meyer used to call the Other-One. As a scientific analysis grows more effective, we no longer explain that behavior in terms of inner events. The world within the skin of the Other-One loses its preferred status." See Brand Blanshard, "The Problem of Consciousness: A Debate with B. F. Skinner," *Philosophy and Phenomenological Research* 27, no. 3 (1967): 317–37.

43. Max Meyer, "The Present Status of the Problem of the Relation Between Mind and Body," *Journal of Philosophy, Psychology and Scientific Methods* 9, no. 14 (1912): 371, https://doi.org/10.2307/2013335.

44. Esper, "Max Meyer," 114. In Meyer's thinking, this reduction to "organism" was inherently humanistic in that it stresses the commonalities among persons and even species. All of us sleep and wake, eat and drink, dance, laugh, cry, reproduce, and die.

45. See Max Planck, "Phantom Problems in Science," in *Scientific Autobiography and Other Papers* (New York: Philosophical Library, 1949), 52–79, 75. In 1946 Planck's long-held views on the unity of science and scientific reasoning would be summarized in his paper on "phantom problems" in science, including the "mind-body problem" and the "problem of free will." Planck viewed the mind-body controversy, like all phantom problems in philosophy and science, as a failure to specify "the viewpoint of the observation" and adhere to it consistently (italics mine). He argued that the "internal" or "psychological" viewpoint and the "external" or "physiological" viewpoint were too frequently

confused: "What you feel, think, want, only you can know as firsthand information. Other people can conclude it only indirectly, from your words, conduct, actions and mannerisms. When such physical manifestations are entirely absent, they have no basis whatever to enable them to know your momentary mental state." The external viewpoint was thus the only one admissible "as the basis of our scientific observation of volitional processes." The establishment of this "external viewpoint" as the basis for the scientific study of human behavior is critical to our story, and Planck's paper suggests the influence that his thinking exerted on Meyer (and even the possibility that Planck had read Meyer's 1921 work). In any case, the resonance between Planck's argument and Meyer's is plain.

46. Max Friedrich Meyer, *Psychology of the Other-One* (Missouri Book Company, 1921), 146, http://archive.org/details/cu31924031214442.
47. Meyer, *Psychology*, 147.
48. Meyer, 402, 406.
49. Meyer, 411–12, 420.
50. Meyer, 402.
51. Meyer, 404.
52. B. F. Skinner, *The Behavior of Organisms: An Experimental Analysis* (Acton, MA: Copley, 1991), 3.
53. Skinner, *The Behavior of Organisms*, 4–6.
54. B. F. Skinner, *Science and Human Behavior*, (Kindle Edition: Free Press, 2012), 228–29.
55. B. F. Skinner, *Beyond Freedom & Dignity* (Indianapolis: Hackett, 2002), 163.
56. Skinner, *Beyond Freedom & Dignity*, 19–20.
57. Skinner, 21, 44, 58.
58. Skinner, *Science and Human Behavior*, 20.
59. Skinner, *Beyond Freedom & Dignity*, 4–5.
60. Skinner, 5–6 (italics mine).
61. Skinner, 59.
62. "Gambling Is a Feature of Capitalism—Not a Bug," *Prospect*, April 2017, http://www.prospectmagazine.co.uk/magazine/gambling-is-a-feature-of-capitalism-not-a-bug; Natasha Dow Schüll, *Addiction by Design: Machine Gambling in Las Vegas* (Princeton, NJ: Princeton University Press, 2014); Howard J. Shaffer, "Internet Gambling & Addiction," Harvard Medical School: Division on Addictions, Cambridge Health Alliance, January 16, 2004; Michael Kaplan, "How Vegas Security Drives Surveillance Tech Everywhere," *Popular Mechanics*, January 1, 2010, http://www.popularmechanics.com/technology/how-to/computer-security/4341499; Adam Tanner, *What Stays in Vegas: The World of Personal Data—Lifeblood of Big Business—and the End of Privacy as We Know It* (New York: PublicAffairs, 2014); Chris Nodder, *Evil by Design: Interaction Design to Lead Us into Temptation* (Indianapolis: Wiley, 2013); Julian Morgans, "Your Addiction to Social Media Is No Accident," *Vice*, May 18, 2017, https://

www.vice.com/en_us/article/vv5jkb/the-secret-ways-social-media-is-built-for
-addiction; "Reasons for Playing Slot Machines Rather Than Table Games in the
U.S.," *Statista*, 2017, https://www.statista.com/statistics/188761/reasons-for
-playing-slot-machines-more-than-table-games-in-the-us.

63. Skinner, *Science and Human Behavior*, 105–6, 282 (italics mine). For insight
into behavior technologies in casinos, see Tanner, *What Stays in Vegas*.

64. Skinner, *Science and Human Behavior*, 105–6.

65. Skinner, 21.

66. Skinner, 282.

67. Skinner, 282.

68. Some may be surprised that George Orwell characterized *1984* to his friend
Anthony Powell as "a utopia written as a novel." See Robert McCrum, "1984:
The Masterpiece That Killed George Orwell," *Guardian*, May 9, 2009, http://
www.theguardian.com/books/2009/may/10/1984-george-orwell. B. F. Skinner
characterized *Walden Two* as "a novel about a Utopian community." B. F.
Skinner, *Walden Two* (Indianapolis: Hackett, 2005), vi.

69. See Skinner's foreword to the 1976 edition of *Walden Two*: B. F. Skinner, *Walden
Two* (New York: Macmillan, 1976).

70. This clip of Mumford's is offered by British documentarian Adam Curtis and
drawn from a longer BBC documentary made in 1968, *Towards Tomorrow: A
Utopia*: https://www.bbc.co.uk/programmes/p0295vz8

71. George Orwell, *1984* (Boston: Houghton Mifflin Harcourt, 2017), 548, 551.

72. Orwell, *1984*, 637–38.

73. Skinner, *Walden Two*, 275–76.

74. Skinner, 137, 149.

CHAPTER THIRTEEN

1. B. F. Skinner, *Walden Two* (Indianapolis: Hackett, 2005), 242–43.

2. B. F. Skinner, "To Know the Future," *Behavior Analyst* 13, no. 2 (1990): 104.

3. Skinner, "*To Know the Future*," 106.

4. Hannah Arendt, *Essays in Understanding* (New York: Schocken, 1994), 319.

5. Hannah Arendt, *The Human Condition* (Chicago: University of Chicago Press,
1998), 322.

6. Hannah Arendt, *The Origins of Totalitarianism* (New York: Schocken, 2004), 620.

7. "Trust," *Our World in Data*, August 3, 2017, https://ourworldindata.org/trust.

8. "Public Trust in Government: 1958–2017," *Pew Research Center for the People
and the Press*, May 3, 2017, http://www.people-press.org/2017/05/03/public
-trust-in-government-1958-2017.

9. Peter P. Swire, "Privacy and Information Sharing in the War on Terrorism,"
Villanova Law Review 51, no. 4 (2006): 951. See also Kristen E. Eichensehr,

"Public-Private Cybersecurity," *Texas Law Review* 95, no. 3 (2017), https://texaslawreview.org/public-private-cybersecurity.

10. Joseph Menn, "Facebook, Twitter, Google Quietly Step Up Fight Against Terrorist Propaganda," *Sydney Morning Herald*, December 7, 2015, http://www.smh.com.au/technology/technology-news/facebook-twitter-google-quietly-step-up-fight-against-terrorist-propaganda-20151206-glgvj2.html.

11. Menn, "Facebook, Twitter, Google." See also Jim Kerstetter, "Daily Report: Tech Companies Pressured on Terrorist Content," *Bits Blog*, December 8, 2015, https://bits.blogs.nytimes.com/2015/12/08/daily-report-tech-companies-pressured-on-terrorist-content; Mark Hosenbell and Patricia Zengerle, "Social Media Terrorist Activity Bill Returning to Senate," *Reuters*, December 7, 2015, http://www.reuters.com/article/us-usa-congress-socialmedia-idUSKBN0TQ 2E520151207.

12. Dave Lee, "'Spell-Check for Hate' Needed, Says Google's Schmidt," *BBC News*, December 7, 2015, http://www.bbc.com/news/technology-35035087.

13. Danny Yadron, "Agenda for White House Summit with Silicon Valley," *Guardian*, January 7, 2016, https://www.theguardian.com/technology/2016/jan/07/white-house-summit-silicon-valley-tech-summit-agenda-terrorism; Danny Yadron, "Revealed: White House Seeks to Enlist Silicon Valley to 'Disrupt Radicalization,'" *Guardian*, January 8, 2016, http://www.theguardian.com/technology/2016/jan/07/white-house-social-media-terrorism-meeting-facebook-apple-youtube.

14. Kashmir Hill, "The Government Wants Silicon Valley to Build Terrorist-Spotting Algorithms. But Is It Possible?" *Fusion*, January 14, 2016, http://fusion.net/story/255180/terrorist-spotting-algorithm.

15. Stefan Wagstyl, "Germany to Tighten Security in Wake of Berlin Terror Attack," *Financial Times*, January 11, 2017, https://www.ft.com/content/bf7972f4-d759-11e6-944b-e7eb37a6aa8e.

16. John Mannes, "Facebook, Microsoft, YouTube and Twitter Form Global Internet Forum to Counter Terrorism," *TechCrunch*, June 26, 2017, http://social.techcrunch.com/2017/06/26/facebook-microsoft-youtube-and-twitter-form-global-internet-forum-to-counter-terrorism; "Partnering to Help Curb the Spread of Terrorist Content Online," Google, December 5, 2016, http://www.blog.google:443/topics/google-europe/partnering-help-curb-spread-terrorist-content-online.

17. "Five Country Ministerial 2017: Joint Communiqué," June 28, 2017, https://www.publicsafety.gc.ca/cnt/rsrcs/pblctns/fv-cntry-mnstrl-2017/index-en.aspx.

18. "European Council Conclusions on Security and Defence, 22/06/2017," June 22, 2017, http://www.consilium.europa.eu/en/press/press-releases/2017/06/22/euco-security-defence.

19. "G20 Leaders' Statement on Countering Terrorism—European Commission Press Release," July 7, 2017, http://europa.eu/rapid/press-release_STATEMENT-17-1955_en.htm; Jamie Bartlett, "Terrorism Adds the Backdrop to the Fight

for Internet Control," *Financial Times,* June 6, 2017, https://www.ft.com
/content/e47782fa-4ac0-11e7-919a-1e14ce4af89b.

20. Spencer Ackerman and Sam Thielman, "US Intelligence Chief: We Might Use
the Internet of Things to Spy on You," *Guardian,* February 9, 2016, http://www
.theguardian.com/technology/2016/feb/09/internet-of-things-smart-home
-devices-government-surveillance-james-clapper.

21. Matt Olsen, Bruce Schneier, and Jonathan Zittrain, "Don't Panic: Making
Progress on the 'Going Dark' Debate," Berkman Klein Center for Internet &
Society at Harvard, February 1, 2016, 13.

22. Haley Sweetland Edwards, "Alexa Takes the Stand," *Time,* May 15, 2017;
Tom Dotan and Reed Albergotti, "Amazon Echo and the Hot Tub Murder,"
Information, December 27, 2016, https://www.theinformation.com/amazon
-echo-and-the-hot-tub-murder.

23. Parmy Olson, "Fitbit Data Now Being Used in the Courtroom," *Forbes,*
November 16, 2014, http://www.forbes.com/sites/parmyolson/2014/11/16
/fitbit-data-court-room-personal-injury-claim; Kate Crawford, "When Fitbit Is
the Expert Witness," *Atlantic,* November 19, 2014, http://www.theatlantic.com
/technology/archive/2014/11/when-fitbit-is-the-expert-witness/382936; Ms.
Smith, "Cops Use Pacemaker Data to Charge Man with Arson, Insurance
Fraud," CSO , January 30, 2017, http://www.csonline.com/article/3162740
/security/cops-use-pacemaker-data-as-evidence-to-charge-homeowner-with
-arson-insurance-fraud.html.

24. Jonah Engel Bromwich, Mike Isaac, and Daniel Victor, "Police Use Surveillance
Tool to Scan Social Media, A.C.L.U. Says," *New York Times,* October 11, 2016,
http://www.nytimes.com/2016/10/12/technology/aclu-facebook-twitter
-instagram-geofeedia.html.

25. Jennifer Levitz and Zusha Elinson, "Boston Plan to Track Web Draws Fire,"
Wall Street Journal, December 5, 2016.

26. Lee Fang, "The CIA Is Investing in Firms That Mine Your Tweets and Instagram
Photos," *Intercept,* April 14, 2016, https://theintercept.com/2016/04/14/in
-undisclosed-cia-investments-social-media-mining-looms-large.

27. Ashley Vance and Brad Stone, "Palantir, the War on Terror's Secret Weapon,"
Bloomberg.com, September 22, 2011, http://www.bloomberg.com/news
/articles/2011-11-22/palantir-the-war-on-terrors-secret-weapon; Ali Winston,
"Palantir Has Secretly Been Using New Orleans to Test Its Predictive Policing
Technology," *Verge,* February 27, 2018.

28. Rogier Creemers, "China's Chilling Plan to Use Social Credit Ratings to Keep
Score on Its Citizens," *CNN.com,* October 27, 2015, https://www.cnn.com
/2015/10/27/opinions/china-social-credit-score-creemers/index.html.

29. Mara Hvistendahl, "Inside China's Vast New Experiment in Social Ranking,"
Wired, December 14, 2017, https://www.wired.com/story/age-of-social-credit.

30. Hvistendahl, "Inside China's Vast New Experiment."

31. Amy Hawkins, "Chinese Citizens Want the Government to Rank Them," *Foreign Policy*, May 24, 2017, https://foreignpolicy.com/2017/05/24/chinese -citizens-want-the-government-to-rank-them.

32. Zhixin Feng et al., "Social Trust, Interpersonal Trust and Self-Rated Health in China: A Multi-level Study," *International Journal for Equity in Health* 15 (November 8, 2016), https://doi.org/10.1186/s12939-016-0469-7.

33. Arthur Kleinman et al., *Deep China: The Moral Life of the Person* (Berkeley: University of California Press, 2011); Mette Halskov Hansen, *iChina: The Rise of the Individual in Modern Chinese Society*, ed. Rune Svarverud (Copenhagen: Nordic Institute of Asian Studies, 2010); Yunxiang Yan, *The Individualization of Chinese Society* (Oxford: Bloomsbury Academic, 2009).

34. Hawkins, "Chinese Citizens Want the Government to Rank Them."

35. Hvistendahl, "Inside China's Vast New Experiment."

36. Hvistendahl.

37. Masha Borak, "China's Social Credit System: AI-Driven Panopticon or Fragmented Foundation for a Sincerity Culture?" *TechNode*, August 23, 2017, https://technode.com/2017/08/23/chinas-social-credit-system-ai-driven-panop ticon-or-fragmented-foundation-for-a-sincerity-culture.

38. "China Invents the Digital Totalitarian State," *Economist*, December 17, 2016, http://www.economist.com/news/briefing/21711902-worrying-implications-its -social-credit-project-china-invents-digital-totalitarian.

39. Shi Xiaofeng and Cao Yin, "Court Blacklist Prevents Millions from Flying, Taking High-Speed Trains," *Chinadaily.com*, February 14, 2017, http://www .chinadaily.com.cn/china/2017-02/14/content_28195359.htm.

40. "China Moving Toward Fully Developed Credit Systems," *Global Times*, June 6, 2017, http://www.globaltimes.cn/content/1052634.shtml.

41. Hvistendahl, "Inside China's Vast New Experiment."

42. Yaxing Yao, Davide Lo Re, and Yang Wang, "Folk Models of Online Behavioral Advertising," in *Proceedings of the 2017 ACM Conference on Computer Supported Cooperative Work and Social Computing*, CSCW '17 (New York: ACM, 2017), 1957–69, https://doi.org/10.1145/2998181.2998316.

43. "China Invents the Digital Totalitarian State."

44. Christopher Lunt, United States Patent: 9100400—Authorization and authentication based on an individual's social network, issued August 4, 2015, http://patft.uspto.gov/netacgi/nph-Parser?Sect1=PTO1&Sect2=HITOFF&d =PALL&p=1&u=%2Fnetahtml%2FPTO%2Fsrchnum.htm&r=1&f=G&l=50&s1 =9100400.PN.&OS=PN/9100400&RS=PN/9100400. The critical paragraph in the patent reads as follows: "In a fourth embodiment of the invention, the service provider is a lender. When an individual applies for a loan, the lender examines the credit ratings of members of the individual's social network who are connected to the individual through authorized nodes. If the average credit rating of these members is at least a minimum credit score, the lender continues to process the loan application. Otherwise, the loan application is rejected."

45. Christer Holloman, "Your Facebook Updates Now Determine Your Credit Score," *Guardian,* August 28, 2014, http://www.theguardian.com/media-net work/media-network-blog/2014/aug/28/social-media-facebook-credit-score -banks; Telis Demos and Deepa Seetharaman, "Facebook Isn't So Good at Judging Your Credit After All," *Wall Street Journal,* February 24, 2016, http:// www.wsj.com/articles/lenders-drop-plans-to-judge-you-by-your-facebook -friends-1456309801; Yanhao Wei et al., "Credit Scoring with Social Network Data" (SSRN Scholarly Paper, Rochester, NY: Social Science Research Network, July 1, 2014), https://papers.ssrn.com/abstract=2475265; Daniel Bjorkegren and Darrell Grissen, "Behavior Revealed in Mobile Phone Usage Predicts Loan Repayment" (SSRN Scholarly Paper, Rochester, NY: Social Science Research Network, July 13, 2015), https://papers.ssrn.com/abstract=2611775.

46. Creemers, "China's Chilling Plan."

47. See Dan Strumpf and Wenxin Fan, "Who Wants to Supply China's Surveillance State? The West," *Wall Street Journal,* November 1, 2017, https://www.wsj.com /articles/who-wants-to-supply-chinas-surveillance-state-the-west-1509540111.

48. Carl J. Friedrich, "The Problem of Totalitarianism—an Introduction," in *Totalitarianism,* ed. Carl J. Friedrich (New York: Grosset & Dunlap, 1964), 1–2.

CHAPTER FOURTEEN

1. Mark Weiser, "The Computer for the 21st Century," *Scientific American,* July 1999, 89.

2. Satya Nadella, "Build 2017," Build Conference 2017, Seattle, May 10, 2017, https://ncmedia.azureedge.net/ncmedia/2017/05/Build-2017-Satya-Nadella -transcript.pdf.

3. Eric Schmidt, "Alphabet's Eric Schmidt: We Should Embrace Machine Learning—Not Fear It," *Newsweek,* January 10, 2017, http://www.newsweek .com/2017/01/20/google-eric-schmidt-embrace-machine-learning-not -fear-it-540369.html.

4. Richard Waters, "FT Interview with Google Co-founder and CEO Larry Page," *Financial Times,* October 31, 2014, http://www.ft.com/intl/cms/s/2/3173f19e -5fbc-11e4-8c27-00144feabdc0.html#axzz3JjXPNno5.

5. Marcus Wohlsen, "Larry Page Lays Out His Plan for Your Future," *Wired,* March 2014, https://www.wired.com/2014/03/larry-page-using-google-build -future-well-living.

6. Waters, "FT Interview with Google Co-founder"; Vinod Khosla, "Fireside Chat with Google Co-founders, Larry Page and Sergey Brin," Khosla Ventures, July 3, 2014, http://www.khoslaventures.com/fireside-chat-with-google-co -founders-larry-page-and-sergey-brin.

7. Miguel Helft, "Fortune Exclusive: Larry Page on Google," *Fortune,* December 11, 2012, http://fortune.com/2012/12/11/fortune-exclusive-larry-page-on -google.

8. Khosla, "Fireside Chat."

9. Larry Page, "2013 Google I/O Keynote," Google I/O, May 15, 2013, http://www
.pcworld.com/article/2038841/hello-larry-googles-page-on-negativity-laws
-and-competitors.html.

10. "Facebook's (FB) CEO Mark Zuckerberg on Q4 2014 Results—Earnings Call
Transcript," *Seeking Alpha,* January 29, 2015, https://seekingalpha.com
/article/2860966-facebooks-fb-ceo-mark-zuckerberg-on-q4-2014-results
-earnings-call-transcript.

11. See Ashlee Vance, "Facebook: The Making of 1 Billion Users," *Bloomberg.com,*
October 4, 2012, http://www.bloomberg.com/news/articles/2012-10-04/face
book-the-making-of-1-billion-users.

12. "Facebook's (FB) CEO Mark Zuckerberg on Q4 2014 Results."

13. "Facebook (FB) Mark Elliot Zuckerberg on Q1 2016 Results—Earnings Call
Transcript," *Seeking Alpha,* April 28, 2016, https://seekingalpha.com/article
/3968783-facebook-fb-mark-elliot-zuckerberg-q1-2016-results-earnings-call
-transcript.

14. Mark Zuckerberg, "Building Global Community," Facebook, February 16,
2017, https://www.facebook.com/notes/mark-zuckerberg/building-global
-community/10154544292806634.

15. Mark Zuckerberg, "Facebook CEO Mark Zuckerberg's Keynote at F8 2017
Conference (Full Transcript)," April 19, 2017, https://singjupost.com
/facebook-ceo-mark-zuckerbergs-keynote-at-f8-2017-conference-full
-transcript.

16. Johann Wolfgang von Goethe, *"The Sorcerer's Apprentice," German Stories at
Virginia Commonwealth University,* 1797, *http://germanstories.vcu.edu/goethe
/zauber_e4.html.*

17. Frank E. Manuel and Fritzie P. Manuel, *Utopian Thought in the Western World*
(Cambridge, MA: Belknap Press, 1979), 20.

18. Manuel and Manuel, *Utopian Thought,* 23 (italics mine).

19. Todd Bishop and Nat Levy, "With $256 Billion, Apple Has More Cash Than
Amazon, Microsoft and Google Combined," *GeekWire,* May 2, 2017, https://
www.geekwire.com/2017/256-billion-apple-cash-amazon-microsoft-google
-combined.

20. Manuel and Manuel, *Utopian Thought in the Western World,* 9.

21. Zuckerberg, "Building Global Community."

22. "Facebook CEO Mark Zuckerberg's Keynote at F8 2017 Conference."

23. Nadella, "Build 2017."

24. Satya Nadella, Chen Qiufan, and Ken Liu, "The Partnership of the Future,"
Slate, June 28, 2016, http://www.slate.com/articles/technology/future_tense
/2016/06/microsoft_ceo_satya_nadella_humans_and_a_i_can_work_together
_to_solve_society.html.

25. Nadella, "Build 2017."

26. Nadella.
27. Nadella (italics mine).
28. Nadella (italics mine).
29. Nadella.
30. Elad Yom-Tov et al., User behavior monitoring on a computerized device, US9427185 B2, filed June 20, 2013, and issued August 30, 2016, http://www .google.com/patents/US9427185.
31. B. F. Skinner, *Walden Two* (Indianapolis: Hackett, 2005), 195–96.
32. Eric Schmidt and Sebastian Thrun, "Let's Stop Freaking Out About Artificial Intelligence," *Fortune,* June 28, 2016, http://fortune.com/2016/06/28/artificial -intelligence-potential.
33. Schmidt and Thrun, "Let's Stop Freaking Out."

CHAPTER FIFTEEN

1. Alex Pentland, "Alex Pentland Homepage—Honest Signals, Reality Mining, and Sensible Organizations," February 2, 2016, http://web.media.mit.edu/~sandy; "Alex Pentland—Bio," World Economic Forum, February 28, 2018, https://www .weforum.org/agenda/authors/alex-pentland; *Edge Video,* "The Human Strategy: A Conversation with Alex 'Sandy' Pentland," October 30, 2017, https://www.edge.org/conversation/alex_sandy_pentland-the-human-strategy.
2. Talks at Google, Sandy Pentland: "Social Physics: How Good Ideas Spread," YouTube.com, March 7, 2014, https://www.youtube.com/watch?v=HMB10 ttu-Ow.
3. Maria Konnikova, "Meet the Godfather of Wearables," *Verge,* May 6, 2014, http://www.theverge.com/2014/5/6/5661318/the-wizard-alex-pentland-father -of-the-wearable-computer.
4. "Alex Pentland," *Wikipedia,* July 22, 2017, https://en.wikipedia.org/w/index. php?title=Alex_Pentland&oldid=791778066; Konnikova, "Meet the Godfather"; Dave Feinleib, "3 Big Data Insights from the Grandfather of Google Glass," *Forbes,* October 17, 2012, http://www.forbes.com/sites/davefeinleib/2012 /10/17/3-big-data-insights-from-the-grandfather-of-google-glass.
5. The term *social physics* originates in the positivist philosophy of Auguste Comte, who preceded Planck in his programmatic vision of a scientific approach to the study of society that would equal the precision of the natural sciences. In the 1830s Comte wrote the following: "Now that the human mind has founded celestial physics, terrestrial physics ... and organic physics ... it only remains to complete the system of observational sciences by the foundation of social physics." See Auguste Comte, *Introduction to Positive Philosophy,* ed. Frederick Ferré (Indianapolis: Hackett, 1988), 13.

    Nearly two hundred years later, Pentland's theory and research in social physics have made him the focus of articles in the *New York Times,* the *Harvard Business Review,* and the *New Yorker,* as well as a prominent speaker on the global circuit, from the UN and the World Economic Forum to corporations and international conferences. At Microsoft and Google he has been featured as the "presiding genius" of the "Big Data revolution," whose "groundbreaking experiments" and "remarkable discoveries" have made his work "the bedrock of a whole new scientific field." With the publication of his book *Social Physics*, Pentland was introduced at the popular Digital-Life-Design Conference in 2014 by the well-known social media analyst Clay Shirky, who opined that Pentland's Human Dynamics Lab "has done more to explain human behavior in groups in the last ten years than any other institution in the world."

6. Konnikova, "Meet the Godfather."
7. Konnikova.
8. Tanzeem Choudhury and Alex Pentland, "The Sociometer: A Wearable Device for Understanding Human Networks" (white paper, Computer Supported Cooperative Work—Workshop on Ad Hoc Communications and Collaboration in Ubiquitous Computing Environments), November 2, 2002.
9. Choudhury and Pentland.
10. Nathan Eagle and Alex Pentland, "Reality Mining: Sensing Complex Social Systems," *Personal and Ubiquitous Computing* 10, no. 4 (2006): 255, https://doi .org/10.1007/s00779-005-0046-3.
11. Alex Pentland, "'Reality Mining' the Organization," *MIT Technology Review,* March 31, 2004, https://www.technologyreview.com/s/402609/reality-mining -the-organization.
12. Eagle and Pentland, "Reality Mining."
13. Kate Greene, "TR10: Reality Mining," *MIT Technology Review,* February 19, 2008, http://www2.technologyreview.com/news/409598/tr10-reality-mining; Alex Pentland, *Social Physics: How Good Ideas Spread—the Lessons from a New Science* (Brunswick, NJ: Scribe, 2014), 217–18.
14. Pentland, *Social Physics*, 2–3.
15. Greene, "TR10."
16. Alex Pentland, "The Data-Driven Society," *Scientific American* 309 (October 2013): 78–83, https://doi.org/doi:10.1038/scientificamerican1013-78.
17. Pentland, "'Reality Mining' the Organization."
18. Nathan Eagle and Alex Pentland, Combined short range radio network and cellular telephone network for interpersonal communications, MIT ID: 10705T, US US7877082B2, filed May 6, 2004, and issued September 19, 2014, https:// patents.google.com/patent/US20150006207A1/en.
19. See Ryan Singel, "When Cell Phones Become Oracles," *Wired,* July 25, 2005, https://www.wired.com/2005/07/when-cell-phones-become-oracles.
20. Pentland, "'Reality Mining' the Organization."
21. D. O. Olguin et al., "Sensible Organizations: Technology and Methodology for Automatically Measuring Organizational Behavior," *IEEE Transactions on Systems, Man, and Cybernetics, Part B (Cybernetics)* 39, no. 1 (2009): 43–55, https://doi.org/10.1109/TSMCB.2008.2006638.

22. Taylor Soper, "MIT Spinoff Tenacity Raises $1.5M to Improve Workplace Productivity with 'Social Physics,'" *GeekWire*, February 10, 2016, https://www.geekwire.com/2016/tenacity-raises-1-5m; Ron Miller, "Endor Emerges from MIT Research with Unique Predictive Analytics Tech," *TechCrunch*, March 8, 2017, http://social.techcrunch.com/2017/03/08/endor-emerges-from-mit-research-with-unique-predictive-analytics-tech; Rob Matheson, "Watch Your Tone," *MIT News*, January 20, 2016, http://news.mit.edu/2016/startup-cogito-voice-analytics-call-centers-ptsd-0120.

23. Ben Waber, *People Analytics: How Social Sensing Technology Will Transform Business and What It Tells Us About the Future of Work* (Upper Saddle River, NJ: FT Press, 2013).

24. Ron Miller, "New Firm Combines Wearables and Data to Improve Decision Making," *TechCrunch*, February 24, 2015, http://social.techcrunch.com/2015/02/24/new-firm-combines-wearables-and-data-to-improve-decision-making.

25. Miller, "New Firm"; Alexandra Bosanac, "How 'People Analytics' Is Transforming Human Resources," *Canadian Business*, October 26, 2015, http://www.canadianbusiness.com/innovation/how-people-analytics-is-transforming-human-resources.

26. Pentland, "The Data-Driven Society."

27. "Alex Pentland Homepage"; Endor.com, December 23, 2017; "Endor—Careers," http://www.endor.com/careers; "Endor—Social Physics," http://www.endor.com/social-physics.

28. "Yellow Pages Acquires Sense Networks," Yellow Pages, January 6, 2014, http://corporate.yp.com/yp-acquires-sense-networks.

29. Alison E. Berman, "MIT's Sandy Pentland: Big Data Can Be a Profoundly Humanizing Force in Industry," *Singularity Hub*, May 16, 2016, https://singularityhub.com/2016/05/16/mits-sandy-pentland-big-data-can-be-a-profoundly-humanizing-force-in-industry.

30. Berman, "MIT's Sandy Pentland."

31. Alex Pentland, "Society's Nervous System: Building Effective Government, Energy, and Public Health Systems," *MIT Open Access Articles*, October 2011, http://dspace.mit.edu/handle/1721.1/66256.

32. Pentland, "Society's Nervous System," 3.

33. Pentland, 6.

34. Pentland, 2–4.

35. Pentland, 3.

36. Pentland, 10.

37. Pentland, 8 (italics mine).

38. Pentland.

39. Pentland, *Social Physics*, 10–11.

40. Pentland, 12 (italics mine).

41. Pentland, 245.

42. Pentland, 7 (italics mine).
43. B. F. Skinner, *Beyond Freedom & Dignity* (Indianapolis: Hackett, 2002), 175.
44. B. F. Skinner, *Walden Two* (Indianapolis: Hackett, 2005), 241.
45. Skinner, *Walden Two*, 162.
46. Skinner, 239.
47. Pentland, *Social Physics*, 19.
48. Pentland, 143, 18.
49. Pentland, 153.
50. Skinner, *Walden Two*, 275.
51. Skinner, 252.
52. Skinner, 255–56.
53. Skinner, 218–19.
54. Pentland, *Social Physics*, 191.
55. Pentland, 2–3.
56. Pentland, 6–7.
57. Pentland, 172 (italics mine).
58. Pentland, 38.
59. Skinner, *Walden Two*, 92–93 (italics mine).
60. Pentland, *Social Physics*, 69.
61. Pentland, 184.
62. Pentland, 152.
63. Pentland, 190.
64. Pentland, 46.
65. Alex Pentland, "The Death of Individuality: What Really Governs Your Actions?" *New Scientist* 222, no. 2963 (2014): 30–31, https://doi.org/10.1016/S0262-4079(14)60684-9.
66. Skinner, *Beyond Freedom*, 155–56.
67. Pentland, *Social Physics*, 191, 203–4.
68. Pentland, "The Death of Individuality."
69. Skinner, *Beyond Freedom*, 200, 205.
70. Skinner, 211.
71. Talks at Google, *Sandy Pentland: "Social Physics: How Good Ideas Spread."*
72. Noam Chomsky, "The Case Against B. F. Skinner," *New York Review of Books*, December 30, 1971.
73. Pentland, *Social Physics*, 189.
74. Pentland, 190.
75. Alex Pentland, "Reality Mining of Mobile Communications: Toward a New Deal on Data," in *Global Information Technology Report, World Economic Forum & INSEAD* (World Economic Forum, 2009), 75–80.
76. Harvard Business Review Staff, "With Big Data Comes Big Responsibility," *Harvard Business Review*, November 1, 2014, https://hbr.org/2014/11/with-big-data-comes-big-responsibility.

77. "Who Should We Trust to Manage Our Data?" World Economic Forum, accessed August 9, 2018, https://www.weforum.org/agenda/2015/10/who -should-we-trust-manage-our-data/.

78. Primavera De Filippi and Benjamin Loveluck, "The Invisible Politics of Bitcoin: Governance Crisis of a Decentralized Infrastructure," *Internet Policy Review* 5, no. 3 (September 30, 2016).

79. Staff, "With Big Data Comes Big Responsibility."

CHAPTER SIXTEEN

1. "The World UNPLUGGED," *The World UNPLUGGED*, https://theworld unplugged.wordpress.com.

2. For an insightful account, see Katherine Losse, *The Boy Kings: A Journey into the Heart of the Social Network* (New York: Free Press, 2012).

3. "Confusion," *The World UNPLUGGED*, February 26, 2011, https:// theworldunplugged.wordpress.com/emotion/confusion.

4. "College Students Spend 12 Hours/Day with Media, Gadgets," *Marketing Charts*, November 30, 2009, https://www.marketingcharts.com/television-11195.

5. Andrew Perrin and Jingjing Jiang, "About a Quarter of U.S. Adults Say They Are 'Almost Constantly' Online," *Pew Research Center*, March 14, 2018, http://www .pewresearch.org/fact-tank/2018/03/14/about-a-quarter-of-americans-report -going-online-almost-constantly; Monica Anderson and Jingjing Jiang, "Teens, Social Media & Technology 2018," *Pew Research Center*, May 31, 2018, http://www .pewinternet.org/2018/05/31/teens-social-media-technology-2018/.

6. Jason Dorsey, "Gen Z—Tech Disruption: 2016 National Study on Technology and the Generation After Millennials," *Center for Generational Kinetics*, 2016, http://3pur2814p18t46fuop22hvvu.wpengine.netdna-cdn.com/wp-content /uploads/2017/01/Research-White-Paper-Gen-Z-Tech-Disruption-c-2016 -Center-for-Generational-Kinetics.pdf.

7. Sarah Marsh, "Girls Suffer Under Pressure of Online 'Perfection,' Poll Finds," *Guardian*, August 22, 2017, http://www.theguardian.com/society/2017 /aug/23/girls-suffer-under-pressure-of-online-perfection-poll-finds.

8. For an early and insightful theoretical discussion of the internet as a zone of personal objectification, see Julie E. Cohen, "Examined Lives: Informational Privacy and the Subject as Object" (SSRN Scholarly Paper, Rochester, NY: Social Science Research Network, August 15, 2000), https://papers.ssrn.com/abstract=233597.

9. Sarah Marsh and *Guardian* readers, "Girls and Social Media: 'You Are Expected to Live Up to an Impossible Standard,'" *Guardian*, August 22, 2017, http://www .theguardian.com/society/2017/aug/23/girls-and-social-media-you-are -expected-to-live-up-to-an-impossible-standard.

10. See Marsh, "Girls and Social Media."

11. "Millennials Check Their Phones More Than 157 Times per Day," *New York*, May 31, 2016, https://socialmediaweek.org/newyork/2016/05/31/millennials -check-phones-157-times-per-day (italics mine).

12. Natasha Dow Schüll, *Addiction by Design: Machine Gambling in Las Vegas* (Princeton, NJ: Princeton University Press, 2014), 166–67.

13. Schüll, *Addiction by Design*, 160.

14. Natasha Dow Schüll, "Beware: 'Machine Zone' Ahead," *Washington Post*, July 6, 2008, http://www.washingtonpost.com/wp-dyn/content/article/2008/07/04 /AR2008070402134.html (italics mine).

15. See Schüll, *Addiction by Design*, 174.

16. Alex Hern, "'Never Get High on Your Own Supply'—Why Social Media Bosses Don't Use Social Media," *Guardian*, January 23, 2018, http://www.theguardian .com/media/2018/jan/23/never-get-high-on-your-own-supply-why-social -media-bosses-dont-use-social-media.

17. Jessica Contrera, "This Is What It's Like to Grow Up in the Age of Likes, Lols and Longing," *Washington Post*, May 25, 2016, http://www.washingtonpost .com/sf/style/2016/05/25/13-right-now-this-is-what-its-like-to-grow-up-in-the -age-of-likes-lols-and-longing.

18. Granville Stanley Hall, *Adolescence: Its Psychology and Its Relations to Physiology, Anthropology, Sociology, Sex, Crime, Religion and Education* (Memphis, TN: General Books, 2013), 1:3.

19. Hall, *Adolescence*, 1:84.

20. Erik H. Erikson, *Identity and the Life Cycle* (New York: W. W. Norton, 1994), 126–27. See also Erik H. Erikson, *Identity: Youth and Crisis* (New York: W. W. Norton, 1994), especially 128–35.

21. For an introduction to this concept, see Jeffrey Jensen Arnett, *Emerging Adulthood: The Winding Road from the Late Teens Through the Twenties* (Oxford: Oxford University Press, 2006).

22. See, for example, Laurence Steinberg and Richard M. Lerner, "The Scientific Study of Adolescence: A Brief History," *Journal of Early Adolescence* 24, no. 1 (2004): 45–54, https://doi.org/10.1177/0272431603260879; Arnett, *Emerging Adulthood*; Daniel Lapsley and Ryan D. Woodbury, "Social Cognitive Development in Emerging Adulthood," in *The Oxford Handbook of Emerging Adulthood* (Oxford: Oxford University Press, 2015); Wim Meeus, "Adolescent Psychosocial Development: A Review of Longitudinal Models and Research," *Developmental Psychology* 52, no. 12 (2016): 1969–93, https://doi.org/10.1037 /dev0000243; Jeffrey Jensen Arnett et al., *Debating Emerging Adulthood: Stage or Process?* (Oxford: Oxford University Press, 2011).

23. Dan P. McAdams, "Life Authorship in Emerging Adulthood," in *The Oxford Handbook of Emerging Adulthood* (Oxford: Oxford University Press, 2015), 438.

24. Lapsley and Woodbury, "Social Cognitive Development," 152.

25. Lapsley and Woodbury, 155. Academic discussions of the individuation -attachment balance frequently turn on questions of culture. How universal are these developmental insights? A passage from Lapsley and Woodbury's review addresses this question in way that I find balanced and reasonable:

> How individuation plays out in different ethnoracial groups, in different cultural settings, and within national boundaries or in cross-national samples, are all matters of empirical inquiry. But the tension between agency and communion is a basic duality of human existence (Bakan, 1966) in our view. How it is calibrated may well show variability across cultures. Some societies may prioritize communion, but agency is not thereby neglected. Other societies may prioritize agency, but the yearning for attachment, communion, and bonding is never absent. Moreover, how agency-communion is manifested will vary within the life course of the self-same individual, depending on relational status, developmental priorities, or life circumstances. However the compromise is struck between agency and communion, emerging adulthood is the developmental period during which the hard bargaining will have to take place, with important implications for later adjustment in adulthood.

The reference in this extract is to David Bakan, *The Duality of Human Existence: Isolation and Communion in Western Man* (Boston: Beacon, 1966).

26. Robert Kegan, *The Evolving Self: Problem and Process in Human Development* (Cambridge, MA: Harvard University Press, 1982), 96. See the discussion on 95–100.

27. Erikson, *Identity: Youth and Crisis*, 130.

28. Kegan, *The Evolving Self*, 19.

29. Lapsley and Woodbury, "Social Cognitive Development," 152.

30. danah boyd, *It's Complicated: The Social Lives of Networked Teens* (New Haven, CT: Yale University Press, 2014), 8.

31. Chris Nodder, *Evil by Design: Interaction Design to Lead Us into Temptation* (Indianapolis: Wiley, 2013), xv.

32. Lapsley and Woodbury, "Social Cognitive Development," 152.

33. Nodder, *Evil by Design*, 5.

34. "What's the History of the Awesome Button (That Eventually Became the Like Button) on Facebook?" *Quora*, September 19, 2017, https://www.quora.com /Whats-the-history-of-the-Awesome-Button-that-eventually-became-the-Like -button-on-Facebook.

35. See John Paul Titlow, "How Instagram Learns from Your Likes to Keep You Hooked," *Fast Company*, July 7, 2017, https://www.fastcompany.com/4043 4598/how-instagram-learns-from-your-likes-to-keep-you-hooked.

36. Adam Alter, *Irresistible: The Rise of Addictive Technology and the Business of Keeping Us Hooked* (New York: Penguin, 2017), 128.

37. Josh Constine, "How Facebook News Feed Works," *TechCrunch*, September 6, 2016, http://social.techcrunch.com/2016/09/06/ultimate-guide-to-the-news-feed.

38. Michael Arrington, "Facebook Users Revolt, Facebook Replies," *TechCrunch*, http://social.techcrunch.com/2006/09/06/facebook-users-revolt-facebook-replies.

39. Victor Luckerson, "Here's How Your Facebook News Feed Actually Works," *Time*, July 9, 2015, http://time.com/collection-post/3950525/facebook-news-feed-algorithm.

40. See Constine, "How Facebook News Feed Works." The quotation is from Will Oremus, "Who Controls Your Facebook Feed," *Slate*, January 3, 2016, http://www.slate.com/articles/technology/cover_story/2016/01/how_facebook_s_news_feed_algorithm_works.html.

41. Constine, "How Facebook News Feed Works."

42. See Luckerson, "Here's How Your Facebook News Feed Actually Works."

43. See Oremus, "Who Controls Your Facebook Feed."

44. Alessandro Acquisti, Laura Brandimarte, and George Loewenstein, "Privacy and Human Behavior in the Age of Information," *Science* 347, no. 6221 (2015): 509–14, https://doi.org/10.1126/science.aaa1465.

45. Jerry Suls and Ladd Wheeler, "Social Comparison Theory," in *Theories of Social Psychology*, ed. Paul A. M. Van Lange, Arie W. Kruglanski, and E. Tory Higgins, vol. 2 (Thousand Oaks, CA: Sage, 2012), 460–82.

46. David R. Mettee and John Riskind, "Size of Defeat and Liking for Superior and Similar Ability Competitors," *Journal of Experimental Social Psychology* 10, no. 4 (1974): 333–51; See also T. Mussweiler and K. Rütter, "What Friends Are For! The Use of Routine Standards in Social Comparison," *Journal of Personality and Social Psychology* 85, no. 3 (2003): 467–81.

47. Suls and Wheeler, "Social Comparison Theory."

48. K. Hennigan and L. Heath, "Impact of the Introduction of Television on Crime in the United States: Empirical Findings and Theoretical Implications," *Journal of Personality and Social Psychology* 42, no. 3 (1982): 461–77; Hyeseung Yang and Mary Beth Oliver, "Exploring the Effects of Television Viewing on Perceived Life Quality: A Combined Perspective of Material Value and Upward Social Comparison," *Mass Communication and Society* 13, no. 2 (2010): 118–38.

49. Amanda L. Forest and Joanne V. Wood, "When Social Networking Is Not Working," *Psychological Science* 23, no. 3 (2012): 295–302; Lin Qiu et al., "Putting Their Best Foot Forward: Emotional Disclosure on Facebook," *Cyberpsychology, Behavior, and Social Networking* 15, no. 10 (2012): 569–72.

50. Jiangmeng Liu et al., "Do Our Facebook Friends Make Us Feel Worse? A Study of Social Comparison and Emotion," *Human Communication Research* 42, no. 4 (2016): 619–40, https://doi.org/10.1111/hcre.12090.

51. Andrew K. Przybylski et al., "Motivational, Emotional, and Behavioral Correlates of Fear of Missing Out," *Computers in Human Behavior* 29, no. 4 (2013): 1841–48, https://doi.org/10.1016/j.chb.2013.02.014.

52. Qin-Xue Liu et al., "Need Satisfaction and Adolescent Pathological Internet Use: Comparison of Satisfaction Perceived Online and Offline," *Computers in Human Behavior* 55 (February 2016): 695–700, https://doi.org/10.1016/j .chb.2015.09.048; Dorit Alt, "College Students' Academic Motivation, Media Engagement and Fear of Missing Out," *Computers in Human Behavior* 49 (August 2015): 111–19, https://doi.org/10.1016/j.chb.2015.02.057; Roselyn J. Lee-Won, Leo Herzog, and Sung Gwan Park, "Hooked on Facebook: The Role of Social Anxiety and Need for Social Assurance in Problematic Use of Facebook," *Cyberpsychology, Behavior, and Social Networking* 18, no. 10 (2015): 567–74, https://doi.org/10.1089/cyber.2015.0002; Jon D. Elhai et al., "Fear of Missing Out, Need for Touch, Anxiety and Depression Are Related to Problematic Smartphone Use," *Computers in Human Behavior* 63 (October 2016): 509–16, https://doi.org/10.1016/j.chb.2016.05.079.

53. Nina Haferkamp and Nicole C. Krämer, "Social Comparison 2.0: Examining the Effects of Online Profiles on Social-Networking Sites," *Cyberpsychology, Behavior and Social Networking* 14, no. 5 (2011): 309–14, https://doi.org/10 .1089/cyber.2010.0120. See also Helmut Appel, Alexander L. Gerlach, and Jan Crusius, "The Interplay Between Facebook Use, Social Comparison, Envy, and Depression," *Current Opinion in Psychology* 9 (June 2016): 44–49, https:///doi .org/10.1016/j.copsyc.2015.10.006.

54. Ethan Kross et al., "Facebook Use Predicts Declines in Subjective Well-Being in Young Adults," *PLoS ONE* 8, no. 8 (2013): e69841, https://doi.org/10.1371 /journal.pone.0069841.

55. Hanna Krasnova et al., "Envy on Facebook: A Hidden Threat to Users' Life Satisfaction?" *Wirtschaftsinformatik Proceedings 2013* 92 (January 1, 2013), http://aisel.aisnet.org/wi2013/92; Christina Sagioglou and Tobias Greitemeyer, "Facebook's Emotional Consequences: Why Facebook Causes a Decrease in Mood and Why People Still Use It," *Computers in Human Behavior* 35 (June 2014): 359–63, https://doi.org/10.1016/j.chb.2014.03.003.

56. Edson C. Tandoc Jr., Patrick Ferruci, and Margaret Duffy, "Facebook Use, Envy, and Depression Among College Students: Is Facebooking Depressing?" *Computers in Human Behavior* 43 (February 2015): 139–46.

57. Adriana M. Manago et al., "Facebook Involvement, Objectified Body Consciousness, Body Shame, and Sexual Assertiveness in College Women and Men," *Springer* 72, nos. 1–2 (2014): 1–14, https://doi.org/10.1007/s11199-014-0441-1.

58. Jan-Erik Lönnqvist and Fenne große Deters, "Facebook Friends, Subjective Well-Being, Social Support, and Personality," *Computers in Human Behavior* 55 (February 2016): 113–20, https://doi.org/10.1016/j.chb.2015.09.002; Daniel C. Feiler and Adam M. Kleinbaum, "Popularity, Similarity, and the Network

Extraversion Bias," *Psychological Science* 26, no. 5 (2015): 593–603, https://doi .org/10.1177/0956797615569580.

59. Brian A. Primack et al., "Social Media Use and Perceived Social Isolation Among Young Adults in the U.S.," *American Journal of Preventive Medicine* 53, no. 1 (2017): 1–8, https://doi.org/10.1016/j.amepre.2017.01.010; Taylor Argo and Lisa Lowery, "The Effects of Social Media on Adolescent Health and Well -Being," *Journal of Adolescent Health* 60, no. 2 (2017): S75–76, https://doi .org/10.1016/j.jadohealth.2016.10.331; Elizabeth M Seabrook, Margaret L. Kern, and Nikki S. Rickard, "Social Networking Sites, Depression, and Anxiety: A Systematic Review," *JMIR Mental Health* 3, no. 4 (2016): e50, https://doi .org/10.2196/mental.5842.

60. Holly B. Shakya and Nicholas A. Christakis, "Association of Facebook Use with Compromised Well-Being: A Longitudinal Study," *American Journal of Epidemiology*, January 16, 2017, https://doi.org/10.1093/aje/kww189.

61. Bernd Heinrich, *The Homing Instinct* (Boston: Houghton Mifflin Harcourt, 2014), 298–99.

62. Erving Goffman, *The Presentation of Self in Everyday Life* (New York: Anchor, 1959), 112–32.

63. There is an extensive literature on this topic, but two articles that specifically reference "chilling effects" in social media are Sauvik Das and Adam Kramer, "Self-Censorship on Facebook," in *Proceedings of the Seventh International AAAI Conference on Weblogs and Social Media*, 2013; Alice E. Marwick and danah boyd, "I Tweet Honestly, I Tweet Passionately: Twitter Users, Context Collapse, and the Imagined Audience," *New Media & Society* 13, no. 1 (2011): 114–33.

64. Shoshana Zuboff, file note, November 9, 2017, Queen's University, Kingston, Ontario.

65. Ben Marder, Adam Joinson, Avi Shankar, and David Houghton, "The Extended 'Chilling' Effect of Facebook: The Cold Reality of Ubiquitous Social Networking," *Computers in Human Behavior* 60 (July 1, 2016): 582–92, https:// doi.org/10.1016/j.chb.2016.02.097.

66. Stanley Milgram and Thomas Blass, *The Individual in a Social World: Essays and Experiments*, 3rd ed. (London: Pinter & Martin, 2010), xxi–xxiii.

CHAPTER SEVENTEEN

1. Gaston Bachelard, *The Poetics of Space* (Boston: Beacon, 1994), 6.

2. Bachelard, *The Poetics of Space*, 7.

3. Bachelard, 91.

4. Philip Marfleet, "Understanding 'Sanctuary': Faith and Traditions of Asylum," *Journal of Refugee Studies* 24, no. 3 (2011): 440–55, https://doi.org/10.1093/jrs /fer040.

5. John Griffiths Pedley, *Sanctuaries and the Sacred in the Ancient Greek World* (New York: Cambridge University Press, 2005), 97.

6. H. Bianchi, *Justice as Sanctuary* (Eugene, OR: Wipf & Stock, 2010). See also Norman Maclaren Trenholme and Frank Thilly, *The Right of Sanctuary in England: A Study in Institutional History,* vol. 1 (Columbia: University of Missouri, 1903).

7. Linda McClain, "Inviolability and Privacy: The Castle, the Sanctuary, and the Body," *Yale Journal of Law & the Humanities* 7, no. 1 (1995): 203, http://digitalcommons.law.yale.edu/yjlh/vol7/iss1/9.

8. Darhl M. Pedersen, "Psychological Functions of Privacy," *Journal of Environmental Psychology* 17, no. 2 (1997): 147–56, https://doi.org/10.1006/jevp.1997.0049. For valuable related discussions in legal scholarship, see Daniel J. Solove, "'I've Got Nothing to Hide' and Other Misunderstandings of Privacy," *San Diego Law Review* 44 (July 12, 2007): 745; Julie E. Cohen, "What Privacy Is For" (SSRN Scholarly Paper, Rochester, NY: Social Science Research Network, November 5, 2012), https://papers.ssrn.com/abstract=2175406.

9. Anita L. Allen, *Unpopular Privacy: What Must We Hide?* Studies in Feminist Philosophy (New York: Oxford University Press, 2011), 4.

10. Orin S. Kerr, "Searches and Seizures in a Digital World," *Harvard Law Review* 119, no. 2 (2005): 531–85; Elizabeth B. Wydra, Brianne J. Gorod, and Brian R. Frazelle, "Timothy Ivory Carpenter v. United States of America—On Writ of Certiorari to the United States Court of Appeals for the Sixth Circuit—Brief of Scholars of the History and Original Meaning of the Fourth Amendment as Amici Curiae in Support of Petitioner," Supreme Court of the United States, August 14, 2017; David Gray, *The Fourth Amendment in an Age of Surveillance* (New York: Cambridge University Press, 2017); David Gray, "The Fourth Amendment Categorical Imperative," *Michigan Law Review,* 2017, http://michiganlawreview.org/the-fourth-amendment-categorical-imperative.

11. See Jennifer Daskal, "The Un-territoriality of Data," *Yale Law Journal* 125, no. 2 (2015): 326–98.

12. Andrew Guthrie Ferguson, "The Internet of Things and the Fourth Amendment of Effects," *California Law Review,* August 3, 2015, 879–80, https://papers.ssrn.com/abstract=2577944.

13. Lisa Van Dongen and Tjerk Timan, "Your Smart Coffee Machine Knows What You Did Last Summer: A Legal Analysis of the Limitations of Traditional Privacy of the Home Under Dutch Law in the Era of Smart Technology" (SSRN Scholarly Paper, Rochester, NY: Social Science Research Network, September 1, 2017), https://papers.ssrn.com/abstract=3090340.

14. For a clear explanation of "consent" under the GDPR, see Sally Annereau, "Understanding Consent Under the GDPR," Global Data Hub, November 2016, https://globaldatahub.taylorwessing.com/article/understanding-consent-under-the-gdpr..

15. McCann FitzGerald and Ruairí Madigan, "GDPR and the Internet of Things: 5 Things You Need to Know," *Lexology*, May 26, 2016, http://www.lexology.com /library/detail.aspx?g=ba0b0d12-bae3-4e93-b832-85c15620b877.

16. Daphne Keller, "The Right Tools: Europe's Intermediary Liability Laws and the 2016 General Data Protection Regulation" (SSRN Scholarly Paper, Rochester, NY: Social Science Research Network, March 22, 2017), https:// papers.ssrn.com/abstract=2914684; Sandra Wachter, "Normative Challenges of Identification in the Internet of Things: Privacy, Profiling, Discrimination, and the GDPR" (SSRN Scholarly Paper, Rochester, NY: Social Science Research Network, December 6, 2017), https://papers.ssrn.com/abstract=3083554; Tal Zarsky, "Incompatible: The GDPR in the Age of Big Data" (SSRN Scholarly Paper, Rochester, NY: Social Science Research Network, August 8, 2017), https://papers.ssrn.com/abstract=3022646; Anna Rossi, "Respected or Challenged by Technology? The General Data Protection Regulation and Commercial Profiling on the Internet" (SSRN Scholarly Paper, Rochester, NY: Social Science Research Network, July 13, 2016), https://papers.ssrn .com/abstract=2852739; Viktor Mayer-Schönberger and Yann Padova, "Regime Change? Enabling Big Data Through Europe's New Data Protection Regulation," *Columbia Science & Technology Law Review* 315 (2016): 315–35.

17. Paul-Olivier Dehaye, e-mail message to DCMS Committee, March 6, 2018, http://data.parliament.uk/writtenevidence/committeeevidence.svc /evidencedocument/digital-culture-media-and-sport-committee/fake-news /written/80117.html.

18. Paul-Olivier Dehaye, e-mail message to DCMS Committee, March 7, 2018, http://data.parliament.uk/writtenevidence/committeeevidence.svc/evidence document/digital-culture-media-and-sport-committee/fake-news/written /80117.html (italics mine).

19. For more on Hive's data and architecture, see the Facebook audit performed in 2011–2012 by the Irish Data Protection Commissioner, following the efforts of privacy activist Max Schrems, who challenged Facebook's accumulation of personal data on EU citizens: "Facebook Audit," Data Protection Commission —Ireland, July 3, 2018, https://www.dataprotection.ie/docs/Facbook-Audit /1290.htm.

20. "How Can I Download a Copy of My Facebook Data?" Facebook, https://www .facebook.com/help/1701730696756992; "What to Look for in Your Facebook Data—and How to Find It," *Wired*, April 26, 2018, https://www.wired.com /story/download-facebook-data-how-to-read.

21. John Paul Titlow, "How Instagram Learns from Your Likes to Keep You Hooked," *Fast Company*, July 7, 2017, https://www.fastcompany.com/40434598 /how-instagram-learns-from-your-likes-to-keep-you-hooked; Lilian Edwards and Michael Veale, "Slave to the Algorithm? Why a 'Right to an Explanation' Is Probably Not the Remedy You Are Looking For" (SSRN Scholarly Paper,

Rochester, NY: Social Science Research Network, May 23, 2017), https://papers.ssrn.com/abstract=2972855; Michael Veale, Reuben Binns, and Jef Ausloos, "When Data Protection by Design and Data Subject Rights Clash," *International Data Privacy Law*, April 26, 2018, https://doi.org/10.1093/idpl/ipy002; Dimitra Kamarinou, Christopher Millard, and Jatinder Singh, "Machine Learning with Personal Data" (SSRN Scholarly Paper, Rochester, NY: Social Science Research Network, November 7, 2016), https://papers.ssrn.com/abstract=2865811.

22. Andrew Tutt, "An FDA for Algorithms," *Administrative Law Review* 69, no. 83 (2017), https://papers.ssrn.com/abstract=2747994.

23. Personal communication.

24. For an excellent discussion of these power dynamics in historical perspective, see Robin Mansell, "Bits of Power: Struggling for Control of Information and Communication Networks," *Political Economy of Communication* 5, no. 1 (2017), 2–29, especially 16.

25. Laura Nader, "The Life of the Law—a Moving Story," *Valparaiso University Law Review* 36, no. 3 (2002): 658.

26. For more on NOYB, see its informative website: "Noyb.Eu | My Privacy Is None of Your Business," https://noyb.eu. See also Hannah Kuchler, "Max Schrems: The Man Who Took on Facebook—and Won," *Financial Times*, April 5, 2018.

27. 2010–2012: Kit Seeborg, "Facebook Q4 2012 Quarterly Earnings," January 31, 2013, https://www.slideshare.net/kitseeborg/fb-q412-investordeck/4-Daily_Active_Users_DAUsMillions_of; 2013–2014: "Facebook Q4 2014 Results," investor.fb.com, August 4, 21018, http://files.shareholder.com/downloads/AMDA-NJ5DZ/3907746207x0x805520/2D74EDCA-E02A-420B-A262-BC096264BB93/FB_Q414EarningsSlides20150128.pdf, 3; 2015–2017: Deborah Crawford et al., "Facebook, Inc. (FB)—Fourth Quarter and Full Year 2016 Results," February 1, 2017, https://s21.q4cdn.com/399680738/files/doc_financials/2017/Q4/Q4-2017-Earnings-Presentation.pdf, 2.

28. United States and Canada: "Facebook: Quarterly Revenue in U.S. and Canada from 1st Quarter 2010 to 2nd Quarter 2018," Statista, 2018, https://www.statista.com/statistics/223280/facebooks-quarterly-revenue-in-the-us-and-canada-by-segment/#0; Europe: "Facebook: Quarterly Revenue in Europe from 1st Quarter 2010 to 2nd Quarter 2018," Statista, 2018, https://www.statista.com/statistics/223279/facebooks-quarterly-revenue-in-europe/#0.

29. "Global Stats," *statcounter.com*, http://gs.statcounter.com.

30. See Daisuke Wakabayashi and Adam Satariano, "How Looming Privacy Regulations May Strengthen Facebook and Google," *New York Times*, April 24, 2018, https://www.nytimes.com/2018/04/23/technology/privacy-regulation-facebook-google.html.

31. "Recommendations for Implementing Transparency, Consent and Legitimate Interest Under the GDPR," Centre for Information Policy Leadership, Hunton and Williams LLP, GDPR Implementation Project, May 19, 2017.

32. "Exclusive: Facebook to Put 1.5 Billion Users Out of Reach of New EU Privacy Law," *Reuters*, April 19, 2018, https://www.reuters.com/article/us-facebook-privacy-eu-exclusive/exclusive-facebook-to-change-user-terms-limiting-effect-of-eu-privacy-law-idUSKBN1HQ00P.

33. Elizabeth E. Joh, "Privacy Protests: Surveillance Evasion and Fourth Amendment Suspicion," *Arizona Law Review* 55, no. 4 (2013): 997–1029; Jeffrey L. Vagle, "Furtive Encryption: Power, Trust, and the Constitutional Cost of Collective Surveillance," *Indiana Law Journal* 90, no. 1 (2015), http://papers.ssrn.com/abstract=2550934.

34. "How to Be Invisible: 15 Anti-surveillance Gadgets & Wearables," *WebUrbanist*, November 28, 2016, http://weburbanist.com/2016/11/28/how-to-be-invisible-15-anti-surveillance-designs-installations. For another relevant article, see "The Role of Hackers in Countering Surveillance and Promoting Democracy," April 29, 2018, https://search-proquest-com.ezproxy.cul.columbia.edu/docview/1719239523?pq-origsite=gscholar.

35. See Zach Sokol, "Hide from Surveillance by Wearing a Mask of This Artist's Face," *Creators*, May 7, 2014, https://creators.vice.com/en_us/article/pgqp87/hide-from-surveillance-by-wearing-a-mask-of-this-artists-face.

36. See "Backslash," Backslash.com, August 4, 2018, http://www.backslash.cc.

37. Mehrdad Hessar et al., "Enabling On-Body Transmissions with Commodity Devices," *UBICOMP 16*, September 12–16, 2016, Heidelberg, Germany.

38. See Adam Harvey, "Stealth Wear—Anti-drone Fasion," ah projects, December 3, 2012, https://ahprojects.com/projects/stealth-wear.

39. See Benjamin Grosser, "Projects," Benjamin Grosser, August 3, 2018, https://bengrosser.com/projects.

40. Cade Metz, "The Unsettling Performance That Showed the World Through AI's Eyes," *Wired*, April 30, 2017, https://www.wired.com/2017/04/unsettling-performance-showed-world-ais-eyes/; Thu-Huong Ha, "Ai Weiwei's New Show Exposes the Creepy Consequences of Our Obsession with Posing for the Camera," *Quartz*, April 29, 2018, https://qz.com/1000684/ai-weiwei-herzog-de-meuron-artwork-hansel-gretel-exposes-the-creepy-consequences-of-our-obsession-with-posing-for-the-camera.

CHAPTER EIGHTEEN

1. Adam Smith, *The Wealth of Nations*, ed. Edwin Cannan (New York: Modern Library, 1994), 485.

2. Friedrich August von Hayek, *The Collected Works of Friedrich August Hayek*, ed. William Warren Bartley (Chicago: University of Chicago Press, 1988), 1:14.

3. Friedrich Hayek, "The Use of Knowledge in Society," in *Individualism and Economic Order* (Chicago: University of Chicago Press, 1980). See the discussion on 85–89.

4. Hayek, "The Use of Knowledge," 89 (italics mine).

5. Ashlee Vance, "Facebook: The Making of 1 Billion Users," *Bloomberg.com*, http://www.bloomberg.com/news /articles/2012-10-04/facebook-the-making-of-1-billion-users.

6. Tom Simonite, "What Facebook Knows," *MIT Technology Review*, June 13, 2012, https://www.technologyreview.com/s/428150/what-facebook-knows.

7. See Vance, "Facebook: The Making of 1 Billion Users."

8. Derek Thompson, "Google's CEO: 'The Laws Are Written by Lobbyists,'" *Atlantic*, October 1, 2010, https://www.theatlantic.com/technology/archive/2010/10/googles-ceo-the-laws-are-written-by-lobbyists/63908.

9. Satya Nadella, "Satya Nadella: Build 2017," *News Center*, May 10, 2017, https://news.microsoft.com/speeches/satya-nadella-build-2017.

10. Smith, *The Wealth of Nations*, 939–40.

11. These data are drawn from my own compilation of General Motors market capitalization and employment data from 1926 to 2008, Google from 2004 to 2016, and Facebook from 2012 to 2016. All market capitalization values are adjusted for inflation to 2016 dollars, as per the Consumer Price Index from Federal Reserve Economic Data, Economic Research Division, Federal Reserve Bank of St. Louis. The sources used to compile these data include Standard & Poor's Capital IQ (Google Market Capitalization and Headcount), Wharton Research Data Services—CRSP (General Motors Market Capitalization), Standard & Poor's Compustat (General Motors Headcount), Thomas Eikon (Facebook Market Capitalization), Company Annual Reports (General Motors Headcount), and SEC Filings (Facebook Headcount).

12. Opinion Research Corporation, "Is Big Business Essential for the Nation's Growth and Expansion?" ORC Public Opinion Index (August 1954); Opinion Research Corporation, "Which of These Comes Closest to Your Impression of the Business Setup in This Country?" ORC Public Opinion Index (January 1955); Opinion Research Corporation, "Now Some Questions About Large Companies. Do You Agree or Disagree on Each of These?... Large Companies Are Essential for the Nation's Growth and Expansion," ORC Public Opinion Index (June 1959). A 1951 report found that the American public lauded big business for job creation, its effectiveness as a mass producer, the development and improvement of products, payment of big taxes, and support of education. See "Poll Finds Public on Industry's Side," *New York Times*, July 15, 1951. A 1966 Harris poll reported 44 percent of Americans crediting the federal government with the nation's prosperity and 34 percent crediting big business. In 1968, when CEO pay was about 24 times that of the average worker, 64 percent of Americans said that business leadership was the best it had ever been. See Louis Harris & Associates, "Which Two or Three Best Describe Most Business Corporation Leaders in the Country?" (April 1966); Louis Harris & Associates, "Compared with What We Have Produced in the Past in This Country, Do You Feel That Our Present Leadership in the Field of Business Is Better, Worse or About the Same as We Have Produced in the Past?" (June

1968). For more background, see also Louis Galambos, *The Public Image of Big Business in America, 1880–1940: A Quantitative Study in Social Change* (Baltimore, MD: Johns Hopkins University Press, 1975).

13. See Alfred D. Chandler, "The Enduring Logic of Industrial Success," *Harvard Business Review*, March 1, 1990, https://hbr.org/1990/03/the-enduring-logic-of-industrial-success; Susan Helper and Rebecca Henderson, "Management Practices, Relational Contracts, and the Decline of General Motors," *Journal of Economic Perspectives* 28, no. 1 (2014): 49–72, https://doi.org/10.1257/jep.28.1.49.

14. David H. Autor et al., "The Fall of the Labor Share and the Rise of Superstar Firms" (SSRN Scholarly Paper, Rochester, NY: Social Science Research Network, May 22, 2017), https://papers.ssrn.com/abstract=2971352. See also Michael Chui and James Manyika, "Competition at the Digital Edge: 'Hyperscale' Businesses," *McKinsey Quarterly*, March 2015.

15. One hundred more data centers are expected to be online by late 2018. Microsoft invested $20 billion in 2017, and in 2018 Facebook announced plans to invest $20 billion in a new hyperscale data center in Atlanta. According to one industry report, hyperscale firms are also building the world's networks, especially subsea cables, which means that "a large portion of the global internet traffic is now running through private networks owned or operated by hyperscalers." In 2016 Facebook and Google teamed up to build a new subsea cable between the US and Hong Kong, described as the highest-capacity transpacific route to date. See João Marges Lima, "Hyperscalers Taking Over the World at an Unprecedented Scale," *Data Economy*, April 11, 2017, https://data-economy.com/hyperscalers-taking-world-unprecedented-scale; João Marges Lima, "Facebook, Google Partners in 12,800Km Transpacific Cable Linking US, China," *Data Economy*, October 13, 2016, https://data-economy.com/facebook-google-partners-in-12800km-transpacific-cable-linking-us-china; João Marges Lima, "Facebook Could Invest up to $20bn in a Single Hyperscale Data Centre Campus," *Data Economy*, January 23, 2018, https://data-economy.com/facebook-invest-20bn-single-hyperscale-data-centre-campus.

16. T. H. Breen, *The Marketplace of Revolution: How Consumer Politics Shaped American Independence* (New York: Oxford University Press, 2005), 22.

17. Breen, *The Marketplace of Revolution*, 222.

18. Breen, XVI–XVII.

19. Breen, 235–39.

20. See Breen, 20.

21. See Breen, 299.

22. Breen, 325.

23. Daron Acemoglu and James A. Robinson, *Why Nations Fail: The Origins of Power, Prosperity, and Poverty* (New York: Crown Business, 2012).

24. Acemoglu and Robinson, *Why Nations Fail*, 313–14. Historian Jack Goldstone observes that the magnitude of Britain's parliamentary reforms defused the

pressure for more violent change, creating a more durable and prosperous democracy. Like Acemoglu and Robinson, he concludes that "national decay" is typically associated with a social pattern in which elites do not identify their interests with those of the public, suggesting the danger of precisely the kind of structural independence enjoyed by surveillance capitalists. See Jack A. Goldstone, *Revolution and Rebellion in the Early Modern World* (Berkeley: University of California Press, 1993), 481, 487; see also Barrington Moore, *Social Origins of Dictatorship and Democracy: Lord and Peasant in the Making of the Modern World* (Boston: Beacon, 1993), 3–39.

25. Michel Crozier, Samuel P. Huntington, and Joji Watanuki, "The Crisis of Democracy: Report on the Governability of Democracies to the Trilateral Commission," 1975, http://trilateral.org/download/doc/crisis_of_democracy .pdf.

26. Ryan Mac, Charlie Warzel, and Alex Kantrowitz, "Growth at Any Cost: Top Facebook Executive Defended Data Collection in 2016 Memo—and Warned That Facebook Could Get People Killed," *Buzzfeed,* March 29, 2018, https:// www.buzzfeed.com/ryanmac/growth-at-any-cost-top-facebook-executive -defended-data?utm_term=.stWyyGQnb#.cnkEEaN0v.

27. Nicholas Thompson and Fred Vogelstein, "Inside the Two Years That Shook Facebook—and the World," *Wired,* February 12, 2018, https://www.wired.com /story/inside-facebook-mark-zuckerberg-2-years-of-hell.

28. Hunt Allcott and Matthew Gentzkow, "Social Media and Fake News in the 2016 Election," *Journal of Economic Perspectives* 31, no. 2 (2017): 211–36.

29. "Nielsen/Netratings Reports Topline U.S. Data for July 2007," Nielsen /Netratings, July 2007.

30. Consumer Watchdog, "Liars and Loans: How Deceptive Advertisers Use Google," February 2011; Jay Greene, "Feds Shut Down High-Tech Mortgage Scammers," *CBSNews.com,* November 16, 2011.

31. US Department of Justice, "Google Forfeits $500 Million Generated by Online Ads & Prescription Drug Sales by Canadian Online Pharmacies," https://www .justice.gov/opa/pr/google-forfeits-500-million-generated-online-ads -prescription-drug-sales-canadian-online.

32. Michela Del Vicario et al., "The Spreading of Misinformation Online," *Proceedings of the National Academy of Sciences* 113, no. 3 (2016): 554–59; Solomon Messing and Sean J. Westwood, "How Social Media Introduces Biases in Selecting and Processing News Content," Pew Research Center, April 8, 2012.

33. See Paul Mozur and Mark Scott, "Fake News in U.S. Election? Elsewhere, That's Nothing New," *New York Times,* November 17, 2016, http://www.nytimes .com/2016/11/18/technology/fake-news-on-facebook-in-foreign-elections -thats-not-new.html.

34. Catherine Buni, "The Secret Rules of the Internet," *Verge,* April 13, 2016, https:// www.theverge.com/2016/4/13/11387934/internet-moderator-history-youtube -facebook-reddit-censorship-free-speech.

35. Madeleine Varner and Julia Angwin, "Facebook Enabled Advertisers to Reach 'Jew Haters,'" *ProPublica*, September 14, 2017, https://www.propublica.org /article/facebook-enabled-advertisers-to-reach-jew-haters.

36. See the instructive discussion in Buni, "The Secret Rules of the Internet"; Nick Hopkins, "Revealed: Facebook's Internal Rulebook on Sex, Terrorism and Violence," *Guardian*, May 21, 2017, https://www.theguardian.com/news/2017 /may/21/revealed-facebook-internal-rulebook-sex-terrorism-violence?utm _source=esp&utm_medium=Email&utm_campaign=GU+Today+USA +-+Collections+2017&utm_term=227190&subid=17990030&CMP=GT_US _collection; Nick Hopkins, "Facebook Moderators: A Quick Guide to Their Job and Its Challenges," *Guardian*, May 21, 2017, https://www.theguardian.com /news/2017/may/21/facebook-moderators-quick-guide-job-challenges; Kate Klonick, "The New Governors: The People, Rules, and Processes Governing Online Speech," *Harvard Law Review* 131 (March 20, 2017), https://papers.ssrn .com/abstract=2937985.

37. Michael Nunez, "Facebook's Fight Against Fake News Was Undercut by Fear of Conservative Backlash," *Gizmodo*, November 14, 2016, http://gizmodo .com/facebooks-fight-against-fake-news-was-undercut-by-fear-1788808204.

38. Varner and Angwin, "Facebook Enabled Advertisers to Reach 'Jew Haters.'"

39. Alex Kantrowitz, "Google Allowed Advertisers to Target 'Jewish Parasite,' 'Black People Ruin Everything,'" *BuzzFeed*, September 15, 2017, https://www.buzzfeed .com/alexkantrowitz/google-allowed-advertisers-to-target-jewish-parasite -black.

40. Jack Nicas, "Big Brands Boost Fake News Sites," *Wall Street Journal*, December 9, 2016; Olivia Solon, "Google's Bad Week: YouTube Loses Millions as Advertising Row Reaches US," *Guardian*, March 25, 2017, http://www.the guardian.com/technology/2017/mar/25/google-youtube-advertising-extremist -content-att-verizon; Alexi Mostrous, "YouTube Hate Preachers Share Screens with Household Names," *Times*, March 17, 2017, https://www.thetimes.co.uk /article/youtube-hate-preachers-share-screens-with-household-names-kdmpm kkjk; Alexi Mostrous, "Advertising Giant Drops Google in Storm Over Extremist Videos," *Times*, March 18, 2017, http://www.thetimes.co.uk/article /advertising-giant-drops-google-in-storm-over-extremist-videos-2klgvv8d5.

41. Shannon Bond, "Trade Group Warns on Google Ad Backlash 'Crisis,'" *Financial Times*, March 24, 2017, https://www.ft.com/content/0936a49e -b521-369e-9d22-c194ed1c0d48; Matthew Garrahan, "AT&T Pulls Some Ads from Google After YouTube Controversy," *Financial Times*, March 22, 2017, https://www.ft.com/content/254d330d-f3d1-3ac2-ab8d-761083d6976a; Sapna Maheshwari and Daisuke Wakabayashi, "AT&T and Johnson & Johnson Pull Ads from YouTube," *New York Times*, March 22, 2017, https://www.nytimes .com/2017/03/22/business/atampt-and-johnson-amp-johnson-pull-ads-from -youtube-amid-hate-speech-concerns.html; Rob Davies, "Google Braces for Questions as More Big-Name Firms Pull Adverts," *Guardian*, March 19, 2017,

https://www.theguardian.com/technology/2017/mar/19/google-braces-for
-questions-as-more-big-name-firms-pull-adverts.

42. Olivia Solon, "Facebook's Fake News: Mark Zuckerberg Rejects 'Crazy Idea'
That It Swayed Voters," Guardian, November 11, 2016,, https://www.the
guardian.com/technology/2016/nov/10/facebook-fake-news-us-election
-mark-zuckerberg-donald-trump.

43. Guy Chazan, "Berlin Looks at Fines for Facebook with Fake News Law,"
Financial Times, December 16, 2016; Guy Chazan, "Germany Cracks Down on
Social Media Over Fake News," *Financial Times,* March 14, 2017, https://www.ft
.com/content/c10aa4f8-08a5-11e7-97d1-5e720a26771b; Jim Pickard, "Amber
Rudd Urges Action from Internet Groups on Extremist Content," *Financial
Times,* March 26, 2017, https://www.ft.com/content/f652c9bc-120d-11e7-80f4
-13e067d5072c; Alexandra Topping, Mark Sweney, and Jamie Grierson, "Google
Is 'Profiting from Hatred' Say MPs in Row Over Adverts," *Guardian,* March 17,
2017, http://www.theguardian.com/technology/2017/mar/17/google-is
-profiting-from-hatred-say-mps-in-row-over-adverts; Sabrina Siddiqui, "'From
Heroes to Villains': Tech Industry Faces Bipartisan Backlash in Washington,"
Guardian, September 26, 2017, http://www.theguardian.com/us-news/2017
/sep/26/tech-industry-washington-google-amazon-apple-facebook; Nancy
Scola and Josh Meyer, "Google, Facebook May Have to Reveal Deepest Secrets,"
Politico, October 1, 2017, http://politi.co/2yBtppQ; Paul Lewis, "Senator Warns
YouTube Algorithm May Be Open to Manipulation by 'Bad Actors,'" *Guardian,*
February 5, 2018, http://www.theguardian.com/technology/2018/feb/05
/senator-warns-youtube-algorithm-may-be-open-to-manipulation-by-bad
-actors.

44. Madhumita Murgia and David Bond, "Google Apologises to Advertisers
for Extremist Content on YouTube," *Financial Times,* March 20, 2017; Sam
Levin, "Mark Zuckerberg: I Regret Ridiculing Fears Over Facebook's Effect on
Election," *Guardian,* September 27, 2017, http://www.theguardian.com
/technology/2017/sep/27/mark-zuckerberg-facebook-2016-election-fake
-news; Robert Booth and Alex Hern, "Facebook Admits Industry Could Do
More to Combat Online Extremism," *Guardian,* September 20, 2017, http://
www.theguardian.com/technology/2017/sep/20/facebook-admits-industry
-could-do-more-to-combat-online-extremism; Scott Shane and Mike Isaac,
"Facebook to Turn Over Russian-Linked Ads to Congress," *New York Times,*
September 21, 2017, https://www.nytimes.com/2017/09/21/technology/face
book-russian-ads.html; David Cohen, "Mark Zuckerberg Seeks Forgiveness in
Yom Kippur Facebook Post," *Adweek,* October 2, 2017, http://www.adweek
.com/digital/mark-zuckerberg-yom-kippur-facebook-post; "Exclusive
Interview with Facebook's Sheryl Sandberg," *Axios,* October 12, 2017, https://
www.axios.com/exclusive-interview-facebook-sheryl-sandberg-2495538841
.html; Kevin Roose, "Facebook's Frankenstein Moment," *New York Times,*

September 21, 2017, https://www.nytimes.com/2017/09/21/technology/face book-frankenstein-sandberg-ads.html.

45. David Cohen, "Mark Zuckerberg Seeks Forgiveness in Yom Kippur Facebook Post."

46. Roose, "Facebook's Frankenstein Moment."

47. Booth and Hern, "Facebook Admits Industry Could Do More to Combat Online Extremism."

48. See Murgia and Bond, "Google Apologises to Advertisers."

49. Mark Bergen, "Google Is Losing to the 'Evil Unicorns,'" *Bloomberg Businessweek,* November 27, 2017.

50. On modest adjustments, see Mike Isaac, "Facebook and Other Tech Companies Seek to Curb Flow of Terrorist Content," *New York Times,* December 5, 2016, http://www.nytimes.com/2016/12/05/technology/facebook-and-other -tech-companies-seek-to-curb-flow-of-terrorist-content.html; Daisuke Wakabayashi, "Google Cousin Develops Technology to Flag Toxic Online Comments," *New York Times,* February 23, 2017, https://www.nytimes.com /2017/02/23/technology/google-jigsaw-monitor-toxic-online-comments.html; Sapna Maheshwari, "YouTube Revamped Its Ad System. AT&T Still Hasn't Returned," *New York Times,* February 12, 2018, https://www.nytimes.com /2018/02/12/business/media/att-youtube-advertising.html; Madhumita Murgia, "Google Reveals Response to YouTube Ad Backlash," *Financial Times,* March 21, 2017, https://www.ft.com/content/46475974-0e30-11e7-b030-76895439 4623; Heather Timmons, "Google Executives Are Floating a Plan to Fight Fake News on Facebook and Twitter," *Quartz,* https://qz.com/1195872/google-face book-twitter-fake-news-chrome; Elizabeth Dwoskin and Hamza Shaban, "Facebook Will Now Ask Users to Rank News Organizations They Trust," *Washington Post,* January 19, 2018, https://www.washingtonpost.com/news /the-switch/wp/2018/01/19/facebook-will-now-ask-its-users-to-rank-news -organizations-they-trust; Hamza Shaban, "Mark Zuckerberg Vows to Remove Violent Threats from Facebook," *Washington Post,* August 16, 2017, https:// www.washingtonpost.com/news/the-switch/wp/2017/08/16/mark -zuckerberg-vows-to-remove-violent-threats-from-facebook. On moves to quash meaningful reform, see Hannah Albarazi, "Zuckerberg Votes Against Shareholder Push for Fake News Transparency," *CBS SFBayArea,* June 2, 2017, http://sanfrancisco.cbslocal.com/2017/06/02/zuckerberg-shareholder-fake -news-transparency; Ethan Baron, "Google Parent Alphabet Gender-Pay Proposal Dead on Arrival," *Mercury News,* June 7, 2017.

51. "Facebook Reports First Quarter 2018 Results."

52. Adam Mosseri, "News Feed FYI: Bringing People Closer Together," *Facebook Newsroom,* January 11, 2018, https://newsroom.fb.com/news/2018/01/news -feed-fyi-bringing-people-closer-together.

53. Sapna Maheshwari, "As Facebook Changes Its Feed, Advertisers See Video Ambitions," *New York Times,* January 21, 2018, https://www.nytimes.com /2018/01/21/business/media/facebook-video-advertising.html.

54. Thomas Paine, *The Life and Works of Thomas Paine,* ed. William M. Van der Weyde (New Rochelle, NY: Thomas Paine Historical Society, 1925), 6:97.

55. Hannah Arendt, *Between Past and Future: Eight Exercises in Political Thought* (New York: Penguin, 2006), 99.

56. Mark Zuckerberg, "Building Global Community," February 16, 2017, https:// www.facebook.com/notes/mark-zuckerberg/building-global-community /10154544292806634.

57. Karissa Bell, "Zuckerberg Removed a Line About Monitoring Private Messages from His Facebook Manifesto," *Mashable,* February 16, 2017, http://mashable .com/2017/02/16/mark-zuckerberg-manifesto-ai.

58. Heather Kelly, "Mark Zuckerberg Explains Why He Just Changed Facebook's Mission," *CNNMoney,* June 22, 2017, http://money.cnn.com/2017/06/22 /technology/facebook-zuckerberg-interview/index.html.

59. Pippa Norris, "Is Western Democracy Backsliding? Diagnosing the Risks," Harvard Kennedy School, March 2017, https://www.hks.harvard.edu /publications/western-democracy-backsliding-diagnosing-risks; Erik Voeten, "Are People Really Turning Away from Democracy?" (SSRN Scholarly Paper, Rochester, NY: Social Science Research Network, December 8, 2016), https:// papers.ssrn.com/abstract=2882878; Amy C. Alexander and Christian Welzel, "The Myth of Deconsolidation: Rising Liberalism and the Populist Reaction," *Journal of Democracy,* April 28, 2017, https://www.journalofdemocracy.org /sites/default/files/media/Journal%20of%20Democracy%20Web%20 Exchange%20-%20Alexander%20and%20Welzel.pdf; Ronald Inglehart, "The Danger of Deconsolidation: How Much Should We Worry?" *Journal of Democracy* 27, no. 3 (2016), https://www.journalofdemocracy.org/article /danger-deconsolidation-how-much-should-we-worry; Roberto Stefan Foa and Yascha Mounk, "The Signs of Deconsolidation," *Journal of Democracy* 28, no. 1 (2017); Ronald Inglehart and Christian Welzel, "Democracy's Victory Is Not Preordained. Inglehart and Welzel Reply," *Foreign Affairs* 88, no. 4 (2009): 157–59; Roberto Stefan Foa, "The End of the Consolidation Paradigm—a Response to Our Critics," *Journal of Democracy,* April 28, 2017.

60. Bart Bonikowski, "Three Lessons of Contemporary Populism in Europe and the United States," *Brown Journal of World Affairs* 23, no. 1 (2016); Bart Bonikowski and Paul DiMaggio, "Varieties of American Popular Nationalism," *American Sociological Review* 81, no. 5 (2016): 949–80; Theda Skocpol and Vanessa Williamson, *The Tea Party and the Remaking of Republican Conservatism* (New York: Oxford University Press, 2016), 74–75.

61. Richard Wike et al., "Globally, Broad Support for Representative and Direct Democracy," *Pew Research Center's Global Attitudes Project,* October 16, 2017,

http://www.pewglobal.org/2017/10/16/globally-broad-support-for-representa tive-and-direct-democracy.

62. As democracy scholar and author of the "democratic recession" thesis Larry Diamond puts it, "It is hard to overstate how important the vitality and self-confidence of U.S. democracy has been to the global expansion of democracy.... Apathy and inertia in Europe and the United States could significantly lower the barriers to new democratic reversals and to authoritarian entrenchment in many more states." See Larry Diamond, "Facing Up to the Democratic Recession," *Journal of Democracy* 26, no. 1 (2015): 141–55, https:// doi.org/10.1353/jod.2015.0009.

63. Naomi Klein, *The Shock Doctrine: The Rise of Disaster Capitalism* (New York: Picador, 2007); Erik Olin Wright, *Envisioning Real Utopias* (London: Verso, 2010); Wendy Brown, *Edgework: Critical Essays on Knowledge and Politics* (Princeton, NJ: Princeton University Press, 2005); Gerald F. Davis, *Managed by the Markets: How Finance Re-shaped America* (New York: Oxford University Press, 2011).

64. Immanuel Wallerstein et al., *Does Capitalism Have a Future?* (Oxford: Oxford University Press, 2013); Erik Olin Wright, *Envisioning Real Utopias* (London: Verso, 2010); Naomi Klein, *This Changes Everything: Capitalism Vs. the Climate* (New York: Simon & Schuster, 2015); Wendy Brown, *Edgework: Critical Essays on Knowledge and Politics* (Princeton, NJ: Princeton University Press, 2005); Davis, *Managed by the Markets;* Wolfgang Streeck, "On the Dismal Future of Capitalism," *Socio-Economic Review* 14, no. 1 (2016): 164–70; Craig Calhoun, "The Future of Capitalism," *Socio-Economic Review* 14, no. 1 (2016): 171–76; Polly Toynbee, "Unfettered Capitalism Eats Itself," *Socio-Economic Review* 14, no. 1 (2016): 176–79; Amitai Etzioni, "The Next Industrial Revolution Calls for a Different Economic System," *Socio-Economic Review* 14, no. 1 (2016): 179–83.

65. See, for example, Nicolas Berggruen and Nathan Gardels, *Intelligent Governance for the 21st Century: A Middle Way Between West and East* (Cambridge: Polity, 2013).

66. Hannah Arendt, *The Origins of Totalitarianism* (New York: Schocken, 2004), 615.

67. Theodor Adorno, "Education after Auschwitz," in *Critical Models: Interventions and Catchwords* (New York: Columbia University Press, 1966).

68. Thomas Piketty, *Capital in the Twenty-First Century* (Cambridge, MA: Belknap Press, 2014), 571.

69. Piketty, *Capital in the Twenty-First Century,* 573. For a wise and elegant defense of democracy, see also Wendy Brown, *Undoing the Demos: Neoliberalism's Stealth Revolution* (New York: Zone, 2015).

70. Roger W. Garrison, "Hayek and Friedman," in *Elgar Companion to Hayekian Economics,* ed. Norman Barry (Northampton, MA: Edward Elgar, 2014).

71. Friedrich Hayek, interview by Robert Bork, November 4, 1978, Center for Oral History Research, University of California, Los Angeles, http://oralhistory .library.ucla.edu.

72. Zygmunt Bauman, *Liquid Modernity* (Cambridge, MA: Polity, 2000); Fernand Braudel, *The Structures of Everyday Life* (New York: Harper & Row, 1981), 1:620.

73. Piketty, *Capital in the Twenty-First Century*, 614–15.

74. Roberto M. Unger, *Free Trade Reimagined: The World Division of Labor and the Method of Economics* (Princeton, NJ: Princeton University Press, 2007), 8, 41 (italics mine).

75. Paine, *The Life and Works*, 6:172.

76. Hannah Arendt, "A Reply" [to Eric Voegelin's review of *Origins of Totalitarianism*], *Review of Politics*, 15 (1953): 79.

77. George Orwell, *In Front of Your Nose 1945–1950: The Collected Essays, Journalism and Letters of George Orwell*, vol. 4, ed. Sonia Orwell and Ian Angus (New York: Harcourt, Brace, and World, 1968), 160–81 (italics mine).

78. Orwell, *In Front of Your Nose.*

79. Hannah Arendt, "What Is Freedom?" in *Between Past and Future: Eight Exercises in Political Thought* (New York: Penguin, 1993), 169.

INDEX

Note: Page numbers with *f* refer to figures.

SHOSHANA ZUBOFF is the author of three books, each of which signaled the start of a new epoch in technological society. In the late 1980s, *In the Age of the Smart Machine* foresaw how computers would revolutionize the modern workplace. At the dawn of the twenty-first century, *The Support Economy* predicted the rise of a digitally distributed capitalism of services tailored to the individual. Today, *The Age of Surveillance Capitalism* reveals a world in which technology users are no longer customers but the raw material for an entirely new industrial system. Dr. Zuboff is the Charles Edward Wilson Professor Emerita at Harvard Business School and Faculty Associate at the Berkman Center for Internet and Society at Harvard Law School.